Finite Difference Methods in Heat Transfer

Finite Difference Methods in Heat Transfer

Second Edition

M. Necati Özişik
Helcio R.B. Orlande
Marcelo José Colaço
Renato Machado Cotta

CRC Press
Taylor & Francis Group
Boca Raton London New York

CRC Press is an imprint of the
Taylor & Francis Group, an **informa** business

CRC Press
Taylor & Francis Group
6000 Broken Sound Parkway NW, Suite 300
Boca Raton, FL 33487-2742

© 2017 by Taylor & Francis Group, LLC
CRC Press is an imprint of Taylor & Francis Group, an Informa business

No claim to original U.S. Government works

Printed on acid-free paper

International Standard Book Number-13: 978-1-4822-4345-1 (Hardback)

This book contains information obtained from authentic and highly regarded sources. Reasonable efforts have been made to publish reliable data and information, but the author and publisher cannot assume responsibility for the validity of all materials or the consequences of their use. The authors and publishers have attempted to trace the copyright holders of all material reproduced in this publication and apologize to copyright holders if permission to publish in this form has not been obtained. If any copyright material has not been acknowledged please write and let us know so we may rectify in any future reprint.

Except as permitted under U.S. Copyright Law, no part of this book may be reprinted, reproduced, transmitted, or utilized in any form by any electronic, mechanical, or other means, now known or hereafter invented, including photocopying, microfilming, and recording, or in any information storage or retrieval system, without written permission from the publishers.

For permission to photocopy or use material electronically from this work, please access www.copyright.com (http://www.copyright.com/) or contact the Copyright Clearance Center, Inc. (CCC), 222 Rosewood Drive, Danvers, MA 01923, 978-750-8400. CCC is a not-for-profit organization that provides licenses and registration for a variety of users. For organizations that have been granted a photocopy license by the CCC, a separate system of payment has been arranged.

Trademark Notice: Product or corporate names may be trademarks or registered trademarks, and are used only for identification and explanation without intent to infringe.

Visit the Taylor & Francis Web site at
http://www.taylorandfrancis.com

and the CRC Press Web site at
http://www.crcpress.com

Printed and bound in the United States of America by Sheridan

To Teresa, Fernanda, and Arthur Orlande (HRBO)

To José Colaço (*in memorium*), Alice B. Colaço, and Roberta Viegas (MJC)

To Carolina, Bianca (*in memorium*), Victor, Clara, and Gabriel (RMC)

Contents

Preface xv
Preface—First Edition xix

1. **Basic Relations** 1
   1.1 Classification of Second-Order Partial Differential Equations 2
       1.1.1 Physical Significance of Parabolic, Elliptic, and
             Hyperbolic Systems 4
   1.2 Parabolic Systems 5
   1.3 Elliptic Systems 7
       1.3.1 Steady-State Diffusion 7
       1.3.2 Steady-State Advection–Diffusion 7
       1.3.3 Fluid Flow 8
   1.4 Hyperbolic Systems 8
   1.5 Systems of Equations 9
       1.5.1 Characterization of System of Equations 10
       1.5.2 Wave Equation 11
   1.6 Boundary Conditions 12
   1.7 Uniqueness of the Solution 15
   Problems 18

2. **Discrete Approximation of Derivatives** 23
   2.1 Taylor Series Formulation 24
       2.1.1 Finite Difference Approximation of First Derivative 25
       2.1.2 Finite Difference Approximation of Second
             Derivative 27
       2.1.3 Differencing via Polynomial Fitting 28
       2.1.4 Finite Difference Approximation of Mixed Partial
             Derivatives 29
       2.1.5 Changing the Mesh Size 31
       2.1.6 Finite Difference Operators 33
   2.2 Control Volume Approach 34
   2.3 Boundary and Initial Conditions 38
       2.3.1 Discretization of Boundary Conditions with
             Taylor Series 40
             2.3.1.1 Boundary Condition of the First Kind 41
             2.3.1.2 Boundary Conditions of the Second and
                     Third Kinds 41

		2.3.2	Discretization of Boundary Conditions with Control Volumes...42
			2.3.2.1 Boundary Condition of the First Kind...............43
			2.3.2.2 Boundary Condition of the Second Kind..........44
			2.3.2.3 Boundary Condition of the Third Kind.............44
	2.4	Errors Involved in Numerical Solutions..46
		2.4.1	Round-Off Errors...46
		2.4.2	Truncation Error..46
		2.4.3	Discretization Error...47
		2.4.4	Total Error..47
		2.4.5	Stability...48
		2.4.6	Consistency..48
	2.5	Verification and Validation..49
		2.5.1	Code Verification..50
		2.5.2	Solution Verification..54
	Problems..58
	Notes..62

3. Methods of Solving Systems of Algebraic Equations........................65
3.1 Reduction to Algebraic Equations 65
3.2 Direct Methods...70
  3.2.1 Gauss Elimination Method 71
  3.2.2 Thomas Algorithm..72
3.3 Iterative Methods...75
  3.3.1 Gauss–Seidel Iteration..75
  3.3.2 Successive Overrelaxation...79
  3.3.3 Red-Black Ordering Scheme 81
  3.3.4 LU Decomposition with Iterative Improvement.................83
  3.3.5 Biconjugate Gradient Method...83
3.4 Nonlinear Systems...84
Problems...88

4. One-Dimensional Steady-State Systems 97
4.1 Diffusive Systems...97
  4.1.1 Slab..97
  4.1.2 Solid Cylinder and Sphere..98
  4.1.3 Hollow Cylinder and Sphere...105
  4.1.4 Heat Conduction through Fins 110
    4.1.4.1 Fin of Uniform Cross Section 112
    4.1.4.2 Finite Difference Solution 113
4.2 Diffusive–Advective Systems 116
  4.2.1 Stability for Steady-State Systems..118
  4.2.2 Finite Volume Method..119
    4.2.2.1 Interpolation Functions...121
Problems...124

5. One-Dimensional Transient Systems 129
5.1 Diffusive Systems 129
    5.1.1 Simple Explicit Method 130
        5.1.1.1 Prescribed Potential at the Boundaries 131
        5.1.1.2 Convection Boundary Conditions 132
        5.1.1.3 Prescribed Flux Boundary Condition 133
        5.1.1.4 Stability Considerations 134
        5.1.1.5 Effects of Boundary Conditions on Stability ... 136
        5.1.1.6 Effects of r on Truncation Error 137
        5.1.1.7 Fourier Method of Stability Analysis 138
    5.1.2 Simple Implicit Method 146
        5.1.2.1 Stability Analysis 147
    5.1.3 Crank–Nicolson Method 148
        5.1.3.1 Prescribed Heat Flux Boundary Condition 151
    5.1.4 Combined Method 152
        5.1.4.1 Stability of Combined Method 154
    5.1.5 Cylindrical and Spherical Symmetry 155
    5.1.6 Application of Simple Explicit Method 156
        5.1.6.1 Solid Cylinder and Sphere 156
        5.1.6.2 Stability of Solution 160
        5.1.6.3 Hollow Cylinder and Sphere 162
    5.1.7 Application of Simple Implicit Scheme 164
        5.1.7.1 Solid Cylinder and Sphere 164
        5.1.7.2 Hollow Cylinder and Sphere 165
    5.1.8 Application of Crank–Nicolson Method 166
5.2 Advective–Diffusive Systems 169
    5.2.1 Purely Advective (Wave) Equation 169
        5.2.1.1 Upwind Method 170
        5.2.1.2 MacCormack's Method 172
        5.2.1.3 Warming and Beam's Method 173
    5.2.2 Advection–Diffusion Equation 179
        5.2.2.1 Simple Explicit Scheme 179
        5.2.2.2 Implicit Finite Volume Method 182
5.3 Hyperbolic Heat Conduction Equation 185
    5.3.1 Finite Difference Representation of Hyperbolic Heat Conduction Equation 186
Problems 190

6. Transient Multidimensional Systems 207
6.1 Simple Explicit Method 207
    6.1.1 Two-Dimensional Diffusion 208
    6.1.2 Two-Dimensional Transient Convection–Diffusion 213
        6.1.2.1 FTCS Differencing 213
        6.1.2.2 Upwind Differencing 214
        6.1.2.3 Control Volume Approach 215

6.2	Combined Method	219
6.3	ADI Method	220
6.4	ADE Method	224
6.5	An Application Related to the Hyperthermia Treatment of Cancer	228
Problems		243
Notes		251

7. Nonlinear Diffusion 253

7.1	Lagging Properties by One Time Step	254
7.2	Use of Three-Time-Level Implicit Scheme	256
	7.2.1 Internal Nodes	257
	7.2.2 Limiting Case R = 0 for Cylinder and Sphere	258
	7.2.3 Boundary Nodes	259
7.3	Linearization	261
	7.3.1 Stability Criterion	263
7.4	False Transient	264
	7.4.1 Simple Explicit Scheme	266
	7.4.2 Simple Implicit Scheme	267
	7.4.3 A Set of Diffusion Equations	267
7.5	Applications in Coupled Conduction and Radiation in Participating Media	268
	7.5.1 One-Dimensional Problem with Diffusion Approximation	268
	7.5.2 Solution of the Three-Dimensional Equation of Radiative Transfer	272
Problems		285

8. Multidimensional Incompressible Laminar Flow 291

8.1	Vorticity-Stream Function Formulation	291
	8.1.1 Vorticity and Stream Function	292
	8.1.2 Finite Difference Representation of Vorticity-Stream Function Formulation	295
	8.1.2.1 Vorticity Transport Equation	296
	8.1.2.2 Poisson's Equation for Stream Function	297
	8.1.2.3 Poisson's Equation for Pressure	298
	8.1.3 Method of Solution for ω and ψ	298
	8.1.3.1 Solution for a Transient Problem	298
	8.1.3.2 Solution for a Steady-State Problem	299
	8.1.4 Method of Solution for Pressure	301
	8.1.5 Treatment of Boundary Conditions	302
	8.1.5.1 Boundary Conditions on Velocity	302
	8.1.5.2 Boundary Conditions on ψ	303
	8.1.5.3 Boundary Condition on ω	304

Contents

		8.1.5.4	Boundary Conditions on Pressure	307
		8.1.5.5	Initial Condition	307
	8.1.6	Energy Equation		308
8.2	Primitive Variables Formulation			309
	8.2.1	Determination of the Velocity Field: The SIMPLEC Method		314
	8.2.2	Treatment of Boundary Conditions		320
		8.2.2.1	Pressure	321
		8.2.2.2	Momentum and Energy Equations	325
8.3	Two-Dimensional Steady Laminar Boundary Layer Flow			329
Problems				332

9. Compressible Flow 339
9.1 Quasi-One-Dimensional Compressible Flow 339
    9.1.1 Solution with MacCormack's Method 342
    9.1.2 Solution with WAF-TVD Method 348
9.2 Two-Dimensional Compressible Flow 354
Problems 358

10. Phase Change Problems 361
10.1 Mathematical Formulation of Phase Change Problems 363
    10.1.1 Interface Condition 364
    10.1.2 Generalization to Multidimensions 365
    10.1.3 Dimensionless Variables 366
    10.1.4 Mathematical Formulation 367
10.2 Variable Time Step Approach for Single-Phase Solidification 368
    10.2.1 Finite Difference Approximation 369
        10.2.1.1 Differential Equation 370
        10.2.1.2 Boundary Condition at $x = 0$ 370
        10.2.1.3 Interface Conditions 371
    10.2.2 Determination of Time Steps 371
        10.2.2.1 Starting Time Step Δt_0 371
        10.2.2.2 Time Step Δt_1 371
        10.2.2.3 Time Step Δt_n 372
10.3 Variable Time Step Approach for Two-Phase Solidification 374
    10.3.1 Finite Difference Approximation 376
        10.3.1.1 Equation for the Solid Phase 376
        10.3.1.2 Boundary Condition at $x = 0$ 376
        10.3.1.3 Equation for the Liquid Phase 377
        10.3.1.4 Interface Conditions 377
    10.3.2 Determination of Time Steps 377
        10.3.2.1 Starting Time Step Δt_0 378
        10.3.2.2 Time Step Δt_1 379
        10.3.2.3 Time Steps Δt_n, $(2 \leq n \leq N - 4)$ 380

		10.3.2.4	Time Step Δt_{N-3}	380
		10.3.2.5	Time Step Δt_{N-2}	381
		10.3.2.6	Time Step Δt_{N-1}	382

10.4 Enthalpy Method 383
    10.4.1 Explicit Enthalpy Method: Phase Change with
          Single Melting Temperature 385
          10.4.1.1 Algorithm for Explicit Method.......................386
          10.4.1.2 Interpretation of Enthalpy Results..................387
          10.4.1.3 Improved Algorithm for Explicit Method.....388
    10.4.2 Implicit Enthalpy Method: Phase Change with
          Single Melting Temperature 389
          10.4.2.1 Algorithm for Implicit Method.......................390
    10.4.3 Explicit Enthalpy Method: Phase Change over a
          Temperature Range 392
10.5 Phase Change Model for Convective–Diffusive Problems..........392
    10.5.1 Model for the Passive Scalar Transport Equation..........395
    10.5.2 Model for the Energy Equation...398
Problems...409

11. Numerical Grid Generation 411
11.1 Coordinate Transformation Relations..413
    11.1.1 Gradient 415
    11.1.2 Divergence 415
    11.1.3 Laplacian..416
    11.1.4 Normal Derivatives..416
    11.1.5 Tangential Derivatives...417
11.2 Basic Ideas in Simple Transformations 419
11.3 Basic Ideas in Numerical Grid Generation and Mapping 422
11.4 Boundary Value Problem of Numerical Grid Generation 429
11.5 Finite Difference Representation of Boundary
     Value Problem of Numerical Grid Generation 436
11.6 Steady-State Heat Conduction in Irregular Geometry 439
11.7 Steady-State Laminar Free Convection in Irregular
     Enclosures—Vorticity-Stream Function Formulation...................445
    11.7.1 The Nusselt Number...454
    11.7.2 Results 455
11.8 Transient Laminar Free Convection in Irregular
     Enclosures—Primitive Variables Formulation 457
11.9 Computational Aspects for the Evaluation of Metrics 461
    11.9.1 One-Dimensional Advection–Diffusion Equation..........461
    11.9.2 Two-Dimensional Heat Conduction in a
          Hollow Sphere 465
Problems...469
Notes..473

12. Hybrid Numerical–Analytical Solutions 477
    12.1 Combining Finite Differences and Integral Transforms 479
          12.1.1 The Hybrid Approach 480
          12.1.2 Hybrid Approach Application: Transient Forced
                    Convection in Channels 481
    12.2 Unified Integral Transforms 489
          12.2.1 Total Transformation 491
          12.2.2 Partial Transformation 493
          12.2.3 Computational Algorithm 497
          12.2.4 Test Case 501
    12.3 Convective Eigenvalue Problem 505
    Problems 517

Appendix A. Subroutine Gauss 527

Appendix B. Subroutine Trisol 529

Appendix C. Subroutine SOR 531

Appendix D. Subroutine BICGM2 533

Appendix E. Program to Solve Example 10.1 541

Bibliography 545

Index 565

Preface

Computational methods for the solution of heat transfer problems have evolved significantly since the publication of the first edition of Professor M. Necati Özişik's book, *Finite Difference Methods in Heat Transfer*. In the meantime, the application of computational techniques to analysis and design in thermal engineering has become a common practice. Nowadays, several commercial computer codes, which can be applied to problems involving single or coupled heat transfer modes, multiple scales, and multiple physical phenomena, are available. Therefore, as the computational methods have gone beyond academic research to become practical engineering tools, understanding their fundamentals and limitations has become mandatory for engineering undergraduate students and practitioners.

Professor Özişik's finite difference book was published in 1994 and was intended as a textbook for advanced senior or graduate courses. It covered the basic discretization techniques and presented applications to transient and steady-state problems in conduction and convection heat transfer. This second edition of *Finite Difference Methods in Heat Transfer* includes new and updated methods, applications, examples, and proposed problems. Chapters and sessions have been rearranged with the objective of making the text clearer and more organized, without losing the identity and the most important contents of the original book. As in the first edition, required mathematical derivations have been kept to a minimum, but the methods are sufficiently detailed in the book to be readily applied by the reader. The modifications introduced in the second edition include, among others, the concepts of verification and validation; detailed discretizations of boundary conditions; new methods for the solution of linear systems of equations; new model equations, such as for quasi-one-dimensional compressible flow; algorithms for pressure-velocity coupling in laminar incompressible flows, such as the SIMPLEC method; modern methods for the solution of hyperbolic systems based on solution features; the discretization and solution of the equation of radiative transfer; phase change problems with convection in the liquid phase and with a solute dispersed in the base material; numerical grid generation in cylindrical and polar coordinates; and recent advancements in the hybrid analytical–numerical generalized integral transform technique (GITT). Seventy-two new proposed problems have been included in the second edition to allow the reader to have practical experience with the added topics.

Although Chapters 5 through 9 have been reorganized in terms of the physical phenomena for which the finite difference method is intended, the same structure of the original book has been kept with respect to the basic and advanced materials covered. Chapters 1 through 6 contain basic material that can be used in a senior-level undergraduate course or an entry-level graduate course for the student's first contact with discretization techniques by finite differences and their application to elliptic, parabolic, and hyperbolic problems. More involved material is covered in Chapters 7 through 12, which can be used in an advanced course of finite differences at the graduate level.

Chapter 1 introduces the classification of second-order partial differential equations and presents examples of elliptic, parabolic, and hyperbolic physical equations. The discretization of partial derivatives using the Taylor series approach is presented in Chapter 2. In that chapter, the discretization of partial differential equations by the finite control volume approach is also introduced. (We keep the same notation as the original book, where finite volume was considered a discretization approach instead of a method by itself.) The discretization of boundary conditions of the first, second, and third kinds is also described in Chapter 2, both for Taylor series and control volume discretization approaches. The concepts of verification and validation of computer codes are discussed in Chapter 2, and techniques for code and solution verification are presented, based on the ASME verification and validation standard.[*] Direct and iterative methods for the solution of systems of algebraic equations, which result from the finite difference discretization of the original continuous problems, are presented in Chapter 3. Chapters 4 through 6 deal with the application of finite difference methods to one-dimensional steady-state problems, one-dimensional transient problems, and transient multidimensional problems, respectively. Discretization schemes specific for each kind of problem are presented in detail in these chapters, which are organized in a sequence of increasing difficulty for a better understanding by the reader. Steady-state multidimensional diffusion problems are considered as a special case of the transient problems in the notes of Chapter 6.

Nonlinear diffusion problems are addressed in Chapter 7, while Chapters 8 and 9 deal with incompressible and compressible flows, respectively. The solution of phase change problems is examined in Chapter 10, where both diffusive and advective–diffusive problems are considered. The elliptic numerical grid generation technique is described in Chapter 11, together with the transformation relations for several operators. Although this technique generates structured grids, which lack the geometrical flexibility of unstructured grids, it readily allows for the extension of the methods presented in the previous chapters to problems in irregular geometries. Finally, Chapter 12 presents the generalized integral transform technique as an alternative hybrid numerical–analytical approach, which utilizes the finite difference method in its numerical portion, as covered in earlier chapters.

---

[*] ASME V&V 20-2009 (2009), Standard for Verification and Validation in Computational Fluid Dynamics and Heat Transfer, ASME, New York.

This book is accompanied by software for elliptic numerical grid generation and solution of transient diffusion problems in two-dimensional geometries, including doubly connected regions. Boundary conditions are allowed to vary pointwise, and the temperature profile can be obtained at some specified time for the entire domain or for some point in the domain as a function of time. The software and its user guides can be downloaded from: http://colaco.freeshell.org/books/Ozisik/.

Writing the new edition of a book is a difficult task, especially when the single author of the previous edition cannot be consulted about the convenience of adding new authors and modifying his original work.* It is necessary to update and include new material without losing the characteristics of the original work, which was not conceived or written by the authors responsible for the new edition. Under these circumstances, we consider completely inappropriate the situation, rather common nowadays, where the names of new edition authors, who were not authors of a previous edition, appear ahead of the name of the deceased author of the original work. Therefore, we prefer that this second edition of *Finite Difference Methods in Heat Transfer* always be considered as Professor Özişik's book, which we contributed to by updating and enlarging his original material. Professor Özişik dedicated his life to education and research in heat transfer and was an iconic worldwide leader in thermal sciences. He published more than 300 research papers in international journals and conferences. He was the author of 11 books, most of them are best sellers that were reedited several times and published in different languages.

The invitation by Mr. Jonathan W. Plant, Executive Editor for Mechanical, Aerospace & Nuclear Engineering from CRC Press/Taylor & Francis Group, to write this second edition of *Finite Difference Methods in Heat Transfer*, was greatly appreciated. It was a great honor to receive an invitation to work on this prestigious book, originally conceived and written by our former friend and advisor. We are also indebted to Professor Afshin J. Ghajar, the Heat Transfer Series Editor for CRC Press/Taylor & Francis, for his encouragement and support as we pursued this challenging task.

The revision work was originally divided in closely equal parts among the three new coauthors (HRBO, MJC, and RMC). However, RMC was nominated as president of the National Commission of Nuclear Energy in Brazil, a government key position, which drastically reduced the time available for him to complete his share of work. Therefore, HRBO and MJC offered to take on more than half of the work still attributed to RMC, which allowed for the timely conclusion of the new edition. RMC is deeply grateful to his coauthors and editors for their help and understanding.

---

* Professor M. Necati Özişik passed away on October 4, 2008.

We are also thankful for the cooperation of the CRC Press/Taylor & Francis Group editorial staff. We would like to express our deepest gratitude for the continuous financial support provided by agencies of the Brazilian government, including CNPq, CAPES, FAPERJ, and ANP/PRH37.

Helcio R.B. Orlande
Marcelo José Colaço
Renato Machado Cotta

Preface—First Edition

In recent years, with ever growing availability of high speed, large capacity computers, the interest in the use of numerical methods, such as the finite difference and finite element methods, for solving problems governed by differential equations has increased significantly. Many complicated engineering problems can now be solved with computers at a very little cost in a very short time. Therefore, practicing engineers and engineering students are interested more than ever in strengthening their background in the use of numerical techniques.

The finite difference and finite element methods are now two universally used approaches for solving linear and nonlinear differential equations governing engineering problems. Depending on the nature of the problem each method has its advantages. Finite difference methods are simple to formulate, can readily be extended to two- or three-dimensional problems, are easy to learn and apply. The finite element method has the flexibility in dealing with problems involving irregular geometry. However, with the advent of numerical grid generation technique, the finite difference method now possesses the geometrical flexibility of the finite element method while maintaining the simplicity of the conventional finite difference technique.

This book presents the finite difference techniques and their use in the solution of engineering problems governed by differential equations, with particular emphasis on applications in the areas of heat transfer and fluid flow. A variety of topics covered include the steady-state and transient problems of heat conduction, duct flow, convective-diffusive systems, nonlinear diffusion, hyperbolic heat conduction, melting–solidification, theory and application of the numerical grid generation techniques, and the hybrid method of solution.

The order of coverage of the material gradually advances from the elementary treatment of the subject given in Chapters 1 through 5, to more general applications presented in Chapters 6 through 10; finally more recent topics such as the numerical grid generation technique and the hybrid method of solution are treated in Chapters 11 and 12. Therefore, this book can be used for self-study and is useful as a text for a senior-level elective or first semester graduate-level course.

The required mathematics background needed to follow the material in this book is limited to elementary calculus except for Chapters 11 and 12, for which familiarity with partial differential equations is needed.

Chapter 1 gives a classification of second-order partial differential equations and discusses the physical significance of parabolic-, hyperbolic-, and elliptic-type partial differential equations in relation to the physical problem governed by them. Chapter 2 presents a step-by-step description of discretization procedure by using Taylor series and the control-volume approach. Chapter 3 describes the salient features of various algorithms for solving systems of algebraic equations resulting from the discretization of partial differential equations governing the physical problem. Chapter 4 is an introduction to the application of finite differences with examples chosen from one-dimensional steady-state diffusive and convective-diffusive systems. Chapter 5 presents various finite difference schemes, such as the explicit, implicit, combined, etc. methods for solving one-dimensional time-dependent diffusion problems. Chapter 6 describes the application of finite difference techniques for the solution of typical two-dimensional problems of diffusion and convection.

Chapter 7 illustrates the use of finite difference methods for solving two-dimensional steady-state diffusion, velocity, and temperature fields. Chapter 8 is concerned with the solution of hyperbolic heat conduction problems while Chapter 9 presents the methods of solving nonlinear diffusion problems. Chapter 10 is devoted to the solution of one-dimensional melting or solidification problems for materials having a single phase-change temperature as well as phase-change taking place over a temperature range. Chapter 11 presents the fundamentals of numerical grid generation technique to transform an irregular physical region into a regular computational domain. The problem is solved in the regular domain by the application of the standard finite-differences and the results are transformed back to the physical domain by numerical coordinate transformation. The application is illustrated by representative examples. Finally, Chapter 12 presents a hybrid approach that combines the generalized integral transform technique and finite differences for the solution of multidimensional transient and steady-state heat transfer problems containing variable coefficients in the differential equation and/or boundary conditions, as well as nonlinear diffusion–convection type problems. The application of the hybrid method is illustrated with problems chosen in the area of transient forced convection in a parallel-plate duct, and pertinent recent references are cited.

I would like to acknowledge valuable discussion with Dr. Woo-Seung Kim during the preparation of this manuscript.

I dedicate this book to my wife, Gül, without her patience and understanding this project would not have been realized.

M. Necati Özişik

1
Basic Relations

Numerical methods are useful for solving fluid dynamics, heat and mass transfer problems, and other partial differential equations of mathematical physics when such problems cannot be handled by exact analysis techniques because of nonlinearities, complex geometries, and complicated boundary conditions. The development of high-speed digital computers significantly enhanced the use of numerical methods in various branches of science and engineering. Many complicated problems can now be solved at a very little cost and in a very short time with the available computing power.

Presently, the *finite difference method* (FDM), the *finite volume method* (FVM), and the *finite-element method* (FEM) are widely used for the solution of partial differential equations of heat, mass, and momentum transfer. Extensive amounts of literature exist on the application of these methods for the solution of such problems. Each method has its advantages depending on the nature of the physical problem to be solved, but there is no best method for all problems. For instance, the dimension of the problem is an important factor that deserves some consideration because an efficient method for one-dimensional problems may not be so efficient for two- or three-dimensional problems. FDMs are simple to formulate and can readily be extended to two- or three-dimensional problems. Furthermore, FDM is very easy to learn and apply for the solution of partial differential equations encountered in the modeling of engineering problems for simple geometries. For problems involving irregular geometries in the solution domain, the FEM is known for having more flexibility because the region near the boundary can readily be divided into subregions. A major drawback of FDM used to be its difficulty to handle effectively the solution of problems over arbitrarily-shaped complex geometries because of interpolation between the boundaries and the interior points, in order to develop finite difference expressions for nodes next to the boundaries. More recently, with the advent of numerical grid generation approaches, the FDM has become comparable to FEM in dealing with irregular geometries, while still maintaining the simplicity of the standard FDM.

In this book, we are concerned with the use of FDMs for the solution of heat, mass, and momentum transport problems encountered in engineering applications. Despite the simplicity of the finite difference representation of governing partial differential equations, it requires considerable experience and knowledge to select the appropriate scheme for a specific problem in hand. The type of partial differential equations, the number of physical dimensions, the type of coordinate system involved, whether the governing equations and boundary conditions are linear or nonlinear, and whether the problem is steady-state or transient are among the factors that affect the type of numerical scheme chosen from a large number of available possibilities. The tailoring of a numerical method for a specific problem in hand is an important first step in the numerical solution with a FDM. In this chapter, we present a classification of partial differential equations encountered in the mathematical formulation of heat, mass, and momentum transfer problems, and we discuss the physical significance of such a classification in relation to the numerical solution of the problem.

1.1 Classification of Second-Order Partial Differential Equations

In the solution of partial differential equations with finite differences, the choice of a particular finite differencing scheme also depends on the type of partial differential equation considered. Generally, partial differential equations are classified into three categories: *elliptic, parabolic,* and *hyperbolic*. To illustrate, we consider the following general second-order partial differential equation with two independent variables x, y, as presented by Forsythe and Wasow (1967):

$$A\frac{\partial^2 \phi}{\partial x^2} + B\frac{\partial^2 \phi}{\partial x \partial y} + C\frac{\partial^2 \phi}{\partial y^2} + D\frac{\partial \phi}{\partial x} + E\frac{\partial \phi}{\partial y} + F\phi + G(x,y) = 0. \tag{1.1}$$

Here, we assume a linear equation (this restriction is not essential), that is, the coefficients A, B, C, D, E, and F, and the known source term G are functions of the two independent variables x, y but not of the dependent variable ϕ.

For heat and mass transfer and fluid flow problems, the generalized dependent variable ϕ denotes a specific dependent variable such as temperature, concentration, velocity component, density, or pressure. The mathematical character of the partial differential equation (1.1) depends on the coefficients of the higher order terms, A, B, and C. The partial differential equation (1.1) at a point (x_0, y_0) is called:

$$\text{Elliptic, if } B^2 - 4AC < 0, \tag{1.2a}$$

$$\text{Parabolic, if } B^2 - 4AC = 0, \tag{1.2b}$$

$$\text{Hyperbolic, if } B^2 - 4AC > 0. \tag{1.2c}$$

For example, the steady-state heat conduction equation with no energy generation and constant properties is shown as follows:

$$\frac{\partial^2 T}{\partial x^2} + \frac{\partial^2 T}{\partial y^2} = 0, \tag{1.3a}$$

where the two-dimensional Laplace's equation is *elliptic* as verified by setting A = 1, B = 0, and C = 1. The steady-state heat conduction equation with energy generation is

$$\frac{\partial^2 T}{\partial x^2} + \frac{\partial^2 T}{\partial y^2} + \frac{1}{k} g(x, y) = 0, \tag{1.3b}$$

which is Poisson's equation and also is *elliptic*.

The one-dimensional time-dependent heat conduction equation

$$\frac{\partial^2 T}{\partial x^2} = \frac{1}{\alpha} \frac{\partial T}{\partial t}, \tag{1.4}$$

is *parabolic*, as can be verified by letting the independent variable t be represented by y and setting A = 1, B = 0, and C = 0.

The second-order wave equation

$$\frac{\partial^2 \phi}{\partial x^2} = \frac{1}{c^2} \frac{\partial^2 \phi}{\partial t^2}, \tag{1.5}$$

where t is the time, x is the space variable, and c is the wave propagation speed, is *hyperbolic*, as can be verified by letting the independent variable t be represented by *y* and setting A = 1, B = 0, and C = $-\frac{1}{c^2}$.

The non-Fourier heat conduction equation

$$\frac{\partial^2 T}{\partial x^2} = \frac{1}{c^2} \frac{\partial^2 T}{\partial t^2} + \frac{1}{\alpha} \frac{\partial T}{\partial t}, \tag{1.6}$$

which is a second-order damped wave equation, is also *hyperbolic*.

For simplicity, we consider the partial differential equation (1.1) in only two independent variables (x, y). The extension to three or more independent variables is straightforward.

For example, the three-dimensional, steady-state heat conduction equation

$$\frac{\partial^2 T}{\partial x^2} + \frac{\partial^2 T}{\partial y^2} + \frac{\partial^2 T}{\partial z^2} + \frac{1}{k} g(x, y, z) = 0, \tag{1.7}$$

is *elliptic*.

The two-dimensional, transient heat conduction equation

$$\frac{\partial^2 T}{\partial x^2} + \frac{\partial^2 T}{\partial y^2} + \frac{1}{k}g(x,y,t) = \frac{1}{\alpha}\frac{\partial T}{\partial t}, \qquad (1.8)$$

is *parabolic*.

1.1.1 Physical Significance of Parabolic, Elliptic, and Hyperbolic Systems

In the foregoing discussion, we considered a purely mathematical criterion given by equations (1.2a)–(1.2c) to classify the second-order partial differential equation (1.1) into parabolic, elliptic, and hyperbolic categories. We now discuss the physical significance of such a classification in the computational and physical aspects.

Consider, for example, the steady-state heat conduction equation (1.3a) or (1.3b), which has second-order partial derivatives in both x and y variables. The conditions at any given location are influenced by changes in conditions at both sides of that location, whether the changes are in the x variable or in the y variable. Thus, the steady-state heat conduction equation is elliptic in both x and y space coordinates and is simply called *elliptic*. The main characteristic of the elliptic equation is that it models a steady-state or equilibrium diffusion process and requires the specification of appropriate boundary conditions at all boundaries.

Now let us consider the one-dimensional, time-dependent heat conduction equation (1.4), which has a second-order partial derivative in the x variable and a first-order partial derivative in the time variable. The conditions at any given location x are influenced by changes in conditions at both sides of that location; hence, the equation is regarded *elliptic* in the x variable. However, in the time variable t, the conditions at any instant are influenced only by changes taking place in conditions at times *earlier* than that time; hence, the equation is parabolic in time and is called *parabolic*. Note that the equation is called *parabolic* if there exists at least one coordinate (i.e., time or space) in which the conditions at any given location (i.e., time or space) are influenced by changes in conditions only at one side (i.e., earlier time or upstream) of that location. The main characteristic of the parabolic equation is that it models a transient state or evolution diffusion process and requires the specification of appropriate boundary conditions at all boundaries plus an initial condition at the starting point of the evolution process. As the temperature field at any time is not affected by the temperature field at *later* times, one starts with a given *initial* temperature field and *marches* forward to compute the temperature fields at successive time steps.

In the case of a *hyperbolic* equation, such as the hyperbolic heat conduction equation (1.6), it exhibits a wavelike propagation of the temperature field with a finite speed, in contrast to the infinite speed of propagation associated with the parabolic heat conduction equation (1.4). Therefore, the solution of

Basic Relations

hyperbolic equations with finite differences requires special considerations and special schemes. The main characteristic of the hyperbolic equation (1.6) is that it models a transient state or evolution propagation process and requires the specification of appropriate boundary conditions at all boundaries plus initial conditions at the starting point of the evolution process—both for the potential and its first derivative in time.

1.2 Parabolic Systems

There is a wide variety of engineering problems that are governed by partial differential equations of the parabolic type. Instead of testing them each time in order to determine whether an equation is parabolic or not, it is convenient to introduce a sufficiently general parabolic differential equation from which numerous other parabolic equations of transport phenomena can be obtained as special cases.

The *general parabolic* conservation equation governing the behavior of some *unknown potential* ϕ in the two-dimensional rectangular coordinates system (y and z space variables) can be written in *conservative form* as*

$$\frac{\partial}{\partial t}(\gamma\phi) + \frac{\partial}{\partial y}(\beta v\phi) + \frac{\partial}{\partial z}(\beta w\phi) = \frac{\partial}{\partial y}\left(\Gamma\frac{\partial\phi}{\partial y}\right) + \frac{\partial}{\partial z}\left(\Gamma\frac{\partial\phi}{\partial z}\right) + S, \quad (1.9)$$

where ϕ is the general dependent variable, Γ is the generalized diffusion coefficient, β and γ are specified coefficients, and S is the specified volumetric source term. In addition, v and w are the velocity components in the y and z directions, respectively. This can readily be generalized for the three space variables, x, y, and z.

The first term on the left-hand side of equation (1.9) represents the *unsteady* term, and the second and third terms represent *advection*. On the right-hand side of this equation, the first and second terms represent *diffusion*, and the last term represents the *volumetric source*.

The physical significances of the dependent variable ϕ, the coefficients β, γ, Γ, and the source term S depends on the physical nature of the problem considered. The following examples illustrate this matter:

---

* The *conservative form* implies that the coefficients of the derivative terms are either constant or, if variable, that their derivatives do not appear in the equation. For example, in the case of a heat conduction equation with space-dependent thermal conductivity, the conservative form is given as

$$\rho C_p \frac{\partial T}{\partial t} = \frac{\partial}{\partial x}\left(k\frac{\partial T}{\partial x}\right),$$

and the nonconservative form as

$$\rho C_p \frac{\partial T}{\partial t} = k\frac{\partial^2 T}{\partial x^2} + \frac{\partial k}{\partial x}\frac{\partial T}{\partial x}.$$

(1) For two-dimensional, transient heat conduction with constant heat capacity, we set:

$\phi \equiv T$, temperature
$\gamma \equiv \rho C_p$, heat capacity
$\beta \equiv 0$ no motion
$\Gamma \equiv k$, thermal conductivity
$S \equiv g$, volumetric energy generation rate

Then equation (1.9) becomes

$$\rho C_p \frac{\partial T}{\partial t} = \frac{\partial}{\partial y}\left(k\frac{\partial T}{\partial y}\right) + \frac{\partial}{\partial z}\left(k\frac{\partial T}{\partial z}\right) + g. \tag{1.10}$$

(2) For two-dimensional, transient, convection–diffusion with constant heat capacity, we set:

$\phi \equiv T$, temperature
$\gamma \equiv \rho C_p$, heat capacity
$\beta \equiv \rho C_p$, heat capacity
$\Gamma \equiv k$, thermal conductivity
$S \equiv g$, volumetric energy generation rate

Then equation (1.9) takes the following form:

$$\rho C_p \left[\frac{\partial T}{\partial t} + \frac{\partial}{\partial y}(vT) + \frac{\partial}{\partial z}(wT)\right] = \frac{\partial}{\partial y}\left(k\frac{\partial T}{\partial y}\right) + \frac{\partial}{\partial z}\left(k\frac{\partial T}{\partial z}\right) + g. \tag{1.11}$$

(3) For two-dimensional, transient, mass advection–diffusion without mass retention effects, we set:

$\phi \equiv C$, mass concentration
$\beta = \gamma = 1$
$\Gamma \equiv D$, diffusion coefficient
$S \equiv S^*$, mass generation rate per unit volume

Then equation (1.9) becomes

$$\frac{\partial}{\partial t}(C) + \frac{\partial}{\partial y}(vC) + \frac{\partial}{\partial z}(wC) = \frac{\partial}{\partial y}\left(D\frac{\partial C}{\partial y}\right) + \frac{\partial}{\partial z}\left(D\frac{\partial C}{\partial z}\right) + S^*. \tag{1.12}$$

In the foregoing examples, we considered applications of equation (1.9) for different cases in heat and mass transfer. The momentum equations, for example, are readily obtained from equation (1.9) by proper interpretation of the coefficients β, γ, and the source term S.

1.3 Elliptic Systems

The problems of steady-state diffusion, convection–diffusion, and some fluid flow problems are governed by partial differential equations that are *elliptic*. To illustrate this matter, we present the governing differential equations for each of these three types of problems in the two-dimensional rectangular coordinates system. The generalization to three dimensions is straightforward.

1.3.1 Steady-State Diffusion

The problems of steady-state heat or mass diffusion are governed by the *Poisson equation*, which is *elliptic* and can be written in the following form:

$$\frac{\partial}{\partial x}\left(\Gamma \frac{\partial \phi}{\partial x}\right) + \frac{\partial}{\partial y}\left(\Gamma \frac{\partial \phi}{\partial y}\right) + S = 0, \tag{1.13}$$

where

ϕ = unknown scalar potential that can represent the temperature T or the mass concentration C,

Γ = generalized diffusion coefficient that can represent the thermal conductivity k or the mass diffusion coefficient D,

S = specified volumetric source term that can be energy generation rate g or mass production rate S^*, per unit volume.

For example, for heat conduction, equation (1.13) becomes

$$\frac{\partial}{\partial x}\left(k \frac{\partial T}{\partial x}\right) + \frac{\partial}{\partial y}\left(k \frac{\partial T}{\partial y}\right) + g = 0, \tag{1.14}$$

and for mass diffusion, it takes the form

$$\frac{\partial}{\partial x}\left(D \frac{\partial C}{\partial x}\right) + \frac{\partial}{\partial y}\left(D \frac{\partial C}{\partial y}\right) + S^* = 0. \tag{1.15}$$

1.3.2 Steady-State Advection–Diffusion

The problems of steady-state heat or mass transfer involving advection and diffusion are also governed by *elliptic* partial differential equations that can be written in the form

$$\frac{\partial}{\partial x}(\beta u \phi) + \frac{\partial}{\partial y}(\beta v \phi) = \frac{\partial}{\partial x}\left(\Gamma \frac{\partial \phi}{\partial x}\right) + \frac{\partial}{\partial y}\left(\Gamma \frac{\partial \phi}{\partial y}\right) + S, \tag{1.16}$$

where the unknown potential ϕ, the generalized diffusion coefficient Γ, and the source term S have been defined previously. The coefficient β is taken as

$\beta = \rho C_p$ for heat transfer and $\beta = 1$ for mass transfer. In addition, u and v are velocity components in the x and y directions, respectively.

For example, for heat transfer in an incompressible flow, equation (1.16) becomes

$$\rho C_p \left[\frac{\partial}{\partial x}(uT) + \frac{\partial}{\partial y}(vT) \right] = \frac{\partial}{\partial x}\left(k\frac{\partial T}{\partial x}\right) + \frac{\partial}{\partial y}\left(k\frac{\partial T}{\partial y}\right) + g, \qquad (1.17a)$$

and for mass transfer, it takes the form

$$\frac{\partial}{\partial x}(uC) + \frac{\partial}{\partial y}(vC) = \frac{\partial}{\partial x}\left(D\frac{\partial C}{\partial x}\right) + \frac{\partial}{\partial y}\left(D\frac{\partial C}{\partial y}\right) + S^*. \qquad (1.17b)$$

We note that equation (1.13) is obtainable from equation (1.16) by setting $\beta = 0$ in the latter.

1.3.3 Fluid Flow

The equations governing the steady-state subsonic forced compressible flow of an isothermal Newtonian fluid in the two-dimensional rectangular coordinates system are given by

x-momentum

$$\frac{\partial}{\partial x}(\rho uu) + \frac{\partial}{\partial y}(\rho vu) = -\frac{\partial p}{\partial x} + \frac{\partial}{\partial x}\left(\mu\frac{\partial u}{\partial x}\right) + \frac{\partial}{\partial y}\left(\mu\frac{\partial u}{\partial y}\right), \qquad (1.18)$$

y-momentum

$$\frac{\partial}{\partial x}(\rho uv) + \frac{\partial}{\partial y}(\rho vv) = -\frac{\partial p}{\partial y} + \frac{\partial}{\partial x}\left(\mu\frac{\partial v}{\partial x}\right) + \frac{\partial}{\partial y}\left(\mu\frac{\partial v}{\partial y}\right), \qquad (1.19)$$

and the continuity equation

$$\frac{\partial}{\partial x}(\rho u) + \frac{\partial}{\partial y}(\rho v) = 0, \qquad (1.20)$$

where μ is the viscosity, p is the pressure, and ρ is the density. These are a set of three nonlinear coupled *elliptic* partial differential equations for the three unknowns, u, v, and p. The system is closed when the proper boundary conditions are specified and the equation of state is provided that relates ρ to T and p.

1.4 Hyperbolic Systems

Problems governed by *hyperbolic* partial differential equations are encountered in a number of applications in heat and fluid flow. For example, transient heat conduction associated with laser pulses of extremely short duration,

Basic Relations

extremely high rates of change of temperature or heat fluxes, or extremely low temperatures approaching absolute zero may be governed by the hyperbolic heat conduction equation (1.6) instead of by the customarily used parabolic heat conduction equation (1.4).

The simplest hyperbolic equation is the first-order linear wave equation given by

$$\frac{1}{c}\frac{\partial u}{\partial t} + \frac{\partial u}{\partial x} = 0, \quad c > 0, \tag{1.21}$$

which governs the wave propagation in the x direction with a speed c. Other examples of hyperbolic systems include the classical, second-order linear wave equation, which can be obtained from equation (1.21),

$$\frac{\partial^2 u}{\partial t^2} = c^2 \frac{\partial^2 u}{\partial x^2}, \quad c > 0, \tag{1.22}$$

for the propagation of sound waves and the hyperbolic heat conduction equation

$$\tau \frac{\partial^2 T}{\partial t^2} + \frac{\partial T}{\partial t} = \alpha \frac{\partial^2 T}{\partial x^2}, \tag{1.23a}$$

where the *relaxation time* τ is defined as

$$\tau = \frac{\alpha}{c^2}. \tag{1.23b}$$

Equation (1.23a) has resulted by combining the non-Fourier heat flux model,

$$\tau \frac{\partial q}{\partial t} + q = -k \frac{\partial T}{\partial x}, \tag{1.24}$$

with the energy equation

$$-\frac{\partial q}{\partial x} = \rho C_p \frac{\partial T}{\partial t}, \tag{1.25}$$

in order to eliminate the heat flux q. We note that for $\tau = 0$, equation (1.23a) reduces to the classical diffusion equation.

1.5 Systems of Equations

In many engineering applications, the physical processes are governed by a system of equations rather than by a single equation. When such equations are to be solved numerically, it is often convenient to combine them into a compact vector form. In some situations, a higher-order partial differential equation can be converted into a system of first-order equations.

Consider a system of first-order partial differential equations in the two independent variables (x,t) expressed in the vector form as

$$\frac{\partial \mathbf{T}}{\partial t} + \frac{\partial \mathbf{F}(\mathbf{T})}{\partial x} + \mathbf{H} = 0, \qquad (1.26)$$

where $\mathbf{T}$ is the unknown component vector, $\mathbf{F}(\mathbf{T})$ is a given vector that is a function of $\mathbf{T}$, and $\mathbf{H}$ is a given source term vector.

For simplicity, we choose a system consisting of two equations; then various vectors in equation (1.26) are defined as

$$\mathbf{T} = \begin{bmatrix} T_1 \\ T_2 \end{bmatrix}, \mathbf{F}(\mathbf{T}) = \begin{bmatrix} F_1(\mathbf{T}) \\ F_2(\mathbf{T}) \end{bmatrix}, \mathbf{H} = \begin{bmatrix} H_1 \\ H_2 \end{bmatrix}, \qquad (1.27a,b,c)$$

$$\frac{\partial \mathbf{T}}{\partial t} = \begin{bmatrix} \frac{\partial T_1}{\partial t} \\ \frac{\partial T_2}{\partial t} \end{bmatrix}, \frac{\partial \mathbf{F}}{\partial x} = \begin{bmatrix} \frac{\partial F_1}{\partial x} \\ \frac{\partial F_2}{\partial x} \end{bmatrix}. \qquad (1.28a,b)$$

The generalization to a system of more than two equations is a straightforward matter.

Equation (1.26) can be expressed in the quasilinear form as

$$\frac{\partial \mathbf{T}}{\partial t} + \mathbf{A} \frac{\partial \mathbf{T}}{\partial x} + \mathbf{H} = 0, \qquad (1.29)$$

where the Jacobian matrix $\mathbf{A}$, for the case of two equations, is given by

$$\mathbf{A} = \frac{\partial \mathbf{F}(\mathbf{T})}{\partial \mathbf{T}} = \begin{bmatrix} \frac{\partial F_1}{\partial T_1} & \frac{\partial F_1}{\partial T_2} \\ \frac{\partial F_2}{\partial T_1} & \frac{\partial F_2}{\partial T_2} \end{bmatrix}. \qquad (1.30)$$

1.5.1 Characterization of System of Equations

An understanding of the behavior of a system of equations, namely, whether it is hyperbolic or elliptic, is important in the selection of an appropriate finite difference scheme for its solution. The system given by equation (1.29) is *hyperbolic* if the eigenvalues of the coefficient matrix $\mathbf{A}$ are all real and distinct and *elliptic* if they are all complex (Richtmyer and Morton 1967; Zahmanoglou and Thoe 1976).

We now generalize this system [equation (1.29)] for the case of two independent spatial variables (x, y) expressed in the form

$$\frac{\partial \mathbf{T}}{\partial t} + \mathbf{A} \frac{\partial \mathbf{T}}{\partial x} + \mathbf{B} \frac{\partial \mathbf{T}}{\partial y} + \mathbf{H} = 0, \qquad (1.31)$$

Basic Relations

where **A** and **B** are the coefficient matrices. Richtmyer and Morton (1967) identify the system [equation (1.31)] as being hyperbolic in the x direction if the eigenvalues of **A** are real and distinct. Similarly, the behavior of the system [equation (1.31)] is said to be hyperbolic in the y direction if the eigenvalues of the matrix **B** are real and distinct. Therefore, it is possible for the system to exhibit hyperbolic behavior in (x, t) and elliptic behavior in (y, t) or vice versa, depending on the nature of the coefficient matrices **A** and **B**.

1.5.2 Wave Equation

Consider the wave equation (1.22) written as

$$\frac{\partial^2 u}{\partial t^2} - c^2 \frac{\partial^2 u}{\partial x^2} = 0. \tag{1.32a}$$

For a constant c, let

$$u_1 = \frac{\partial u}{\partial t} \text{ and } u_2 = c\frac{\partial u}{\partial x}, \tag{1.32b,c}$$

so that equation (1.32a) can be split-up into two first-order equations in the form

$$\frac{\partial u_1}{\partial t} - c\frac{\partial u_2}{\partial x} = 0, \tag{1.33a}$$

$$\frac{\partial u_2}{\partial t} - c\frac{\partial u_1}{\partial x} = 0, \tag{1.33b}$$

and the resulting two equations can be expressed in the matrix form as

$$\frac{\partial \mathbf{U}}{\partial t} + \mathbf{A}\frac{\partial \mathbf{U}}{\partial x} = 0, \tag{1.34a}$$

where

$$\mathbf{U} = \begin{bmatrix} u_1 \\ u_2 \end{bmatrix} \text{ and } \mathbf{A} = \begin{bmatrix} 0 & -c \\ -c & 0 \end{bmatrix} \tag{1.34b,c}$$

The wave equation (1.32a) is hyperbolic; therefore, the aforementioned split-up form should also retain its hyperbolic character. The eigenvalues of the matrix **A** are determined from

$$\det\{\mathbf{A} - \lambda \mathbf{I}\} = 0 \tag{1.35a}$$

where **I** is the identity matrix. This result leads to

$$\lambda^2 - c^2 = 0 \tag{1.35b}$$

which gives the eigenvalues as $\lambda_1 = c$ and $\lambda_2 = -c$. Thus, as expected, the system is hyperbolic because the eigenvalues are real and distinct.

1.6 Boundary Conditions

To identify the various types of *linear* boundary conditions, we introduce the following definitions:

$$\text{boundary condition of the first kind: } \phi = \text{prescribed} \qquad (1.36a)$$

$$\text{boundary condition of the second kind: } \Gamma \frac{\partial \phi}{\partial \mathbf{n}} = \text{prescribed} \qquad (1.36b)$$

$$\text{boundary condition of the third kind: } \Gamma \frac{\partial \phi}{\partial \mathbf{n}} + h\phi = \text{prescribed} \qquad (1.36c)$$

where $\frac{\partial}{\partial \mathbf{n}}$ denotes differentiation along the outward drawn normal to the boundary surface, that is,

$$\frac{\partial \phi}{\partial \mathbf{n}} = \nabla \phi \cdot \mathbf{n}. \qquad (1.37)$$

The boundary conditions of the first, second, and third kinds are also commonly referred to as Dirichlet, Neumann, and Robin boundary conditions, respectively.

When the right-hand sides of equations (1.36a)–(1.36c) vanish, boundary conditions are said to be *homogeneous*.

Let the vector **n** be represented in Cartesian coordinates as

$$\mathbf{n} = (l\mathbf{i} + m\mathbf{j} + n\mathbf{k}), \qquad (1.38)$$

where **i**, **j**, and **k** denote the unit vectors along the positive x, y, and z directions, respectively, and l, m, and n are the direction cosines of the vector **n**. Therefore, in Cartesian coordinates, $\frac{\partial \phi}{\partial \mathbf{n}}$ reduces to

$$\frac{\partial \phi}{\partial \mathbf{n}} = l\frac{\partial \phi}{\partial x} + m\frac{\partial \phi}{\partial y} + n\frac{\partial \phi}{\partial z}. \qquad (1.39)$$

Equation (1.39) can be further simplified if the boundary coincides with a surface of constant x, y, or z coordinates. Figure 1.1 illustrates such a case, for boundaries at x = 0 and x = L, where **n** = −**i** and **n** = **i**, respectively. Therefore, for this case, we have

$$\frac{\partial \phi}{\partial \mathbf{n}} = -\frac{\partial \phi}{\partial x} \quad \text{at } x=0, \qquad (1.40a)$$

$$\frac{\partial \phi}{\partial \mathbf{n}} = \frac{\partial \phi}{\partial x} \quad \text{at } x=L. \qquad (1.40b)$$

Basic Relations

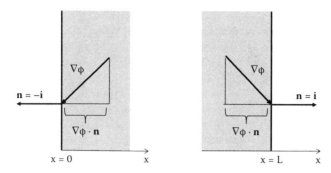

FIGURE 1.1
Illustration for the normal derivative.

Example 1.1

In a heat conduction problem, derive the boundary condition at a differential element dA of a surface that is remotely heated by a heat flux q_{sup} and exchange heat by convection and radiation with the surroundings. No heat is generated at the surface. The surface exchanges heat by convection with a heat transfer coefficient h to a gas that is at the temperature T_∞. The gas is supposed to be transparent to radiation. The emissivity of the surface is ε and it exchanges heat by radiation with a surrounding surface at the temperature T_{surr}. The figure shown here illustrates the physical situation under analysis.

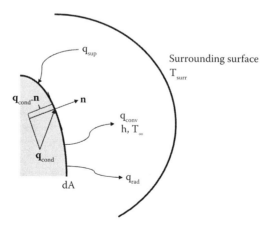

Solution

The energy balance at the surface, by taking into account that a surface has no mass and therefore does not accumulate heat, is written as

$$(\mathbf{q}_{cond} \cdot \mathbf{n} + q_{sup})dA = (q_{conv} + q_{rad})dA, \tag{a}$$

where **n** is the unit vector normal to the surface. The conduction heat flux vector can be obtained from Fourier's Law as

$$\mathbf{q}_{cond} = -k\nabla T, \qquad (b)$$

so that

$$\mathbf{q}_{cond} \cdot \mathbf{n} = -k\nabla T \cdot \mathbf{n} = -k\frac{\partial T}{\partial \mathbf{n}}. \qquad (c)$$

Note that $\mathbf{q}_{cond} \cdot \mathbf{n}$ was assumed as a positive quantity in the energy balance [Equation (a)]. By also writing the convective and the radiative heat fluxes as positive quantities, in accordance with the provided information that heat is lost (and not gained) by convection and radiation, we have, respectively,

$$q_{conv} = h(T - T_\infty), \qquad (d)$$

$$q_{rad} = \varepsilon\sigma(T^4 - T_{surr}^4), \qquad (e)$$

where $\sigma = 5.67 \times 10^{-8}$ W/m²K⁴ is the Stefan–Boltzmann constant (Özişik 1985). Equations (c)–(e) are then substituted into Equation (a), which is rearranged to yield

$$k\frac{\partial T}{\partial \mathbf{n}} + hT + \varepsilon\sigma T^4 = q_{sup} + hT_\infty + \varepsilon\sigma T_{surr}^4, \qquad (f)$$

which is a nonlinear boundary condition because of radiation.

If it is assumed that $T \approx T_{surr}$, radiation can be linearized in the form (Özişik 1985)

$$q_{rad} = h_{rad}(T - T_{surr}), \qquad (g)$$

where

$$h_{rad} = 4\varepsilon\sigma T_{surr}^3 \qquad (h)$$

is the heat transfer coefficient for radiation. By using Equation (g) instead of Equation (e) for the radiative flux, the energy balance given by Equation (a) becomes

$$k\frac{\partial T}{\partial \mathbf{n}} + h_{comb}T = q_{sup} + hT_\infty + h_{rad}T_{surr}, \qquad (i)$$

which is a linear boundary condition of the third kind, in the same form as equation (1.36c). In Equation (i), h_{comb} is the combined heat transfer coefficient due to convection and linearized radiation, that is,

$$h_{comb} = h + h_{rad}. \qquad (j)$$

Basic Relations

Equation (i) shows that a third kind boundary condition is obtained as long as there is heat transfer by convection and/or linearized radiation at the surface.

When the heat flux q_{sup} is imposed by an electrical resistance in direct contact with the surface, and not remotely such as illustrated by the accompanying figure, there is no heat transfer by convection and by radiation over the surface. Thus, Equation (i) reduces to

$$k\frac{\partial T}{\partial \mathbf{n}} = q_{sup}, \qquad (k)$$

which is a linear boundary condition of the second kind, in the same form as equation (1.36b). Whereas a boundary condition of the third kind involves convective and/or linearized radiation, a boundary condition of the second kind is obtained by an imposed heat flux over the surface, without convection and radiation heat transfer.

The boundary condition of the first kind [see equation (1.36a)],

$$T = T_\infty, \qquad (l)$$

can be obtained as a special case of equation (i), when the heat transfer coefficient is very large, such as for convection with phase change (e.g., boiling or condensation) and $q_{sup} = 0$. Alternatively, the boundary condition of the first kind results in cases where the surface is in direct contact with a thermal reservoir maintained at T_∞.

1.7 Uniqueness of the Solution

The study of uniqueness and existence of solutions for a given system of equations is not frequently considered in engineering simulations. However, it is instructive to examine some simple situations in order to illustrate the implications of such matter.

Consider a steady-state heat conduction problem with energy generation in a finite, closed domain given by

$$\nabla^2 T + \frac{1}{k} g = 0 \text{ in the region} \qquad (1.41a)$$

$$\frac{\partial T}{\partial \mathbf{n}} = 0 \text{ on all boundaries,} \qquad (1.41b)$$

where $\frac{\partial}{\partial \mathbf{n}}$ denotes the derivative along the outward drawn normal to the boundary surface. Just by physical considerations, we conclude that such a problem cannot have a steady-state solution because the energy generated in the medium has no way to escape as all boundaries are insulated; the temperature is bound to increase continuously with time.

Let us now consider another steady-state heat conduction problem in a finite domain, with no energy generation in the medium but with all boundaries subjected to prescribed heat fluxes. The mathematical formulation of this problem is given by

$$\nabla^2 T = 0 \text{ in the region} \tag{1.42a}$$

$$k \frac{\partial T}{\partial \mathbf{n}} = f \text{ on all boundaries,} \tag{1.42b}$$

where f is a function of the boundary position. Again, by physical reasoning, we conclude that this problem cannot have a steady-state solution unless the amount of heat entering the medium through part of the boundary surfaces is equal to the amount of heat leaving the domain through the rest of the boundary surfaces. Even so, if such a condition is satisfied, the steady-state solution for the problem is unique only to within an additive constant, that is, $T(\mathbf{r}) + c$, where the arbitrary constant, c, vanishes both in the differential equation and in the boundary condition given by equations (1.42a) and (1.42b).

For a number of physical, nonlinear boundary value problems, multiple solutions exist or no solution exists (Kubicek and Hlavacek 1983). For a nonlinear boundary value problem, it is difficult to prove rigorously the existence of a solution. There are physical problems that do not possess a solution for particular values of the parameters. In many engineering problems, nonlinearities are frequently caused by chemical reactions, radiation effects, dependence of the rate, equilibrium, and transport coefficients on concentration and temperature, as well as by viscous energy dissipation. A strong, exothermic, autocatalytic reaction or radiation effects may give rise to multiple steady-state solutions. Consider, for example, the nonlinear boundary value problem governing the explosion of a solid explosive material given in the form

$$\frac{d^2 y}{dR^2} + \frac{1}{R}\frac{dy}{dR} = -\delta\, e^y \quad 0 < R < 1 \tag{1.43a}$$

subject to the boundary conditions

$$\frac{dy}{dR} = 0 \text{ at } R = 0, \tag{1.43b}$$

$$y = 0 \text{ at } R = 1. \tag{1.43c}$$

Basic Relations

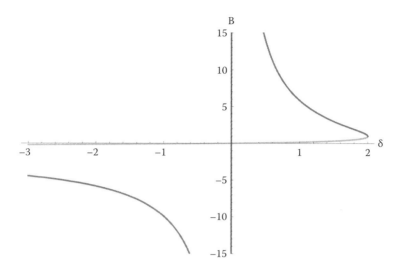

FIGURE 1.2
Roots of equation (1.44b).

The solution of this nonlinear problem is given in the form (Kubicek and Hlavacek 1983)

$$y = \ln \frac{8B/\delta}{(B\ R^2 + 1)^2}, \qquad (1.44a)$$

where the constant B is determined from

$$\frac{8B/\delta}{(B+1)^2} = 1, \qquad (1.44b)$$

which gives the following two roots:

$$B_1 = -\frac{-4 + 2\sqrt{4 - 2\delta} + \delta}{\delta} \qquad (1.45a)$$

$$B_2 = \frac{4 + 2\sqrt{4 - 2\delta} - \delta}{\delta}. \qquad (1.45b)$$

It can be seen from Figure 1.2 that, for $\delta < 2$, equation (1.44b) has two distinct real roots; hence, the problem [equation (1.43)] has two solutions for this range of δ. For $\delta = 2$, there is only one root, $B = 1$; hence, the problem has only one solution. Finally, for $\delta > 2$, the problem possesses no solution.

PROBLEMS

1.1. Consider the following differential equation:

$$x\frac{\partial^2 T}{\partial x^2} + \frac{\partial^2 T}{\partial y^2} = 0.$$

This equation can be elliptic, hyperbolic, or parabolic depending on whether $x > 0$, $x < 0$, or $x = 0$, respectively. Explain when this equation is elliptic, hyperbolic, and parabolic.

1.2. Consider the following partial differential equation:

$$\frac{\partial T}{\partial x} + A\frac{\partial T}{\partial y} - B\frac{\partial^2 T}{\partial y^2} = 0,$$

where A and B are constants. Specify whether this equation is elliptic, hyperbolic, or parabolic.

1.3. Consider the following partial differential equation:

$$A\frac{\partial^2 T}{\partial x^2} + B\frac{\partial^2 T}{\partial x \partial y} + C\frac{\partial^2 T}{\partial y^2} = 0.$$

Determine the nature of this equation for the following cases:
  i. $A = 1$, $B = 3$, and $C = 2$
  ii. $A = 1$, $B = -2$, and $C = 1$
  iii. $A = 1$, $B = 3$, and $C = 3$

1.4. Determine the nature of the following partial differential equations:

(i) $\dfrac{\partial^2 T}{\partial x^2} + \dfrac{\partial^2 T}{\partial y^2} + \dfrac{1}{k}g(x,y) = 0$

(ii) $\dfrac{\partial^2 T}{\partial x^2} - \dfrac{\partial^2 T}{\partial x \partial y} + \dfrac{\partial T}{\partial y} = 4$

1.5. Consider slug flow (i.e., uniform flow velocity) for forced convection inside a circular tube. The energy equation in dimensionless form, including the effects of axial heat conduction in the fluid, is given by

$$\frac{\partial T}{\partial X} = \frac{1}{R}\frac{\partial}{\partial R}\left(R\frac{\partial T}{\partial R}\right) + \frac{1}{(\text{Pe})^2}\frac{\partial^2 T}{\partial X^2},$$

where R and X are radial and axial dimensionless coordinates, respectively, and Pe is the Peclet number. Discuss the nature of this equation for the values of the Peclet number being finite and Pe → ∞.

1.6. Consider Burger's equation given by

$$\frac{\partial u}{\partial t} + u \frac{\partial u}{\partial x} = \upsilon \frac{\partial^2 u}{\partial x^2},$$

where u is the velocity, t the time, x the coordinate, and υ the kinematic viscosity of the fluid. Discuss the nature of this equation.

1.7. For extremely short times (i.e., picosecond or shorter) or at temperatures near absolute zero, the effects of finite speed of propagation become important and heat conduction is governed by the following equation:

$$\frac{1}{c^2} \frac{\partial^2 T}{\partial t^2} + \frac{1}{\alpha} \frac{\partial T}{\partial t} = \frac{\partial^2 T}{\partial x^2},$$

where c is the wave propagation speed. In the standard heat conduction, c is regarded as infinite, and this equation reduces to the usual heat conduction equation. Discuss the nature of this equation.

1.8. Determine the nature of the following two-dimensional energy equation for flow in a parallel plates channel:

$$\rho C_p \left[\frac{\partial T}{\partial t} + \frac{\partial}{\partial x}(uT) + \frac{\partial}{\partial y}(vT) \right] = \frac{\partial}{\partial x}\left(k \frac{\partial T}{\partial x}\right) + \frac{\partial}{\partial y}\left(k \frac{\partial T}{\partial y}\right) + g,$$

where T is the temperature, t is the time, x and y are the axial and transversal coordinates, respectively, u and v are the velocity components, and g is the energy generation rate.

1.9. A plane wall, confined to the region $0 \leq x \leq L$, is subjected to a heat supply at a rate of q_o W/m² at the boundary surface x = 0 and dissipates heat by convection with a heat transfer coefficient h_∞ W/(m² °C) into an ambient air at temperature T_∞ from the boundary surface at x = L. Write the boundary conditions at x = 0 and x = L.

1.10. Consider a two-dimensional heat conduction problem in a rectangular shape confined to the region $0 \leq x \leq a$, $0 \leq y \leq b$. Write the mathematical formulation of boundary conditions for the following cases:

i. Boundary at x = 0: Heat is removed at a constant rate of q_o W/m².

ii. Boundary at x = a: Heat is dissipated by convection with a heat transfer coefficient h_a into the ambient air at constant temperature T_∞.

iii. Boundary at y = 0: Kept insulated.

iv. Boundary at y = b: Heat is supplied into the solid at a rate of q_b W/m².

1.11. A spherical shell has an inside radius $R = r_1$ and outside radius $R = r_2$. At the inside surface, it is heated electrically at a rate of q_1 W/m², and at the outside surface, heat is dissipated by convection with a heat transfer coefficient h_2 into an ambient air at a constant temperature T_∞. Write the boundary conditions.

1.12. A copper bar of radius $R = b$ is initially at a uniform temperature T_i. The heating of the rod begins at time $t = 0$ by the passage of electric current that generates heat throughout the rod at a constant rate of g_o W/m³. The rod dissipates heat by free convection with a heat transfer coefficient $h = C(T - T_\infty)^{1/4}$ into the ambient fluid at constant temperature T_∞. Assuming constant thermal conductivity k for the solid and one-dimensional time-dependent problem, write the mathematical formulation of this heat conduction problem.

1.13. Show that the following three different forms of the differential operator in the spherical system are equivalent:

$$\frac{1}{R^2}\frac{d}{dR}\left(R^2\frac{dT}{dR}\right) = \frac{1}{R}\frac{d^2}{dR^2}(RT) = \frac{d^2T}{dR^2} + \frac{2}{R}\frac{dT}{dR}$$

1.14. Determine the nature of the following convection–diffusion equation:

$$\rho C_p\left[\frac{\partial}{\partial x}(uT) + \frac{\partial}{\partial y}(vT)\right] = \frac{\partial}{\partial x}\left(k\frac{\partial T}{\partial x}\right) + \frac{\partial}{\partial y}\left(k\frac{\partial T}{\partial y}\right) + g.$$

1.15. Set up the mathematical formulation of the following heat conduction problems:

i. A slab in $0 \leq x \leq L$ is initially at a temperature $F(x)$. For times $t > 0$, the boundary at $x = 0$ is kept insulated and the boundary at $x = L$ dissipates heat by convection into a medium at zero temperature.

ii. A semi-infinite region $0 \leq x < \infty$ is initially at a temperature $F(x)$. For times $t > 0$, heat is generated in the medium at a constant rate of g_o W/m³, while the boundary at $x = 0$ is kept at zero temperature.

iii. A solid cylinder $0 \leq R \leq b$ is initially at a temperature $F(R)$. For times $t > 0$, heat is generated in the medium at a rate of $g(R)$, W/m³, while the boundary at $R = b$ dissipates heat by convection into a medium at zero temperature.

iv. A solid sphere $0 \leq R \leq b$ is initially at temperature $F(R)$. For times $t > 0$, heat is generated in the medium at a rate of $g(R)$, W/m³, while the boundary at $R = b$ is kept at a uniform temperature T_o.

Basic Relations

1.16. A one-dimensional unsteady inviscid compressible flow is described by the equations

$$\frac{\partial}{\partial t}(\rho) + \frac{\partial}{\partial x}(\rho u) = 0,$$

$$\frac{\partial}{\partial t}(\rho u) + \frac{\partial}{\partial x}(p + \rho u^2) = 0,$$

$$\frac{\partial}{\partial t}(E_t) + \frac{\partial}{\partial x}[(E_t + p)u] = 0,$$

where ρ is density, u is velocity, p is pressure, and E_t is the total energy. Write these equations in the matrix system form.

1.17. The Laplace equation $\frac{\partial^2 u}{\partial x^2} + \frac{\partial^2 u}{\partial y^2} = 0$ is an elliptic equation. Split up this equation into two first-order equations, examine the eigenvalues of the coefficient matrix, and show that the eigenvalues are both complex (i.e., $\lambda_1 = +i$ and $\lambda_2 = -i$), which is consistent with the elliptic nature of the problem.

1.18. Determine whether the following system of equations is elliptic or hyperbolic:

$$\frac{\partial \mathbf{T}}{\partial x} + \mathbf{A}\frac{\partial \mathbf{T}}{\partial y} = 0,$$

where

$$\mathbf{T} = \begin{bmatrix} T_1 \\ T_2 \end{bmatrix}, \quad \mathbf{A} = \begin{bmatrix} 0 & -1 \\ 1 & 0 \end{bmatrix}.$$

2

Discrete Approximation of Derivatives

When a partial differential equation is solved analytically over a given region, subject to appropriately specified initial and boundary conditions, the resulting solution satisfies the differential equation at every point within the domain of interest, which may contain time and several spatial independent variables. On the other hand, in numerical methods for the solution of partial differential equations—such as the finite difference technique, which is the subject of this book, the problem domain is *discretized* so that the values of the unknown dependent variables are considered only at a finite number of nodal points instead of at every point over the domain.

Consider, as an example, a differential problem to be solved in a one-dimensional spatial domain of length L, which involves only the independent variable x in Cartesian coordinates. In the finite difference method, the domain $0 \leq x \leq L$ is discretized—let us say in equal segments of length Δx, where $\Delta x = L/M$, thus generating a grid (or mesh) of $(M + 1)$ nodes in the spatial domain, as illustrated by Figure 2.1. Therefore, $(M + 1)$ algebraic equations are developed by discretizing the governing differential equations and the boundary conditions for the problem. The problem of solving the ordinary or partial differential equations over the problem domain is then transformed to the task of development of a set of algebraic equations and their solution by a suitable algorithm.

This seemingly simple approach is complicated by the fact that the nature of the resulting set of algebraic equations depends on the character of the partial differential equations governing the physical problem, that is, whether they are parabolic, elliptic, or hyperbolic. Furthermore, there are numerous discretization schemes; hence, one must choose the one that is the most appropriate for the nature of the problem.

Two basic approaches commonly used to discretize the derivatives in partial differential equations include: (i) the use of *Taylor series expansion* and (ii) the use of *control volumes*. These two approaches are usually referenced as the finite difference method and the finite volume method, respectively. Both discretization approaches will be addressed in this book and applied to different problems of interest in heat transfer and fluid flow.

In this chapter, we present discretization of partial differential equations by Taylor series expansion and by control volumes, and we discuss the types of errors involved in the discretization process and during the solution of the resulting system of algebraic equations.

Several classical references on the fundamentals of discretization and finite difference methods include Richtmeyer and Morton (1967), Smith (1978),

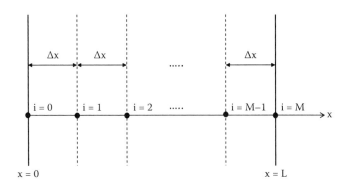

FIGURE 2.1
Example of discretization of a one-dimensional spatial domain.

Pletcher et al. (2012), Berezin and Zhidkov (1965), Roache (1976), Patankar (1980), and Jaluria and Torrance (1986).

2.1 Taylor Series Formulation

The idea of the finite difference representation of a derivative can be introduced by recalling the definition of the derivative of a general function F(x, y) at a point $x = x_0$, $y = y_0$

$$\frac{\partial F(x,y)}{\partial x} = \lim_{\Delta x \to 0} \frac{F(x_0 + \Delta x, y_0) - F(x_0, y_0)}{\Delta x} \quad (2.1)$$

Clearly, if the function F(x, y) is continuous, the right-hand side of equation (2.1) can be a reasonable approximation of $\partial F/\partial x$ for a *sufficiently small* but finite Δx.

A formal basis for developing finite difference approximation of derivatives is through the use of Taylor series expansion. Consider Taylor series expansion of a function f(x) about a point x_0 in the *forward* (i.e., positive x) and *backward* (i.e., negative x) directions given, respectively, by

$$f(x_0 + \Delta x) = f(x_0) + \frac{df}{dx}\bigg|_{x=x_0} \Delta x + \frac{d^2 f}{dx^2}\bigg|_{x=x_0} \frac{(\Delta x)^2}{2!} + \frac{d^3 f}{dx^3}\bigg|_{x=x_0} \frac{(\Delta x)^3}{3!} + \cdots \quad (2.2)$$

$$f(x_0 - \Delta x) = f(x_0) - \frac{df}{dx}\bigg|_{x=x_0} \Delta x + \frac{d^2 f}{dx^2}\bigg|_{x=x_0} \frac{(\Delta x)^2}{2!} - \frac{d^3 f}{dx^3}\bigg|_{x=x_0} \frac{(\Delta x)^3}{3!} + \cdots \quad (2.3)$$

These two expressions form the basis for developing finite difference approximations for the first derivative df/dx, at x_0. Rearranging equations (2.2) and (2.3), the *forward* and *backward* finite difference approximations for the first derivative, respectively, become

$$\left.\frac{df}{dx}\right|_{x=x_0} = \frac{f(x_0 + \Delta x) - f(x_0)}{\Delta x} + 0(\Delta x) \quad \text{(forward)} \tag{2.4}$$

$$\left.\frac{df}{dx}\right|_{x=x_0} = \frac{f(x_0) - f(x_0 - \Delta x)}{\Delta x} + 0(\Delta x) \quad \text{(backward)} \tag{2.5}$$

where the "order of" notation "$0(\Delta x)$" characterizes the *truncation error* associated with the finite difference approximation. This represents the difference between the derivative and its finite difference representation. For equation (2.4), it is given by

$$0(\Delta x) \equiv \left.\frac{d^2 f}{dx^2}\right|_{x=x_0} \frac{(\Delta x)}{2} + \left.\frac{d^3 f}{dx^3}\right|_{x=x_0} \frac{(\Delta x)^2}{6} + \cdots \tag{2.6}$$

Subtracting equation (2.3) from equation (2.2), the *central difference* approximation is determined as

$$\left.\frac{df}{dx}\right|_{x=x_0} = \frac{f(x_0 + \Delta x) - f(x_0 - \Delta x)}{2\Delta x} - 0[(\Delta x)^2] \quad \text{(central)} \tag{2.7}$$

where

$$0[(\Delta x)^2] \equiv \left.\frac{d^3 f}{dx^3}\right|_{x=x_0} \frac{(\Delta x)^2}{6} + \left.\frac{d^5 f}{dx^5}\right|_{x=x_0} \frac{(\Delta x)^4}{120} + \cdots \tag{2.8}$$

An analysis of the truncation error associated with various finite difference representations given here reveals that the central difference approximation is *second order* in Δx; hence, it is a more accurate approximation than the forward and backward differences.

In the aforementioned developments, only two grid points were used for finite difference approximation of the first derivative. However, there are situations in which more grid points need to be retained in finite difference approximation of derivatives in order to improve the accuracy of the representation. In the following section, we summarize finite difference approximations for the first and second derivatives with two-, three-, and four-point formulae.

2.1.1 Finite Difference Approximation of First Derivative

Let i be the grid point at x_0 and f_i the function f evaluated at x_0. Then, the notations $i + 1$ and $i - 1$ refer, respectively, to the grid points at $x_0 + \Delta x$ and $x_0 - \Delta x$. Similarly, the notations $i + 2$ and $i - 2$ refer to the grid points at $x_0 + 2\Delta x$ and $x_0 - 2\Delta x$, respectively, and so on. Using this notation, we present below two-, three-, and four-point formulae for the first derivative.

Two-point formulae:

$$f'_i = \frac{f_{i+1} - f_i}{\Delta x} + 0(\Delta x) \quad \text{(forward)} \qquad (2.9a)$$

$$f'_i = \frac{f_i - f_{i-1}}{\Delta x} + 0(\Delta x) \quad \text{(backward)} \qquad (2.9b)$$

$$f' = \frac{f_{i+1} - f_{i-1}}{2\Delta x} + 0[(\Delta x)^2] \quad \text{(central)} \qquad (2.9c)$$

These three formulae can be rewritten more compactly as a single equation in the form

$$f'_i = \frac{(1-\varepsilon)f_{i+1} + 2\varepsilon f_i - (1+\varepsilon)f_{i-1}}{2\Delta x} \qquad (2.10)$$

where

$$\varepsilon = \begin{cases} -1 & \text{for forward} \\ 0 & \text{for central} \\ +1 & \text{for backward} \end{cases}$$

Three-point formulae:

$$f'_i = \frac{1}{2\Delta x}(-3f_i + 4f_{i+1} - f_{i+2}) + 0[(\Delta x)^2] \qquad (2.11a)$$

$$f'_i = \frac{1}{2\Delta x}(f_{i-2} - 4f_{i-1} + 3f_i) + 0[(\Delta x)^2] \qquad (2.11b)$$

Four-point formulae:

$$f'_i = \frac{1}{6\Delta x}(-11f_i + 18f_{i+1} - 9f_{i+2} + 2f_{i+3}) + 0[(\Delta x)^3] \qquad (2.12a)$$

$$f'_i = \frac{1}{6\Delta x}(-2f_{i-1} - 3f_i + 6f_{i+1} - f_{i+2}) + 0[(\Delta x)^3] \qquad (2.12b)$$

$$f'_i = \frac{1}{6\Delta x}(f_{i-2} - 6f_{i-1} + 3f_i + 2f_{i+1}) + 0[(\Delta x)^3] \qquad (2.12c)$$

Three- or four-point formulae, as given here, are useful to represent a first derivative at a node i on the boundary by using more than two grid points inside the domain, in order to improve the accuracy of approximation.

Discrete Approximation of Derivatives

Example 2.1

Let T_0 be the temperature of the grid point on the boundary and T_1, T_2, T_3, ... be the temperatures at the neighboring grid points along the positive x-direction (see Figure 2.1). The heat flux at the boundary $x = 0$ is to be determined from its definition given by $q_w = -k(\partial T/\partial x)_{x=0}$. Represent the derivative of temperature at $x = 0$ with finite differences using approximations of order $0(\Delta x)$, $0[(\Delta x)^2]$, and $0[(\Delta x)^3]$.

Solution

The forward differencing schemes must be used because the grid points $i = 1, 2, 3, \ldots$ in relation to the boundary node $i = 0$ are located along the positive x-direction. Forward finite difference representations, accurate to the order $0(\Delta x)$, $0[(\Delta x)^2]$, and $0[(\Delta x)^3]$, obtained from equations (2.9a), (2.11a), and (2.12a), respectively, are given by

$$\frac{dT}{dx} = \frac{T_1 - T_0}{\Delta x} \tag{a}$$

$$\frac{dT}{dx} = \frac{1}{2\Delta x}(-3T_0 + 4T_1 - T_2) \tag{b}$$

$$\frac{dT}{dx} = \frac{1}{6\Delta x}(-11T_0 + 18T_1 - 9T_2 + 2T_3) \tag{c}$$

2.1.2 Finite Difference Approximation of Second Derivative

The Taylor series expansions given by equations (2.2) and (2.3) can be used to develop finite difference approximations for the second derivative. To obtain the *central* finite difference approximation for the second derivative, equations (2.2) and (2.3) are added, the resulting expression is solved for $(d^2f/dx^2)_{x_0}$, and the result, written with the abbreviated notation, is given by

$$f''_i = \frac{f_{i-1} - 2f_i + f_{i+1}}{(\Delta x)^2} + 0[(\Delta x)^2] \quad \text{(central)} \tag{2.13a}$$

where

$$0[(\Delta x)^2] \equiv \frac{(\Delta x)^2}{12} f''''_i + \cdots$$

To develop *forward* and *backward* finite difference approximations for the second derivatives, the functions $f(x_0 + 2\Delta x)$ and $f(x_0 - 2\Delta x)$ are expanded in Taylor series. The function $f'(x_0)$ is eliminated between the expansion of $f(x_0 + 2\Delta x)$ and the expansion given by equation (2.2), and the resulting expression is solved for $(d^2f/dx^2)_{x_0}$. The *forward* finite difference approximation for the second derivative is determined as

$$f''_i = \frac{f_i - 2f_{i+1} + f_{i+2}}{(\Delta x)^2} + 0(\Delta x) \quad \text{(forward)} \tag{2.13b}$$

Similarly, the function $f'(x_0)$ is eliminated between the expansion of $f(x_0 - 2\Delta x)$ and the expansion given by equation (2.3), and the resulting expression is solved for $(d^2f/dx^2)_{x_0}$. The *backward* finite difference approximation of the second derivative is determined as

$$f''_i = \frac{f_{i-2} - 2f_{i-1} + f_i}{(\Delta x)^2} + 0(\Delta x) \quad \text{(backward)} \tag{2.13c}$$

where

$$0(\Delta x) = \Delta x f'''_i + \ldots$$

The finite difference approximations for the second derivative given here utilize three grid points. Approximations utilizing more than three points can also be developed; we list some of such representations as follows:

$$f''_i = \frac{2f_i - 5f_{i+1} + 4f_{i+2} - f_{i+3}}{(\Delta x)^2} + 0[(\Delta x)^2] \tag{2.14a}$$

$$f''_i = \frac{-f_{i-3} + 4f_{i-2} - 5f_{i-1} + 2f_i}{(\Delta x)^2} + 0[(\Delta x)^2] \tag{2.14b}$$

Example 2.2

Discretize the steady-state one-dimensional heat conduction equation, with constant thermal conductivity, in Cartesian coordinates

$$\frac{\partial^2 T}{\partial x^2} + \frac{1}{k} g(x) = 0 \tag{a}$$

by using central finite differences of second order.

Solution
By applying equation (2.13a) and the notation $g_i = g(i\Delta x)$, equation (a) reduces to

$$\frac{T_{i+1} - 2T_i + T_{i-1}}{(\Delta x)^2} + \frac{g_i}{k} = 0 \tag{b}$$

2.1.3 Differencing via Polynomial Fitting

Finite difference expressions can also be developed by representing the function f in the form of a polynomial and evaluating the coefficients in terms of

Discrete Approximation of Derivatives

the function values at the neighboring nodes. For example, consider the representation of f(x) by fitting the parabola in the form

$$f(x) = ax^2 + bx + c \tag{2.15a}$$

to the nodes $x = 0$, Δx and $2\Delta x$. Therefore,

$$f'(x) = 2ax + b \text{ and } f'(0) = b \tag{2.15b}$$

and we can write

$$f(0) = c, \ f(\Delta x) = a(\Delta x)^2 + b\Delta x + c, \text{ and } f(2\Delta x) = 4a(\Delta x)^2 + 2b\Delta x + c \tag{2.15c,d,e}$$

Solving equations (2.15c,d,e) for b yields

$$f'_i = \frac{1}{2\Delta x}(-3f_i + 4f_{i+1} - f_{i+2}) \tag{2.16}$$

which is identical to equation (2.11a). This approach is particularly useful in developing finite difference expressions for nonuniform values of Δx as well as for calculating the gradients needed for determining the heat or mass flux at the wall.

2.1.4 Finite Difference Approximation of Mixed Partial Derivatives

Often, it may be necessary to represent mixed partial derivatives, such as $\partial^2 f/\partial x \partial y$, in finite differences. The finite difference approximation can be developed by the successive application of finite differencing of the first derivative in the x and y variables.

For illustration purposes, we consider finite difference approximation of the mixed partial derivative $\partial^2 f/\partial x \partial y$ and use the central difference formula [equation (2.9c)] to discretize the first derivative for both the x and y variables. We write

$$\frac{\partial}{\partial x}\left(\frac{\partial f}{\partial y}\right) = \frac{1}{2\Delta x}\left(\left.\frac{\partial f}{\partial y}\right|_{i+1,j} - \left.\frac{\partial f}{\partial y}\right|_{i-1,j}\right) + 0(\Delta x)^2 \tag{2.17a}$$

where the subscripts i and j denote the grid points associated with the discretization in the x and y variables, respectively (see Figure 2.2). By applying the central difference formula once more to discretize the partial derivatives with respect to the y variable on the right-hand side of equation (2.17a), we obtain

$$\frac{\partial}{\partial x}\left(\frac{\partial f}{\partial y}\right) = \frac{1}{2\Delta x}\left(\frac{f_{i+1,j+1} - f_{i+1,j-1}}{2\Delta y} - \frac{f_{i-1,j+1} - f_{i-1,j-1}}{2\Delta y}\right) + 0[(\Delta x)^2, (\Delta y)^2] \tag{2.17b}$$

which is the finite difference approximation of the mixed partial derivative $\partial^2 f/\partial x \partial y$ using central differences for both x and y variables. The order of

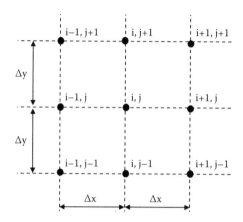

FIGURE 2.2
Discretization in two dimensions in Cartesian coordinates.

differentiation is immaterial if the derivatives are continuous; that is, $\partial^2 f/\partial x \partial y$ and $\partial^2 f/\partial y \partial x$ are equal.

In the aforementioned illustration, we applied central differences for both derivatives in x and y. If all possible combinations of forward, backward, and central differences are considered, nine different cases arise for finite difference approximation of $\partial^2 f/\partial x \partial y$. Table 2.1 lists the finite difference approximations for each of these nine different cases. The order of the truncation errors associated with each of the nine cases shown in Table 2.1 can be verified by Taylor series expansion in two variables. For example, to verify the result $0[(\Delta x)^2,(\Delta y)^2]$ shown for Case #9 of Table 2.1, we expand each of the following in Taylor series

$$f_{i+1,j+1} \equiv f(x + \Delta x, y + \Delta y)$$

$$f_{i+1,j-1} \equiv f(x + \Delta x, y - \Delta y)$$

$$f_{i-1,j+1} \equiv f(x - \Delta x, y + \Delta y)$$

$$f_{i-1,j-1} \equiv f(x - \Delta x, y - \Delta y)$$

and substitute the resulting expansions into the difference formula given by equation (2.17b), which uses central differencing in both the x and y variables. After cancellations and simplifications, it can be shown that the leading truncation error term is given by

$$\text{TE} = -\frac{(\Delta x)^2}{6}\frac{\partial^4 f}{\partial x^3 \partial y} - \frac{(\Delta y)^2}{6}\frac{\partial^4 f}{\partial x \partial y^3} - \cdots \qquad (2.18)$$

Thus, the finite differencing given by equation (2.17b) is accurate to $0[(\Delta x)^2, (\Delta y)^2]$.

Discrete Approximation of Derivatives

TABLE 2.1
Finite Difference Approximation of Mixed Partial Derivative $\partial^2 f/\partial x \partial y$

Case No.	Difference Scheme[a] x	y	Finite Difference Approximation	Order of the Error
1	F	F	$\dfrac{1}{\Delta x}\left(\dfrac{f_{i+1,j+1}-f_{i+1,j}}{\Delta y} - \dfrac{f_{i,j+1}-f_{i,j}}{\Delta y}\right)$	$0[\Delta x, \Delta y]$
2	F	B	$\dfrac{1}{\Delta x}\left(\dfrac{f_{i+1,j}-f_{i+1,j-1}}{\Delta y} - \dfrac{f_{i,j}-f_{i,j-1}}{\Delta y}\right)$	$0[\Delta x, \Delta y]$
3	F	C	$\dfrac{1}{\Delta x}\left(\dfrac{f_{i+1,j+1}-f_{i+1,j-1}}{2\Delta y} - \dfrac{f_{i,j+1}-f_{i,j-1}}{2\Delta y}\right)$	$0[\Delta x, (\Delta y)^2]$
4	B	F	$\dfrac{1}{\Delta x}\left(\dfrac{f_{i,j+1}-f_{i,j}}{\Delta y} - \dfrac{f_{i-1,j+1}-f_{i-1,j}}{\Delta y}\right)$	$0[\Delta x, \Delta y]$
5	B	B	$\dfrac{1}{\Delta x}\left(\dfrac{f_{i,j}-f_{i,j-1}}{\Delta y} - \dfrac{f_{i-1,j}-f_{i-1,j-1}}{\Delta y}\right)$	$0[\Delta x, \Delta y]$
6	B	C	$\dfrac{1}{\Delta x}\left(\dfrac{f_{i,j+1}-f_{i,j-1}}{2\Delta y} - \dfrac{f_{i-1,j+1}-f_{i-1,j-1}}{2\Delta y}\right)$	$0[\Delta x, (\Delta y)^2]$
7	C	F	$\dfrac{1}{2\Delta x}\left(\dfrac{f_{i+1,j+1}-f_{i+1,j}}{\Delta y} - \dfrac{f_{i-1,j+1}-f_{i-1,j}}{\Delta y}\right)$	$0[(\Delta x)^2, \Delta y]$
8	C	B	$\dfrac{1}{2\Delta x}\left(\dfrac{f_{i+1,j}-f_{i+1,j-1}}{\Delta y} - \dfrac{f_{i-1,j}-f_{i-1,j-1}}{\Delta y}\right)$	$0[(\Delta x)^2, \Delta y]$
9	C	C	$\dfrac{1}{2\Delta x}\left(\dfrac{f_{i+1,j+1}-f_{i+1,j-1}}{2\Delta y} - \dfrac{f_{i-1,j+1}-f_{i-1,j-1}}{2\Delta y}\right)$	$0[(\Delta x)^2, (\Delta y)^2]$

[a] B = backward difference, C = central difference, F = forward difference.

2.1.5 Changing the Mesh Size

In most engineering applications, one will often have some idea of the general shape of the solution, especially of the locations where it will exhibit sudden changes. Therefore, to obtain higher resolution in the region where the solution gradients are expected to be large, it is desirable to use a finer mesh over that particular region, while still using a coarse discretization on regions where the gradient is expected to be small (and not excessively increasing the number of grid points where the solution needs to be calculated). To illustrate this matter, we consider the simplest situation involving a change in mesh spacing only in one direction at some point in the region.

Figure 2.3 shows a change of the mesh size from Δx_1 to Δx_2 at some node i. A Taylor series expansion about the node i can be used to develop finite difference approximation for the first and second derivatives at node i.

1. **Approximation of First Derivative**: The function f(x) is expanded about the node i in *forward* and *backward* Taylor series, respectively, as

$$f_{i+1} = f_i + \Delta x_2 \left.\dfrac{df}{dx}\right|_i + \dfrac{(\Delta x_2)^2}{2!}\left.\dfrac{d^2 f}{dx^2}\right|_i + \dfrac{(\Delta x_2)^3}{3!}\left.\dfrac{d^3 f}{dx^3}\right|_i + 0[(\Delta x_2)^4] \quad (2.19a)$$

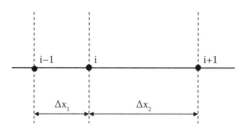

FIGURE 2.3
Change of the mesh size from Δx_1 to Δx_2 at the node i.

$$f_{i-1} = f_i - \Delta x_1 \left.\frac{df}{dx}\right|_i + \frac{(\Delta x_1)^2}{2!}\left.\frac{d^2f}{dx^2}\right|_i - \frac{(\Delta x_1)^3}{3!}\left.\frac{d^3f}{dx^3}\right|_i + 0[(\Delta x_1)^4] \quad (2.19b)$$

Equation (2.19b) is subtracted from equation (2.19a), and the resulting expression is solved for the first derivative $\left.\frac{df}{dx}\right|_i$.

$$\left.\frac{df}{dx}\right|_i = \frac{f_{i+1} - f_{i-1}}{\Delta x_2 + \Delta x_1} - \frac{1}{2}\frac{(\Delta x_2)^2 - (\Delta x_1)^2}{\Delta x_2 + \Delta x_1}\left.\frac{d^2f}{dx^2}\right|_i + 0[(\Delta x)^2] \quad (2.20)$$

where $0[(\Delta x)^2]$ means the *largest* of $0[(\Delta x_1)^2]$ or $0[(\Delta x_2)^2]$.

Then, the finite difference approximation of the first derivative at the node i where the mesh size is changed from Δx_1 to Δx_2 becomes

$$\left.\frac{df}{dx}\right|_i = \frac{f_{i+1} - f_{i-1}}{\Delta x_2 + \Delta x_1} \quad (2.21a)$$

and equation (2.20) implies that the finite difference expression [equation (2.21a)] is *second-order* accurate only if $\Delta x_2 \rightarrow \Delta x_1$. Then we have

$$0\left[\frac{(\Delta x_2)^2 - (\Delta x_1)^2}{\Delta x_2 + \Delta x_1}\right] \leq 0[(\Delta x_1)^2] \quad (2.21b)$$

We note that, if the mesh size varies from Δx_1 to Δx_2 abruptly, say $\Delta x_2 \cong 2\Delta x_1$, then the accuracy of the differencing at i deteriorates to *first order*.

2. **Approximation of Second Derivative**: To obtain a difference approximation for the second derivative at the node i, equation (2.19b) is multiplied by $(\Delta x_2/\Delta x_1)^2$, and the resulting expression is added to equation (2.19a) to give

$$f_{i+1} + \varepsilon^2 f_{i-1} = (1+\varepsilon^2)f_i + (1-\varepsilon)\Delta x_2 \frac{df}{dx}\bigg|_i + (\Delta x_2)^2 \frac{d^2f}{dx^2}\bigg|_i \quad (2.22a)$$
$$+ \frac{1}{6}(\Delta x_2 - \Delta x_1)(\Delta x_2)^2 \frac{d^3f}{dx^3}\bigg|_i + 0[(\Delta x)^4]$$

where
$$\varepsilon = \frac{\Delta x_2}{\Delta x_1} \quad (2.22b)$$

and $0[(\Delta x)^4]$ means the largest of $0[(\Delta x_1)^4]$ or $0[(\Delta x_2)^4]$. The finite difference approximation for the second derivative is obtained by solving equation (2.22a) for $\frac{d^2f}{dx^2}\big|_i$ to yield

$$\frac{d^2f}{dx^2}\bigg|_i = \frac{f_{i+1} - (1+\varepsilon^2)f_i + \varepsilon^2 f_{i-1}}{(\Delta x_2)^2} - \frac{1-\varepsilon}{\Delta x_2}\left(\frac{f_{i+1}-f_{i-1}}{\Delta x_2 + \Delta x_1}\right) + 0[(\Delta x_2 - \Delta x_1)] \quad (2.23)$$

This expression is accurate to second order only if $\Delta x_2 \to \Delta x_1$; otherwise, it is first-order accurate. These results imply that, unless the mesh spacing is changed slowly, the truncation error deteriorates.

2.1.6 Finite Difference Operators

Finite difference operators are often used in order to express the finite difference expressions in compact forms, and different notations have been used by Smith (1978), Peyret and Taylor (1983), and others. Here we present some of the commonly used difference operators with particular emphasis on the notations proposed by Peyret and Taylor (1983).

Let indice i refer to the grid points selected along the x-axis. The forward, backward, and central difference approximations of the first derivative about the node i can be expressed with operator notation as

$$\text{forward} \quad \frac{df}{dx}\bigg|_i = \frac{f_{i+1} - f_i}{\Delta x} \equiv \Delta_x^+ f_i \quad (2.24a)$$

$$\text{backward} \quad \frac{df}{dx}\bigg|_i = \frac{f_i - f_{i-1}}{\Delta x} \equiv \Delta_x^- f_i \quad (2.24b)$$

$$\text{central} \quad \frac{df}{dx}\bigg|_i = \frac{f_{i+1} - f_{i-1}}{2\Delta x} \equiv \Delta_x^0 f_i \quad (2.24c)$$

The central finite difference approximation for the *second derivative* is represented by

$$\frac{d^2f}{dx^2}\bigg|_i = \frac{f_{i+1} - 2f_i + f_{i-1}}{(\Delta x)^2} \equiv \Delta_{xx} f_i \quad (2.24d)$$

It can readily be verified that the following relationships exist among different operators

$$\frac{1}{2}(\Delta_x^+ + \Delta_x^-) = \Delta_x^0 \tag{2.25a}$$

$$\Delta_x^+ - \Delta_x^- = \Delta x\, \Delta_{xx} \tag{2.25b}$$

$$\Delta_x^+ \Delta_x^- = \Delta_{xx} \tag{2.25c}$$

For example, the finite difference approximation to the two-dimensional steady-state heat conduction equation

$$\frac{\partial^2 T}{\partial x^2} + \frac{\partial^2 T}{\partial y^2} + \frac{1}{k}g(x,y) = 0 \tag{2.26}$$

can be represented with the aforementioned operator notation as

$$(\Delta_{xx} + \Delta_{yy})T_{i,j} + \frac{1}{k}g_{i,j} = 0 \tag{2.27}$$

where $T(i\Delta x, j\Delta y) \equiv T_{i,j}$ and $g(i\Delta x, j\Delta y) \equiv g_{i,j}$.

Therefore, the finite difference approximation of equation (2.26) is given by

$$\frac{T_{i+1,j} - 2T_{i,j} + T_{i-1,j}}{(\Delta x)^2} + \frac{T_{i,j+1} - 2T_{i,j} + T_{i,j-1}}{(\Delta y)^2} + \frac{g_{i,j}}{k} = 0 \tag{2.28}$$

2.2 Control Volume Approach

In the previous section, the Taylor series approach was used as a purely mathematical procedure to develop the finite difference approximation to the derivatives. In the alternative *control volume* approach, the finite difference equations are developed by constraining the partial differential equation to a finite control volume and conserving the specific physical quantity such as mass, momentum, or energy over this control volume. The basic concept thus consists of writing the conservation equation of interest over a small volume surrounding a grid point.

To develop the control volume statement for a small finite region, it is instructive to work from the partial differential equation governing the specific physical quantity. For illustration purposes, we consider the transient heat conduction equation with energy generation given in the form

$$\rho C_p \frac{\partial T(\mathbf{r},t)}{\partial t} = -\nabla \cdot \mathbf{q}(\mathbf{r},t) + g(\mathbf{r},t) \tag{2.29}$$

where g(r,t) is the volumetric energy generation rate, **r** is the position vector, and the heat flux vector **q** is related to the temperature T by Fourier's law

$$\mathbf{q}(\mathbf{r},t) = -k\nabla T(\mathbf{r},t) \tag{2.30}$$

We integrate equation (2.29) over a small fixed volume V, that is,

$$\int \rho C_p \frac{\partial T}{\partial t} dV = -\int \nabla \cdot \mathbf{q}\, dV + \int g\, dV \tag{2.31}$$

The integral on the left-hand side can be removed by utilizing the mean value theorem for integrals. Similarly, the integral for the source term g is removed. The volume integral over the divergence of the heat flux vector is transformed to a surface integral by means of the divergence theorem. Then equation (2.31) becomes

$$V\rho C_p \frac{\partial \overline{T}}{\partial t} = -\int \mathbf{q}\cdot\mathbf{n}\, dS + V\overline{g} \tag{2.32a}$$

where S is the surface area of the control volume V, while $\overline{T}$ and $\overline{g}$ are suitable averages over the control volume of temperature and the energy generation rate, respectively. Introducing the heat flux vector **q** from equation (2.30) into equation (2.32a), we find

$$V\rho C_p \frac{\partial \overline{T}}{\partial t} = \int k\frac{\partial T}{\partial n} dS + V\overline{g} \tag{2.32b}$$

where

$$\nabla T \cdot \mathbf{n} = \frac{\partial T}{\partial n} \tag{2.32c}$$

Here **n** and $\frac{\partial}{\partial n}$ are the outward drawn normal unit vectors at the surface of the control volume and the derivative along this direction, respectively.

Equation (2.32a) or (2.32b) is representative of the principle of conservation of energy over the finite control volume V. They state that the rate of energy entering the control volume through its boundary surfaces S, plus the rate of energy generated in the volume element, is equal to the rate of increase of stored energy in the control volume. Furthermore, since fluxes are conserved in transport between adjacent control volumes, the conservation principle is also satisfied for an assembly of control volumes. That is, the numerical solution will satisfy both the *local* and *global* conservation properties; hence, the formulation given by equations (2.32a–c) is *fully conserving*.

In this development of the control volume conservation equations (2.32a–c), our starting point was the diffusion equation, which was integrated over the control volume. Clearly, an alternative approach is to recall the fact that the diffusion equation is usually derived from the conservation of energy over a control volume and, hence, to directly apply the appropriate conservation principle to a control volume.

Here, we have developed a control volume energy conservation equation for the physical phenomena involving transient heat conduction. Similar conservation expressions can be developed for the conservation of mass or momentum and include situations involving convective transport.

Once the control volume conservation equation is available, the corresponding finite difference equation over the control volume is readily obtained by discretizing the derivative terms in this conservation equation.

The control volume approach for the development of finite difference equations has distinct advantages of being readily applicable to multidimensional problems, complicated boundary conditions, and situations involving variable mesh and variable physical properties. On the other hand, the accuracy estimates with the control volume approach are difficult compared to that with the Taylor series expansion method, which readily provides information on the order of the truncation error involved.

When applying the control volume approach to develop the finite difference equations, the finite difference nodes must be established first and then the control volumes must be identified. To illustrate this matter, we present in Figure 2.4 a one-dimensional domain with grid points, a control volume around node i, and the values of the potential T at the grid points. In this illustration, the nodes are placed with equal spacing Δx, but unequal spacing poses no difficulty. A typical internal grid point is identified as i, and the value of the potential at this point is T_i. The following simple example will illustrate the basic concepts in the application of the control volume approach to develop finite difference equations.

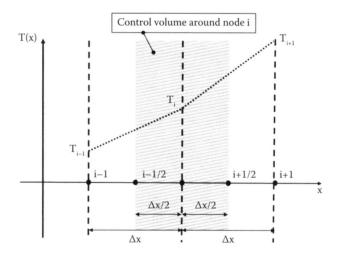

FIGURE 2.4
Control volume for one-dimensional situation.

Example 2.3

Use the control volume approach to discretize the steady-state one-dimensional heat conduction equation in Cartesian coordinates

$$\frac{\partial}{\partial x}\left(k\frac{\partial T}{\partial x}\right) + g(x) = 0 \qquad (a)$$

Solution

The integration of equation (a) in a control volume such as the one presented in Figure 2.4 around node i gives

$$\left[\left(k\frac{dT}{dx}\right)_{i+1/2} - \left(k\frac{dT}{dx}\right)_{i-1/2}\right]S + \bar{g}_i V = 0 \qquad (b)$$

where $\bar{g}_i$ is a suitable average of $g(x)$ over the control volume associated with the node i. For the one-dimensional problem considered here, unit lengths are taken along the y and z directions. Hence, the surface area at the interfaces of the control volume is also unitary, that is, $S = 1$ and V = volume of the control volume = $S\Delta x = \Delta x$. The subscript $i+1/2$ refers to the interface location between the nodes i and $i+1$, similarly $i-1/2$ refers to the interface location between the nodes i and $i-1$.

To develop the finite difference equation for the node i in equation (a), the derivatives in the diffusive fluxes at $i+1/2$ and $i-1/2$ are discretized by assuming a piecewise linear profile for the temperature between the neighboring nodes, as illustrated in Figure 2.4, and the aforementioned values of S and V are introduced. The finite difference equation for the internal node i then becomes

$$k_{i+1/2}\frac{T_{i+1} - T_i}{\Delta x} - k_{i-1/2}\frac{T_i - T_{i-1}}{\Delta x} + \Delta x \bar{g}_i = 0 \qquad (c)$$

Example 2.3 illustrates an interesting characteristic of the control volume approach for discretization, which results from the use of the conservative form of the governing equation: it naturally deals with spatial dependent thermophysical properties. Note in equation (c) that the local values of thermal conductivities at the nodes $i+1/2$ and $i-1/2$ appeared in the approximations of the diffusive fluxes at the surfaces of the control volume. Such values can be suitably approximated with the local thermophysical properties in the neighboring nodes. For example, for the surface at $i+1/2$, one can write:

$$k_{i+1/2} = \frac{k_{i+1} + k_i}{2} \qquad (2.33a)$$

or, alternatively,

$$k_{i+1/2} = \frac{2k_{i+1}k_i}{k_{i+1} + k_i} \qquad (2.33b)$$

which are the arithmetic and the harmonic means of k_{i+1} and k_i, respectively. The arithmetic mean does not appropriately cope with sharp variations in the values of the thermal conductivity across the control volume surface as the harmonic mean does (Patankar 1980). Therefore, harmonic means are recommended instead of arithmetic means.

The reader can recognize that equation (c) of Example 2.3 reduces to equation (b) of Example 2.2 in the case of constant thermal conductivity, that is, $k_{i-1/2} = k_{i+1/2} = k$.

2.3 Boundary and Initial Conditions

The mathematical formulation of well-posed problems, depending on the character of the governing equation (that is, if it is parabolic, elliptic, or hyperbolic; see Chapter 1), requires, in broad terms, the specifications of boundary conditions at the surface of the volume as well as of initial conditions. The initial conditions are required only for the solution of transient problems, and they are specified as the spatial distributions of the dependent variables and of their time derivatives (if necessary). For example, in the case of a transient heat conduction problem in a volume V, the initial condition is given by the temperature distribution in V at a time that is arbitrarily designated as time zero (that is, the time that the governing equation is integrated from). Therefore, we can write for this example

$$T(\mathbf{r},t) = F(\mathbf{r}) \text{ at } t = 0, \quad \text{for } \mathbf{r} \in V \qquad (2.34)$$

where $F(\mathbf{r})$ is a known function. The discretization of the initial condition is then straightforward and simply consists of the evaluation of $F(\mathbf{r})$ at each finite difference node. The discretization of boundary conditions is not that simple and will be further discussed in this section, for both the Taylor series and the control volume discretization approaches.

Linear boundary conditions for the mathematical formulation of physical problems can be classified as (see also Section 1.6): (i) first kind, where the value of the dependent variable is specified; (ii) second kind, where the normal derivative of the dependent variable is specified; and (iii) third kind, where a linear combination of the dependent variable and its normal derivative is specified. Obviously, the specification of the boundary condition is strongly related to the physics of the problem. For example, for the energy conservation equation in conduction or multi mode heat transfer, the boundary condition results from the application of the energy conservation principle at the surface of the body and ultimately represents how the body exchanges heat with the surroundings (Özişik 1993). A general case was analyzed in Example 1.1, which could be reduced to first-, second-, and third-kind boundary conditions.

In order to illustrate the discretization of boundary conditions, we consider the one-dimensional steady-state heat conduction problem in Cartesian coordinates in the domain presented by Figure 2.1, which is modeled by (see also Examples 2.2 and 2.3)

$$\frac{d}{dx}\left(k\frac{dT}{dx}\right) + g(x) = 0 \text{ in } 0 < x < L \quad (2.35a)$$

or by

$$\frac{d^2T}{dx^2} + \frac{1}{k}g(x) = 0 \text{ in } 0 < x < L \quad (2.35b)$$

for constant thermal conductivity.

Nine different combinations of boundary conditions can be specified for this problem, depending on its physics. For the sake of brevity in the discretization procedure, the boundary conditions at $x = 0$ and $x = L$ will be considered here as being of the same type.

For the *first-kind boundary conditions* we have:

$$T = \theta_0 \quad \text{at } x = 0 \quad (2.36a)$$

$$T = \theta_L \quad \text{at } x = L \quad (2.36b)$$

while for the *second-kind boundary conditions*, with imposed heat flux q_0 at $x = 0$ and heat flux q_L imposed at $x = L$, we have:

$$-k\frac{dT}{dx} = q_0 \quad \text{at } x = 0 \quad (2.37a)$$

$$k\frac{dT}{dx} = q_L \quad \text{at } x = L \quad (2.37b)$$

The *third-kind boundary conditions* are presented for a case of convective heat transfer at $x = 0$ and $x = L$, with the heat transfer coefficient and surrounding temperature h_0 and $T_{\infty,0}$, respectively, at $x = 0$. Similarly, at $x = L$, the heat transfer coefficient and the surrounding temperature are given by h_L and $T_{\infty,L}$, respectively. Thus,

$$-k\frac{dT}{dx} + h_0 T = h_0 T_{\infty,0} \quad \text{at } x = 0 \quad (2.38a)$$

$$k\frac{dT}{dx} + h_L T = h_L T_{\infty,L} \quad \text{at } x = L \quad (2.38b)$$

TABLE 2.2

Coefficients of equations (2.39) for Boundary Conditions of the Second and Third Kind

	x = 0		x = L	
Boundary Condition	a_0	f_0	a_L	f_L
Second kind	0	q_0	0	q_L
Third kind	h_0	$h_0 T_{\infty,0}$	h_L	$h_L T_{\infty,L}$

For simplicity in the analysis, equation (2.37) and (2.38) will be written in the following general form

$$-k\frac{dT}{dx} + a_0 T = f_0 \quad \text{at } x = 0 \tag{2.39a}$$

$$k\frac{dT}{dx} + a_L T = f_L \quad \text{at } x = L \tag{2.39b}$$

where the coefficients a and f are given by Table 2.2 for boundary conditions of the second and third kinds.

2.3.1 Discretization of Boundary Conditions with Taylor Series

The discretized form of equation (2.35b), by using central finite difference approximation, is given by (see Example 2.2):

$$T_{i-1} - 2T_i + T_{i+1} + \frac{g_i}{k}(\Delta x)^2 = 0 \tag{2.40}$$

Equation (2.40) is valid for the interior nodes inside the domain, that is, for $1 \leq i \leq M - 1$, as illustrated by Figure 2.5. The equations for the nodes at the boundaries are now written, depending on the type of boundary condition that is applied.

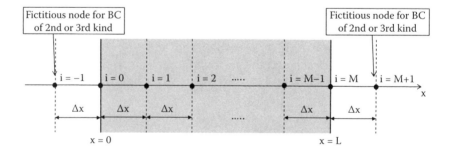

FIGURE 2.5
Discretization of the domain.

2.3.1.1 Boundary Condition of the First Kind

By writing equation (2.40) for the node i = 1, we obtain

$$T_0 - 2T_1 + T_2 + \frac{g_1}{k}(\Delta x)^2 = 0 \qquad (2.41a)$$

where the temperature at the node i = 0 is obtained from the boundary condition [equation (2.36a)] and then equation (2.41a) becomes

$$-2T_1 + T_2 + \frac{g_1}{k}(\Delta x)^2 + \theta_0 = 0 \qquad (2.41b)$$

Similarly, for i = M−1 equation (2.40) becomes

$$T_{M-2} - 2T_{M-1} + T_M + \frac{g_{M-1}}{k}(\Delta x)^2 = 0 \qquad (2.42a)$$

and by substituting the boundary condition [equation (2.36b)], that is, $T_M = \theta_L$, we obtain

$$T_{M-2} - 2T_{M-1} + \frac{g_{M-1}}{k}(\Delta x)^2 + \theta_L = 0 \qquad (2.42b)$$

Therefore, the system formed by equation (2.41b) for i = 1, equation (2.40) for i = 2,...,M−2, and equation (2.42b) for i = M−1 contains M−1 equations and M−1 unknowns given by temperatures at nodes i = 1,...,M−1. The solution of this system of M−1 equations provides the finite difference solution for the problem. Such a solution, as well all the others that will be examined throughout this book, is highly dependent on the mesh spacing because the approximation error of the finite difference approximation is a function of Δx, as presented earlier in Section 2.1. For the present case, the solution is of second order in Δx because the truncation error in equations (2.40), (2.41b), and (2.42b) is $0[(\Delta x)^2]$ [see equation (2.13a)].

2.3.1.2 Boundary Conditions of the Second and Third Kinds

Although the system resulting from the case with boundary conditions of the first kind at x = 0 and x = L was of size (M−1), the system that results for boundary conditions of the second or third kinds, both at x = 0 and x = L, is of size M + 1. Such is the case because the temperatures at these two boundaries (nodes i = 0 and i = M) are unknown when the boundary conditions of the second or third kinds are applied.

A very simple approach for the discretization of boundary conditions of the second and third kinds, which keeps the truncation error of second order in Δx, is to use the concept of fictitious nodes. This concept will be illustrated in this section by utilizing the general forms of equation (2.39) for the boundary conditions of the second and third kinds.

The approach of fictitious nodes starts by writing at the boundary nodes, $i = 0$ and $i = M$, the discretized form of the governing equation [given by equation (2.40), in this case]. We obtain

$$T_{-1} - 2T_0 + T_1 + \frac{g_0}{k}(\Delta x)^2 = 0 \quad \text{at } i = 0 \tag{2.43a}$$

$$T_{M-1} - 2T_M + T_{M+1} + \frac{g_M}{k}(\Delta x)^2 = 0 \quad \text{at } i = M \tag{2.43b}$$

Equations (2.43a,b) involve the temperatures at the fictitious nodes $i = -1$ and $i = M + 1$ that are outside the domain of interest (see Figure 2.5). The temperatures T_{-1} and T_{M+1} are eliminated from equation (2.43a,b) by using the boundary conditions equation (2.39a,b). The derivatives in equation (2.39a,b) are discretized with central differences that are of the second order in Δx, at $i = 0$ and $i = M$, respectively,

$$-k\frac{T_1 - T_{-1}}{2\Delta x} + a_0 T_0 = f_0 \quad \text{at } i = 0 \tag{2.44a}$$

$$k\frac{T_{M+1} - T_{M-1}}{2\Delta x} + a_L T_M = f_L \quad \text{at } i = M \tag{2.44b}$$

which are solved for the temperatures at the fictitious nodes and substituted into equation (2.43a,b) to yield

$$-2\left(1 + \frac{\Delta x}{k}a_0\right)T_0 + 2T_1 + \frac{g_0}{k}(\Delta x)^2 + 2\frac{\Delta x}{k}f_0 = 0 \quad \text{at } i = 0 \tag{2.45a}$$

$$2T_{M-1} - 2\left(1 + \frac{\Delta x}{k}a_L\right)T_M + \frac{g_M}{k}(\Delta x)^2 + 2\frac{\Delta x}{k}f_L = 0 \quad \text{at } i = M \tag{2.45b}$$

Equation (2.45a) for $i = 0$, equation (2.40) for $1 \leq i \leq M-1$, and equation (2.45b) for $i = M$ give the system of $M + 1$ equations and $M + 1$ unknowns to be solved for the cases of boundary conditions of the second or third kinds at $x = 0$ and $x = L$, where the coefficients a_0, a_L, f_0, and f_L are given by Table 2.2.

2.3.2 Discretization of Boundary Conditions with Control Volumes

The discretization of equation (2.35a) with the control volume approach results in (see Example 2.3):

$$\left[\left(k\frac{dT}{dx}\right)_{i+1/2} - \left(k\frac{dT}{dx}\right)_{i-1/2}\right] + \overline{g}_i \Delta x = 0 \tag{2.46}$$

where the nodes i correspond to the central position of the finite control volume. For equation (2.46), we consider M identical control volumes of length Δx, as illustrated by Figure 2.6.

Discrete Approximation of Derivatives

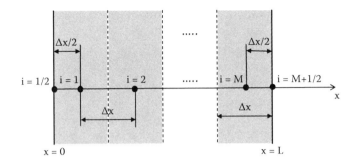

FIGURE 2.6
Domain discretization with finite volumes.

By using central finite difference approximation for the temperature gradients, equation (2.46) becomes

$$k_{i-1/2}T_{i-1} - (k_{i-1/2} + k_{i+1/2})T_i + k_{i+1/2}T_{i+1} + (\Delta x)^2 \bar{g}_i = 0 \qquad (2.47)$$

Equation (2.47) directly applies for volumes that do not have surfaces that coincide with the boundaries of the domain (at the positions $i - 1/2$ and $i + 1/2$), that is, it is valid for $2 \leq i \leq M-1$. For the derivation of the discretized equations of control volumes $i = 1$ and $i = M$, the fluxes at the positions $i - 1/2$ and $i + 1/2$ in equation (2.46), respectively, are substituted from the boundary conditions as described in the following.

2.3.2.1 Boundary Condition of the First Kind

At $x = 0$, the flux at $i = 1/2$ is approximated by forward differences in half of the control volume as follows

$$\left(k\frac{dT}{dx}\right)_{1/2} = k_{1/2}\frac{T_1 - T_{1/2}}{\frac{\Delta x}{2}} \qquad (2.48)$$

where, from the boundary condition equation (2.36a), $T_{1/2} = \theta_0$. By substituting equation (2.48) and discretizing the flux at $i = 3/2$ by central differences, equation (2.47) becomes:

$$-(2k_{1/2} + k_{3/2})T_1 + k_{3/2}T_2 + (\Delta x)^2 \bar{g}_1 + 2k_{1/2}\theta_0 = 0 \quad \text{for } i = 1 \qquad (2.49)$$

Similarly, at $x = L$, the flux at $i = M + 1/2$ is approximated by backward differences, that is,

$$\left(k\frac{dT}{dx}\right)_{M+1/2} = k_{M+1/2}\frac{T_{M+1/2} - T_M}{\frac{\Delta x}{2}} \qquad (2.50)$$

where, from the boundary condition equation (2.36b), $T_{M+1/2} = \theta_L$. By substituting equation (2.50) and discretizing the flux at $i = M-1/2$ by central differences, equation (2.46) becomes

$$k_{M-1/2}T_{M-1} - (2k_{M+1/2} + k_{M-1/2})T_M + (\Delta x)^2 \bar{g}_M + 2k_{M+1/2}\theta_L = 0 \text{ for } i = M \quad (2.51)$$

Therefore, the system formed by equation (2.49) for $i = 1$, equation (2.47) for $i = 2,\ldots,M-1$, and equation (2.51) for $i = M$ is composed of M equations with M unknowns, which are the temperatures at the center of each control volume used in the discretization of the domain.

2.3.2.2 Boundary Condition of the Second Kind

For the control volume discretization, boundary conditions of the second kind are the easiest to apply because the fluxes at the surfaces of the control volumes naturally appear in equation (2.46). At control volume surfaces that coincide with the boundaries of the domain, the fluxes are directly substituted from the boundary conditions, given in this case by equations (2.37a,b). Therefore, by writing equation (2.46) for $i = 1$ and substituting equation (2.37a), we obtain

$$\left[\left(k\frac{dT}{dx}\right)_{3/2} + q_0\right] + \bar{g}_1 \Delta x = 0 \quad (2.52a)$$

Similarly, we have for $i = M$

$$\left[q_L - \left(k\frac{dT}{dx}\right)_{M-1/2}\right] + \bar{g}_M \Delta x = 0 \quad (2.52b)$$

which are now rewritten by applying central approximations for the gradients. The following equations result:

$$-k_{3/2}T_1 + k_{3/2}T_2 + (\Delta x)^2 \bar{g}_1 + \Delta x\, q_0 = 0 \text{ for } i = 1 \quad (2.53a)$$

$$k_{M-1/2}T_{M-1} - k_{M-1/2}T_M + (\Delta x)^2 \bar{g}_M + \Delta x\, q_L = 0 \text{ for } i = M \quad (2.53b)$$

Equations (2.53a,b), together with equation (2.47) for $i = 2,\ldots,M-1$, form the system of equations for the case with boundary conditions of the second kind at $x = 0$ and $x = L$.

2.3.2.3 Boundary Condition of the Third Kind

At $x = 0$, the discretization of the boundary condition [equation (2.38a)] with forward finite differences gives

$$-k_{1/2}\frac{T_1 - T_{1/2}}{\frac{\Delta x}{2}} + h_0 T_{1/2} = h_0 T_{\infty,0} \quad (2.54)$$

Discrete Approximation of Derivatives 45

which is then solved for $T_{1/2}$ as

$$T_{1/2} = \frac{h_0 T_{\infty,0} \Delta x + 2k_{1/2} T_1}{h_0 \Delta x + 2k_{1/2}} \quad (2.55)$$

Equation (2.55) is substituted into the definition of the convective flux at the boundary $x = 0$, that is, $q_0 = h_0(T_{\infty,0} - T_{1/2})$, thus resulting in

$$q_0 = h_0^*(T_{\infty,0} - T_1) \quad (2.56a)$$

where h_0^* is a modified heat transfer coefficient, which takes into account conduction from the boundary to the center of the control volume in addition to convection at the surface, and is given by

$$h_0^* = \frac{2k_{1/2} h_0}{h_0 \Delta x + 2k_{1/2}} \quad (2.56b)$$

By now substituting equation (2.56a) into equation (2.52a), we obtain the discretized equation for the control volume at $i = 1$ in terms of the modified heat transfer coefficient h_0^*, that is,

$$-(k_{3/2} + h_0^* \Delta x) T_1 + k_{3/2} T_2 + (\Delta x)^2 \bar{g}_1 + h_0^* T_{\infty,0} \Delta x = 0 \quad \text{for } i = 1 \quad (2.57)$$

Analogous operations are now performed in order to obtain the discretized equation for the control volume at $i = M$ with third-kind boundary condition at $x = L$. The temperature at the boundary $x = L$ is found as:

$$T_{M+1/2} = \frac{h_L T_{\infty,L} \Delta x + 2k_{M+1/2} T_M}{h_L \Delta x + 2k_{M+1/2}} \quad (2.58)$$

and the discretized equation for $i = M$ is obtained as

$$k_{M-1/2} T_{M-1} - (k_{M-1/2} + h_L^* \Delta x) T_M + (\Delta x)^2 \bar{g}_M + h_L^* T_{\infty,L} \Delta x = 0 \quad \text{for } i = M \quad (2.59)$$

where the modified heat transfer coefficient at $i = M$ is given by

$$h_L^* = \frac{2k_{M+1/2} h_L}{h_L \Delta x + 2k_{M+1/2}} \quad (2.60)$$

Therefore, for third-kind boundary conditions at $x = 0$ and $x = L$, the system of algebraic equations that result from the discretization with finite control volumes is formed by equation (2.57), equation (2.47) written for $i = 2,\ldots,$ $M-1$, and equation (2.59).

We note that the systems of algebraic equations for boundary conditions of the second and of the third kinds resulting from the control volume discretization do not involve the temperatures at the surfaces of the body, which are also unknown. The temperatures at the surfaces $x = 0$ and $x = L$ can be calculated after the solution of such algebraic systems, by discretizing the boundary conditions, equation (2.37) or (2.38). For boundary conditions of the third kind, the temperatures at the boundaries $x = 0$ and $x = L$ are obtained from equations (2.55) and (2.58), respectively. For boundaries conditions of the second kind, such temperatures are given, respectively, by:

$$T_{1/2} = T_1 + \frac{q_0 \Delta x}{2 k_{1/2}} \qquad (2.61a)$$

$$T_{M+1/2} = T_M + \frac{q_L \Delta x}{2 k_{M+1/2}} \qquad (2.61b)$$

2.4 Errors Involved in Numerical Solutions

In the solution of differential equations with finite differences, a variety of schemes are available for the discretization of derivatives and the solution of the resulting system of algebraic equations. In many situations, questions arise regarding the round-off and truncation errors involved in the numerical computations as well as the consistency, stability, and the convergence of the finite difference scheme. Such matters will be discussed in various chapters throughout this book as the occasions arise. Here, we present a brief description of the significance of these terminologies.

2.4.1 Round-Off Errors

Computations are rarely made in exact arithmetic. This means that real numbers are represented in "floating point" form and, as a result, errors are caused due to the rounding off of the real numbers. Even though modern computers can represent numbers to several decimal places, in extreme cases, such errors (called "round-off' errors) can accumulate and become a main source of error in the solution.

2.4.2 Truncation Error

In finite difference representation of derivatives with Taylor's series expansion, the higher order terms are neglected by truncating the series; the error caused as a result of such truncation is called the *truncation error*.

Discrete Approximation of Derivatives

For example, in *forward* differencing of the first derivative to the order Δx, as given by equation (2.4), the term

$$0(\Delta x) = -\frac{1}{2}\Delta x\, f'(x_0) - \frac{1}{6}(\Delta x)^2\, f'''(x_0) + \ldots \qquad (2.62)$$

represents the truncation error and the lowest order term on the right-hand side (i.e., Δx) gives the order of the method.

The truncation error identifies the difference between the exact solution of a differential equation and its finite difference solution without the round-off error, that is

$$\begin{pmatrix} \text{Exact solution} \\ \text{of PDE} \end{pmatrix} - \begin{pmatrix} \text{Solution of finite} \\ \text{difference equation} \\ \text{without the} \\ \text{round-off error} \end{pmatrix} = \begin{pmatrix} \text{Truncation} \\ \text{error} \end{pmatrix} \qquad (2.63)$$

Consider, for example, the Laplacian operator governing the steady-state heat conduction in a solid

$$L(T) \equiv \frac{\partial^2 T}{\partial x^2} + \frac{\partial^2 T}{\partial y^2} = 0 \qquad (2.64a)$$

and its finite difference approximation given by

$$L_{FD}(T) \equiv \frac{T_{i-1,j} - 2T_{i,j} + T_{i+1,j}}{(\Delta x)^2} + \frac{T_{i,j-1} - 2T_{i,j} + T_{i,j+1}}{(\Delta y)^2} \qquad (2.64b)$$

Then we write

$$L(T) - L_{FD}(T) = \text{Truncation error} = 0[(\Delta x)^2, (\Delta y)^2] \qquad (2.65)$$

that identifies the error resulting from the discretization of the governing partial differential equation (2.64a).

2.4.3 Discretization Error

This term is commonly used to identify the error due to the truncation error in the finite difference representation of the governing differential equation and boundary conditions.

2.4.4 Total Error

This is involved in finite difference calculations consisting of the discretization error plus the round-off error. The discretization error increases with increasing mesh size, while the round-off error decreases with increasing mesh size. Therefore, the total error is expected to exhibit a minimum as the mesh size is decreased.

2.4.5 Stability

In the numerical solution of differential equations with finite differences, errors are introduced at almost every stage of the calculations. The solution scheme is said to be stable if the errors involved in numerical computations are not amplified without bounds as the numerical calculations progresses.

2.4.6 Consistency

The discretized form obtained by finite differences may not represent the original continuous partial differential problem but a different one, even when the mesh size tends to zero. If such a problem happens, the discretized form is *inconsistent* or *incompatible* with the original differential problem.

Therefore, although the concept of consistency relates the discretized equation to the actual partial differential equation, stability relates the numerical solution of the discretized equation to the exact solution of the discretized equation. The concept of *convergence* relates the numerical solution of the discretized equation to the exact solution of the partial differential equation. A convergent discretization scheme is one such that the numerical solution of the discretized equation approaches the exact solution of the partial differential problem, as the spatial and time mesh are refined, that is, $\Delta x \to 0$ and $\Delta t \to 0$.

As expected, the concepts of stability, consistency, and convergence are related, as stated by Lax's Equivalence Theorem: For a well-posed linear initial value problem and a consistent discretization scheme, stability is the necessary and sufficient condition for convergence (Richtmyer and Morton 1967). Lax's theorem only applies to smooth solutions (Laney 1998), and its proof is limited to linear problems. Therefore, nonlinear problems with solution discontinuities, such as in compressible flows, require stability criteria that are heuristic and associated with the physics of the problem. Nonlinear stability criteria are also focused on not allowing spurious oscillations or overshoots on the solution to propagate, such as in the linear case. Nonlinear stability conditions are discussed in Chapter 9, in the analysis of compressible inviscid flows. For further details on nonlinear stability conditions, the reader is referred to Laney (1998).

Example 2.4

Determine the truncated leading error term in finite difference approximation with central differences of the following differential operator.

$$L(T) \equiv \frac{d^2T(x)}{dx^2} + A\frac{dT(x)}{dx} + BT(x) = 0 \tag{a}$$

Solution
Utilizing the results given in Table 2.3, the truncated leading error term is determined as

TABLE 2.3
Various Differencing Schemes and the Truncated Leading Error Terms

Derivative	Finite Difference Form	Truncated Leading Error Terms[a]
$\dfrac{df(x)}{dx}$	$\dfrac{f(x+\Delta x)-f(x)}{\Delta x}$ (forward)	$-\dfrac{\Delta x}{2}f'' - \dfrac{(\Delta x)^2}{6}f'''$
$\dfrac{df(x)}{dx}$	$\dfrac{f(x)-f(x-\Delta x)}{\Delta x}$ (backward)	$+\dfrac{\Delta x}{2}f'' - \dfrac{(\Delta x)^2}{6}f'''$
$\dfrac{df(x)}{dx}$	$\dfrac{f(x+\Delta x)-f(x-\Delta x)}{2\Delta x}$ (central)	$-\dfrac{(\Delta x)^2}{6}f'''$
$\dfrac{d^2f(x)}{dx^2}$	$\dfrac{f(x-\Delta x)-2f(x)+f(x+\Delta x)}{(\Delta x)^2}$ (central)	$-\dfrac{(\Delta x)^2}{12}f''''$

[a] Primes denote differentiation with respect to x.

$$TE = -\frac{(\Delta x)^2}{12}T'''' - A\frac{(\Delta x)^2}{6}T''' \qquad (b)$$

In a finite difference representation of a problem, one is concerned with the discretization of the governing differential equation as well as the boundary conditions associated with it. If anywhere in the discretization procedure the model is accurate only to the first order, then the accuracy of calculations is accurate only to the first order even if the model is second-order accurate everywhere else.

2.5 Verification and Validation

Verification and validation aim at assessing the accuracy of computational simulations. In the last few years, the subjects of verification and validation (V&V) have become of great interest and intensely researched. In fact, several international societies have special committees and conferences dedicated to V&V, and the number of journals specifically related to these subjects has increased. Indeed, even standards are now available where the basic principles and procedures are established. The concepts of V&V are briefly introduced in this section, which is based on the American Society of Mechanical Engineers' Standard for Verification and Validation in Computational Fluid Dynamics and Heat Transfer (ASME 2009).

Validation is defined as "the process of determining the degree to which a model is an accurate representation of the real world from the perspective of the intended uses of the model" (ASME 2009). By the word "model" it is understood to mean the mathematical formulation that is used to represent a physical problem as well as its computational solution. The words "real world" mean that validation necessarily involves the comparison between computational and experimental results of the physical problem under study. Uncertainties present both in the measured data and in the computational results are therefore required for the validation process.

In general, uncertainty refers to a statistical model for the errors, in the form of a probability distribution function, while the error is itself a realization of this probability distribution function. Although readers are generally familiar with experimental uncertainties, the subject of computational simulation under uncertainty is recent, and different methods have been proposed for such a purpose (Xiu 2010). Uncertainties in the computational simulation of real problems are due to: (i) simplification hypotheses used to derive the mathematical formulation of the physical problem; (ii) discretization (truncation) errors, as discussed earlier; and (iii) values used for the input parameters (such as the thermophysical properties appearing in the formulation).

The process of verification must be performed before the process of validation. Verification includes *code verification* and *solution verification*. In the verification of the computational code, an analysis is made to establish if it actually solves the mathematical model for which it is intended. The estimation of the accuracy of the computational solution is the objective of the solution verification. Therefore, among the three sources of uncertainties mentioned earlier, verification is specifically aimed at evaluating the discretization errors.

This section deals only with verification. In fact, verification might be sufficient for engineering design and analysis where no experimental data are available and validation cannot be performed.

2.5.1 Code Verification

Code verification involves the comparison of the computational solution of the code under verification to a *benchmark* solution. Benchmark solutions are ideally analytical and exact from the computational point of view, that is, with negligible round-off errors. However, the benchmark solution can also be obtained from another previously verified code. Exact analytical solutions for heat conduction problems can be found (Özişik 1993) by using the method of separation of variables and the classical integral transform technique (CITT). Moreover, the CITT has been formalized for different classes of diffusion problems (Mikhailov and Özişik 1984b). The basis of the CITT has been extended for broader classes of problems with the generalized integral transform technique (GITT; see Cotta 1990, 1993, 1994a, 1994b). The GITT has a hybrid numerical–analytical character that gives enough flexibility for treating problems with variable coefficients, nonlinearities, convective-diffusive behavior, in irregular domains, and so on and allows user controlled accuracy to generate benchmark solutions. The GITT is the subject of Chapter 12 in this book.

The so-called method of manufactured solutions can also be used to generate analytical solutions for the mathematical problem implemented in the

Discrete Approximation of Derivatives

code under verification, as will be described next (ASME 2009). Consider the mathematical formulation of the physical problem that is implemented in the code that is being verified, written in the following operator form:

$$L[f(\mathbf{r},t)] = 0 \quad \text{for} \quad \mathbf{r} \in V, t > 0 \tag{2.66}$$

where **r** is the position vector in the coordinate system in which the problem is formulated, t is the time variable, V is the spatial domain, and f(r,t) is a solution of the problem. Hence, the operator L[.] can be even nonlinear. The operator L[.], given by equation (2.66), does not include initial and boundary conditions for the problem.

An arbitrary analytical expression M(r,t) is now selected. Note that this expression can be completely disconnected from the physics of the problem and even from the class of fundamental solutions for the problem; it is only required that it be analytical and easy to calculate. Because M(r,t) was arbitrarily selected, when it is substituted into equation (2.66), one generally obtains

$$L[M(\mathbf{r},t)] = Q(\mathbf{r},t) \neq 0 \quad \text{for} \quad \mathbf{r} \in V, t > 0 \tag{2.67}$$

where Q(r,t) is a source-function that results from the fact that M(r,t) does not necessarily satisfy equation (2.66). Note that Q(r,t) = 0 if M(r,t) is a solution of L[f(r,t)] = 0.

The process of code verification with the manufactured solution M(r,t) thus consists of comparing the computational solution of equation (2.67), and not of equation (2.66), to the analytical solution M(r,t). We note that M(r,t) must contain constants to be determined by satisfying the boundary and initial conditions of equation (2.66).

Example 2.5

Generate a manufactured solution for the problem given by

$$\frac{d^2 T}{dx^2} + \frac{1}{k} g(x) = 0 \quad \text{in } 0 < x < L \tag{a}$$

$$T = \theta_0 \quad \text{at } x = 0 \tag{b}$$

$$T = \theta_L \quad \text{at } x = L \tag{c}$$

Solution

The manufactured solution is proposed in the following form:

$$M(x) = a + b \sin(x) \tag{d}$$

By substituting (d) into

$$L[M(x)] = \frac{d^2 M}{dx^2} + \frac{1}{k} g(x) \tag{e}$$

we obtain

$$Q(x) = -b\sin(x) + \frac{1}{k}g(x) \tag{f}$$

The constants a and b in equation (d) are obtained by applying the boundary conditions (b,c). We obtain:

$$M(x) = \theta_0 + (\theta_L - \theta_0)\frac{\sin(x)}{\sin(L)} \tag{g}$$

Equation (g) is, therefore, the solution of the following problem

$$\frac{d^2T}{dx^2} + \frac{1}{k}g(x) = Q(x) \quad \text{in } 0 < x < L \tag{h}$$

$$T = \theta_0 \quad \text{at } x = 0 \tag{i}$$

$$T = \theta_L \quad \text{at } x = L \tag{j}$$

which is basically the original problem with a modified heat source term; that is, we can write equation (h) as

$$\frac{d^2T}{dx^2} + \frac{1}{k}g^*(x) = 0 \quad \text{in } 0 < x < L \tag{k}$$

where

$$g^*(x) = g(x) - \frac{Q(x)}{k} \tag{l}$$

Hence, the finite difference code developed for the solution of equations (a–c) can be straightforwardly applied to the solution of equations (h–j), in order to allow for the code verification by using the manufactured solution given by equation (d). We note that an analytical benchmark solution for equations (a–c) can be easily obtained for code verification. On the other hand, the objective of this example was to introduce the method of manufactured solutions, which can now be used by the reader to verify the codes that he or she develops for more complicated problems.

The comparison of the numerical solution with the benchmark solution used for code verification is usually performed through a grid convergence analysis. In a grid convergence analysis, the grid size is systematically reduced in order to analyze the discretization error between the numerical and benchmark solutions

$$\varepsilon_h(\mathbf{r},t) = N_h(\mathbf{r},t) - M(\mathbf{r},t) \tag{2.68}$$

where $N_h(\mathbf{r},t)$ is the numerical solution and h is the grid size used in this calculation.

Discrete Approximation of Derivatives

Partial differential problems in space and time variables involve grid sizes in each spatial coordinate and in time. Usually, the grid convergence is separately examined for each of these variables, and the grid should be successively reduced by a factor of at least 1.3. On the other hand, complicated problems in complicated geometries might require that the grid be simultaneously refined in all spatial variables. For these cases, the grid size, h, can be obtained, for example, in Cartesian coordinates as:

$$h = (\Delta x \, \Delta y \, \Delta z)^{1/3} \tag{2.69}$$

It is expected that the discretization error decreases as the grid size is reduced and takes the form

$$\varepsilon_h(\mathbf{r}, t) = C h^p + \text{(higher order terms)} \tag{2.70}$$

where C is a constant and p is the order of the error (p = 1 for first-order error, p = 2 for second-order error, etc.). Therefore, by performing the code verification through a grid convergence analysis, not only the code is verified but also the order of the error can be examined and compared to that which is theoretically expected (see Section 2.1).

If the error given by equation (2.70) is not satisfactorily reduced as the grid becomes smaller, the code cannot be accepted for its intended purpose of simulation. Sources of mistakes that must be carefully examined can be many, including those in the analytical development of the discretized forms of governing equations, boundary conditions, and initial conditions; in writing the computational code and the required solution algorithms; as well as in inappropriate discretization procedures. If none of these mistakes are found, effects of code parameters on the solution must be examined, such as tolerances for iterative procedures eventually used and even if the grid is sufficiently converged.

Example 2.6

Perform a grid convergence analysis and calculate the order of the discretization error for the first derivative of the function $f(x) = e^{-x}$ at $x = 1$, by using forward and central finite differences.

Solution

The analytical (exact) first derivative of $f(x) = e^{-x}$ is $df(x)/dx = -e^{-x}$. These functions are shown by Figure 2.7. By using the forward and central finite differences given by equations (2.9a) and (2.9c), respectively, we can write

$$d_1 = \frac{e^{-(x+h)} - e^{-x}}{h} + O(h) \tag{a}$$

$$d_2 = \frac{e^{-(x+h)} - e^{-(x-h)}}{2h} + O(h^2) \tag{b}$$

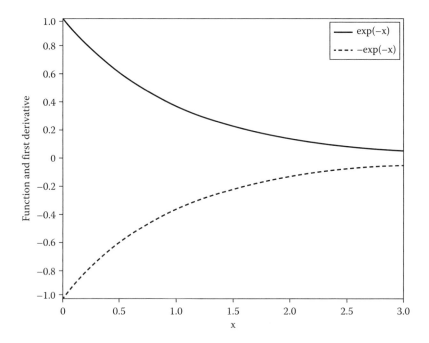

FIGURE 2.7
Function $f(x) = e^{-x}$ and its exact first derivative.

Equations (a) and (b) were calculated for 10 values of h, starting at h = 0.25 and sequentially halved. The discretization errors between the numerical derivatives and the exact derivative at x = 1 are presented by figure 2.8. This figure shows that the discretization errors are reduced as h decreases and that the errors for the central difference approximation are smaller and reduce faster than those for the forward approximation, as expected. By adjusting equation (2.70) to the discretization errors, their orders are calculated as 0.99 for forward approximation and 2.00 for central approximation, which are in excellent agreement with the theoretical values of the finite difference formulae equations (2.9a) and (2.9c), respectively.

2.5.2 Solution Verification

Solution verification is performed after code verification. Although code verification is performed by computing the error between the computational solution and a benchmark solution, in solution verification an estimate of the discretization error is obtained by grid refinement applied to a case of interest for which a benchmark solution is not available. Therefore, solution verification must be performed for every simulation where the code has not been verified.

Discrete Approximation of Derivatives

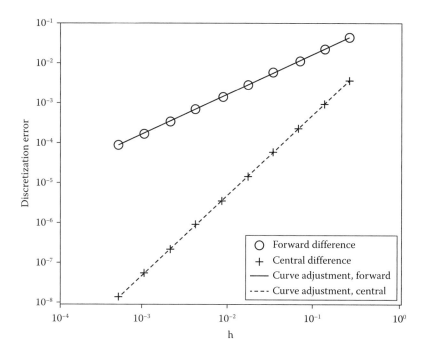

FIGURE 2.8
Discretization error for the forward and central difference approximations.

Solution verification is performed by applying the following five steps (ASME 2009):

Step 1: Define a representative grid size, h, for the problem of interest. For a steady-state one-dimensional problem in Cartesian coordinates, clearly h = Δx. For three-dimensional problems, h can be computed from equation (2.69).

Step 2: Select (at least) three different values of h, such that $h_1 < h_2 < h_3$, for which the computational solution is asymptotic, and let

$$r_{21} = \frac{h_2}{h_1}; \quad r_{32} = \frac{h_3}{h_2} \quad (2.71a,b)$$

The values of r_{21} and r_{32} should be larger than 1.3. Then, for each of the grid sizes h_1, h_2, and h_3, compute the numerical solution of the problem, and, if of interest, compute dependent variables resulting from such a solution (e.g., the Nusselt number in a convective problem). Denote the variable that will be used for the estimation of the discretization error as φ_i, which is computed with grid size h_i. Therefore, by using equation (2.70) for the discretization error, we can write

$$\varphi_1 = \varphi_{exact} + Ch_1^p \tag{2.72a}$$

$$\varphi_2 = \varphi_{exact} + Ch_2^p \tag{2.72b}$$

$$\varphi_3 = \varphi_{exact} + Ch_3^p \tag{2.72c}$$

Step 3: Calculate the order p of the method that is actually observed with the code by using the results obtained for each grid size. In order to calculate p, we define

$$\delta_{21} = \varphi_2 - \varphi_1 \tag{2.73a}$$

$$\delta_{32} = \varphi_3 - \varphi_2 \tag{2.73b}$$

and note that, by using equations (2.72a–c), we have

$$\frac{\delta_{21}}{\delta_{32}} = \frac{1}{r_{21}^p}\left(\frac{r_{21}^p - 1}{r_{32}^p - 1}\right) \tag{2.74}$$

Equation (2.74) is then solved for p. However, if $r_{21} = r_{32}$, p can be computed explicitly as

$$p = \frac{1}{\ln(r_{21})}\ln\left(\frac{\delta_{32}}{\delta_{21}}\right) \tag{2.75}$$

where the variable φ is assumed monotonic.

Step 4: With the order p, estimate the exact value of the variable φ at the finest grid. By substituting equations (2.72a,b) into equation (2.73a), we obtain

$$\delta_{21} = C(h_2^p - h_1^p) \tag{2.76}$$

and, therefore,

$$C = \frac{\delta_{21}}{h_1^p(r_{21}^p - 1)} \tag{2.77}$$

With equations (2.73a) and (2.77), equation (2.72a) can be written as

$$\varphi_{exact}^1 = \frac{r_{21}^p \varphi_1 - \varphi_2}{r_{21}^p - 1} \tag{2.78}$$

where the superscript 1 was used to denote that this value estimated for φ_{exact} was obtained with the finest grid, that is, with the grid size h_1.

Step 5: Calculate and report the following error estimates:

The absolute discrepancy between the solutions with the two most refined meshes,

$$\delta_{21}^a = |\varphi_2 - \varphi_1| \quad (2.79a)$$

the relative discrepancy between the solutions with the two most refined meshes, unless $\varphi_1 = 0$,

$$\tilde{\delta}_{21}^a = \left| \frac{\varphi_2 - \varphi_1}{\varphi_1} \right| \quad (2.79b)$$

and the relative error,

$$\tilde{\delta}_1 = \left| \frac{\varphi_{\text{exact}}^1 - \varphi_1}{\varphi_{\text{exact}}^1} \right| \quad (2.79c)$$

where φ_{exact}^1 is given by equation (2.78).

It is also recommended that the so-called grid convergence index (GCI) be reported, which represents an estimate of the solution uncertainty at the 95% confidence level. The GCI is given by

$$\text{GCI} = \frac{F_s}{r_{21}^p - 1} \left| \frac{\varphi_2 - \varphi_1}{\varphi_1} \right| \quad (2.79d)$$

where F_s is a safety factor recommended as 1.25.

Example 2.7

Perform solution verification of the approximation of the first derivative of the function $f(x) = e^{-x}$ at $x = 1$ by using forward finite differences.

Solution

The approximation by forward finite differences for the first derivative of the function $f(x) = e^{-x}$ is given by equation (a) in Example 2.6. We consider the following values of grid size: $h_1 = 0.01$, $h_2 = 0.02$, and $h_3 = 0.04$, which are clearly in a region where the solution behavior is asymptotic (notice in Figure 2.8 that these values of grid size correspond to a region where the error reduces with converged order 1). For the selected grid sizes, $r_{21} = r_{32} = 2$. Table 2.4 shows the values of the first derivative computed with equation (a) of Example 2.6 for such values of h, as well as the discrepancies between the solutions at successive levels of refinements.

TABLE 2.4
Solutions and Discrepancies for Different Grid Sizes

h	φ	δ
0.01	−0.3660	—
0.02	−0.3642	0.0018
0.04	−0.3606	0.0036

Equation (2.75) is used to compute the observed order of the error, which is p = 0.9856. By applying the definitions from equations (2.78) and (2.79) we obtain

$\varphi^1_{exact} = -0.3679$
$\delta^a_{21} = 0.0018$
$\tilde{\delta}^a_{21} = 0.0050$
$\tilde{\delta}_1 = 0.0051$
GCI = 0.0063

Therefore, the error between the estimate φ^1_{exact} and the exact solution is expected to be smaller than the solution uncertainty of 0.0063 at the 95% confidence level. Such is actually the case, where the derivative calculated with the exact solution $df(x)/dx = -e^{-x}$ is 0.3679. In fact, the estimate of the exact value φ^1_{exact} perfectly agrees with the actual value with four decimal places.

PROBLEMS

2.1. Using a Taylor's series expansion, show that a forward difference representation of df/dx, which is accurate to the order of $O(h^3)$, is given in subscript notation as

$$f'_i = \frac{2f_{i+3} - 9f_{i+2} + 18f_{i+1} - 11f_i}{6h} + O(h^3)$$

2.2. Consider the function $f(x) = 2e^x$. Using a mesh size $\Delta x \equiv h = 0.1$, determine $f'(x)$ at $x = 2$ with the forward formula equation (2.9a) accurate to $O(h)$ and the central difference formula equation (2.9c) accurate to $O(h^2)$ and compare the results with the exact value.

2.3. The values of f(x) at equally spaced (i.e., h = 1) x locations are given by

x	1	2	3	4	5
f(x)	25	30	27	17	−9

Calculate the first derivatives f'(1), f'(3), and f'(5) by using the second-order accurate—that is, $O(h^2)$ formula—for the first derivatives.

Discrete Approximation of Derivatives

2.4. Consider the function $f(x) = \sin(10\pi x)$. By using a mesh size $\Delta x = h = 0.2$, evaluate $f'(0)$ using the forward difference representation of the first derivatives given by

$$f'_i = \frac{f_{i+1} - f_i}{h} + 0(h)$$

and

$$f'_i = \frac{1}{2h}[-3f_i + 4f_{i+1} - f_{i+2}] + 0(h^2)$$

Compare the results obtained by finite differencing with that obtained from the exact analytic evaluation. Explain the reason for the difference between the exact and numerical results.

2.5. Consider the function $f(x) = x^{1/3}$. Evaluate $f'(0)$ using forward differencing schemes given by equations (2.9a) and (2.11a) accurate to $0(h)$ and $0(h^2)$, respectively, with a mesh size $\Delta x = h = 1$. Compare these results with the exact analytic evaluation of $f'(0)$. Try to explain the discrepancy between the exact and the finite difference results by taking the finite difference representation with different values of h, each one smaller than the one before.

2.6. Consider the function $f(x) = e^x$. Calculate $f'(1)$ with finite differences using forward and central difference formulae given, respectively, by equations (2.9a,c), by taking a step size $\Delta x = h = 0.1$. Compare these results with the exact analytic answer and explain the reason for the higher accuracy of central difference formula in comparison to forward difference formula.

2.7. Develop the finite difference approximation for the Laplacian of T in Cartesian coordinates given by

$$\frac{\partial^2 T}{\partial x^2} + \frac{\partial^2 T}{\partial y^2}$$

by using the second-order accurate central differencing formulae and give the resulting truncation errors.

2.8. By using the control volume approach, develop finite difference approximation for the following three-dimensional heat conduction equation

$$\frac{\partial^2 T}{\partial x^2} + \frac{\partial^2 T}{\partial y^2} + \frac{\partial^2 T}{\partial z^2} + \frac{1}{k} g(x, y, z) = 0$$

about an internal node "i,j,p."

2.9. Repeat Problem 2.8 using Taylor series expansion and central differencing scheme for the discretization of the second derivatives. Determine the order of accuracy of the truncation error.

2.10. By using the control volume approach, develop finite difference approximation about an internal node for the following three-dimensional heat conduction equation with spatially variable thermal conductivity

$$\frac{\partial}{\partial x}\left(k\frac{\partial T}{\partial x}\right) + \frac{\partial}{\partial y}\left(k\frac{\partial T}{\partial y}\right) + \frac{\partial}{\partial z}\left(k\frac{\partial T}{\partial z}\right) + g(x,y,z) = 0$$

2.11. Using Taylor series expansion in two variables, verify the finite difference approximation and the order of the truncation errors for the Case Nos. 1, 2, and 3 shown in Table 2.1.

2.12. Heat transfer by convection from the tube wall kept at a uniform temperature to a flow with uniform velocity and negligible radial variation in fluid temperature can be represented with the energy equation expressed in the dimensionless form as

$$\frac{d^2 T}{dx^2} + \frac{dT}{dx} - 2 = 0$$

Develop finite difference approximation for this equation by using the second-order accurate central difference formula for both derivatives and give the truncation error.

2.13. Write second-order accurate finite difference approximation at an interior node for the following equation

$$Pe\frac{d\theta}{dx} = \frac{d^2\theta}{dx^2} - Bi\theta$$

Discrete Approximation of Derivatives 61

2.14. Consider steady-state heat conduction in a slab of thickness L in which the boundary at x = 0 is thermally insulated and the boundary at x = L exchanges heat by convection with the surrounding medium at the temperature T_∞ with a heat transfer coefficient h. The internal heat generation is null. The formulation for this problem is given by

$$\frac{d^2T(x,t)}{dx^2} = 0 \quad \text{in } 0 < x < L$$
$$-k\frac{dT}{dx} = 0 \quad \text{at } x = 0$$
$$k\frac{dT}{dx} + hT = hT_\infty \quad \text{at } x = L$$

Use second-order central differences for the discretization of the differential equation for the problem and apply the concept of the fictitious node to obtain the discretized equations for the boundary nodes.

2.15. Use the finite control volume approach to obtain the discretized equations for Problem 2.14.

2.16. Consider steady-state heat conduction in a slab of thickness L, in which the boundary at x = 0 is maintained at the temperature θ_0 and the boundary at x = L is heated by a constant heat flux q_0. Heat is generated uniformly in the slab with a volumetric rate, g W/m³. The formulation for this problem is given by

$$\frac{d^2T(x,t)}{dx^2} + \frac{g}{k} = 0 \quad \text{in } 0 < x < L$$
$$T = \theta_0 \quad \text{at } x = 0$$
$$k\frac{dT}{dx} = q_L \quad \text{at } x = L$$

Use second-order central differences for the discretization of the differential equations for the problem and apply the concept of the fictitious node to obtain the discretized equations for the boundary nodes.

2.17. Use the finite control volume approach to obtain the discretized equations for Problem 2.16.

2.18. Develop a manufactured solution for Problem 2.14 in the form M(x)= a + b cos(x). Also, develop the formulation for the problem that satisfies the obtained manufactured solution.

2.19. Develop a manufactured solution for Problem 2.16 in the form M(x) = a + b sin(x). Also, develop the formulation for the problem that satisfies the obtained manufactured solution.

2.20. Perform the solution verification steps for the first derivative of the function $f(x) = 2e^x$ at $x = 2$, by using the difference formulae given by equations (2.9a–c), (2.11a,b), and (2.12a–c). Explain the behavior observed with the different finite difference approximations.

2.21. Repeat Problem 2.20 for the functions $f(x) = \sin(x)$ and $f(x) = \cos(x)$, at $x = \pi/2$.

2.22. Repeat Problem 2.20 for the function $f(x) = \tan(x)$, at $x = \pi/4$ and at $x = 15\pi/32$.

NOTES

Discretization Formulae*

1. Forward Difference Formula for the k-th Derivative

$$\left.\frac{d^k y}{dx^k}\right|_{x=0} \equiv y_0^{(k)} = C(w_0 y_0 + w_1 y_1 + \ldots + w_{n-1} y_{n-1}) + E$$

where various coefficients are given by

Derivative	w_0	w_1	w_2	w_3	w_4	n	C	E
y_0'	−1	1				2	$\frac{1}{h}$	$-\frac{1}{2}hy^{ii}$
	−3	4	−1			3	$\frac{1}{2h}$	$\frac{1}{3}h^2 y^{iii}$
	−11	18	−9	2		4	$\frac{1}{6h}$	$-\frac{1}{4}h^3 y^{iv}$
	−25	48	−36	16	−3	5	$\frac{1}{12h}$	$\frac{1}{5}h^4 y^{v}$
y_0''	1	−2	1			3	$\frac{1}{h^2}$	$-hy^{iv}$
	2	−5	4	−1		4	$\frac{1}{h^3}$	$\frac{11}{12}h^2 y^{iv}$
	35	−104	114	−56	11	5	$\frac{1}{12h^2}$	$-\frac{5}{6}h^3 y^{v}$
y_0'''	−1	3	−3	1		4	$\frac{1}{h^3}$	$-\frac{3}{2}h y^{iv}$
	−5	18	−24	14	−3	5	$\frac{1}{2h^3}$	$\frac{7}{4}h^2 y^{v}$
y_0^{IV}	1	−4	6	−4	1	5	$\frac{1}{h^4}$	$-2h\, y^{v}$

Example: A four-point forward difference formula for the first derivative

$$\left.\frac{dy}{dx}\right|_{x=0} = \frac{1}{6h}(-11y_0 + 18y_1 - 9y_2 + 2y_3) - \frac{1}{4}h^3 y^{IV}$$

---

* From D. V. Griffith and I. M. Smith, *Numerical Methods for Engineers*, CRC Press, 1992, Boca Raton, FL. With permission.

2. Central Difference Formula for the k-th Derivative

$$\left.\frac{d^k y}{dx^k}\right|_{x=0} \equiv y_0^{(k)} = C(w_{-m}y_{-m} + w_{-m+1}y_{-m+1} + \cdots$$
$$+ w_0 y_0 + \cdots + w_{m-1}y_{m-1} + w_m y_m) + E$$

where various coefficients are given by

Derivative	w_{-3}	w_{-2}	w_{-1}	w_0	w_1	w_2	w_3	n	C	E
			−1	0	1			3	$\frac{1}{2h}$	$-\frac{1}{6}h^2 y^{iii}$
y_0'		1	−8	0	8	−1		5	$\frac{1}{12h}$	$\frac{1}{30}h^4 y^v$
	−1	9	−45	0	45	−9	1	7	$\frac{1}{60h}$	$-\frac{1}{140}h^6 y^{vii}$
			1	−2	1			3	$\frac{1}{h^2}$	$-\frac{1}{12}h^2 y^{iv}$
y_0''		−1	16	−30	16	−1		5	$\frac{1}{12h^2}$	$\frac{1}{90}h^4 y^{iv}$
	2	−27	270	−490	270	−27	2	7	$\frac{1}{180h^2}$	$-\frac{1}{560}h^6 y^{viii}$
y_0'''		−1	2	0	−2	1		5	$\frac{1}{2h^3}$	$-\frac{1}{4}h^2 y^v$
	1	−8	13	0	−13	8	−1	7	$\frac{1}{8h^3}$	$\frac{7}{120}h^4 y^{vii}$
y_0^{IV}		1	−4	6	−4	1		5	$\frac{1}{h^4}$	$-\frac{1}{6}h^2 y^{vi}$
	−1	12	−39	56	−39	12	−1	7	$\frac{1}{6h^4}$	$\frac{7}{240}h^4 y^{viii}$

Example: A three point central difference formula for the second derivative becomes

$$\left.\frac{d^2 y}{dx^2}\right|_{x=0} = \frac{1}{h^2}(y_{-1} - 2y_0 + y_1) - \frac{1}{12}h^6 y^{IV}$$

3. **Backward Difference Formula for the k-th Derivative**

$$\left.\frac{d^k y}{dx^k}\right|_{x=0} \equiv y_0^{(k)} = C(w_0 y_0 + w_{-1} y_{-1} + \ldots + w_{-n-1} y_{-n-1}) + E$$

where various coefficients are given by

Derivative	w_{-4}	w_{-3}	w_{-2}	w_{-1}	w_0	n	C	E
y_0'				-1	1	2	$\frac{1}{h}$	$\frac{1}{2} h y^{ii}$
			1	-4	3	3	$\frac{1}{2h}$	$\frac{1}{3} h^2 y^{iii}$
		-2	9	-18	11	4	$\frac{1}{6h}$	$\frac{1}{4} h^3 y^{iv}$
	3	-16	36	-48	25	5	$\frac{1}{12h}$	$\frac{1}{5} h^4 y^{v}$
y_0''			1	-2	1	3	$\frac{1}{h^2}$	$h y^{iii}$
		-1	4	-5	2	4	$\frac{1}{h^2}$	$\frac{11}{12} h^2 y^{iv}$
	11	-56	114	-104	35	5	$\frac{1}{12h^2}$	$\frac{5}{6} h^3 y^{v}$
y_0'''		-1	3	-3	1	4	$\frac{1}{h^3}$	$\frac{3}{2} h y^{iv}$
	3	-14	24	-18	5	5	$\frac{1}{2h^3}$	$\frac{7}{4} h^2 y^{v}$
y_0^{IV}	1	-4	6	-4	1	5	$\frac{1}{h^4}$	$2h\, y^{v}$

Example: A three point backward difference formula for the second derivative becomes

$$\left.\frac{d^2 y}{dx^2}\right|_{x=0} = \frac{1}{h^2}(y_{-2} - 2y_{-1} + y_0) + h\, y^{iii}$$

3
Methods of Solving Systems of Algebraic Equations

In the previous chapter, we described discrete approximation of differential equations using both the Taylor series expansion and the control volume approach. Thus, the problem governed by a single or a set of differential equations and boundary conditions can be approximated by a system of algebraic equations. One needs to examine the nature of the resulting system of equations because the proper choice of the computer subroutine for solving sets of algebraic equations is strongly affected by the following considerations:

1. Whether the problem is linear or nonlinear;
2. Whether the coefficient matrix is tridiagonal, full, or sparse (i.e., large percentage of entries are zero);
3. Whether the number of operations involved in the algorithm is so large as to give rise to excessive accumulation of roundoff errors;
4. Whether the coefficient matrix is "diagonally dominant";
5. Whether the coefficient matrix is ill-conditioned (i.e., small changes in the input data, such as those introduced by roundoff errors produce large changes in the solution).

Therefore, the objective of this chapter is to illustrate, with a simple example, the basic steps in the transformation of a problem governed by a differential equation and some specified boundary conditions into a system of algebraic equations and then present an overview of various methods for solving systems of algebraic equations as well as discuss their advantages and limitations.

3.1 Reduction to Algebraic Equations

A large variety of finite difference schemes is available for discretizing the derivatives in differential equations; the choice depends on the nature of the governing differential equation and its boundary conditions. These matters will be discussed in greater detail in the following chapters. Here, our objective is to illustrate the basic steps in the transformation of a differential equation and its boundary conditions into a set of algebraic equations. We consider the following simple example.

Energy is generated in a slab of thickness L at a rate of g(x) W/m³, while it is dissipated from the boundary surfaces at x = 0 and x = L by convection into ambients at temperatures $T_{\infty,0}$ and $T_{\infty,L}$, with heat transfer coefficients h_0 and h_L, respectively. The mathematical formulation of this problem for the steady state is given by

$$\frac{d^2 T(x)}{dx^2} + \frac{1}{k}g(x) = 0 \quad \text{in} \quad 0 < x < L \tag{3.1a}$$

$$-k\frac{dT(x)}{dx} + h_0 T(x) = h_0 T_{\infty,0} \quad \text{at } x = 0 \tag{3.1b}$$

$$k\frac{dT(x)}{dx} + h_L T(x) = h_L T_{\infty,L} \quad \text{at } x = L \tag{3.1c}$$

The basic steps in the transformation of this problem by finite differences into a set of algebraic equations for the temperatures T_i at a finite number of grid points i = 0,1,...,M, chosen over the solution domain of the problem are as follows.

1. The domain $0 \leq x \leq L$ is divided into M equal subregions each of thickness $\Delta x = L/M$ as illustrated in Figure 3.1. (Grids of unequal size can also be used.)
2. The differential equation (3.1a) is discretized by a suitable finite difference scheme at the internal grid points i = 1,2,...,M−1. Here, we use the classical second-order accurate central-difference formula given by equation (2.13a) to discretize the second derivative. The differential equation (3.1a) reduces to

$$\frac{T_{i-1} - 2T_i + T_{i+1}}{(\Delta x)^2} + \frac{1}{k}g_i = 0$$

with a truncation error $0[(\Delta x)^2]$. This result is rearranged in the form

$$T_{i-1} - 2T_i + T_{i+1} + G_i = 0, \quad i = 1,2,3,...,(M-1) \tag{3.2a}$$

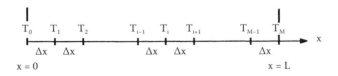

FIGURE 3.1
One-dimensional finite difference grid.

where

$$G_i = \frac{(\Delta x)^2 g_i}{k} \qquad (3.2b)$$

The system (3.2) provides M−1 algebraic equations, but it contains M + 1 unknown grid-point temperatures $T_0, T_1, \ldots, T_M$. Two additional relations needed for making the number of equations equal to the number of unknowns are obtained by discretizing the boundary conditions as described next.

3. The boundary conditions given by equations (3.1b,c) need to be discretized because they contain the first derivative of temperature. If the *forward* and *backward* differencing formulae given by equations (2.9a,b) are used, the results are first-order accurate, that is, $0(\Delta x)$. It is desirable to use a second-order accurate formula in order to be consistent with the second-order accuracy of the discretized differential equation.

A second-order accurate formula for the first derivative is given by equation (2.9c); but to implement this formula at the boundary grid points i = 0 and i = M, additional grid points are needed to the left and to the right of the boundary nodes i = 0 and i = M, respectively. Therefore, *fictitious nodes* located at a distance Δx to the left and right of the boundaries at x = 0 and x = L, at *fictitious temperatures* T_{-1} and T_{M+1}, respectively, are considered as illustrated in Figure 3.2. Then the application of the central difference formula [equation (2.9c)] to the boundary conditions [Equations (3.1b) and (3.1c), respectively] gives

$$-k \frac{T_1 - T_{-1}}{2\Delta x} + h_0 T_0 = h_0 T_{\infty,0} \qquad (3.3a)$$

$$k \frac{T_{M+1} - T_{M-1}}{2\Delta x} + h_L T_M = h_L T_{\infty,L} \qquad (3.3b)$$

To eliminate the fictitious temperatures T_{-1} and T_{M+1} from equations (3.3a,b), two additional relations are obtained by

FIGURE 3.2
Fictitious nodes at fictitious temperatures T_{-1} and T_{M+1}.

evaluating the difference equation (3.2a) for $i = 0$ and $i = M$, to give, respectively,

$$T_{-1} - 2T_0 + T_1 + \frac{(\Delta x)^2 g_0}{k} = 0 \tag{3.4a}$$

$$T_{M-1} - 2T_M + T_{M+1} + \frac{(\Delta x)^2 g_M}{k} = 0 \tag{3.4b}$$

The elimination of T_{-1} and T_{M+1} between equations (3.3) and (3.4) results in the following two finite difference equations

$$2T_1 - 2\beta_0 T_0 + (2\gamma_0 + G_0) = 0 \quad \text{at } x = 0, (i = 0) \tag{3.5a}$$

$$2T_{M-1} - 2\beta_L T_M + (2\gamma_L + G_M) = 0 \quad \text{at } x = L, (i = M) \tag{3.5b}$$

where

$$\left.\begin{array}{l} \beta_0 \equiv 1 + \dfrac{\Delta x \, h_0}{k}, \quad \gamma_0 \equiv \dfrac{\Delta x (h_0 T_{\infty,0})}{k} \\[6pt] \beta_L \equiv 1 + \dfrac{\Delta x \, h_L}{k}, \quad \gamma_L \equiv \dfrac{\Delta x (h_L T_{\infty,L})}{k} \\[6pt] G_0 \equiv \dfrac{(\Delta x)^2 g_0}{k}, \quad G_M \equiv \dfrac{(\Delta x)^2 g_M}{k} \end{array}\right\} \tag{3.6}$$

Equations (3.2) and (3.5) provide $M + 1$ algebraic equations for the determination of $M + 1$ unknown node temperatures T_i ($i = 0,1,2,\ldots,M$). These equations are summarized here

$$2T_1 - 2\beta_0 T_0 = -(2\gamma_0 + G_0), \quad i = 0 \tag{3.7}$$

$$T_{i-1} - 2T_i + T_{i+1} = -G_i, \quad i = 1,2,\ldots,M-1 \tag{3.8}$$

$$2T_{M-1} - 2\beta_L T_M = -(2\gamma_L + G_M), \quad i = M \tag{3.9}$$

4. The sets of equations (3.7) – (3.9) are expressed in the matrix form

$$\mathbf{A T = B} \tag{3.10}$$

where

$$\mathbf{A} = \begin{bmatrix} -2\beta_0 & 2 & 0 & \cdots & 0 & 0 & 0 \\ 1 & -2 & 1 & & 0 & 0 & 0 \\ 0 & 1 & -2 & 1 & 0 & 0 & 0 \\ \vdots & & & & & & \vdots \\ 0 & & & & 1 & -2 & 1 \\ 0 & 0 & \cdots & \cdots & 0 & 2 & -2\beta_L \end{bmatrix} = \text{known coefficient matrix}$$

$$\tag{3.11a}$$

Methods of Solving Systems of Algebraic Equations

$$T = \begin{bmatrix} T_0 \\ T_1 \\ \cdot \\ \cdot \\ T_M \end{bmatrix} = \text{unknown vector}, \quad B = \begin{bmatrix} -(G_0 + 2\gamma_0) \\ -G_1 \\ \cdot \\ \cdot \\ -G_{M-1} \\ -(G_M + 2\gamma_L) \end{bmatrix} = \text{known vector}$$

(3.11b)

Thus, the problem of solving the one-dimensional steady-state heat conduction given by equations (3.1a–c) is transformed to the solution of a set of algebraic equations (3.10) for the determination of M+1 node temperatures, T_i, i = 0,1,...,M.

For the one-dimensional problem considered here, the coefficient matrix **A** is a *tridiagonal* matrix. Depending on the nature of the problem, the dimensions, and the finite difference scheme used, a multidiagonal, a full, or a sparse matrix may result.

In the aforementioned illustration, we used fictitious nodes in order to develop a second-order accurate finite difference scheme to discretize the boundary conditions. The same equations can also be developed by applying the control volume approach for a volume element about the boundary node as demonstrated in Example 3.1 given here.

If temperature is prescribed at any boundary, then the node temperature at that boundary is known and no additional finite difference equation is needed for that boundary.

Example 3.1

Show that the second-order accurate finite difference approximations for the boundary conditions given by equations (3.5a,b) are also obtainable with the control volume approach by writing an energy balance for a volume element of thickness $\Delta x/2$ about the boundary nodes i = 0 and i = M.

Solution

Consider control volumes of thickness $\Delta x/2$ next to the boundary surfaces at x = 0 and x = L as illustrated in the accompanying figure.

The steady-state energy conservation principle for each of these control volumes is stated as

$$\begin{pmatrix} \text{Rate of heat} \\ \text{gain by} \\ \text{convection} \end{pmatrix} + \begin{pmatrix} \text{Rate of heat} \\ \text{gain by} \\ \text{conduction} \end{pmatrix} + \begin{pmatrix} \text{Rate of} \\ \text{energy} \\ \text{generation} \end{pmatrix} = 0$$

The application of this conservation equation for the boundary nodes about i = 0 and i = M, respectively, gives

$$h_0(T_{\infty,0} - T_0) + k\frac{T_1 - T_0}{\Delta x} + \frac{\Delta x}{2}g_0 = 0 \quad \text{at } x = 0 \text{ (i.e., } i = 0)$$

$$h_L(T_{\infty,L} - T_M) + k\frac{T_{M-1} - T_M}{\Delta x} + \frac{\Delta x}{2}g_M = 0 \quad \text{at } x = L \text{ (i.e., } i = M)$$

These results are rearranged as

$$2T_1 - \left(2 + \frac{2\Delta x h_0}{k}\right)T_0 + \left[\frac{2\Delta x h_0}{k}T_{\infty,0} + \frac{(\Delta x)^2 g_0}{k}\right] = 0$$

$$2T_{M-1} - \left(2 + \frac{2\Delta x h_L}{k}\right)T_M + \left[\frac{2\Delta x h_L}{k}T_{\infty,L} + \frac{(\Delta x)^2 g_M}{k}\right] = 0$$

which are the same as those given by equations (3.5a,b).

So far, we have illustrated the basic steps in the transformation of a partial differential equation and its boundary conditions into a system of algebraic equations. The methods of solving such a system of algebraic equations can be put into one of the two categories:

1. The *direct* methods in which a finite number of operations is involved in the solution
2. The *iterative* techniques in which answers become progressively more accurate as the number of iterations is increased, provided that the convergence criterion related to the *diagonal dominance* of the coefficient matrix is satisfied

In the remainder of this chapter, we present an overview of the *direct* and *iterative* methods of solving systems of algebraic equations and discuss the implications of *nonlinear systems*.

3.2 Direct Methods

Generally, the direct methods are preferred for systems having banded matrix coefficients and for problems involving relatively simple geometries

Methods of Solving Systems of Algebraic Equations　　　　　　　　　　71

and boundary conditions. They are very efficient but require large computer storage and give rise to the accumulation of round-off errors. There is a wealth of literature on the subject of solving systems of simultaneous algebraic equations because of the importance of this subject in scientific computing. Here, we present a brief discussion of some of these direct methods.

One of the most elementary methods of solving a set of algebraic equations is by employing *Cramer's rule*. The method is not practical for large number of equations because it involves a large number of operations. To solve a set of N equations, the number of basic operations needed is of the order of $0(N^4)$, which implies that doubling the number of equations to be solved increases the computer time on the order of 2^4 or 16 times.

3.2.1 Gauss Elimination Method

This is a direct method commonly used for solving simultaneous algebraic equations. In this method, the coefficient matrix is transformed into an upper triangular matrix by systematic application of some algebraic operations under which the solution to the system of equations remains invariant. Two principal operations applied include

1. Multiplication or division of any equation by a constant
2. Replacement of any equation by the sum (or difference) of that equation with any other equation

Once the system is transformed into an upper diagonal form, the solution starts from the last equation and proceeds upward by back substitutions. We illustrate the procedure with the following simple example involving only three unknowns: T_1, T_2, and T_3.

$$a_{11}T_1 + a_{12}T_2 + a_{13}T_3 = d_1 \tag{3.12a}$$

$$a_{21}T_1 + a_{22}T_2 + a_{23}T_3 = d_2 \tag{3.12b}$$

$$a_{31}T_1 + a_{32}T_2 + a_{33}T_3 = d_3 \tag{3.12c}$$

We choose the first equation as the "pivot" equation and use it to eliminate T_1 from the second and third equations. We obtain

$$a_{11}T_1 + a_{12}T_2 + a_{13}T_3 = d_1 \tag{3.13a}$$

$$0 + a_{22}^*T_2 + a_{23}^*T_3 = d_2^* \tag{3.13b}$$

$$0 + a_{32}^*T_2 + a_{33}^*T_3 = d_3^* \tag{3.13c}$$

where $a_{22}^* = a_{22} - (a_{21}/a_{11})a_{12}$, $a_{23}^* = a_{23} - (a_{21}/a_{11})a_{13}$, $d_2^* = d_2 - (a_{21}/a_{11})d_1$, $a_{32}^* = a_{32} - (a_{31}/a_{11})a_{12}$, $a_{33}^* = a_{33} - (a_{31}/a_{11})a_{13}$, and $d_3^* = d_3 - (a_{31}/a_{11})d_1$.

To eliminate T_2 from the third equation, the second equation is used as the "pivot" equation. Then the system (3.13) takes the triangular form

$$a_{11}T_1 + a_{12}T_2 + a_{13}T_3 = d_1 \qquad (3.14)$$

$$a_{22}^*T_2 + a_{23}^*T_3 = d_2^* \qquad (3.15)$$

$$a_{33}'T_3 = d_3' \qquad (3.16)$$

where $a_{33}' = a_{33}^* - (a_{32}^*/a_{22}^*)a_{23}^*$ and $d_3' = d_3^* - (a_{32}^*/a_{22}^*)d_2^*$.

The unknowns T_i are immediately determined from this system by starting from the last equation and by back substitution. We obtain

$$T_3 = d_3'/a_{33}' \qquad (3.17)$$

$$T_2 = (d_2^* - a_{23}^*T_3)/a_{22}^* \qquad (3.18)$$

$$T_1 = (d_1 - a_{13}T_3 - a_{12}T_2)/a_{11} \qquad (3.19)$$

This procedure can be readily generalized to a system of N equations.

The number of multiplications involved in the solution of a system of N algebraic equations with a full matrix by using Gauss elimination varies as N^3, which is much less than N^4 needed for solution with Cramer's method. In Appendix A at the end of the book, we present a subroutine, written in Fortran, for Gauss elimination.

3.2.2 Thomas Algorithm

In the case of a tridiagonal system of algebraic equations, such as the one encountered in the solution of one-dimensional heat conduction problems, the Gauss elimination method can be further simplified by taking advantage of the zeros of the tridiagonal coefficient matrix. This modified procedure, generally referred to as the *Thomas algorithm*, is an extremely efficient method for solving such equations (Thomas 1949). Consider a system of N algebraic equations having a tridiagonal coefficient matrix given by equation (3.20). To solve this system of equations, the matrix of the coefficients is put into upper diagonal form by the elimination process described here:

$$\begin{bmatrix} b_1 & c_1 & 0 & 0 & \ldots & 0 & 0 \\ a_2 & b_2 & c_2 & 0 & & 0 & 0 \\ 0 & a_3 & b_3 & c_3 & & 0 & 0 \\ \cdot & & & & & \cdot & \\ \cdot & & & & & \cdot & \\ \cdot & & & & & \cdot & \\ 0 & 0 & & a_{N-1} & b_{N-1} & c_{N-1} \\ 0 & 0 & 0 & \ldots & 0 & a_N & b_N \end{bmatrix} \begin{bmatrix} T_1 \\ T_2 \\ T_3 \\ \cdot \\ \cdot \\ \cdot \\ T_{N-1} \\ T_N \end{bmatrix} = \begin{bmatrix} d_1 \\ d_2 \\ d_3 \\ \cdot \\ \cdot \\ \cdot \\ d_{N-1} \\ d_N \end{bmatrix} \qquad (3.20)$$

1. The first equation (row) is chosen as the "pivot," multiplied by "a_2/b_1," and subtracted from the second equation (row) to eliminate a_2. The resulting second equation is equivalent to

$$\text{replacing "}b_2\text{" by } \left(b_2 - \frac{a_2}{b_1}c_1\right)$$

$$\text{replacing "}d_2\text{" by } \left(d_2 - \frac{a_2}{b_1}d_1\right)$$

2. The modified second equation is chosen as the "pivot," and a similar approach is followed to eliminate a_3. The resulting third equation is equivalent to

$$\text{replacing "}b_3\text{" by } \left(b_3 - \frac{a_3}{b_2}c_2\right)$$

$$\text{replacing "}d_3\text{" by } \left(d_3 - \frac{a_3}{b_2}d_2\right)$$

3. The procedure is continued until a_N is eliminated from the last equation. Thus, the general procedure for upper diagonalizing equation (3.20) is stated as

$$\text{Replace "}b_i\text{" by } \left(b_i - \frac{a_i}{b_{i-1}}c_{i-1}\right) \quad \text{for } i = 2, 3, \ldots, N \quad (3.21a)$$

$$\text{Replace "}d_i\text{" by } \left(d_i - \frac{a_i}{b_{i-1}}d_{i-1}\right) \quad \text{for } i = 2, 3, \ldots, N \quad (3.21b)$$

Once the triangular form is achieved by the aforementioned procedure, the unknown T_i's are determined by back substitution, starting from the last equation and working backward

$$T_N = \frac{d_N}{b_N} \quad (3.22a)$$

$$T_i = \frac{d_i - c_i T_{i+1}}{b_i}, \quad i = N-1, N-2, \ldots, 1. \quad (3.22b)$$

In Appendix B, we present a program written in FORTRAN for the *Thomas algorithm*.

Using the Thomas algorithm, the number of basic arithmetic operations for solving a tridiagonal set is of the order N, in contrast to $0(N^3)$ operations required for solving with Gauss elimination. Therefore, not only are the computation times much shorter, but the round-off errors also are significantly reduced.

Example 3.2

Finite difference approximation of steady-state heat conduction in a slab with energy generation and prescribed heat flux at one boundary, using four nodes, results in the following system with a tridiagonal coefficient matrix.

$$\begin{bmatrix} -1 & 1 & 0 & 0 \\ 1 & -2 & 1 & 0 \\ 0 & 1 & -2 & 1 \\ 0 & 0 & 1 & -2 \end{bmatrix} \begin{bmatrix} T_1 \\ T_2 \\ T_3 \\ T_4 \end{bmatrix} = \begin{bmatrix} -40 \\ -30 \\ -30 \\ -30 \end{bmatrix}$$

Solve this problem using the Thomas algorithm.

Solution

The upper diagonalization procedure defined by equations (3.21a,b) gives

$$b_i' \text{s:} \begin{cases} b_1 = -1 \\ b_2 \equiv b_2 - \dfrac{a_2}{b_1} c_1 = -2 - \dfrac{(1)}{(-1)}(1) = -1 \\ b_3 \equiv b_3 - \dfrac{a_3}{b_2} c_2 = -2 - \dfrac{(1)}{(-1)}(1) = -1 \\ b_4 \equiv b_4 - \dfrac{a_4}{b_3} c_3 = -2 - \dfrac{(1)}{(-1)}(1) = -1 \end{cases}$$

and

$$d_i' \text{s:} \begin{cases} d_1 = -40 \\ d_2 \equiv d_2 - \dfrac{a_2}{b_1} d_1 = -30 - \dfrac{(1)}{(-1)}(-40) = -70 \\ d_3 \equiv d_3 - \dfrac{a_3}{b_2} d_2 = -30 - \dfrac{(1)}{(-1)}(-70) = -100 \\ d_4 \equiv d_4 - \dfrac{a_4}{b_3} d_3 = -30 - \dfrac{(1)}{(-1)}(-100) = -130 \end{cases}$$

The back substitution, defined by equation (3.22), gives the four temperatures as

$$T_i' \text{s:} \begin{cases} T_4 = \dfrac{d_4}{b_4} = \dfrac{-130}{-1} = 130 \\ T_3 = \dfrac{d_3 - c_3 T_4}{b_3} = \dfrac{-100 - (1)(130)}{(-1)} = 230 \\ T_2 = \dfrac{d_2 - c_2 T_3}{b_2} = \dfrac{-70 - (1)(230)}{(-1)} = 300 \\ T_1 = \dfrac{d_1 - c_1 T_2}{b_1} = \dfrac{-40 - (1)(300)}{(-1)} = 340 \end{cases}$$

3.3 Iterative Methods

When the number of equations is very large, the coefficient matrix is sparse but not banded and the computer storage is critical, an iterative method is preferred to the direct method of solution. If the iterative process is convergent, the solution is obtained within a specified accuracy of the exact answer in a finite number of operations. The method is certain to converge for a system having *diagonal dominance*.

Iterative methods have rather simple algorithms, are easy to apply, and are not restricted for use with simple geometries and boundary conditions. They are also preferred when the number of operations in the calculations is so large that the direct methods may prove inadequate because of the accumulation of roundoff errors and computational cost. If the sparse coefficient matrix has an average of p nonzero elements per row, then one iteration requires about pN operations for a matrix of order N, or a total of KpN operations for K iterations. This number should be compared with the N^3 operations required by Gauss elimination. The advantage of the iterative method depends on the number of required iterations for a prescribed accuracy, and is clearly favored by the increase on the size of the system, N.

The *Gauss–Seidel iteration* (often called the Liebman iteration) is one of the efficient procedures for solving large, sparse systems of equations. The convergence can be accelerated by the procedure called *successive overrelaxation* (SOR). Other variations of the Gauss–Seidel procedure have been discussed by Lapidus and Pinder (1982).

In this section, we discuss the Gauss–Seidel iteration, SOR, the *red-black ordering* scheme, and the *biconjugate gradient method*.

3.3.1 Gauss–Seidel Iteration

This is a very simple, efficient *point-iterative* procedure for solving large, sparse systems of algebraic equations. The Gauss–Seidel iteration is based on the idea of successive approximations, but it differs from the standard iteration in that the most recently determined values are used in each round of iterations. Basic steps are as follows:

1. Solve each equation for the *main-diagonal* unknown.
2. Make an initial guess for all the unknowns.
3. Computations begin with the use of the guessed values to compute a first approximation for each of the main-diagonal unknowns solved successively in Step 1. In each computation, wherever possible, the most recently determined values are used and the first round of iterations is completed.
4. The procedure is continued until a specified convergence criterion is satisfied for all the unknowns.

To illustrate the procedure, we consider the following three equations:

$$a_{11}T_1 + a_{12}T_2 + a_{13}T_3 = d_1 \tag{3.23a}$$

$$a_{21}T_1 + a_{22}T_2 + a_{23}T_3 = d_2 \tag{3.23b}$$

$$a_{31}T_1 + a_{32}T_2 + a_{33}T_3 = d_3 \tag{3.23c}$$

where $a_{ii} \neq 0$ for $i = 1$ to 3. Equations are successively solved for the main-diagonal unknowns

$$T_1 = \frac{1}{a_{11}}(d_1 - a_{12}T_2 - a_{13}T_3) \tag{3.24a}$$

$$T_2 = \frac{1}{a_{22}}(d_2 - a_{21}T_1 - a_{23}T_3) \tag{3.24b}$$

$$T_3 = \frac{1}{a_{33}}(d_3 - a_{31}T_1 - a_{32}T_2) \tag{3.24c}$$

Initial guess values are chosen as

$$T_1^{(0)}, \ T_2^{(0)}, \ \text{and} \ T_3^{(0)} \tag{3.25}$$

These guess values are used together with the most recently computed values to complete the first round of iterations as

$$T_1^{(1)} = \frac{1}{a_{11}}(d_1 - a_{12}T_2^{(0)} - a_{13}T_3^{(0)}) \tag{3.26a}$$

$$T_2^{(1)} = \frac{1}{a_{22}}(d_2 - a_{21}T_1^{(1)} - a_{23}T_3^{(0)}) \tag{3.26b}$$

$$T_3^{(1)} = \frac{1}{a_{33}}(d_3 - a_{31}T_1^{(1)} - a_{32}T_2^{(1)}) \tag{3.26c}$$

These first approximations are then used together with the most recently computed values to complete the second round of iterations as

$$T_1^{(2)} = \frac{1}{a_{11}}(d_1 - a_{12}T_2^{(1)} - a_{13}T_3^{(1)}) \tag{3.27a}$$

$$T_2^{(2)} = \frac{1}{a_{22}}(d_2 - a_{21}T_1^{(2)} - a_{23}T_3^{(1)}) \tag{3.27b}$$

$$T_3^{(2)} = \frac{1}{a_{33}}(d_3 - a_{31}T_1^{(2)} - a_{32}T_2^{(2)}) \tag{3.27c}$$

The iteration procedure is continued in a similar manner.

General expressions for the "n+1"th round of iterations of the aforementioned system is written as

$$T_1^{(n+1)} = \frac{1}{a_{11}}[d_1 - a_{12}T_2^{(n)} - a_{13}T_3^{(n)}] \qquad (3.28a)$$

$$T_2^{(n+1)} = \frac{1}{a_{22}}[d_2 - a_{21}T_1^{(n+1)} - a_{23}T_3^{(n)}] \qquad (3.28b)$$

$$T_3^{(n+1)} = \frac{1}{a_{33}}[d_3 - a_{31}T_1^{(n+1)} - a_{32}T_2^{(n+1)}] \qquad (3.28c)$$

In the general case of N equations, the "n + 1" th round of iterations can be written as

$$T_i^{(n+1)} = \frac{1}{a_{ii}}\left\{d_i - \sum_{j=1}^{i-1} a_{ij}T_j^{(n+1)} - \sum_{j=i+1}^{N} a_{ij}T_j^{(n)}\right\} \quad \text{for} \quad i = 1 \text{ to N.} \qquad (3.29)$$

The criterion for convergence can be specified either in terms of *absolute convergence* in the form

$$\left|T_i^{(n+1)} - T_i^{(n)}\right| \leq \varepsilon \qquad (3.30)$$

or as the *relative convergence criterion* in the form

$$\left|\frac{T_i^{(n+1)} - T_i^{(n)}}{T_i^{(n+1)}}\right| \leq \varepsilon \qquad (3.31)$$

which should be satisfied for all T_i. The convergence criterion given by equation (3.31), for $T_i^{(n+1)} \neq 0$, is the safest choice if the magnitudes of T_i cannot be guessed beforehand; but such a testing process requires more computer time than checking with the criterion given by equation (3.30). If the approximate magnitudes of T_i are known beforehand, the criterion given by equation (3.30) is preferred.

The convergence of iterative methods does not depend on the initial guess for the unknowns but does depend on the character of the coefficient matrix. For a convergent system, a good first guess for the unknowns significantly reduces the number of iterations for the specified convergence criterion to be satisfied. The system of equations in which the diagonal elements are the largest elements (in magnitude) in each row is best suited for iterative solution. In situations when this is not the case, equations may be rearranged in order to bring the largest element in each row on the diagonal, if possible. Fortunately, in most heat transfer problems, the diagonal elements of the difference equations happen to be the largest element in each row.

A sufficient condition for convergence is given by

$$|a_{ii}| \geq \sum_{\substack{j=1 \\ i \neq j}}^{N} |a_{ij}| \quad \text{for } i = 1, 2, ..., N \tag{3.32a}$$

and

$$|a_{ii}| > \sum_{\substack{j=1 \\ i \neq j}}^{N} |a_{ij}| \quad \text{for at least one } i \text{ (i.e., row)} \tag{3.32b}$$

This condition requires that, for each equation, the magnitude of the diagonal element be greater than or equal to the sum of the magnitudes of the other coefficients in the equation. However, in practice, convergence might be obtained when the sufficient condition of diagonal dominance, given by equations (3.32a,b), is not satisfied.

Example 3.3

Perform the first three iterations of the Gauss–Seidel method for solving the following system of equations.

$$6T_1 + T_2 + 3T_3 = 17$$

$$T_1 - 10T_2 + 4T_3 = -7$$

$$T_1 + T_2 + 3T_3 = 12$$

Solution

We note that, in each equation, the largest element (in magnitude) is in the diagonal. These equations are solved for the main-diagonal unknowns as

$$T_1 = \frac{1}{6}(17 - T_2 - 3T_3)$$

$$T_2 = \frac{1}{10}(7 + T_1 + 4T_3)$$

$$T_3 = \frac{1}{3}(12 - T_1 - T_2)$$

and the initial guess values are arbitrarily chosen as

$$T_1^{(0)} = T_2^{(0)} = T_3^{(0)} = 1$$

The first round of iterations is determined as

$$T_1^{(1)} = \frac{1}{6}\left[17 - T_2^{(0)} - 3T_3^{(0)}\right] = 2.167$$

Methods of Solving Systems of Algebraic Equations

$$T_2^{(1)} = \frac{1}{10}\left[7 + T_1^{(1)} + 4T_3^{(0)}\right] = 1.317$$

$$T_3^{(1)} = \frac{1}{3}\left[12 - T_1^{(1)} - T_2^{(1)}\right] = 2.839$$

The second round of iterations becomes

$$T_1^{(2)} = \frac{1}{6}\left[17 - T_2^{(1)} - 3T_3^{(1)}\right] = 1.194$$

$$T_2^{(2)} = \frac{1}{10}\left[7 + T_1^{(2)} + 4T_3^{(1)}\right] = 1.955$$

$$T_3^{(2)} = \frac{1}{3}\left[12 - T_1^{(2)} - T_2^{(2)}\right] = 2.950$$

and the third round of iterations gives

$$T_1^{(3)} = \frac{1}{6}\left[17 - T_2^{(2)} - 3T_3^{(2)}\right] = 1.032$$

$$T_2^{(3)} = \frac{1}{10}\left[7 + T_1^{(3)} + 4T_3^{(2)}\right] = 1.999$$

$$T_3^{(3)} = \frac{1}{3}\left[12 - T_1^{(3)} - T_2^{(3)}\right] = 2.989$$

The values obtained with three iterations are sufficiently close to the exact answer $T_1 = 1$, $T_2 = 2$, and $T_3 = 3$. The solution is converging to the exact results.

3.3.2 Successive Overrelaxation

The Gauss–Seidel method described previously generally does not converge sufficiently fast. SOR is a method that can accelerate the convergence. To illustrate the basic idea in this approach, we rearrange equations (3.28a–c) in the form

$$T_1^{(n+1)} = T_1^{(n)} + \left[\frac{1}{a_{11}}\left\{d_1 - a_{11}T_1^{(n)} - a_{12}T_2^{(n)} - a_{13}T_3^{(n)}\right\}\right] \quad (3.33a)$$

$$T_2^{(n+1)} = T_2^{(n)} + \left[\frac{1}{a_{22}}\left\{d_2 - a_{21}T_1^{(n+1)} - a_{22}T_2^{(n)} - a_{23}T_3^{(n)}\right\}\right] \quad (3.33b)$$

$$T_3^{(n+1)} = T_3^{(n)} + \left[\frac{1}{a_{33}}\left\{d_3 - a_{31}T_1^{(n+1)} - a_{32}T_2^{(n+1)} - a_{33}T_3^{(n)}\right\}\right] \quad (3.33c)$$

That is, we added to the right-hand sides of equations (3.28a–c), respectively, the terms $0 = T_1^{(n)} - T_1^{(n)}$, $0 = T_2^{(n)} - T_2^{(n)}$, and $0 = T_3^{(n)} - T_3^{(n)}$ and regrouped them. As the exact solution is approached, $T_i^{(n+1)}$ approaches $T_i^{(n)}$ and the terms inside the brackets become zero identically. Therefore, the terms inside the square brackets can be regarded as correction terms to $T_i^{(n)}$ (i = 1,2,3) for each iteration. In the SOR method, the bracketed terms are multiplied by a factor ω, called the *relaxation parameter*, and equations (3.33a–c) are rewritten as

$$T_1^{(n+1)} = T_1^{(n)} + \frac{\omega}{a_{11}} \{d_1 - a_{11}T_1^{(n)} - a_{12}T_2^{(n)} - a_{13}T_3^{(n)}\} \tag{3.34a}$$

$$T_2^{(n+1)} = T_2^{(n)} + \frac{\omega}{a_{22}} \{d_2 - a_{21}T_1^{(n+1)} - a_{22}T_2^{(n)} - a_{23}T_3^{(n)}\} \tag{3.34b}$$

$$T_3^{(n+1)} = T_3^{(n)} + \frac{\omega}{a_{33}} \{d_3 - a_{31}T_1^{(n+1)} - a_{32}T_2^{(n+1)} - a_{33}T_3^{(n)}\} \tag{3.34c}$$

The values of the relaxation parameter must lie in the range $0 < \omega < 2$ for convergence. The range $0 < \omega < 1$ corresponds to *underrelaxation*, $1 < \omega < 2$ *overrelaxation*, and $\omega = 1$ to *Gauss–Seidel iteration*.

The aforementioned procedure for SOR can be generalized for the case of N equations as

$$T_i^{(n+1)} = T_i^{(n)} + \frac{\omega}{a_{ii}} \left\{ d_i - \sum_{j=1}^{i-1} a_{ij} T_j^{(n+1)} - \sum_{j=i}^{N} a_{ij} T_j^{(n)} \right\}, \quad i = 1 \text{ to } N. \tag{3.35}$$

which is rearranged as

$$T_i^{(n+1)} = \frac{\omega}{a_{ii}} \left\{ d_i - \sum_{j=1}^{i-1} a_{ij} T_j^{(n+1)} - \sum_{j=i+1}^{N} a_{ij} T_j^{(n)} \right\} + (1-\omega) T_i^{(n)} \quad \text{for } i = 1 \text{ to } N$$

$$\tag{3.36a}$$

Note that the terms inside the parentheses in equation (3.36a) are the same as those inside the parentheses on the right-hand side of the Gauss–Seidel iteration given by equation (3.29). Therefore, equation (3.36a) can be expressed in the form

$$T_i^{(n+1)} = \omega \hat{T}_i^{(n)} + (1 - \omega) T_i^{(n)}, \quad i = 1 \text{ to } N \tag{3.36b}$$

where $\hat{T}_i^{(n)} \equiv$ the right-hand side of Gauss–Seidel iteration formula given by equation (3.29). Clearly, the case $\omega = 1$ corresponds to the Gauss–Seidel iteration. The choice of the relaxation parameter affects the speed of convergence, but the determination of the optimal value of ω is a difficult matter. Some numerical experimentation is necessary for selecting a proper value of ω for a given problem. With the proper choice of ω, it may be possible to reduce the computation time by an order of magnitude; therefore, when the number of equations is large and reduction of the computation time is important, some experimentation with different values of ω is worthwhile.

The physical significance of the relaxation parameter ω is as follows. For ω = 1, the Gauss–Seidel computed value of the unknown is stored as the current value. For underrelaxation, 0 < ω < 1, a weighted average of the Gauss–Seidel value and the value from the previous iteration is stored as the current value. For overrelaxation, 1 < ω < 2, the current stored value is essentially extrapolated beyond the Gauss–Seidel value. For ω > 2, the calculations diverge. In Appendix C, we present a subroutine written in Fortran for successive *overrelaxation*.

3.3.3 Red-Black Ordering Scheme

The red-black ordering scheme for solving systems of algebraic equations generated by finite differences allows for the vectorization of the computer code and is used in conjunction with the Gauss–Seidel (or SOR) technique. The method is powerful and generally reduces CPU time significantly when compared to the Gauss–Seidel (or SOR) method. The basis for this approach is the reordering of equations in the Gauss–Seidel (or SOR) algorithm in such a way that the solution to the system can take advantage of the vectorizing capabilities of modern processors.

Consider, for example, the steady-state heat conduction equation with energy generation in the rectangular coordinate system given by

$$\frac{\partial^2 T}{\partial x^2} + \frac{\partial^2 T}{\partial y^2} = f \tag{3.37a}$$

and its finite difference form for a square grid expressed as

$$T_{ij}^{(n+1)} = \frac{1}{4}\left(T_{i+1,j}^{(n)} + T_{i-1,j}^{(n+1)} + T_{i,j+1}^{(n)} + T_{i,j-1}^{(n+1)} - h^2 f_{ij}\right) \tag{3.37b}$$

where $h = \Delta x = \Delta y$. Gauss–Seidel is used, and the computation is assumed to move through the grid points from *left to right* and *bottom to top*. Here, the superscript n denotes the iteration level and h the mesh size. Figure 3.3 shows the iteration procedure for the Gauss–Seidel scheme, which requires two newly updated values and two old values in order to compute $T_{ij}^{(n+1)}$ at the n + 1 iteration level. On the other hand, the iteration procedure for the red-black ordering scheme shown in Figure 3.4 requires most recently solved values of the opposite color in order to compute $T_{ij}^{(n+1)}$ at the n + 1 iteration level.

Such an approach allows the vectorization of the algorithm (Ortega 1989). To implement the Gauss–Seidel (or SOR) iteration according to the red-black

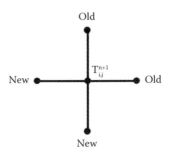

FIGURE 3.3
Gauss–Seidel iteration procedure.

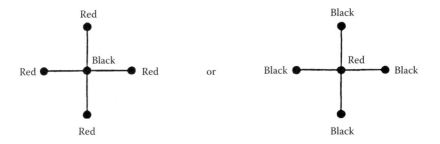

FIGURE 3.4
Red-black iteration scheme.

```
•   •   •   •   •   •   •   •   •   •
R   B   R   B   R   B   R   B   R   B
•   •   •   •   •   •   •   •   •   •
B   R   B   R   B   R   B   R   B   R
•   •   •   •   •   •   •   •   •   •
R   B   R   B   R   B   R   B   R   B
•   •   •   •   •   •   •   •   •   •
B   R   B   R   B   R   B   R   B   R
```

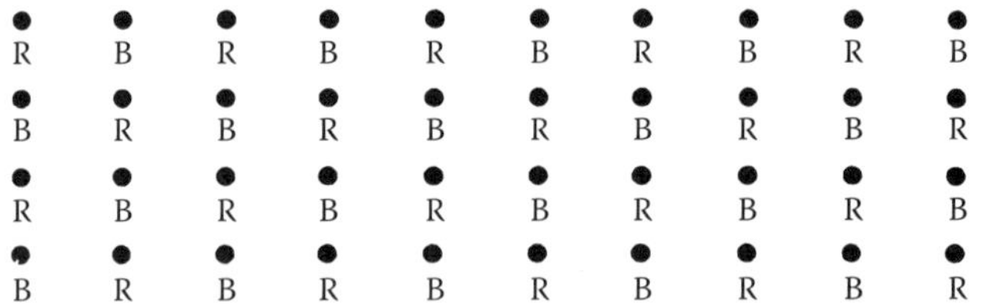

FIGURE 3.5
A checkerboard ordering of grid points.

ordering scheme, the computational grid can be arranged in a checkerboard ordering of all grid points as illustrated in Figure 3.5. In order to accommodate the red-black iteration scheme, the grid points are divided into two classes, red and black, and ordered within each class. The computations are performed using equation (3.36b) by starting with "red" (or black) and using the most recently solved values of the opposite color neighboring nodes to compute T_{ij}^{n+1}. Similar calculations are then performed by moving to the next "red" point on the right by jumping over the black point. The calculations are repeated until all red points from left to right as well as from bottom to top are exhausted. The calculations are then repeated alternately with black and red points until convergence.

3.3.4 LU Decomposition with Iterative Improvement

The iterative Gauss–Seidel (or SOR) method described previously is effective for the solution of a system of algebraic equations if the coefficient matrix is diagonally dominant and sparse. Such a requirement is generally satisfied for most partial differential equations governing heat transfer problems. In the case of a system of algebraic equations with an ill-conditioned, near-singular, or non-diagonally dominant coefficient matrix, the Gauss–Seidel (or SOR) iteration results in the accumulation of roundoff errors, which in turn causes the solution to oscillate and soon diverge. For such cases, the accumulation of large round-off error is alleviated by the use of a powerful matrix-solving routine called *LU decomposition with iterative improvement* (Press et al. 1992, 31–43). Three subroutines, LUDCMP, LUBKSB, and MPROVE, are needed for the implementation of this scheme (Press et al. 1992). The subroutines LUDCMP and LUBKSB are called only once, while MPROVE may be called numerous times to eliminate roundoff errors associated with near-singular matrices and, hence, to considerably improve the solution.

3.3.5 Biconjugate Gradient Method

Another iterative method, which is highly vectorizable, is the biconjugate gradient method. This method performs very well, with a high rate of convergence. The method is fully described in Press et al. (1992) along with the subroutine LINBCG needed for its implementation. A basic algorithm, which uses a preconditioning matrix $\tilde{\mathbf{A}}$ for solving the linear system $\mathbf{Ax} = \mathbf{b}$, can be described in the following steps:

1. Set the iteration counter $\qquad$ $n = 0$
2. Set $\qquad$ $\mathbf{x}^{(n)}$ = initial guess
3. Calculate the residual $\qquad$ $\mathbf{r} = \mathbf{b} - \mathbf{A}\,\mathbf{x}^{(n)}$
4. Calculate the product of the matrix by the residual $\qquad$ $\mathbf{rr} = \mathbf{A}\,\mathbf{r}$
5. Calculate the product of the preconditioner by the right-hand side $\qquad$ $\mathbf{z} = \tilde{\mathbf{A}}\,\mathbf{b}$
6. Calculate the norm of $\mathbf{z}$ $\qquad$ $bnrm = |\mathbf{z}|$
7. Calculate the product of the preconditioner by the residual $\qquad$ $\mathbf{z} = \tilde{\mathbf{A}}\,\mathbf{r}$

8. Increase the iteration counter $\quad n = n + 1$

9. Calculate the product of the transpose of the preconditioner by **rr** $\quad \mathbf{zz} = \tilde{\mathbf{A}}^T \mathbf{rr}$

10. Calculate the inner product of **z** and **rr** $\quad \text{bknum} = \mathbf{z}^T \mathbf{rr}$

11. If n = 1, then
    a. Set $\quad \mathbf{p} = \mathbf{z}$
    b. Set $\quad \mathbf{pp} = \mathbf{zz}$

12. Else
    a. Calculate $\quad \text{bk} = \text{bknum}/\text{bkden}$
    b. Set $\quad \mathbf{p} = \text{bk } \mathbf{p} + \mathbf{z}$
    c. Set $\quad \mathbf{pp} = \text{bk } \mathbf{pp} + \mathbf{zz}$

13. Set $\quad \text{bkden} = \text{bknum}$

14. Calculate the product of the matrix by **p** $\quad \mathbf{z} = \mathbf{A} \mathbf{p}$

15. Calculate the inner product of **z** and **pp** $\quad \text{akden} = \mathbf{z}^T \mathbf{pp}$

16. Set $\quad \text{ak} = \text{bknum}/\text{akden}$

17. Calculate the product of the transpose of **A** by **pp** $\quad \mathbf{zz} = \mathbf{A}^T \mathbf{pp}$

18. Calculate the new solution vector $\quad \mathbf{x} = \mathbf{x} + \text{ak } \mathbf{p}$

19. Calculate $\quad \mathbf{r} = \mathbf{r} - \text{ak } \mathbf{z}$

20. Calculate $\quad \mathbf{rr} = \mathbf{rr} - \text{ak } \mathbf{zz}$

21. Calculate the product of the preconditioner by **r** $\quad \mathbf{z} = \tilde{\mathbf{A}} \mathbf{r}$

22. Calculate the error $\quad \text{error} = |\mathbf{z}^T \mathbf{z}|/\text{bnrm}$

23. If error > tolerance, go back to Step 8.

One of the common choices for $\tilde{\mathbf{A}}$ is to use the Jacobi preconditioner. In this case, the matrix $\tilde{\mathbf{A}}$ is taken as $1/\text{diag}(\mathbf{A})$. Appendix D presents a Fortran code for the biconjugate gradient method, with a Jacobi preconditioner.

3.4 Nonlinear Systems

The boundary value problems become nonlinear due to the nonlinearity of the governing differential equations or of the boundary conditions or both. Most physical problems are actually nonlinear. There is no difficulty in applying the finite difference approximation to discretize a nonlinear problem; but the difficulty is associated with the solution of the resulting system of algebraic equations. Because the system of algebraic equations resulting from linear problems can readily be solved with the available algorithms, generally every effort is made to linearize the nonlinear systems of equations. For example, the simplest and common approach to linearize the difference equations is a procedure known as *lagging the coefficients*, that is, by evaluating the nonlinear coefficients at the previous time level (i.e., lagged). Another approach is the *iterative update* of the coefficients. That is, the coefficients are first evaluated at the previous iteration and then the system is solved for the new values of the unknowns at the n + 1 level. Calculations are then repeated to obtain improved predictions.

Linearization can also be performed for the solution of nonlinear systems, such as in the Newton–Raphson iterative method described here.

The Newton–Raphson method is an algorithm for finding the roots of systems of nonlinear algebraic equations by iteration. Consider, for example, the following system of N algebraic equations

$$F_1(x_1, x_2, \cdots, x_N) = 0$$
$$F_2(x_1, x_2, \cdots, x_N) = 0$$
$$\vdots \qquad \vdots$$
$$F_N(x_1, x_2, \cdots, x_N) = 0$$

(3.38)

We need to find $x_1, x_2, \ldots, x_N$ such that this system of equations is satisfied.

To develop the iteration scheme, the equations are written in the vector form as

$$\mathbf{F}(\mathbf{x}) = 0 \qquad (3.39)$$

and the Taylor series expansion is considered

$$\mathbf{F}(\mathbf{x}^{(n+1)}) = \mathbf{F}(\mathbf{x}^{(n)}) + \left(\frac{\partial \mathbf{F}}{\partial \mathbf{x}}\right)(\mathbf{x}^{(n+1)} - \mathbf{x}^{(n)}) + \cdots \qquad (3.40)$$

We need $\mathbf{F}(\mathbf{x}^{(n+1)}) = 0$. The Taylor series is truncated, and this condition is imposed to obtain

$$\mathbf{F}(\mathbf{x}^{(n)}) + \left(\frac{\partial \mathbf{F}(\mathbf{x}^{(n)})}{\partial \mathbf{x}}\right)(\mathbf{x}^{(n+1)} - \mathbf{x}^{(n)}) = 0 \qquad (3.41)$$

which is solved for $\mathbf{x}^{(n+1)}$

$$\mathbf{x}^{(n+1)} = \mathbf{x}^{(n)} - \left(\frac{\partial \mathbf{F}(\mathbf{x}^{(n)})}{\partial \mathbf{x}}\right)^{-1} \mathbf{F}(\mathbf{x}^{(n)}) \qquad (3.42a)$$

where $(\partial \mathbf{F}/\partial \mathbf{x})$ is the *Jacobian* matrix $\mathbf{J}$ defined as

$$\mathbf{J} \equiv \frac{\partial \mathbf{F}(\mathbf{x}^{(n)})}{\partial \mathbf{x}} = \begin{bmatrix} \dfrac{\partial F_1}{\partial x_1} & \cdots & \dfrac{\partial F_1}{\partial x_N} \\ \vdots & & \vdots \\ \dfrac{\partial F_N}{\partial x_1} & \cdots & \dfrac{\partial F_N}{\partial x_N} \end{bmatrix} \qquad (3.42b)$$

Equations (3.42a,b) define the Newton–Raphson method of iterations, where the superscript (n) denotes the values obtained at the (n)-th iteration and (n + 1) indicates the values at the (n + 1)-th iteration.

Special case N = 2: For illustration purposes, we consider the Newton–Raphson method for the case of two equations given in the form

$$\begin{aligned} F_1(x,y) &= 0 \\ F_2(x,y) &= 0 \end{aligned} \qquad (3.43)$$

The Newton–Raphson method given by equations (3.42a,b) reduces to

$$\begin{bmatrix} x^{(n+1)} \\ y^{(n+1)} \end{bmatrix} = \begin{bmatrix} x^{(n)} \\ y^{(n)} \end{bmatrix} - \mathbf{J}^{-1} \begin{bmatrix} F_1(x^{(n)}, y^{(n)}) \\ F_2(x^{(n)}, y^{(n)}) \end{bmatrix} \qquad (3.44a)$$

where $\mathbf{J}$ is the Jacobian matrix defined as

$$\mathbf{J} = \begin{bmatrix} \dfrac{\partial F_1}{\partial x} & \dfrac{\partial F_1}{\partial y} \\ \dfrac{\partial F_2}{\partial x} & \dfrac{\partial F_2}{\partial y} \end{bmatrix} \qquad (3.44b)$$

If the explicit inverse of the Jacobian matrix $\mathbf{J}$ is introduced, equations (3.44a,b) are rewritten in the form

$$\begin{bmatrix} x^{(n+1)} \\ y^{(n+1)} \end{bmatrix} = \begin{bmatrix} x^{(n)} \\ y^{(n)} \end{bmatrix} - \left(\frac{1}{D} \begin{bmatrix} \dfrac{\partial F_2}{\partial y} & -\dfrac{\partial F_1}{\partial y} \\ -\dfrac{\partial F_2}{\partial x} & \dfrac{\partial F_1}{\partial x} \end{bmatrix} \begin{bmatrix} F_1(x^{(n)}, y^{(n)}) \\ F_2(x^{(n)}, y^{(n)}) \end{bmatrix} \right) \qquad (3.45a)$$

where the determinant D is

$$D = \begin{vmatrix} \dfrac{\partial F_1}{\partial x} & \dfrac{\partial F_1}{\partial y} \\ \dfrac{\partial F_2}{\partial x} & \dfrac{\partial F_2}{\partial y} \end{vmatrix} \qquad (3.45b)$$

Special case of n = 1: For this special case, we have only one equation

$$F(x) = 0 \tag{3.46}$$

and the Newton–Raphson method reduces to

$$x^{(n+1)} = x^{(n)} - \frac{F(x^{(n)})}{F'(x^{(n)})} \tag{3.47}$$

where the prime denotes differentiation with respect to x.

If a good initial guess is made, the Newton–Raphson iteration process converges extremely fast. Iterations are terminated when the computed changes in the values of $|x^{(n+1)} - x^{(n)}|$ become less than some specified tolerance ε.

A good initial guess is essential for the successful convergence of this method. For a single equation, a priori information on the location of the roots is often available. However, in the case of systems of equations, it is quite difficult to find good initial guess values in the neighborhood of the solution; hence, the convergence of the solution poses a serious problem. If the initial guess is far from the exact solution, the matrix J may become ill-conditioned.

The problem of finding good initial guess values for the Newton–Raphson method has been the subject of numerous investigations. Also, as the system becomes large, the assembling and inversion of J becomes more time consuming.

Example 3.4

Two nonlinear algebraic equations are given by

$$F_1(x,y) = x^2 - 2y + 2 = 0$$

$$F_2(x,y) = 2x^2 - y - 5 = 0$$

Write the Newton–Raphson algorithm for solving these two equations and perform the first iteration.

Solution

This is a two-equation system and we apply the explicit form of the algorithm given by equations (3.45a,b). The determinant D becomes

$$D = \begin{vmatrix} 2x^{(n)} & -2 \\ 4x^{(n)} & -1 \end{vmatrix} = 6x^{(n)}$$

where superscript (n) denotes n-th iteration, and we write

$$\begin{bmatrix} \dfrac{\partial F_2}{\partial y} & -\dfrac{\partial F_1}{\partial y} \\ -\dfrac{\partial F_2}{\partial x} & \dfrac{\partial F_1}{\partial x} \end{bmatrix} = \begin{bmatrix} -1 & +2 \\ -4x^{(n)} & 2x^{(n)} \end{bmatrix}$$

Introducing these results into equation (3.90a), we obtain the following algorithm for iterations

$$\begin{bmatrix} x^{(n+1)} \\ y^{(n+1)} \end{bmatrix} = \begin{bmatrix} x^{(n)} \\ y^{(n)} \end{bmatrix} - \left(\frac{1}{6x^{(n)}} \begin{bmatrix} -1 & 2 \\ -4x^{(n)} & 2x^{(n)} \end{bmatrix} \begin{bmatrix} (x^{(n)})^2 - 2y^{(n)} + 2 \\ 2(x^{(n)})^2 - y^{(n)} - 5 \end{bmatrix} \right)$$

To perform the iterations, we need an initial guess. We choose $x^{(0)} = y^{(0)} = 1$; then this expression gives $(x^{(1)}, y^{(1)})$ as

$$\begin{bmatrix} x^{(1)} \\ y^{(1)} \end{bmatrix} = \begin{bmatrix} 1 \\ 1 \end{bmatrix} - \left(\frac{1}{6} \begin{bmatrix} -1 & 2 \\ -4 & 2 \end{bmatrix} \begin{bmatrix} 1 \\ -4 \end{bmatrix} \right) = \begin{bmatrix} 2.5 \\ 3.0 \end{bmatrix}$$

Thus, the first iteration gives

$$x^{(1)} = 2.5, \quad y^{(1)} = 3.0$$

These results are used in the aforementioned expression to determine $x^{(2)}$, $y^{(2)}$, and the procedure is continued until desired convergence is achieved.

PROBLEMS

3.1. Solve the following set of equations using Gauss elimination, Gauss–Seidel iteration, and the biconjugate gradient method. Compare the results and the number of floating-point operations.

$$\begin{bmatrix} 4 & 3 & -10 \\ 7 & -2 & 3 \\ 5 & 18 & 13 \end{bmatrix} \begin{bmatrix} T_1 \\ T_2 \\ T_3 \end{bmatrix} = \begin{bmatrix} -20 \\ 12 \\ 80 \end{bmatrix}$$

3.2. Solve the following set of equations using Gauss elimination, Gauss–Seidel iteration, and the biconjugate gradient method. Compare the results and the number of floating-point operations.

$$\begin{bmatrix} 3 & 2 & -1 \\ 2 & 4 & 1 \\ 3 & 1 & 5 \end{bmatrix} \begin{bmatrix} T_1 \\ T_2 \\ T_3 \end{bmatrix} = \begin{bmatrix} 8 \\ 20 \\ 29 \end{bmatrix}$$

3.3. Carry out, with hand calculations, the first three iterations of the Gauss–Seidel method for the following set of equations.

$$9T_1 + 2T_2 - 3T_3 = 12$$
$$2T_1 - 8T_2 + 2T_3 = -8$$
$$T_1 + 3T_2 + 6T_3 = 32$$

Methods of Solving Systems of Algebraic Equations

3.4. The following equation is to be solved by Gauss–Seidel iteration. However, it does not seem to be suitable because the diagonal elements are not the largest in each row.

$$\begin{bmatrix} 2 & -3 & -5 & 21 \\ 5 & 10 & -15 & 2 \\ 40 & -20 & 3 & -5 \\ 4 & 30 & -10 & -2 \end{bmatrix} \begin{bmatrix} T_1 \\ T_2 \\ T_3 \\ T_4 \end{bmatrix} = \begin{bmatrix} 65 \\ -12 \\ -11 \\ 26 \end{bmatrix}$$

By reordering the equations, remedy this difficulty and then solve by Gauss–Seidel iteration.

3.5. Consider the heat conduction equation for a solid cylinder given in the form

$$\frac{d^2T}{dR^2} + \frac{1}{R}\frac{dT}{dR} + \frac{1}{k}g(R) = 0 \quad \text{in} \quad 0 \leq R < a$$

This equation contains singularity at $R = 0$. Using L'Hospital's rule, show that at $R = 0$, this equation can be replaced by

$$2\frac{d^2T}{dR^2} + \frac{1}{k}g(0) = 0$$

3.6. Consider the nonlinear differential equation

$$\frac{d^2T}{dx^2} = -1 - \sin\left(\frac{dT}{dx}\right)$$

subject to the boundary conditions

$$T(x) = 0 \text{ at } x = 0$$

$$T(x) = 1 \text{ at } x = 1$$

Using the central difference formula for both the first and second derivatives, write the finite difference form of this problem suitable for iterative calculations by dividing the domain $0 \leq x \leq 1$ into 10 equal parts.

3.7. Consider an iron rod of length L = 10 cm long, diameter D = 1 cm, and thermal conductivity k = 50 W/(m°C). One end of the rod is maintained at T_0 = 200°C, the other end at 0°C, while it is exposed to convection from its lateral surfaces into ambient air at 0°C, with a heat transfer coefficient h = 200 W/(m²°C). If we assume one-dimensional, steady-state heat flow, the mathematical formulation of this problem is given by

$$\frac{d^2T(x)}{dx^2} - b^2 T(x) = 0 \quad \text{in } 0 < x < L$$

$$T(x) = 200°C \quad \text{at } x = 0$$

$$T(x) = 0°C \quad \text{at } x = L$$

where

$$b^2 = \frac{ph}{kA} = \frac{4h}{kD}$$

and A = cross-section area of the rod, P = perimeter, D = diameter, k = thermal conductivity.

By dividing the region $0 \le x \le L$ into five equal parts, calculate the temperature distribution along the rod using finite differences. Compare the finite difference solution with the exact solution.

3.8. Consider the following one-dimensional, steady-state heat conduction problem:

$$\frac{d^2T(x)}{dx^2} + \frac{1}{k}g = 0 \quad \text{in } 0 < x < L$$

$$\frac{dT(x)}{dx} = 0 \quad \text{at } x = 0$$

$$k\frac{dT(x)}{dx} + hT(x) = 0 \quad \text{at } x = L$$

Write the finite difference formulation of this heat conduction problem by dividing the region $0 < x < L$ into four equal parts.

3.9. In a parallel-plate fuel element for a gas-cooled nuclear reactor, the heat generation in the fuel element has approximately a cosine distribution. The simplest steady-state model for the temperature distribution in the fuel element may be taken as

$$\frac{d^2T(x)}{dx^2} + \frac{1}{k}g_0 \cos\left(\frac{\pi x}{2L}\right) = 0 \quad \text{in } 0 < x < L$$

$$\frac{dT}{dx} = 0 \quad \text{at } x = 0$$

$$T = 0 \quad \text{at } x = L$$

where L is the half-thickness of the fuel element. By dividing the region into four equal parts, calculate the temperature distribution with finite differences for $k = 12 \text{ W}/(\text{m}°\text{C})$, $L = 5 \times 10^{-3}$ m, and $g_0 = 6 \times 10^8 \text{ W/m}^3$. Compare the numerical results with the exact solution.

3.10. Consider the following one-dimensional, steady-state heat conduction problem

$$\frac{d^2T(x)}{dx^2} + \frac{1}{k}g_0 = 0 \quad \text{in} \quad 0 < x < L$$

$$-k\frac{dT(x)}{dx} = q_0 \quad \text{in} \quad x = 0$$

$$T = 0 \quad \text{at} \quad x = L$$

a. Write the finite difference formulation of this heat conduction problem by dividing the region $0 < x < L$ into four equal parts.
b. Compute the node temperatures for $k = 12 \text{ W}/(\text{m}°\text{C})$, $L = 0.012$ m, $q_0 = 10^5 \text{ W/m}^2$, and $g_0 = 4 \times 10^7 \text{ W/m}^3$.
c. Compare the numerical solution at the nodes with the exact solution.

3.11. Write the finite difference formulation of the heat conduction equation

$$\frac{\partial^2 T}{\partial x^2} + \frac{\partial^2 T}{\partial y^2} = 0$$

for the square region of side L by using a mesh size $\Delta x = \Delta y = L/3$ for the boundary conditions shown in the figure here. Express the resulting equations in matrix form for the unknown node temperatures T_m, $m = 1$ to 4 and calculate the unknown node temperatures.

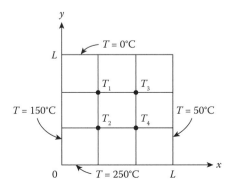

3.12. Write the finite difference formulation of the heat conduction equation

$$\frac{\partial^2 T}{\partial x^2} + \frac{\partial^2 T}{\partial y^2} = 0$$

for the rectangular region $3/4L$ by L by using a mesh size $\Delta x = \Delta y = L/4$ for the boundary conditions shown in the figure here. Express the resulting finite difference equations in matrix form for the unknown node temperatures T_m, $m = 1$ to 6, and calculate these temperatures.

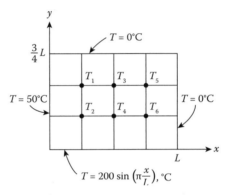

3.13. Write the finite difference formulation of the heat conduction equation

$$\frac{\partial^2 T}{\partial x^2} + \frac{\partial^2 T}{\partial y^2} + \frac{g}{k} = 0$$

for nodes $m = 1, 2, \ldots, 12$ at which the temperatures T_i are unknown, as shown in the figure here. The temperatures f_1, f_2, and f_3 are considered specified. Express these results in matrix form for the unknown node temperatures T_m, $m = 1$ to 12.

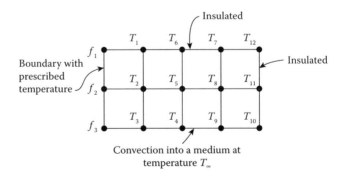

3.14. Consider two-dimensional, steady-state heat conduction in a rectangular region 4 L by 3 L, as shown in the figure here. The boundary surface at x = 0 is insulated; the boundary surface at x = 4 L dissipates heat by convection into an ambient at 300 K; and boundary surfaces at y = 0 and y = 3 L are maintained at 1000 K and 500 K, respectively. The heat transfer coefficient for convection is h = 70 W/(m²°C), and the thermal conductivity of the material is k = 4 W/(m°C). By using a mesh size $\Delta x = \Delta y = 0.1$ m, write T_m, m = 1 to 10 and calculate the node temperatures.

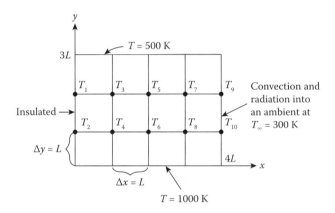

3.15. Solve by finite differences the steady-state temperature distribution in the rectangular region subject to the boundary conditions shown in the figure here by setting L = 4 cm and $T_0 = 100°C$, using a mesh size $\Delta x = \Delta y = 1$ cm.

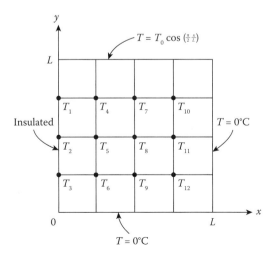

3.16. Solve by finite differences the steady-state temperature distribution in a rectangular region subject to the boundary conditions shown in the figure here by setting L = 4 cm and T_0 = 200°C, using a mesh size $\Delta x = \Delta y = 1$ cm.

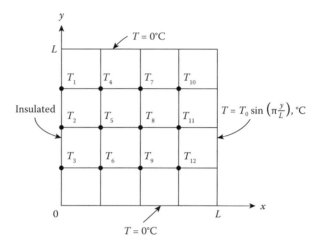

3.17. Solve by finite differences the steady-state temperature distribution in a rectangular region subject to the boundary conditions shown in the figure here by setting a = 6 cm, b = 4 cm and T_0 = 200°C, using a mesh size $\Delta x = \Delta y = 1$ cm.

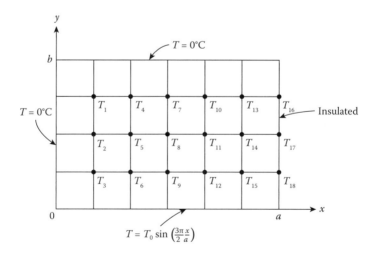

3.18. Consider two-dimensional, steady-state heat conduction in a 3 cm by 5 cm rectangular region subject to the boundary conditions shown in the figure here. Using a mesh size $\Delta x = \Delta y = 1$ cm, develop a finite difference formulation of this heat conduction problem and calculate the node temperatures.

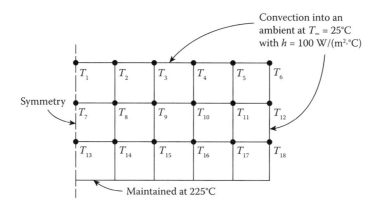

3.19. Write the finite difference formulation for two-dimensional, steady-state heat conduction with energy generation at a constant rate of $g_0 \mathrm{W/m^3}$ for a rectangular region of cross section 0.2 m by 0.3 m for the six nodes m = 1,2,...,6 by setting $\Delta x = \Delta y = 0.1$ m. The boundary conditions are shown in the figure here.

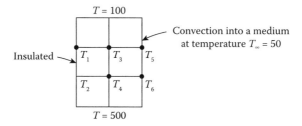

3.20. Consider the following 16 equations having a tridiagonal matrix:

$$\begin{bmatrix} -2 & 1 & 0 & 0 & \cdot & 0 & 0 & 0 & 0 \\ 1 & -2 & 1 & 0 & & & 0 & 0 \\ 0 & 1 & -2 & 1 & & & 0 & 0 \\ 0 & & & & & & & \\ \cdot & & & 1 & -2 & 1 & 0 & 0 \\ \cdot & & & 0 & 1 & -2 & 1 & 0 \\ 0 & & & 0 & 0 & 1 & -2 & 1 \\ 0 & 0 & 0 & 0 & 0 & 0 & 1 & -2 \end{bmatrix} \begin{bmatrix} T_1 \\ T_2 \\ T_3 \\ \cdot \\ \cdot \\ T_{14} \\ T_{15} \\ T_{16} \end{bmatrix} = \begin{bmatrix} -1 \\ -2 \\ -2 \\ \cdot \\ -2 \\ -2 \\ -2 \\ +1 \end{bmatrix}$$

Examine the solution of this system of equations using
a. Thomas algorithm
b. Gauss elimination
c. SOR by trying several different relaxation parameter ω, including ω = 1, which corresponds to Gauss–Seidel iterations
d. The biconjugate gradient method

Examine the number of iterations required for a convergence tolerance $\varepsilon = 10^{-3}$, and compare the results with those obtained by the direct method of solution.

3.21. Consider the following steady-state heat conduction equation in cylindrical coordinates

$$\frac{1}{R}\frac{\partial}{\partial R}\left(R\frac{\partial T}{\partial R}\right) + \frac{\partial^2 T}{\partial z^2} + \frac{1}{k}g(R,z) = 0$$

Using a grid-point notation $R = i\Delta R$, $z = j\Delta z$, write this equation in the finite difference form.

4

One-Dimensional Steady-State Systems

In this chapter, we examine finite difference representation and solution of one-dimensional, steady-state problems in Cartesian, cylindrical, and spherical coordinates. The diffusive and convective systems will also be considered, and the effects of flow on the numerical stability of the resulting finite difference equations will be discussed.

4.1 Diffusive Systems

Consider one-dimensional, constant property, steady-state diffusion in a slab, cylinder, or sphere. The governing energy equation is written in the form

$$\frac{1}{R^p}\frac{d}{dR}\left(R^p\frac{dT}{dR}\right) + \frac{1}{k}g(R) = 0, \quad R \neq 0 \tag{4.1a}$$

or

$$\frac{d^2T}{dR^2} + \frac{p}{R}\frac{dT}{dR} + \frac{1}{k}g(R) = 0, \quad R \neq 0 \tag{4.1b}$$

where $p = \begin{cases} 0 & \text{slab} \\ 1 & \text{cylinder} \\ 2 & \text{sphere} \end{cases}$

and in the case of heat conduction, the term g(R) represents the volumetric energy generation rate (i.e., W/m^3). In the case of mass diffusion, k is replaced by the diffusion coefficient D(cm^2/s), g(R) represents the volumetric mass generation rate (i.e., g/cm^3·s), and T represents the volumetric mass concentration (i.e., g/cm^3).

4.1.1 Slab

Equation 4.1 reduces to

$$\frac{d^2T}{dR^2} + \frac{1}{k}g(R) = 0 \tag{4.1c}$$

The finite difference representation of this equation has been discussed previously.

4.1.2 Solid Cylinder and Sphere

For a solid cylinder and sphere, equation (4.1b) has an apparent singularity at the origin $R = 0$. However, an examination of equation (4.1b) reveals that both R and dT/dR become zero for $R = 0$; hence, we have $\frac{0}{0}$ ratio at the origin. By the application of L'Hospital's rule, it can be shown that this ratio has the following determinate form:

$$\left(\frac{1}{R}\frac{dT}{dR}\right)_{R=0} = \frac{\frac{d}{dR}\left(\frac{dT}{dR}\right)}{\frac{d}{dR}(R)}\bigg|_{R=0} = \frac{d^2T}{dR^2}\bigg|_{R=0} \qquad (4.2)$$

Then equation (4.1b), at $R = 0$, becomes

$$(1+p)\frac{d^2T(R)}{dR^2} + \frac{1}{k}g(R) = 0, \quad R = 0 \qquad (4.3)$$

To approximate such equations in finite differences, a network mesh of size δ, as illustrated in Figure 4.1, is constructed over the region. Then, by using the second-order accurate finite difference formula, the first and the second derivatives are directly discretized. The resulting finite difference approximation to equation (4.1b) becomes

$$\frac{T_{i-1} - 2T_i + T_{i+1}}{\delta^2} + \frac{p}{i\delta}\frac{T_{i+1} - T_{i-1}}{2\delta} + \frac{1}{k}g_i = 0 \qquad (4.4)$$

$$\text{for } i = 1, 2, \ldots, M-1, \quad 0(\delta^2)$$

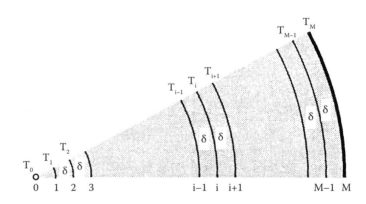

FIGURE 4.1
Nomenclature for finite difference representation for cylindrical and spherical symmetry.

This system provides $M-1$ algebraic equations for the $M+1$ unknown node temperatures $T_1,\ldots,T_{M-1}$.

An additional relationship is obtained by discretizing equation (4.3) at $R = 0$. In order to use a second-order accurate central-difference formula at $R = 0$, a node is needed to the left of the origin $R = 0$. This is achieved by considering a fictitious node "–1" at a fictitious temperature T_{-1} located at a distance δ to the left of the R axis. The resulting finite difference approximation of equation (4.3) at $R = 0$ becomes

$$(1+p)\frac{T_{-1}-2T_0+T_1}{\delta^2} + \frac{1}{k}g_0 = 0, \quad i = 0 \tag{4.5a}$$

where the fictitious temperature T_{-1} is determined by utilizing the symmetry condition at the node $i = 0$; that is

$$\left.\frac{dT}{dR}\right|_{R=0} = \frac{T_{-1}-T_1}{2\delta} = 0, \quad \text{giving } T_{-1} = T_1 \tag{4.5b}$$

Introducing equation (4.5b) into equation (4.5a), the additional finite difference equation is determined as

$$2(1+p)\frac{T_1-T_0}{\delta^2} + \frac{1}{k}g_0 = 0 \quad \text{for } i = 0 \tag{4.6}$$

Equations (4.4) and (4.6) are now rearranged, respectively, as

$$2(1+p)(T_1-T_0) + \frac{\delta^2 g_0}{k} = 0 \quad \text{for } i = 0 \tag{4.7}$$

$$\left(1-\frac{p}{2i}\right)T_{i-1} - 2T_i + \left(1+\frac{p}{2i}\right)T_{i+1} + \frac{\delta^2 g_i}{k} = 0 \quad \text{for } i = 1, 2, \ldots, M-1 \tag{4.8}$$

and $p = \begin{cases} 1 & \text{cylinder} \\ 2 & \text{sphere} \end{cases}$

One more equation is needed to make the number of equations equal to the number of unknowns. It is obtained by considering the boundary condition at the node $i = M$ (i.e., $R = b$). The following possibilities can be considered at the node "M" (see also Chapter 2):

1. The temperature T_b is specified at the boundary $R = b$. Then we have

$$T_M = T_b \tag{4.9}$$

and the system of equations (4.7), (4.8), and (4.9) provides $M + 1$ relations for the determination of $(M + 1)$ unknown node temperatures.

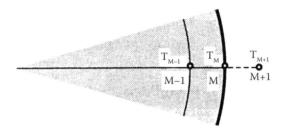

FIGURE 4.2
Fictitious node M + 1 at fictitious temperature T_{M+1}.

2. The boundary condition at R = b is convection into an ambient at a constant temperature $T_{\infty,b}$ with a heat transfer coefficient h_b. The boundary condition is given by

$$k\frac{dT}{dR} + h_b T = h_b T_{\infty,b} \quad \text{at} \quad R = b \qquad (4.10)$$

To discretize this equation about the boundary node M with a second-order central difference formula, an additional node is needed to the right of the node M. This is obtained by considering an extension of the region by a distance δ to the right of the node M, giving rise to a fictitious node M + 1 at a fictitious temperature T_{M+1} as illustrated in Figure 4.2. Then the discretization of Equation 4.10 about the node M with the central difference formula gives

$$k\frac{T_{M+1} - T_{M-1}}{2\delta} + h_b T_M = h_b T_{\infty,b} \qquad (4.11)$$

An additional relationship needed to eliminate T_{M+1} is determined by evaluating equation (4.8) for i = M. We obtain

$$\left(1 - \frac{p}{2M}\right)T_{M-1} - 2T_M + \left(1 + \frac{p}{2M}\right)T_{M+1} + \frac{\delta^2 g_M}{k} = 0 \qquad (4.12)$$

The elimination of T_{M+1} between equations (4.11) and (4.12) gives

$$2T_{M-1} - 2\beta_M T_M + 2\gamma_M + G_M = 0 \quad \text{for} \quad i = M \qquad (4.13a)$$

where

$$\beta_M = 1 + \left(1 + \frac{p}{2M}\right)\frac{\delta h_b}{k} \qquad (4.13b)$$

$$\gamma_M = \left(1 + \frac{p}{2M}\right)\frac{\delta}{k} h_b T_{\infty,b} \qquad (4.13c)$$

$$G_M = \frac{\delta^2 g_M}{k} \tag{4.13d}$$

which is accurate $O(\delta^2)$.

Equations (4.7), (4.8), and (4.13a) provide $M + 1$ relations for the determination of $M + 1$ unknown node temperatures for convection boundary conditions at $R = b$.

3. The boundary condition at $R = b$ is a prescribed heat flux boundary condition. For this case, the steady-state solution does not exist unless the energy generated in the medium equals the total heat removal rate from the boundaries. Even for such a case, the steady-state solution for a solid cylinder or sphere is not unique. Hence, such a situation will not be considered.

Example 4.1

A 10-cm diameter solid steel bar of thermal conductivity $k = 40$ W/m°C is heated electrically by the passage of electric current, which generates energy within the rod at a rate of $g = 4 \times 10^6$ W/m³. Heat is dissipated from the surface of the rod by convection with a heat transfer coefficient $h = 400$ W/m²°C into an ambient at temperature $T_\infty = 20$°C. By dividing the radius into five equal parts, develop the finite difference equations for this heat conduction problem. Compare the finite difference solution with the exact analytic solution for the cases when the *first-order* and the *second-order* accurate differencing are used for the convection boundary condition.

Solution

The problem involves six unknown node temperatures, T_i, $i = 0,1,\ldots,5$, since the region $0 \le r \le b$ is divided into five equal parts. The six finite difference equations needed for their determination are obtained as follows. For the center node, equation (4.7), for $p = 1$, gives

$$4(T_1 - T_0) + \frac{g_0 \delta^2}{k} = 0, \quad i = 0 \tag{a}$$

For the internal nodes, equation (4.8), for $p = 1$, gives

$$\left(1 - \frac{1}{2i}\right) T_{i-1} - 2T_i + \left(1 + \frac{1}{2i}\right) T_{i+1} + \frac{\delta^2 g_i}{k} = 0 \quad \text{for } i = 1, 2, 3, 4 \tag{b}$$

For the convection boundary node $i = 5$, a second-order accurate finite difference equation is obtained from equations (4.13a–d) by setting $M = 5$.

$$2T_4 - 2\beta_5 T_5 + 2\gamma_5 + G_5 = 0, \quad i = 5 \tag{c}$$

where

$$\beta_5 = 1 + \left(1 + \frac{1}{10}\right)\frac{\delta h}{k}$$

$$\gamma_5 = \left(1 + \frac{1}{10}\right)\frac{\delta}{k}hT_\infty, \quad G_5 = \frac{\delta^2 g_5}{k}$$

Thus equations (a), (b), and (c) provide six algebraic equations for determination of the six unknown node temperatures T_i, $i = 0,1,\ldots,5$.

If a first-order finite difference approximation was used for the convection boundary condition, the resulting finite difference equation would be

$$T_5 = \frac{1}{1 + \frac{\delta h}{k}}\left(T_4 + \frac{\delta h}{k}T_\infty\right) \quad \text{for } i = 5 \tag{d}$$

which is less accurate than the second-order accurate finite differencing given by Equation (c). The following numerical values are given:

$$b = 0.05 \text{ m}, M = 5, g_5 = 4 \times 10^6 \text{ W/m}^3$$

$$h = 400 \text{ W/m}^2{}^\circ\text{C}, k = 40 \text{ W/m}^\circ\text{C}, T_\infty = 20^\circ\text{C}$$

Then various quantities are evaluated as

$$\delta = \frac{b}{M} = \frac{0.05}{5} = 0.01\text{m}, \quad \frac{\delta^2 g}{k} = \frac{(0.01)^2(4 \times 10^6)}{40} = 10$$

$$\frac{\delta h}{k} = \frac{(0.01)(400)}{40} = 0.1, \quad \frac{\delta h}{k}T_\infty = (0.1)(20) = 2$$

and the finite difference equations (a) and (b), respectively, become

$$4(T_1 - T_0) + 10 = 0, \quad i = 0 \tag{e}$$

$$\left(1 - \frac{2}{2i}\right)T_{i-1} - 2T_i + \left(1 + \frac{1}{2i}\right)T_{i+1} + 10 = 0, \quad i = 1, 2, 3, 4 \tag{f}$$

For the boundary condition at $i = M = 5$, one can use either the first-order accurate formula (d)

$$T_5 = \frac{1}{1.1}(T_4 + 2), \quad i = 5 \tag{g}$$

or the second-order accurate formula (c)

$$T_4 - 1.11T_5 + 7.2 = 0, \quad i = 5 \tag{h}$$

One-Dimensional Steady-State Systems

Summarizing, equations (e), (f), and (g) or equations (e), (f), and (h) provide six algebraic equations for the determination of six unknown node temperatures.

The exact solution of this problem is given by

$$T(R) = T_\infty + \frac{gb}{2h} + \frac{gb^2}{4k}\left[1 - \left(\frac{R}{b}\right)^2\right]$$

where

$$\frac{gb}{2h} = \frac{(4 \times 10^6)(5 \times 10^{-2})}{2 \times 400} = 250, \quad T_\infty = 20°C,$$

$$\frac{gb^2}{4k} = \frac{(4 \times 10^6)(25 \times 10^{-4})}{4 \times 40} = 62.5$$

Then the exact solution takes the form

$$T(R) = 20 + 250 + 62.5\left[1 - \left(\frac{R}{b}\right)^2\right]$$

Table 4.1 shows a comparison of finite difference solutions with the exact results for the cases when the *first-order* and *second-order* accurate formulae are used for the convection boundary condition. The Gauss elimination method is used to solve the resulting algebraic equations. The numerical results obtained with the second-order accurate formula are in excellent agreement with the exact solution; but the solution with the first-order formula is not so good; it underpredicts temperature from about 7% to 9%. Increasing the number of subdivisions from $M = 5$ to $M = 10$ improves the accuracy of the results with the first-order formula to about 4%.

TABLE 4.1

Comparison of the Results with the Exact Solution for Example 4.1

$\frac{R}{b}$	Exact	First-Order Accurate	M = 5 Second-Order Accurate	M = 10 First-Order Accurate
0.0	332.50	307.50	332.50	320.00
0.2	330.00	305.00	330.00	317.50
0.4	322.50	297.50	322.50	310.00
0.6	310.00	285.00	310.00	297.00
0.8	292.50	267.50	292.50	280.00
1.0	270.00	245.00	270.00	257.50

Example 4.2

Repeat Example 4.1 for the case of a solid sphere.

Solution

The physical problem is exactly the same as Example 4.1, except a solid sphere of diameter D = 10 cm is considered instead of a solid cylinder.

Therefore, the finite difference equations are similar to those given in Example 4.1, except p = 2. Then the final equations become

$$6(T_1 - T_0) + 10 = 0 \quad \text{for } i = 0 \tag{a}$$

$$\left(1 - \frac{1}{i}\right)T_{i-1} - 2T_i + \left(1 + \frac{1}{i}\right)T_{i+1} + 10 = 0 \quad \text{for } i = 1, 2, 3, 4 \tag{b}$$

For the boundary node i = 5, we try both the *first-order* accurate formula

$$T_5 = (T_4 + 2)/1.1 \quad \text{for } i = M = 5 \tag{c}$$

and the *second-order* accurate formula

$$T_4 - 1.12 T_5 + 7.4 = 0 \quad \text{for } i = M = 5 \tag{d}$$

The *exact* solution of this heat conduction problem is given by

$$T(R) = T_\infty + \frac{gb}{3h} + \frac{gb^2}{6k}\left[1 - \left(\frac{R}{b}\right)^2\right]$$

where

$$\frac{gb}{3h} = \frac{(4 \times 10^6)(5 \times 10^{-2})}{3 \times 400} = \frac{500}{3}$$

$$\frac{gb^2}{6k} = \frac{(4 \times 10^6)(25 \times 10^{-4})}{6 \times 40} = \frac{125}{3}, \quad T_\infty = 20°C$$

Then the exact solution takes the form

$$T(R) = 20 + \frac{500}{3} + \frac{125}{3}\left[1 - \left(\frac{R}{b}\right)^2\right]$$

Table 4.2 shows a comparison of the finite difference solutions obtained using the first-order and the second-order accurate formulae with the exact solution. The Gauss elimination is used for solving the resulting system of algebraic equations. The numerical results obtained with the second-order accurate formula are in excellent agreement with the exact solution. However, the accuracy of the numerical results obtained with the *first-order* formula is not so good; it underpredicts temperature from about 7% to 9%. Increasing the number of subdivisions of the region from M = 5 to M = 10 improves the accuracy of the results with the *first-order* formula to about 5%.

TABLE 4.2

Comparison of the Results with the Exact Solution for Example 4.2

$\dfrac{R}{b}$	Exact	First-Order Accurate	M = 5 Second-Order Accurate	M = 10 First-Order Accurate
0.0	228.333	211.667	228.333	220.000
0.2	226.667	210.000	226.667	218.333
0.4	221.667	205.000	221.667	213.333
0.6	213.333	196.667	213.333	205.000
0.8	201.667	185.000	201.667	193.333
1.0	186.667	170.000	186.667	178.333

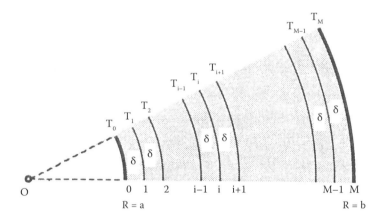

FIGURE 4.3
Nomenclature for finite difference representation for hollow sphere or cylinder.

4.1.3 Hollow Cylinder and Sphere

We now consider heat conduction in a hollow cylinder and sphere of inner radius $R = a$ and outer radius $R = b$. To solve this problem with finite differences, a grid is constructed over the region as illustrated in Figure 4.3. The governing heat conduction equation is given by

$$\frac{d^2T}{dR^2} + \frac{p}{R}\frac{dT}{dR} + \frac{1}{k}g(R) = 0 \quad \text{in } a < R < b \tag{4.14}$$

For finite difference representation of this equation, the region $a \leq R \leq b$ is divided into M subregions each of thickness δ given by

$$\delta = \frac{b-a}{M} \tag{4.15}$$

The differential equation is discretized by using the second-order accurate central-difference formula for both the second and the first derivatives. We obtain

$$\frac{T_{i-1} - 2T_i + T_{i+1}}{\delta^2} + \frac{p}{a + i\delta} \frac{T_{i+1} - T_{i-1}}{2\delta} + \frac{1}{k} g_i = 0 \qquad (4.16)$$

which is rearranged in the form

$$\left[1 - \frac{p}{2\left(\frac{a}{\delta} + i\right)}\right] T_{i-1} - 2T_i + \left[1 + \frac{p}{2\left(\frac{a}{\delta} + i\right)}\right] T_{i+1} + \frac{\delta^2 g_i}{k} = 0 \quad \text{for } i = 1, 2, \ldots, M-1$$

(4.17)

where $p = \begin{cases} 1 & \text{cylinder} \\ 2 & \text{sphere} \end{cases}$

The system of equations (4.17) provides $M - 1$ algebraic equations, but involves $(M + 1)$ unknown node temperatures T_i, $i = 0, 1, 2, \ldots, M$. The additional two relationships are obtained from the boundary conditions at $R = a$ and $R = b$. The following possibilities are considered for the boundary conditions (see also Chapter 2):

1. Temperatures T_a and T_b are prescribed at the boundaries $R = a$ and $R = b$. Then the system of equations (4.17) provides $M - 1$ relations for the determination of $M - 1$ internal node temperatures because $T_0 = T_a$ and $T_M = T_b$ are known.
2. The boundary conditions at $R = a$ and $R = b$ are convection into ambients at temperatures $T_{\infty,a}$ and $T_{\infty,b}$ with heat transfer coefficients h_a and h_b, respectively.

$$-k \frac{dT}{dR} + h_a T = h_a T_{\infty,a}, \quad R = a \qquad (4.18a)$$

$$k \frac{dT}{dR} + h_b T = h_b T_{\infty,b}, \quad R = b \qquad (4.18b)$$

Two additional relations are obtained by discretizing these two boundary conditions. Here, we prefer to use the second-order accurate central-difference formula to discretize these boundary conditions. To apply the central difference formula, the region is assumed to extend by one grid length δ to the left of the boundary node "0" to obtain a fictitious node "-1" at a fictitious temperature T_{-1} and extend by one grid length δ to the right of the boundary node "M" to obtain a fictitious node "M + 1" at a fictitious temperature T_{M+1}. Then the boundary condition equations (4.18a) and (4.18b) are discretized about the nodes "0" and "M," respectively, by using the central difference formula. The resulting expressions contain the unknown fictitious temperatures T_{-1} and T_{M+1}. These unknown temperatures are eliminated by utilizing the expressions obtained from equation (4.17) by evaluating it for i = 0 and i = M. Finally, the second-order accurate finite difference approximation of the boundary conditions [equations (4.18a,b)], respectively, become

$$2T_1 - 2\beta_0 T_0 + 2\gamma_0 + G_0 = 0 \quad \text{for } i = 0 \tag{4.19a}$$

$$2T_{M-1} - 2\beta_M T_M + 2\gamma_M + G_M = 0 \quad \text{for } i = M \tag{4.19b}$$

where

$$\beta_0 = 1 + \left(1 - \frac{p}{2\frac{a}{\delta}}\right)\frac{\delta h_a}{k}, \quad \beta_M = 1 + \left(1 + \frac{p}{2\left(\frac{a}{\delta} + M\right)}\right)\frac{\delta h_b}{k} \tag{4.20a,b}$$

$$\gamma_0 = \left(1 - \frac{p}{2\frac{a}{\delta}}\right)\frac{\delta}{k}(h_a T_{\infty,a}), \quad \gamma_M = \left(1 + \frac{p}{2\left(\frac{a}{\delta} + M\right)}\right)\frac{\delta}{k}(h_b T_{\infty,b})$$

$$\tag{4.20c,d}$$

$$G_0 = \frac{\delta^2 g_0}{k}, \quad G_M = \frac{\delta^2 g_M}{k} \tag{4.20e,f}$$

Summarizing, equations (4.17), (4.19a), and (4.19b) provide M + 1 algebraic equations for the determination of M + 1 unknown node temperatures T_i, i = 0,1,2,…,M.

3. The heat flux is prescribed at any one of the boundaries.

$$-k\frac{dT}{dR} = q_a, \quad R = a \tag{4.21a}$$

or

$$k\frac{dT}{dR} = q_b, \quad R = b \tag{4.21b}$$

Here, positive values of q_a or q_b imply that heat flow is into the medium.

To obtain the finite difference form of the prescribed heat flux boundary conditions [equations (4.21a,b)], we compare them with the convection boundary conditions given by equations (4.18a,b). A term-by-term comparison gives

$$\beta_0 = 1, \quad \beta_M = 1 \qquad (4.22a)$$

$$\gamma_0 = \left(1 - \frac{p}{2\frac{a}{\delta}}\right)\frac{\delta}{k}q_0, \quad \gamma_M = \left(1 + \frac{p}{2\left(\frac{a}{\delta} + M\right)}\right)\frac{\delta}{k}q_b \qquad (4.22b)$$

Therefore, a second-order accurate finite difference approximation for the prescribed heat flux boundary conditions [equations (4.21a,b)] are obtainable from equations (4.19a,b) by making the substitutions given by equations (4.22a,b). We obtain

$$2T_1 - 2T_0 + 2\left(1 - \frac{p}{2\frac{a}{\delta}}\right)\frac{\delta}{k}q_a + \frac{\delta^2 g_0}{k} = 0 \quad \text{for } i = 0 \qquad (4.23a)$$

or

$$2T_{M-1} - 2T_M + 2\left(1 + \frac{p}{2\left(\frac{a}{\delta} + M\right)}\right)\frac{\delta}{k}q_b + \frac{\delta^2 g_M}{k} = 0 \quad \text{for } i = M$$

(4.23b)

The case involving prescribed heat flux at both boundaries does not have a steady-state solution unless the rate of heat generation added to the rate provided to the body on its boundaries equals the total heat removal from the boundaries. Even for such a case, the solution is not unique. For this reason, in the problems of one-dimensional steady-state heat flow in a hollow cylinder or sphere, only one of the boundary conditions will be considered to have a prescribed heat flux.

Example 4.3

Consider steady-state, radial heat conduction in a hollow sphere of inside radius a = 2 cm and outside radius b = 7 cm. Energy is generated at a rate of g = 5 x 10^6 W/m^3 while the inside surface is maintained at a constant temperature T_a = 100°C and the outside surface dissipates heat by convection with a heat transfer coefficient $h_b = 500$ W/m²°C into an ambient at zero temperature. The thermal conductivity of the solid is $k = 50$ W/m°C. By dividing the region "b–a" into five equal parts, develop the finite difference approximation for this heat conduction problem. Compare the finite difference solution with the exact solution for the problem.

Solution
The problem involves five unknown node temperatures T_i, $i = 1,2,...,5$, since the temperature at the inner boundary surface

$$T_0 = T_a = 100°C$$

is known.

The region is divided into five equal parts; therefore, $\delta = 1$ cm. The finite difference equations for the internal nodes are obtained from equation (4.17) by setting $p = 2$

$$\left(1 - \frac{1}{\frac{a}{\delta} + i}\right) T_{i-1} - 2T_i + \left(1 + \frac{1}{\frac{a}{\delta} + i}\right) T_{i+1} + \frac{\delta^2 g_i}{k} = 0 \quad \text{for } i = 1,2,3,4$$

(a)

The finite difference equation for the node $i = M = 5$ at the outer boundary is obtained from equation (4.19b) by setting $M = 5$ and $p = 2$

$$2T_4 - 2\beta_5 T_5 + 2\gamma_5 + G_5 = 0 \quad \text{for } i = 5 \quad \text{(b)}$$

where

$$\beta_5 = 1 + \left(1 + \frac{1}{\frac{a}{\delta} + 5}\right) \frac{\delta h_b}{k}$$

$$\gamma_5 = \left(1 + \frac{1}{\frac{a}{\delta} + 5}\right) \frac{\delta h_b}{k} T_\infty, \quad G_5 = \frac{\delta^2 g_5}{k}$$

The following numerical values are given: $a = 0.02$ m, $b = 0.07$ m, $M = 5$, $g = 5 \times 10^6$ W/m³, $h_b = 500$ W/m²°C, $k = 50$ W/m°C, and $T_\infty = 0°C$. Then various quantities are evaluated as

$$\delta = \frac{b-a}{M} = \frac{0.07-0.02}{5} = 0.01 \text{ m}, \quad \frac{a}{\delta} = \frac{0.02}{0.01} = 2$$

$$\frac{\delta^2 g}{k} = \frac{(0.01)^2 \times 5 \times 10^6}{50} = 10, \quad \frac{\delta h_b}{k} = \frac{(0.01)(500)}{50} = 0.1$$

$$\frac{\delta h_b}{k} T_\infty = (0.1)(0) = 0$$

TABLE 4.3

Comparison of Finite Difference Results with the Exact Solution for Example 4.3

Node Number	Exact	Node Temperature T_i Finite Difference
0	100.00	100.00
1	153.54	153.07
2	172.80	172.11
3	176.36	175.53
4	170.40	169.47
5	157.58	156.58

Then the finite difference equations (a) and (b), respectively, become

$$\left(1 - \frac{1}{2+i}\right)T_{i-1} - 2T_i + \left(1 + \frac{1}{2+i}\right)T_{i+1} + 10 = 0 \quad \text{for } i = 1, 2, 3, 4 \quad \text{(c)}$$

and

$$T_4 - 1.1143 T_5 + 5 = 0 \quad \text{for } i = 5 \quad \text{(d)}$$

Summarizing equations (c) and (d) provide five equations for the five unknown node temperatures T_i, $i = 1, 2, \ldots, 5$, since the temperature $T_0 = 100°C$ is known. The exact solution for this problem is given by

$$T(R) = T_a - \frac{gR^2}{6k} + \frac{C_1}{R} + C_2$$

where

$$C_1 = \frac{ab^2}{ak + bh_b(b-a)} \left\{ h_b(T_a - T_\infty) - \frac{g}{3}\left[b + \frac{h_b}{2k}(b^2 - a^2)\right] \right\}$$

$$C_2 = \frac{ga^2}{6k} - \frac{C_1}{a}$$

and various quantities are defined previously.

Table 4.3 gives a comparison of the finite difference solution with the exact results. Finite difference calculations are performed by using the Gauss elimination. The results are in good agreement with the exact solution. Note that the node $i = 0$ corresponds to the inner boundary where the temperature is prescribed.

4.1.4 Heat Conduction through Fins

The problems of steady-state heat flow through fins or extended surfaces are typical examples of one-dimensional heat conduction where a partial

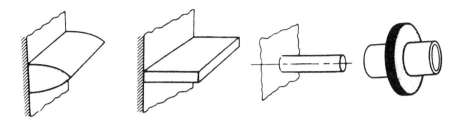

FIGURE 4.4
Typical fin profiles.

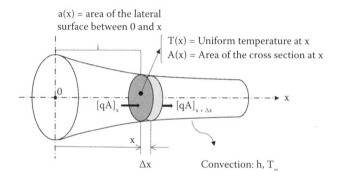

FIGURE 4.5
Nomenclature for the one-dimensional diffusive–convective system with no flow.

lumping formulation is used; that is, gradients are approximated by the boundary conditions in the fin cross section. Figure 4.4 shows typical fin profiles. Applications of fins include, among others, heat transfer in internal combustion engines, automobile radiators, boiler tubes, electrical transformers, electronic equipment cooling, and heat transfer enhancement.

To illustrate the type of governing differential equations associated with this problem, we consider a one-dimensional fin of variable cross section as illustrated in Figure 4.5. The energy equation governing the temperature distribution T(x) is given by Özişik (1977) as

$$\frac{d}{dx}\left[A(x)\frac{d\theta(x)}{dx}\right] - \left(\frac{h}{k}\right)\frac{da(x)}{dx}\theta(x) = 0 \qquad (4.24)$$

where

$A(x)$ = cross-section area normal to the x axis at the location x
$a(x)$ = lateral surface area between $x = 0$ and x
h = heat transfer coefficient
k = thermal conductivity of fin material
$\theta(x) = T(x) - T_\infty$
T_∞ = temperature of the ambient fluid

4.1.4.1 Fin of Uniform Cross Section

We consider a fin of length L and uniform cross section $A(x) = A$ = constant. The fin base at $x = 0$ is kept at constant temperature T_0, and heat is dissipated from its lateral surfaces by convection with a heat transfer coefficient h into an ambient at a constant temperature T_∞. The heat loss from the fin tip is negligible compared to that from the lateral surfaces. The mathematical formulation of this fin problem is given by

$$\frac{d^2\theta(x)}{dx^2} - m^2\theta(x) = 0 \quad \text{in } 0 < x < L \tag{4.25a}$$

$$\theta(x) = \theta_0 \quad \text{at } x = 0 \tag{4.25b}$$

$$\frac{d\theta(x)}{dx} = 0 \quad \text{at } x = L \tag{4.25c}$$

where the differential equation (4.25a) is obtained from equation (4.24) by proper simplifications, and various parameters are defined as

$$m^2 = \frac{Ph}{Ak}, \quad \theta(x) = T(x) - T_\infty, \quad \text{and} \quad \theta_0 = T_0 - T_\infty$$

where P is the perimeter. Exact analytic solution of this problem is given by

$$\frac{\theta(x)}{\theta_0} = \frac{\cosh[m(L-x)]}{\cosh(mL)} \tag{4.26}$$

and the heat flow rate Q through the fin is determined from

$$Q = -Ak \frac{d\theta(x)}{dx}\bigg|_{x=0} = Ak\, m\, \tanh(mL) \tag{4.27}$$

4.1.4.2 Finite Difference Solution

We apply a second-order accurate central differencing scheme to discretize both the differential equation and the boundary conditions of the system [equation (4.25)] and to obtain the following finite difference equations

$$\theta_{i-1} - [2 + (m\Delta x)^2]\theta_i + \theta_{i+1} = 0, \; i = 1,2,\ldots,(M-1) \quad (4.28a)$$

$$2\theta_{M-1} - [2 + (m\Delta x)^2]\theta_M = 0, \; i = M \quad (4.28b)$$

where the base temperature $\theta_0 = T_0 - T_\infty$ and the parameter $m^2 = (Ph/kA)$ are known. The system [equation (4.28)] provides M algebraic equations for the M unknown node temperatures θ_i, $i = 1,2,\ldots,M$.

To evaluate the heat flow rate Q through the fin, we need to discretize equation (4.27), that is

$$Q = -Ak \left. \frac{d\theta(x)}{dx} \right|_{x=0} \quad (4.29a)$$

For improved accuracy, we prefer to use a second-order accurate central-difference formula to discretize the derivative about the node at $x = 0$. The additional node needed for the central difference formula is obtained by assuming the region is extended by one grid length Δx to the left of the node 0 to obtain a fictitious node "-1" at a fictitious temperature T_{-1}. Then the derivative term in equation (4.29a) is discretized by using a second-order accurate central-difference formula. To eliminate the resulting fictitious temperature T_{-1}, an additional expression is obtained by evaluating the finite difference equation (4.28a) for $i = 0$. After eliminating T_{-1}, the finite difference form of equation (4.29a) is determined as

$$Q = \frac{Ak}{\Delta x}(\theta_0 - \theta_1) + \frac{Ph\Delta x}{2}\theta_0 \quad (4.29b)$$

where the first term on the right-hand side represents heat flow by conduction and the second term represents the heat flow from the lateral surface by convection. The latter term arises from the fact that a finite value of Δx is involved in finite differencing instead of $\Delta x \to 0$ used in the mathematical definition of a derivative. Clearly, if very small steps are taken by choosing very small Δx, the contribution of convection becomes negligible and equation (4.29b) approaches the usual definition of derivative.

Equation (4.29b) could also be developed by writing an energy balance equation for a differential volume element of thickness $\Delta x/2$ about the node 0 at the boundary surface $x = 0$.

If first-order accurate finite differencing were used to discretize the derivative in equation (4.29a), the convective term that appears in equation (4.29b) would not be present.

In the following example, we illustrate the finite difference solution of the problem given by equations (4.25a–c) and examine the accuracy of the prediction of heat flow rate Q through the fin.

Example 4.4

An iron rod of length L = 5 cm, diameter D = 2 cm, and thermal conductivity k = 50 W/(m°C) protruding from a wall is exposed to an ambient at T_∞ = 30°C. The heat transfer coefficient between the ambient and the rod surface is h = 100 W/(m²°C), and the base of the rod is kept at a constant temperature T_0 = 330°C. Assuming a one-dimensional steady-state heat flow, calculate the rate of heat loss from the rod into the ambient using finite differences. Compare the finite difference results with the exact analytic solution for the problem.

Solution

The mathematical formulation of this problem is the same as that for the fin problem given by equations (4.25a–c), that is,

$$\frac{d^2\theta(x)}{dx^2} - m^2\theta(x) = 0 \quad \text{in} \quad 0 < x < L$$

$$\theta(x) = \theta_0 \quad \text{at} \quad x = 0$$

$$\frac{d\theta(x)}{dx} = 0 \quad \text{at} \quad x = L$$

where $\theta(x) = T(x) - T_\infty$.

The finite difference approximation of this problem is given by equations (4.28a,b). We consider the rod being divided into five equal subregions. The resulting finite difference equations are obtained from equations (4.28a,b) by setting M = 5.

$$\theta_{i-1} - [2 + (m\Delta x)^2]\theta_i + \theta_{i+1} = 0 \quad \text{for } i = 1 \text{ to } 4$$

$$2\theta_4 - [2 + (m\Delta x)^2]\theta_5 = 0 \quad \text{for } i = 5$$

where

$$\theta_0 = 330 - 30 = 300°C, \quad \Delta x = \frac{L}{M} = \frac{0.05}{5} = 0.01 \text{m}$$

$$m^2 = \frac{Ph}{Ak} = \frac{\pi Dh}{\frac{\pi}{4}D^2 k} = \frac{4h}{Dk} = \frac{4 \times 100}{0.02 \times 50} = 400, \quad (m\Delta x)^2 = 0.04$$

The resulting finite difference equations become

$$i = 1: \; -2.04\theta_1 + \theta_2 = -300$$

$$i = 2: \; \theta_1 - 2.04\theta_2 + \theta_3 = 0$$

$$i = 3: \theta_2 - 2.04\theta_3 + \theta_4 = 0$$

$$i = 4: \theta_3 - 2.04\theta_4 + \theta_5 = 0$$

$$i = 5: 2\theta_4 - 2.04\theta_5 = 0$$

which are expressed in the matrix form as

$$\begin{bmatrix} -2.04 & 1.00 & 0.00 & 0.00 & 0.00 \\ 1.00 & -2.04 & 1.00 & 0.00 & 0.00 \\ 0.00 & 1.00 & -2.04 & 1.00 & 0.00 \\ 0.00 & 0.00 & 1.00 & -2.04 & 1.00 \\ 0.00 & 0.00 & 0.00 & 2.00 & -2.04 \end{bmatrix} \begin{bmatrix} \theta_1 \\ \theta_2 \\ \theta_3 \\ \theta_4 \\ \theta_5 \end{bmatrix} = \begin{bmatrix} -300 \\ 0 \\ 0 \\ 0 \\ 0 \end{bmatrix}$$

This system is solved in order to obtain the finite difference solution, which is in excellent agreement with the analytic solution [equation (4.26)], as shown in Table 4.4.

The heat flow rate through the fin is calculated by using equation (4.29b) as

$$Q = \frac{Ak}{\Delta x}(\theta_0 - \theta_1) + \frac{Ph\Delta x}{2}\theta_0$$

where $\theta_0 = 300$, $\theta_1 = 260.1$, $\dfrac{Ak}{\Delta x} = \left(\dfrac{\pi}{4} \times 0.02^2\right)\left(\dfrac{50}{0.01}\right) = \dfrac{\pi}{2}$

$$\frac{Ph\Delta x}{2} = (\pi \times 0.02)\frac{100 \times 0.01}{2} = 0.0314$$

The heat transfer rate is found to be

$$Q = 1.571\,(300 - 260.1) + (0.0314)(300)$$

$$Q = 62.68 + 9.42 = 72.1 \text{ W}$$

which is very close to the exact result $Q = 71.76$ W.

TABLE 4.4

Comparison of Results for Example 4.4

x/L	θ_i	Exact	Finite Difference
0.2	θ_1	260.0	260.1
0.4	θ_2	230.5	230.6
0.6	θ_3	210.2	210.4
0.8	θ_4	198.3	198.6
1.0	θ_5	194.4	194.7

Note that, if the first-order accurate formula was used to discretize the derivative in equation (4.29a), the convective term would not be included in the calculation of the heat flow rate Q. Then for such a case, the finite difference solution using five subdivisions of the region would involve about 13% error in the prediction of the heat flow rate.

4.2 Diffusive–Advective Systems

Consider a radially averaged temperature of a fluid flowing inside a duct in which streamwise diffusion cannot be neglected in comparison to convection. Thus, the problem of heat transfer involves a mainstream convection as well as diffusion along the flow direction. The solution of the finite difference equations for flow problems of this type requires special considerations because the presence of flow may give rise to numerical instability in the solution of difference equations under certain conditions. To illustrate the implications of this matter, we consider an incompressible flow with constant properties, inside a circular tube of radius R, length L, with a uniform velocity u in the positive x-direction. We assume that the radial variation of fluid temperature is negligible; hence, the temperature field is one-dimensional, that is, T(x). The boundaries at the inlet and outlet are maintained at constant temperatures T_0 (at x = 0) and T_L (at x = L), respectively, while the tube wall is kept at temperature T_0. There is heat transfer with a coefficient h between the fluid and the tube wall. Figure 4.6 illustrates the geometry and the coordinates. By writing an energy balance over a differential control volume of thickness dx and radius R, the one-dimensional energy equation is derived as

$$\frac{d^2T}{dx^2} - \frac{u}{\alpha}\frac{dT(x)}{dx} - \frac{2h}{kR}(T - T_0) = 0 \qquad (4.30)$$

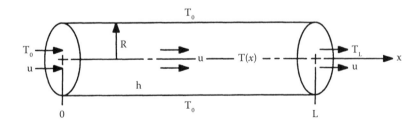

FIGURE 4.6
Geometry and coordinates for one-dimensional steady flow.

One-Dimensional Steady-State Systems

which is written in the form

$$\frac{d^2\theta(x)}{dx^2} - \frac{u}{\alpha}\frac{d\theta(x)}{dx} - \frac{2h}{kR}\theta(x) = 0 \qquad (4.31a)$$

where

$$\theta(x) = \frac{T(x) - T_0}{T_L - T_0} \qquad (4.31b)$$

and α is the thermal diffusivity of the fluid.

The one-dimensional diffusive–advective energy equation (4.31a) is different from the energy equation (4.25a) because the former contains the first derivative of temperature resulting from the flow.

For finite difference approximation of equation (4.31a), the second-order accurate central-difference formula is customarily used to discretize the second derivative of temperature. The use of the central difference formula may look attractive to discretize the first derivative of temperature because it is also second-order accurate. However, it will be shown that the use of the central difference formula to discretize the first derivative of temperature in equation (4.31a) may give rise to numerical instability in the solution of the resulting finite difference equations under certain situations. To alleviate such an instability resulting from the central differencing, the *upwind difference* scheme can be used to discretize the first derivative term. We present the following finite difference representations of equation (4.31a) with both methods and examine the numerical stability of the resulting finite difference equations. Afterward, we present a finite volume method to solve this problem.

Use of Central Differencing: Equation (4.31a) is discretized by using central differencing for both the second and the first derivative terms. We obtain

$$\frac{\theta_{i-1} - 2\theta_i + \theta_{i+1}}{(\Delta x)^2} - \frac{u}{\alpha}\frac{\theta_{i+1} - \theta_{i-1}}{2\Delta x} - \frac{2h}{kR}\theta_i = 0 \qquad (4.32)$$

which is rearranged in the form

$$\left(1 + \frac{Pe_{\Delta x}}{2}\right)\theta_{i-1} - 2\left(1 + \frac{h(\Delta x)^2}{kR}\right)\theta_i + \left(1 - \frac{Pe_{\Delta x}}{2}\right)\theta_{i+1} = 0 \qquad (4.33a)$$

where

$$Pe_{\Delta x} = \frac{u\Delta x}{\alpha} = \text{Mesh Peclet number} \qquad (4.33b)$$

Upwind or Upstream Differencing: We now discretize equation (4.31a) by using central differencing for the second derivative and *upwind differencing*, that is, backward differencing $(\theta_i - \theta_{i-1})/\Delta x$, for the first derivative when $u > 0$, to obtain

$$\frac{\theta_{i-1} - 2\theta_i + \theta_{i+1}}{(\Delta x)^2} - \frac{u}{\alpha}\frac{\theta_i - \theta_{i-1}}{\Delta x} - \frac{2h}{kR}\theta_i = 0 \tag{4.34}$$

which is rearranged as

$$(1 + Pe_{\Delta x})\theta_{i-1} - \left(2 + Pe_{\Delta x} + \frac{2h(\Delta x)^2}{kR}\right)\theta_i + \theta_{i+1} = 0 \tag{4.35a}$$

where

$$Pe_{\Delta x} = \frac{u\Delta x}{\alpha} = \text{Mesh Peclet number} \tag{4.35b}$$

The problem of instability associated with the solution of the finite difference equations (4.33a) and (4.35a) is now discussed.

4.2.1 Stability for Steady-State Systems

The finite difference equations (4.33a) and (4.35a) can be written formally as

$$A\theta_{i-1} + B\theta_i + C\theta_{i+1} = 0 \tag{4.36}$$

The numerical stability of the solution of algebraic equations of this form has been studied by Shih (1984), and his results can be summarized as follows.

Let equation (4.36) be arranged such that

$$B < 0 \tag{4.37}$$

then the cases

$$A > 0, C < 0 \quad \text{and} \quad A < 0, C < 0 \tag{4.38}$$

lead to physically unrealistic situations. The condition that will produce physically meaningful solutions is given by

$$A > 0, C > 0 \quad \text{for } B < 0 \tag{4.39}$$

We now apply the criterion given by equation (4.39) to examine the numerical stability of the solution of the finite difference equations (4.33a) and (4.35a).
The nodal coefficients for equation (4.33a) are given by

$$A = 1 + \frac{Pe_{\Delta x}}{2} > 0 \tag{4.40a}$$

$$B = -2\left(1 + \frac{h(\Delta x)^2}{kR}\right) < 0 \tag{4.40b}$$

One-Dimensional Steady-State Systems

$$C = 1 - \frac{Pe_{\Delta x}}{2} > 0 \quad \text{if } Pe_{\Delta x} < 2 \tag{4.40c}$$

Clearly, the conditions [equations (4.40a,b)] are always satisfied because $Pe_{\Delta x}$ and other parameters are all positive quantities; but the criterion [equation (4.40c)] is satisfied if

$$Pe_{\Delta x} = \frac{u\Delta x}{\alpha} < 2 \tag{4.41}$$

Therefore, the solution of the finite difference equation (4.33a) is numerically stable only for $Pe_{\Delta x} < 2$. This restriction implies that for a given u and α, the step size Δx should be chosen small enough to satisfy the stability criterion given by equation (4.41). This indicates that the presence of flow gives rise to instability in the solution of finite difference equations. The factor 2 in equation (4.41) becomes 4 for the two-dimensional case and 6 for the three-dimensional case.

Now we examine the nodal coefficients for the upwind differencing given by equation (4.35a). We have

$$A = 1 + Pe_{\Delta x} \tag{4.42a}$$

$$B = -\left(2 + Pe_{\Delta x} + \frac{2h(\Delta x)^2}{kR}\right) \tag{4.42b}$$

$$C = 1 \tag{4.42c}$$

Clearly, all the conditions defined by equation (4.39) are satisfied; hence, the difference equation (4.35a) is unconditionally stable. But it is less accurate than equation (4.33a) because the upwind-differencing scheme is only first-order accurate.

To demonstrate the effects of flow velocity on stability, we consider the situation with negligible flow velocity and set $Pe_{\Delta x} = 0$. Then the finite difference equation (4.33a) reduces to

$$\theta_{i-1} - 2\left(1 + \frac{h(\Delta x)^2}{kR}\right)\theta_i + \theta_{i+1} = 0 \tag{4.43}$$

which satisfies all the conditions given by equation (4.39) and hence is unconditionally stable.

4.2.2 Finite Volume Method

In this section, we will briefly introduce the finite volume method for solving diffusive–convective problems. The method will be more detailed when applied to the transient Navier–Stokes equations in Chapter 7.

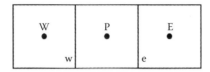

FIGURE 4.7
Control volume for P and its neighbors.

The basic idea is to integrate equation (4.31a) in the control volume around a point P represented in Figure 4.7. In this figure, the capital letters W and E represent the center of the control volume's neighbors to P in the west and east directions, respectively. This figure also shows, in lowercase letters, the interfaces w and e between P and the west and east control volumes, respectively.

Let us rewrite equation (4.31a) as

$$\frac{d^2\theta(x)}{dx^2} - \frac{\rho u C_P}{k}\frac{d\theta(x)}{dx} - \frac{2h}{kR}\theta(x) = 0 \qquad (4.44a)$$

because $\alpha = k/(\rho\, C_p)$. We can also write

$$\frac{k}{C_P}\frac{d^2\theta(x)}{dx^2} - \rho u \frac{d\theta(x)}{dx} - \frac{1}{C_P}\frac{2h}{R}\theta(x) = 0 \qquad (4.44b)$$

Now, considering constant ρ, u, k, and Cp, we have

$$\frac{d}{dx}\left(\frac{k}{C_P}\frac{d\theta}{dx}\right) - \frac{d}{dx}(\rho u \theta) - \frac{2h}{C_P R}\theta = 0 \qquad (4.44c)$$

where, for the sake of simplicity, we dropped the dependence of θ with x. Integrating equation (4.44c) in the control volume around P, we obtain

$$\int_w^e \left[\frac{d}{dx}\left(\frac{k}{C_P}\frac{d\theta}{dx}\right) - \frac{d}{dx}(\rho u \theta) - \frac{2h}{C_P R}\theta\right]dx = 0 \qquad (4.45)$$

Considering the integrand of the third term of this equation as a representative average value within the control volume, the following equation can be obtained

$$\left[\frac{k}{C_P}\frac{d\theta}{dx}\right]_w^e - [\rho u \theta]_w^e - \frac{2h}{C_P R}\theta|_P \Delta x = 0 \qquad (4.46)$$

We can now define the following variables, which represent the mass flow rates per unit length through the east and west boundaries of the control volume P (in this section, the velocity is constant, but in Chapter 7 we will consider a nonuniform velocity u).

$$\dot{M}_e = [\rho u]_e \qquad (4.47a)$$

$$\dot{M}_w = [\rho u]_w \qquad (4.47b)$$

We can also define the following diffusive coefficient (again, in this section, this coefficient is constant, but in Chapter 7 we will consider nonuniform properties)

$$D_{11} = \frac{k}{C_P} \qquad (4.48)$$

Replacing equations (4.47) and (4.48) into equation (4.46), we obtain:

$$\left[D_{11}\frac{d\theta}{dx}\right]_e - \left[D_{11}\frac{d\theta}{dx}\right]_w - [\dot{M}\theta]_e + [\dot{M}\theta]_w - \frac{2h}{C_P R}\theta\bigg|_P \Delta x = 0 \qquad (4.49)$$

4.2.2.1 Interpolation Functions

From equation (4.49), it is possible to verify that the temperature θ, as well as its derivatives, must be calculated at the boundaries of the control volume (locations indicated by lowercase letters w and e). Thus, some interpolation functions are necessary because these quantities are only known at the centers of the control volumes.

The correct choice of the interpolation functions is one of the most critical steps when solving the conservation equations. One of the first attempts to approximate the derivatives at the boundaries of the control volumes was by using the finite difference central differencing scheme. However, such approximation is only valid for low Reynolds (and Peclet) numbers, where diffusion is more important than advection. When the Peclet number increases, this approach presents some numerical oscillations (Patankar and Spalding 1970), as seen earlier. Another approach consists of a first-order upwind scheme. The upwind scheme was originally proposed by Courant et al. (1952). In this approach, the diffusive term is approximated by a central finite difference scheme, while the convective term is calculated assuming that the value of u at the interface is equal to its value at the center of either neighboring volumes, depending on the sign of the velocity. Thus, for example, in the x-direction,

$$u_e = u_P \quad \text{if } u > 0 \qquad (4.50a)$$

$$u_e = u_E \quad \text{if } u_e < 0 \qquad (4.50b)$$

where the subindices refer to the points shown in Figure 4.7.

Spalding (1972) and Patankar and Spalding (1970) proposed a hybrid scheme, which contemplates convective–diffusive and convective-dominant problems. Such schemes use the central finite difference scheme to grid Peclet numbers between −2 and 2, and the upwind scheme elsewhere. Thus, it eliminates the oscillatory behavior of the solution presented by the central scheme. Leonard and Mokhtari (1990) showed that such schemes simulate nonphysical conditions for highly convective problems. Patankar (1980), based on the exponential scheme (Spalding 1972), developed the power law scheme, which is able to obtain good results for grid Peclet numbers up to six. For $Pe_{\Delta x}$ greater than six, the power law scheme tends toward the first-order upwind scheme. The problem with first-order schemes is the numerical viscosity or diffusivity inherent in them, which can result in highly inaccurate results, depending on the range of grid Peclet numbers.

One of the hybrid methods widely employed was developed by Raithby and Torrance (1974) and uses weighted functions to evaluate the diffusive and convective terms. This method is known as the weighted upstream differencing scheme (WUDS). Raithby (1976) also developed an extension of the WUDS, known as the skew WUDS (SWUDS), which extends the applicability of the WUDS to grids where the velocity vector is no longer orthogonal to the grid lines, such as in the vicinity of recirculation zones. Some other studies also tried to extend the WUDS scheme to more general situations.

In problems with high grid Peclet numbers and, especially for three-dimensional problems, where the components of the grid Peclet number (for laminar and turbulent cases) are usually greater than two (Leonard and Mokhtari 1990), there are more robust high-order upwind schemes available. Leonard (1979) introduced two schemes based on quadratic interpolation: the first one, quadratic upstream interpolation for convective kinematics (QUICK), was developed for steady-state problems, while the second one, QUICK with estimated streaming terms (QUICKEST), was used mainly for transient problems. The QUICKEST scheme is explicit, third-order, upwind, and with a dissipative fourth-order error. Third-order upwind schemes can be susceptible to nonphysical oscillations (Gerges and McCorquodale 1997). Although some oscillations are small when compared to other schemes, they can be amplified when solving nonlinear momentum equations. These oscillations, however, can be eliminated with a commonly used technique known as the universal limiter (Leonard 1991; Leonard 1997; Leonard and Mokhtari 1990; Leonard and Niknafs 1991; Park and Kwon 1996; Piperno and Depeyre 1996). Using the concepts of the QUICKEST scheme, Leonard et al. (1995) introduced a third-order upwind scheme for multidimensional problems, uniformly third-order polynomial interpolation algorithm (UTOPIA). Later, Gerges, and McCorquodale (1997) introduced a new third-order upwind scheme that extended the UTOPIA algorithm for cases where the velocity field was no longer orthogonal to the grid lines. This method is known as the skew third-order upwinding scheme (STOUS). Both methods (STOUS and UTOPIA) suffer from small oscillations that can be eliminated using the universal limiter technique.

A more complete review about third-order upwind schemes can be found in Leonard (1997), and a comparison among high-order upwind methods and central finite difference schemes can be found in Tafti (1996). In this book, we will present the WUDS scheme, which is valid for low-to-moderate velocity fluid flows.

Consider again the one-dimensional diffusive–advective problem in the x-direction given by equation (4.49). In the WUDS scheme, the interpolation function is associated with two coefficients α and β that depend on the grid Peclet number and are used as weight functions for the convection and diffusion terms. The value of θ and its derivative at the interface of the control volume are written, using the east face as an example, as (Raithby and Torrance 1974)

$$\theta_e = \left(\frac{1}{2} + \alpha_e\right)\theta_P + \left(\frac{1}{2} - \alpha_e\right)\theta_E \qquad (4.51a)$$

$$\left.\frac{\partial \theta}{\partial x}\right|_e = \beta_e\left(\frac{\theta_E - \theta_P}{\Delta x}\right) \qquad (4.51b)$$

The previous equations can reduce to the

- Central finite difference scheme for $\alpha_e = 0$ and $\beta_e = 1$
- Upwind scheme with $u_e \gg 0$ for $\alpha_e = 0.5$ and $\beta_e = 0$
- Upwind scheme with $u_e \ll 0$ for $\alpha_e = -0.5$ and $\beta_e = 0$

Raithby and Schneider (1988) proposed the following expressions for these two coefficients

$$\alpha = \frac{\text{Pr}_{\Delta x}^2}{10 + 2\text{Pr}_{\Delta x}^2}\text{sign}(\text{Pr}_{\Delta x}) \qquad (4.52a)$$

$$\beta = \frac{1 + 0.005\text{Pr}_{\Delta x}^2}{1 + 0.05\text{Pr}_{\Delta x}^2} \qquad (4.52b)$$

In equation (4.52a), $\text{sign}(\text{Pr}_{\Delta x})$ is the sign function of $\text{Pr}_{\Delta x}$, which is positive for $\text{Pr}_{\Delta x} > 0$ and negative for $\text{Pr}_{\Delta x} < 0$.

Applying equations (4.51a,b) and the corresponding equation for the west interface to equation (4.49), we obtain

$$\dot{M}_e\left[\left(\frac{1}{2}+\alpha_e\right)\theta_P+\left(\frac{1}{2}-\alpha_e\right)\theta_E\right]-\dot{M}_w\left[\left(\frac{1}{2}+\alpha_w\right)\theta_P+\left(\frac{1}{2}-\alpha_w\right)\theta_W\right]$$
$$=D_{11e}\beta_e\left(\frac{\theta_E-\theta_P}{\Delta x}\right)-D_{11w}\beta_w\left(\frac{\theta_P-\theta_W}{\Delta x}\right)-\frac{2h}{C_pR}\theta|_P\Delta x \quad (4.53)$$

or, defining the following coefficients

$$A_e = \dot{M}_e\left(\frac{1}{2}-\alpha_e\right)-\frac{D_{11e}\beta_e}{\Delta x} \quad (4.54)$$

$$A_w = -\dot{M}_w\left(\frac{1}{2}-\alpha_w\right)-\frac{D_{11w}\beta_w}{\Delta x} \quad (4.55)$$

$$A_P = \dot{M}_e\left(\frac{1}{2}+\alpha_e\right)-\dot{M}_w\left(\frac{1}{2}+\alpha_w\right)+D_{11e}\frac{\beta_e}{\Delta x}+D_{11w}\frac{\beta_w}{\Delta x}+\frac{2h}{C_pR}\Delta x \quad (4.56)$$

we can write

$$A_P\theta_P + A_e\theta_E + A_w\theta_W = 0 \quad (4.57)$$

Equation (4.57), written for all volumes of the domain, with proper boundary conditions, can be solved by the techniques presented in Chapter 3.

PROBLEMS

4.1. Consider the following steady-state heat-conduction problem for a solid cylinder

$$\frac{d^2T}{dR^2}+\frac{1}{R}\frac{dT}{dR}+\frac{1}{k}g(R)=0,\ 0<R<1$$

$$\frac{dT}{dR}=0 \text{ at } R=0, \text{ and } T=0 \text{ at } R=1$$

Develop the finite difference equations for this heat conduction problem by dividing the region $0 \le R \le 1$ into 10 equal parts. Use the finite difference formulae presented in Chapter 2, as well as the control volume approach, for the discretization.

4.2. Consider the following steady-state heat conduction problem for a solid sphere

$$\frac{d^2T}{dR^2} + \frac{2}{R}\frac{dT}{dR} + \frac{1}{k}g(R) = 0, \quad 0 < R < 1$$

$$\frac{dT}{dR} = 0 \quad \text{at } R = 0, \quad \text{and } T = 0 \quad \text{at } R = 1$$

Develop the finite difference equations for this heat conduction problem by dividing the region $0 \leq R \leq 1$ into 10 equal parts. Use the finite difference formulae presented in Chapter 2, as well as the control volume approach, for the discretization.

4.3. Consider the following steady-state heat conduction problem for a hollow cylinder

$$\frac{d^2T}{dR^2} + \frac{1}{R}\frac{dT}{dR} + \frac{1}{R}g(R) = 0, \quad 1 < R < 2$$

$$T = 0, \quad R = 1$$

$$\frac{dT}{dR} + HT = 0, \quad R = 2$$

Develop the finite difference equations for this heat conduction problem by dividing the region $1 < R < 2$ into five equal parts. Use the finite difference formulae presented in Chapter 2, as well as the control volume approach, for the discretization.

4.4. Repeat Problem 4.3 for a hollow sphere $1 < R < 2$.

4.5. Consider the following steady-state heat conduction problem for a hollow cylinder

$$\frac{d^2T}{dR^2} + \frac{1}{R}\frac{dT}{dR} + \frac{1}{k}g(R) = 0, \quad 1 < R < 2$$

$$-\frac{dT}{dR} + HT = 0, \quad R = 1$$

$$\frac{dT}{dR} = 0, \quad R = 2$$

Develop the finite difference equations for this heat conduction problem by dividing the region $1 < R < 2$ into five equal parts. Use the finite difference formulae presented in Chapter 2, as well as the control volume approach, for the discretization.

4.6. Consider the following steady-state heat conduction problem for a hollow sphere

$$\frac{d^2T}{dR^2} + \frac{2}{R}\frac{dT}{dR} + \frac{1}{k}g(R) = 0, \quad 1 < R < 2$$

$$\frac{dT}{dR} = 0, \quad R = 1$$

$$\frac{dT}{dR} + HT = 0, \quad R = 2$$

Develop the finite difference equations for this heat conduction problem by dividing the region $1 < R < 2$ into five equal parts. Use the finite difference formulae presented in Chapter 2, as well as the control volume approach, for the discretization.

4.7. Consider the following steady-state heat conduction problem for a hollow cylinder

$$\frac{d^2T}{dR^2} + \frac{1}{R}\frac{dT}{dR} + \frac{1}{k}g(R) = 0, \quad 1 < R < 2$$

$$-\frac{dT}{dR} + H_1 T = 0, \quad R = 1$$

$$\frac{dT}{dR} + H_2 T = 0, \quad R = 2$$

Develop finite difference form of this heat conduction problem by dividing the region $1 \le R \le 2$ into 10 equal parts and using second-order accurate finite differencing for the boundary conditions.

4.8. Repeat Problem 4.7 for a hollow sphere.

4.9. A 0.5-cm-diameter solid-copper sphere of thermal conductivity k = 380 W/(m·°C) has energy generation at a rate of $g_0 = 10^8$ W/m^3. The sphere dissipates heat from its outer surface by convection with a heat transfer coefficient h = 150 W/(m^2·°C) into an ambient at T_∞ = 25°C. Calculate the radial distribution of steady-state temperature by finite differences by dividing the region into five elements, each of radial thickness ΔR = 0.05 cm.

4.10. (a) Consider one-dimensional, steady-state, radial heat conduction in a rod of radius b = 2 cm, thermal conductivity k = 350 W/(m·°C), in which energy is generated at a rate of g = 1.4 × 10^9 W/m^3. The boundary surface at R = b is kept at zero temperature. Dividing the region $0 \le R \le b$ into five equal parts, calculate the radial temperature distribution in the rod by finite differences. Compare the finite difference calculations with the exact solution. (b) Repeat the solution by dividing the region $0 \le R \le b$ into 10 equal parts.

4.11. A 0.5-cm-diameter long copper rod of thermal conductivity k = 380 W/(m·°C) has heat generation at a rate of $g_0 = 10^8$ W/m³ as a result of the passage of electric current. The rod dissipates heat from its outer surface by convection with a heat transfer coefficient h = 150 W/(m²·°C) into an ambient at $T_\infty = 25°C$. The axial variation of temperature is neglected. Calculate the radial distribution of steady-state temperature by finite differences by dividing the region into five circular elements, each of radial thickness $\Delta R = 0.05$ cm.

4.12. (a) Consider one-dimensional, steady-state, radial heat conduction in a solid sphere of radius b = 2 cm, thermal conductivity k = 350 W/(m·°C), in which energy is generated at a rate of $g = 1.4 \times 10^9$ W/m³. The boundary surface at R = b is kept at zero temperature. By dividing the region $0 \leq R \leq b$ into five equal parts, calculate the radial temperature distribution in the sphere by finite differences. Compare the finite difference calculations with the exact solution. (b) Repeat the solution by dividing the region $0 \leq R \leq b$ into 10 equal parts.

4.13. A 12-cm-diameter solid steel bar of thermal conductivity k = 60 W/m°C is heated electrically by the passage of electric current that generates energy at a rate of $g = 10^6$ W/m³. Heat is dissipated from the outer surface of the rod by convection with a heat transfer coefficient h = 500 W/m²°C into an ambient at $T_\infty = 25°C$. By dividing the region $0 \leq R \leq 6$ cm into six equal parts, develop the finite difference form of this heat conduction problem. Compare the finite difference solution with the exact solution for the problem by using the first-order and the second-order accurate finite differencing for the boundary condition.

4.14. Repeat Problem 4.13 for a solid sphere of diameter D = 12 cm.

4.15. Consider the following steady-state heat conduction problem in a plate of thickness L

$$\frac{d^2T(x)}{dx^2} + \frac{1}{k}g = 0 \quad \text{in} \quad 0 < x < L$$

$$-k\frac{dT}{dx} = q_0 \quad \text{at} \quad x = 0$$

$$T = 0 \quad \text{at} \quad x = 1$$

where k = 12 W/m°C, $q_0 = 10^5$ W/m², $g = 4 \times 10^7$ W/m³, and L = 0.012 m. By dividing the region into four equal parts, write the finite difference equations and solve the resulting system of algebraic equations by Thomas algorithm.

5
One-Dimensional Transient Systems

In the preceding chapters, we considered the finite difference representation and solution of one-dimensional steady-state systems. In this chapter, we focus attention on finite difference formulation, solution, and stability considerations of one-dimensional transient systems.

The transient problems have numerous important applications in various branches of science and engineering. Almost all industrial processes experience transients during various stages of operation. For example, thermal transients within a body are generally initiated by sudden variation of the boundary condition or the energy generation in the medium. The start-up or shutdown of nuclear reactors, ovens, furnaces, and so on, are typical examples of the activation of transients as a result of variations in the energy generation rate. The cooling of a hot solid suddenly exposed to a cold ambient is a typical example of transients resulting from a change in boundary conditions.

In this chapter, we will discuss problems dealing with pure diffusion, pure advection, and advection plus diffusion, with parabolic and hyperbolic behaviors.

5.1 Diffusive Systems

Here, for pedagogical purposes, we first focus attention on the one-dimensional heat or mass diffusion problems governed by the linear parabolic differential equation

$$\frac{\partial T(x,t)}{\partial t} = \alpha \frac{\partial^2 T(x,t)}{\partial x^2} \quad (5.1)$$

subjected to appropriate linear boundary conditions, in order to demonstrate the use of various finite difference schemes and the general solution algorithms. For example, 13 different schemes are listed by Richtmyer and Morton (1967) for finite difference representation of the aforementioned equation.

The problems of cylindrical and spherical symmetry are considered separately in order to illustrate the implications of finite difference representation of the differential equation and the boundary conditions for such configurations.

5.1.1 Simple Explicit Method

For simplicity in the analysis, we consider a diffusion problem in a finite region $0 \leq x \leq L$ given in the form

$$\frac{\partial T(x,t)}{\partial t} = \alpha \frac{\partial^2 T(x,t)}{\partial x^2}, \quad 0 < x < L, \quad t > 0 \tag{5.2}$$

The region $0 \leq x \leq L$ is divided into M equal parts of mesh size

$$\Delta x = \frac{L}{M} \tag{5.3}$$

and the differential equation (5.2) is discretized by using the second-order accurate central difference formula for the spatial second derivative and the first-order accurate forward differencing formula for the time derivative to yield

$$\frac{T_i^{n+1} - T_i^n}{\Delta t} = \alpha \frac{T_{i-1}^n - 2T_i^n + T_{i+1}^n}{(\Delta x)^2} + 0[\Delta t, (\Delta x)^2] \tag{5.4a}$$

where

$$T(x,t) = T(i\Delta x, n\Delta t) \equiv T_i^n \tag{5.4b}$$

and, in the explicit method, the spatial derivative was discretized with the values of the function T at time $t_n = n\Delta t$. Equation (5.4a) is rearranged as

$$T_i^{n+1} = rT_{i-1}^n + (1-2r)T_i^n + rT_{i+1}^n \tag{5.5}$$

where

$$r = \frac{\alpha \Delta t}{(\Delta x)^2} \tag{5.6}$$

$n = 0,1,2,\ldots$ and $i = 1,2,\ldots,M-1$, and the truncation error is of order $0[\Delta t,(\Delta x)^2]$.

Equation (5.5) is called the simple explicit form of finite difference approximation of the diffusion equation (5.2) because it involves only one unknown T_i^{n+1} for the time level $n+1$, which can be directly calculated from equation (5.5) when the potentials T_{i-1}^n, T_i^n, and T_{i+1}^n at the previous time level n are available.

Figure 5.1 schematically illustrates the finite difference molecules associated with the simple explicit scheme applied to the simple diffusion equation. Clearly, the system [equation (5.5)], for $i = 1,2,\ldots,M-1$ provides M−1 algebraic relations but contains M + 1 unknown node potentials T_i^{n+1} ($i = 0,1,2,\ldots,M$). Two additional relations needed to make the number of equations equal to the number of unknowns are obtained from the two boundary conditions at $i = 0$ and $i = M$. If the boundary potentials are prescribed, then the

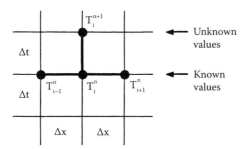

FIGURE 5.1
The finite difference molecules for the simple explicit scheme.

number of equations is equal to the number of unknowns. However, for convection or prescribed heat flux boundary conditions, the boundary potentials are unknown. In such situations, two additional relations are obtained by discretizing the boundary conditions. Next, we describe the solution algorithm and the discretization of the boundary conditions. The reader is also referred to Chapter 2 for details on the discretization of the boundary conditions.

5.1.1.1 Prescribed Potential at the Boundaries

Suppose the potentials (i.e., temperatures) are prescribed at the boundaries $i = 0$ and $i = M$. We have

$$T_0^n = T_a = \text{known} \tag{5.7a}$$

$$T_M^n = T_b = \text{known} \tag{5.7b}$$

Then the system of equations (5.5) provides M−1 explicit relations for the determination of M−1 unknown internal node potentials T_i^{n+1}, $i = 1,2,...,M-1$ because the boundary potentials T_0^n and T_M^n are known for all times. The computational algorithm is as follows:

1. Start the calculations with n = 0. Compute the T_i^1, $i = 1,2,...,M-1$ at the end of the first time step from equation (5.5) because the right-hand side of this equation is known from the initial condition.
2. Set n = 1 and calculate T_i^2, $i = 1,2,...,M-1$ at the end of the second time step from equation (5.5) because the right-hand side of this equation is known from the previous time step.
3. Repeat the procedure for each subsequent time step and continue calculations until a specified time or some specified value of the potential is reached.

5.1.1.2 Convection Boundary Conditions

Consider the boundary surfaces at $x = 0$ and $x = L$ subjected to convection with heat transfer coefficients h_0 and h_L into ambients at temperatures $T_{\infty,0}$ and $T_{\infty,L}$, respectively. We have

$$-k\frac{\partial T}{\partial x} + h_0 T = h_0 T_{\infty,0} = \text{known} \quad \text{at } x = 0 \tag{5.8a}$$

$$k\frac{\partial T}{\partial x} + h_L T = h_L T_{\infty,L} = \text{known} \quad \text{at } x = L \tag{5.8b}$$

where the temperatures at the boundary nodes $i = 0$ and $i = M$ are unknown. Two additional relations are obtained by discretizing these two boundary conditions.

A very simple approach to discretize these boundary conditions is to use forward differencing for equation (5.8a) and backward differencing for equation (5.8b); but the results are only first-order accurate, that is, $0(\Delta x)$. A second-order accurate, that is, $0[(\Delta x)^2]$, differencing of these boundary conditions is possible if central differencing is used to discretize the first derivatives in these boundary conditions. To apply the central differencing, we consider fictitious nodes "-1" at a fictitious temperature T^n_{-1} and "$M + 1$" at a fictitious temperature T^n_{M+1} obtained by extending the region by Δx to the left and right, respectively, as illustrated in Figure 5.2.

Utilizing these fictitious nodes, the central differencing is used to discretize the boundary conditions [equation (5.8)].

$$-k\frac{T^n_1 - T^n_{-1}}{2\Delta x} + h_0 T^n_0 = h_0 T_{\infty,0}, \quad i = 0 \tag{5.9a}$$

$$k\frac{T^n_{M+1} - T^n_{M-1}}{2\Delta x} + h_L T^n_M = h_L T_{\infty,L}, \quad i = M \tag{5.9b}$$

where T^n_{-1} and T^n_{M+1} are the fictitious temperatures at the fictitious nodes "-1" and "$M + 1$". Two additional relations needed to eliminate these fictitious

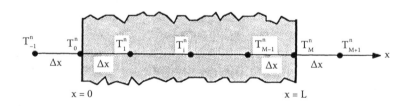

FIGURE 5.2
Fictitious nodes -1 and $M + 1$ at fictitious temperatures T^n_{-1} and T^n_{M+1}.

temperatures are determined by evaluating equation (5.5) for $i = 0$ and $i = M$; we, respectively, obtain

$$T_0^{n+1} = rT_{-1}^n + (1 - 2r)T_0^n + rT_1^n \quad \text{for} \quad i = 0 \tag{5.10a}$$

$$T_M^{n+1} = rT_{M-1}^n + (1 - 2r)T_M^n + rT_{M+1}^n \quad \text{for} \quad i = M \tag{5.10b}$$

Now, T_{-1}^n is eliminated between equations (5.9a) and (5.10a), while T_{M+1}^n is eliminated between equations (5.9b) and (5.10b). We obtain

$$T_0^{n+1} = (1 - 2r\beta_0)T_0^n + 2rT_1^n + 2r\gamma_0 \quad \text{for} \quad i = 0 \tag{5.11a}$$

$$T_M^{n+1} = 2rT_{M-1}^n + (1 - 2r\beta_L)T_M^n + 2r\gamma_L \quad \text{for} \quad i = M \tag{5.11b}$$

where

$$\beta_0 = 1 + \frac{\Delta x\, h_0}{k}, \quad \gamma_0 = \frac{\Delta x\, h_0}{k} T_{\infty,0} \tag{5.12a}$$

$$\beta_L = 1 + \frac{\Delta x\, h_L}{k}, \quad \gamma_L = \frac{\Delta x\, h_L}{k} T_{\infty,L} \tag{5.12b}$$

$$r = \frac{\alpha \Delta t}{(\Delta x)^2} \tag{5.12c}$$

Equations (5.11) and (5.12) are second-order accurate finite difference approximation of the convection boundary conditions given by equations (5.8a) and (5.8b), respectively. The finite difference equations (5.5) together with equations (5.11a) and (5.11b) provide $M + 1$ expressions for the determination of $M + 1$ unknown node temperatures at each time step.

It is to be noted that the finite difference representations given by equations (5.11a) and (5.11b) can also be developed by writing an energy balance for a control volume of thickness $\Delta x/2$ adjacent to the boundary nodes $i = 0$ and $i = M$, respectively.

5.1.1.3 Prescribed Flux Boundary Condition

We now examine the finite difference representation of prescribed heat flux boundary conditions given in the form

$$-k\frac{\partial T}{\partial x} = q_0 = \text{known} \quad \text{at} \quad x = 0 \tag{5.13a}$$

$$k\frac{\partial T}{\partial x} = q_L = \text{known} \quad \text{at} \quad x = L \tag{5.13b}$$

where q_0 and q_L are prescribed heat fluxes (i.e., W/m^2) applied at the boundary surfaces $x = 0$ and $x = L$, respectively.

A comparison of these boundary conditions with the convection boundary conditions [equation (5.8)] reveals that the second-order accurate finite difference approximations of the prescribed heat flux boundary conditions [equation (5.13)] are immediately obtained from equations (5.11a,b) by making

$$\beta_0 = \beta_L = 1, \quad \gamma_0 = \frac{\Delta x}{k} q_0, \quad \gamma_L = \frac{\Delta x}{k} q_L \qquad (5.14\text{a-c})$$

as

$$T_0^{n+1} = (1 - 2r)T_0^n + 2r\, T_1^n + 2r \frac{\Delta x q_0}{k} \quad \text{for} \quad i = 0 \qquad (5.15\text{a})$$

$$T_M^{n+1} = 2r\, T_{M-1}^n + (1 - 2r)T_M^n + 2r \frac{\Delta x q_L}{k} \quad \text{for} \quad i = M \qquad (5.15\text{b})$$

5.1.1.4 Stability Considerations

If the solution of the finite difference equations (5.5) should remain stable (i.e., nondivergent or nonoscillatory), the value of the parameter r, for use in these equations, should be restricted to

$$0 < r \leq \frac{1}{2} \qquad (5.16\text{a})$$

where

$$r = \frac{\alpha \Delta t}{(\Delta x)^2} \qquad (5.16\text{b})$$

This stability criterion implies that, for given values of α and Δx, the magnitude of the time step Δt cannot exceed the limit imposed by equation (5.16a). Figure 5.3 illustrates the effects of the value of r on the stability of finite difference solution with the *explicit* method. Clearly, since the stability criterion is violated for $r = (5/9) > (1/2)$, the solution begins to oscillate and diverge, while the results obtained with $r = (5/11) < (1/2)$ are stable and in good agreement with the exact solution for the problem.

The physical significance of the restriction on the maximum value of r can be illustrated with the following physical argument.

Suppose at any time level n, the temperatures T_{i-1}^n and T_{i+1}^n at the nodes $i-1$ and $i+1$ are equal. Equation (5.5), for this specific case, can be rearranged as

$$T_i^{n+1} = T_{i-1}^n + (1 - 2r)(T_i^n - T_{i-1}^n) \quad \text{for} \quad T_{i-1}^n = T_{i+1}^n \qquad (5.17\text{a})$$

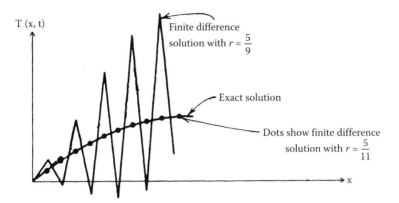

FIGURE 5.3
Effect of the parameter r on the stability of solution in the explicit method.

For illustration purposes, let $T_{i-1}^n = T_{i+1}^n = 0°C$ and $T_i^n = 100°C$. Equation (5.17a) is now used to calculate the temperature T_i^{n+1} of the node i at the next time level n + 1 as

$$T_i^{n+1} = 0 + (1 - 2r)(100 - 0) = (1 - 2r)100 \tag{5.17b}$$

The physical situation requires that the temperature T_i^{n+1} cannot go below the temperature of the two neighboring nodes, 0°C, but neither goes above or stays equal to the temperature at the previous time step, 100°C. An examination of equation (5.17b) reveals that a value of (1 − 2r) less than zero or greater or equal to one violates such a requirement. Therefore, to obtain meaningful results from the solution of the finite difference equation (5.5), the coefficient (1 − 2r) should satisfy the following criterion

$$0 \leq 1 - 2r < 1 \text{ or } r = \frac{\alpha \, \Delta t}{(\Delta x)^2} \leq \frac{1}{2} \tag{5.18a}$$

which is the same as given previously by equations (5.16a,b). Such a restriction on the maximum value of r imposes the following limitation to the maximum size of the time step

$$\Delta t \leq \frac{(\Delta x)^2}{2\alpha} \tag{5.18b}$$

The implications of this restriction for the computation time are as follows. Suppose calculations performed with some mesh size Δx up to a time t^* require N time steps each of size Δt (i.e., $t^* = N\Delta t$). If calculations are to be repeated, say, with a halved mesh size $\Delta x_1 = \frac{1}{2}\Delta x$, for the purpose of improved accuracy, it requires four times as many time steps to reach the same time level t^*. Furthermore, each time level calculation requires twice as long time because there are twice as many nodal points. Thus, for the one-dimensional problem, reducing the mesh size by one-half increases the computer time eight-fold. Therefore, the restriction imposed on the maximum time step by the stability criterion is a major disadvantage of the explicit method.

5.1.1.5 Effects of Boundary Conditions on Stability

The stability criterion given by equations (5.18) is developed by considering the finite difference equations (5.5) for the internal nodes of the region. If the boundary conditions for the problem involve prescribed temperature and/or prescribed heat flux, no additional restrictions are imposed by the boundary conditions, and the stability criterion given by equations (5.18) remains applicable for the solution of the finite difference equations. However, in the case of convection boundary conditions, the second-order accurate finite difference approximation given by equations (5.11a,b) imposes a more severe restriction on the parameter r than that imposed by the criterion $r \leq \frac{1}{2}$.

The stability criteria associated with the finite difference equations (5.11a,b) can be developed by a physical argument similar to that described previously.

Consider the finite difference equation (5.11a) for $\gamma_0 = 0$, which corresponds to convection into an ambient at zero temperature. We obtain

$$T_0^{n+1} = 2rT_1^n + (1 - 2r\beta_0)T_0^n \qquad (5.19a)$$

Suppose at any time level, the temperatures of the nodes 0 and 1 are, respectively, $T_0^n = 100°C$ and $T_1^n = 0°C$. equation (5.19a), to be used for predicting the temperature T_0^{n+1} of the node 0 at the next time level n + 1, becomes

$$T_0^{n+1} = (1 - 2r\beta_0)100 \qquad (5.19b)$$

The physically meaningful situation for the problem requires that the temperature T_0^{n+1} can assume values between 0°C and 100°C but cannot go below the temperature of the neighboring node and of the ambient. An examination of equation (5.19b) reveals that a negative value or a value greater than one of the parameter $(1 - 2r\beta_0)$ violates this requirement. Therefore, to obtain physically meaningful results from the solution of the finite difference equation (5.11a), the following criteria should be satisfied

$$0 \leq 1 - 2r\beta_0 < 1 \text{ or } 0 < r \leq \frac{1}{2\beta_0} = \frac{1}{2 + 2\frac{\Delta x \, h_0}{k}} \qquad (5.20a)$$

Similarly, for the boundary condition [equation (5.11b)], we write

$$0 \leq 1 - 2r\beta_L < 1 \text{ or } 0 < r \leq \frac{1}{2\beta_L} = \frac{1}{2 + 2\frac{\Delta x \, h_L}{k}} \qquad (5.20b)$$

One-Dimensional Transient Systems

Clearly, the stability criteria imposed by equations (5.20a,b) are more restrictive than that of $r \leq \frac{1}{2}$; the smallest r value obtained from equations (5.20a,b) should be used as the stability criterion for the solution of finite difference equations.

Smith (1978) examines the stability criterion for the convection boundary conditions more rigorously by using the matrix method of stability analysis and obtains results that are slightly less restrictive than those given by equations (5.20a,b). That is, Smith's analysis leads to stability criterion in which the terms $2\Delta x h_0/k$ and $2\Delta x h_L/k$ appearing in the denominator of equations (5.20a,b) are replaced by $\Delta x h_0/k$ and $\Delta x h_L/k$, respectively.

5.1.1.6 Effects of r on Truncation Error

For the simple explicit scheme, the value chosen for the parameter r should not exceed the limit imposed by the stability criterion. We now examine the value of r that may produce the minimum truncation error.

The diffusion equation (5.2) is discretized as discussed previously, and the lower order truncation errors are retained.

$$\left(\frac{T_i^{n+1} - T_i^n}{\Delta t} - \frac{\Delta t}{2}\frac{\partial^2 T}{\partial t^2} + \cdots\right) - \alpha\left(\frac{T_{i+1}^n - 2T_i^n + T_{i-1}^n}{(\Delta x)^2} - \frac{(\Delta x)^2}{12}\frac{\partial^4 T}{\partial x^4} + \cdots\right) = 0 \quad (5.21a)$$

Then, the truncation error, TE associated with this discretization scheme is written as

$$TE = -\frac{\Delta t}{2}\frac{\partial^2 T}{\partial t^2} + \alpha\frac{(\Delta x)^2}{12}\frac{\partial^4 T}{\partial x^4} + \text{Higher order terms} \quad (5.21b)$$

The term $\partial^2 T/\partial t^2$ is now manipulated as

$$\frac{\partial^2 T}{\partial t^2} = \frac{\partial}{\partial t}\left(\frac{\partial T}{\partial t}\right) = \frac{\partial}{\partial t}\left(\alpha\frac{\partial^2 T}{\partial x^2}\right) = \alpha\frac{\partial^2}{\partial x^2}\left(\frac{\partial T}{\partial t}\right) = \alpha\frac{\partial^2}{\partial x^2}\left(\alpha\frac{\partial^2 T}{\partial x^2}\right) = \alpha^2\frac{\partial^4 T}{\partial x^4} \quad (5.21c)$$

where we utilized equation (5.2). We introduce equation (5.21c) into equation (5.21b)

$$TE = \alpha\left[-\frac{\alpha \Delta t}{2} + \frac{(\Delta x)^2}{12}\right]\frac{\partial^4 T}{\partial x^4} + \text{Higher order terms} \quad (5.21d)$$

and recall the definition of $r = \alpha \Delta t/(\Delta x)^2$. Then equation (5.21d) is written as

$$TE = \alpha\left(-\frac{\alpha \Delta t}{2} + \frac{\alpha \Delta t}{12r}\right) + \text{Higher order terms} \quad (5.21e)$$

Clearly, for $r = \frac{1}{6}$, the term inside the parenthesis vanishes; hence, the truncation error becomes the lowest.

5.1.1.7 Fourier Method of Stability Analysis

We now present a rather straightforward but more rigorous analysis of the stability of finite difference equations by using the Fourier (or Neumann) method of stability analysis.

Computers cannot perform calculations to infinite accuracy. Therefore, in the numerical solution of finite difference equations with a digital computer, round-off errors are introduced during calculations. The mathematical analysis of stability is concerned with the examination of the growth of errors while the computations are being performed. For an unstable system, the error grows larger without bound, but for a stable system it should not grow without a bound.

Two most commonly used methods of analysis of stability include the *matrix* method and the *Fourier* (or von Neumann) method. In the matrix method, the finite difference representation of the differential equation and the boundary conditions are expressed in matrix form, and the problem of stability is transformed to the examination of the eigenvalues of the coefficient matrix. The errors will not increase exponentially with time, and the system is stable if modules of the largest of the eigenvalue are less than or equal to unity (Fox 1962; Smith 1978). Therefore, the matrix method is capable of including the effects of boundary conditions on stability; but it requires some knowledge of matrix algebra and, in many cases, a closed-form solution for the eigenvalues is not available.

In the Fourier method, the errors are expressed in a finite Fourier series and then the propagation of errors with time are examined. The method does not accommodate the effects of boundary conditions; but it is simple, straightforward, and can readily be extended to multidimensional problems.

Consider the one-dimensional transient heat conduction equation (5.2) expressed in finite difference form by using the explicit method

$$\frac{T_j^{n+1} - T_j^n}{\Delta t} = \alpha \frac{T_{j-1}^n - 2T_j^n + T_{j+1}^n}{(\Delta x)^2} \quad (5.22a)$$

where the subscript j is the discretization index for the space variable (i.e., $x = j\Delta x$) and n for the time variable (i.e., $t = n\Delta t$). The numerical solution of the problem, T_N, can be written as the sum of the *exact solution* of the problem, T_E, and an error term ε in the form

$$T_N = T_E + \varepsilon \quad (5.22b)$$

One-Dimensional Transient Systems

where the numerical solution must satisfy the difference equation (5.22a). Substituting equation (5.22b) into the difference equation (5.22a) and noting that T_E should also satisfy the difference equation, we obtain

$$\frac{\varepsilon_j^{n+1} - \varepsilon_j^n}{\Delta t} = \alpha \frac{\varepsilon_{j-1}^n - 2\varepsilon_j^n + \varepsilon_{j+1}^n}{(\Delta x)^2} \tag{5.23}$$

Numerical errors are introduced at almost every stage of the calculations. Assume that the errors introduced at nodal points along the initial (i.e., t = 0) line could be expressed in a finite Fourier series in terms of sine–cosine or complex exponentials. Here, we prefer to use the latter. To examine the propagation of errors as time increases, one needs to consider only a single term in the series, because the finite difference equations are linear. With these considerations, one examines the propagation of error due to a single term expressed in the form

$$\varepsilon(j\Delta x, n\Delta t) \equiv \varepsilon_j^n = e^{\gamma n \Delta t} e^{i\beta_m j \Delta x} \tag{5.24}$$

where $i = \sqrt{-1}$, β_m are the Fourier modes, γ is in general a complex quantity, $n\Delta t = t$, and $j\Delta x = x$. This equation is expressed in the form

$$\varepsilon_j^n = \xi^n e^{i\beta_m j \Delta x} \tag{5.25a}$$

Similarly, we write

$$\varepsilon_{j+1}^n = \xi^n e^{i\beta_m (j+1) \Delta x} \tag{5.25b}$$

$$\varepsilon_j^{n+1} = \xi^{n+1} e^{i\beta_m j \Delta x} \tag{5.25c}$$

where

$$\xi = e^{\gamma \Delta t} \tag{5.26}$$

For this definition of ξ, the error term ε_j^n will not increase without a bound as t increases if

$$|\xi| \leq 1 \tag{5.27}$$

This criterion applies to linear differential equations with constant coefficients and is necessary and sufficient for two-time level difference equations; but it is not always sufficient for three or more level equations, although it is always necessary (Richtmyer and Morton 1967; Smith 1978).

To determine the stability criterion for the explicit finite difference equation (5.22a), we substitute the error terms given by equations (5.25a–c) into equation (5.23), that is,

$$e^{i\beta_m j \Delta x}(\xi^{n+1} - \xi^n) = \frac{\alpha \Delta t}{(\Delta x)^2} e^{i\beta_m j \Delta x} \xi^n (e^{i\beta_m \Delta x} - 2 + e^{-i\beta_m \Delta x}) \tag{5.28a}$$

and, after cancellation and some rearrangement, we obtain

$$\xi - 1 = 2 \frac{\alpha \Delta t}{(\Delta x)^2} \left(\frac{e^{i\beta_m \Delta x} + e^{-i\beta_m \Delta x}}{2} - 1 \right) \tag{5.28b}$$

Noting that

$$\cos(\beta_m \Delta x) = \frac{e^{i\beta_m \Delta x} + e^{-i\beta_m \Delta x}}{2} \tag{5.29}$$

Equation (5.28b) is written as

$$\xi = 1 - 2r[1 - \cos(\beta_m \Delta x)] \tag{5.30a}$$

where

$$r = \frac{\alpha \Delta t}{(\Delta x)^2} \tag{5.30b}$$

Here the parameter ξ is called the *amplification factor*. Recalling its definition by equation (5.26), the initial errors will not be amplified, and the finite difference calculations will remain stable if the condition $|\xi| \leq 1$ is satisfied for all values of β_m. Applying this restriction to equation (5.30a), we obtain

$$|1 - 2r[1 - \cos(\beta_m \Delta x)]| \leq 1 \tag{5.31a}$$

or

$$-1 \leq \{1 - 2r[1 - \cos(\beta_m \Delta x)]\} \leq 1 \tag{5.31b}$$

which must be satisfied for all possible Fourier modes β_m. The right-hand side of this inequality is satisfied for all possible values of β_m. To satisfy the left-hand side under the most strict conditions, we must have $1 - \cos(\beta_m \Delta x) = 2$. Then, we have

$$-1 \leq (1 - 4r) \text{ or } r = \frac{\alpha \Delta t}{(\Delta x)^2} \leq \frac{1}{2} \tag{5.32a,b}$$

which is the stability criterion for the solution of the explicit finite difference equation (5.22a) or (5.5).

One-Dimensional Transient Systems

The amplification factor ξ defined earlier represents the ratio of the error at time level $n+1$ to that at time level n, that is,

$$\xi = \frac{\varepsilon^{n+1}}{\varepsilon^n} = 1 - 2r(1 - \cos\phi) \qquad (5.33a)$$

where

$$\phi = \beta_m \Delta x \qquad (5.33b)$$

For the specific problem considered here, the amplification factor as given is a real quantity that has no imaginary part and hence involves no phase shift. However, there are situations, such as those encountered in the finite difference representation of the hyperbolic heat conduction equation, wave equation, or the energy equation with convection term, where the expression for the amplification factor has an imaginary part. In such cases, one needs to examine the modulus of the amplification factor $|\xi|$ as well as the phase angle in order to establish the stability criterion.

Example 5.1

Consider the following transient heat conduction problem in a slab given in dimensionless form as

$$\frac{\partial^2 T}{\partial x^2} = \frac{\partial T}{\partial t} \qquad 0 < x < 1, \qquad t > 0$$

$$\frac{\partial T}{\partial x} = 0 \qquad x = 0, \qquad t > 0$$

$$T = 0 \qquad x = 1, \qquad t > 0$$

$$T = 100 \cos\left(\frac{\pi}{2}x\right) \qquad t = 0, \qquad 0 \leq x \leq 1$$

Solve this problem numerically with the explicit finite difference scheme by dividing the region $0 \leq x \leq 1$ into five equal parts by using (a) a first-order accurate and (b) a second-order accurate finite differencing scheme for the boundary condition at $x = 0$. The exact solution of this problem is given by

$$T(x,t) = 100 \cos\left(\frac{\pi}{2}x\right) \exp\left(-\frac{\pi^2}{4}t\right)$$

Compare the temperature of the insulated surface obtained by finite difference solution with the exact solution given here. Use $r = \frac{1}{5}$ for numerical calculations.

Solution

The region $0 \leq x \leq 1$ is divided into five equal parts. The problem involves five unknown node temperatures T_i^n, $i = 0,1,2,3,4$ because the temperature of the boundary surface at $x = 1$ is prescribed as $T_5^n = 0$.

The finite difference equations for the internal nodes are obtained from equation (5.5) by setting $r = \dfrac{1}{5} = 0.2$.

$$T_i^{n+1} = 0.2 T_{i-1}^n + 0.6 T_i^n + 0.2 T_{i+1}^n \quad \text{for} \quad i = 1,2,3,4 \quad \text{and} \quad T_5^i = 0.$$

An additional relation is obtained by discretizing the insulated boundary condition.

1. If a first-order accurate formula is used, the finite difference approximation of the insulated boundary condition becomes

$$\dfrac{T_1^n - T_0^n}{\Delta x} = 0 \quad \text{or} \quad T_0^n = T_1^n \quad \text{for} \quad i = 0$$

2. If a second-order accurate scheme is used, the finite difference approximation of the insulated boundary condition is obtained from equation (5.15a) by setting $q_0 = 0$ and $r = 0.2$ as

$$T_0^{n+1} = 0.6 \, T_0^n + 0.4 \, T_1^n \quad \text{for} \quad i = 0$$

The initial condition for the problem becomes

$$T_i^0 = 100 \cos\left(\dfrac{\pi}{10} i\right), \quad i = 0, 1, 2, \ldots, 5$$

In Table 5.1, we present a comparison of finite difference solutions with the exact results for the temperature for the insulated boundary. Clearly, the finite difference solution utilizing the second-order accurate differencing for the boundary condition is very close to the exact solution, while the finite difference solution utilizing first-order accurate differencing for the boundary condition deviates about 10% to 5% from the exact results.

TABLE 5.1

A Comparison of Finite Difference Solutions with the Exact Results for Example 5.1

		T(0,t)	
		Finite Difference	
Time t	Exact	First-Order Accurate	Second-Order Accurate
0.2	61.0498	55.0210	61.0004
0.4	37.2708	31.4255	37.2105
0.6	22.7537	17.9513	22.6986
0.8	13.8911	10.2544	13.8462
1.0	8.4805	5.8577	8.4463
1.2	5.1773	3.3461	5.1523
1.4	3.1607	1.9114	3.1429
1.6	1.9296	1.0919	1.9172

Example 5.2

Consider the following transient heat conduction problem given in the dimensionless form as

$$\frac{\partial^2 T}{\partial x^2} = \frac{\partial T}{\partial t} \quad 0 < x < 1, \quad t > 0$$

$T = 0, x = 0, t > 0$
$T = 0, x = 1, t > 0$
$T = 10 \sin(2\pi x), \quad t = 0, \quad 0 \leq x \leq 1$

The exact analytic solution of this problem is

$$T(x,t) = 10 \, e^{-4\pi^2 t} \sin(2\pi x)$$

Solve this problem with finite differences using the explicit method taking:

a. $\Delta x = 0.1, r = 0.25 \therefore \Delta t = 0.0025$
b. $\Delta x = 0.1, r = 0.50 \therefore \Delta t = 0.0050$

and compare the temperature at the location $x = 0.3$ with the exact results.

Solution

The finite difference approximation of the differential equation of heat conduction is obtained from equation (5.5) as

$$T_i^{n+1} = rT_{i-1}^n + (1-2r)T_i^n + rT_{i+1}^n, \quad i = 1, 2, \ldots, 9$$

The temperatures at the boundary nodes are specified as

$$T_0^n = 0 \quad \text{and} \quad T_{10}^n = 0$$

and the initial condition becomes

$$T_i^0 = 10 \sin(0.2\pi i), \quad i = 0, 1, 2, \ldots, 10$$

The aforementioned system of algebraic equations can readily be solved for a specified value of r satisfying the stability criterion. Here we consider two cases:

1. $r = 0.25$: $\quad T_i^{n+1} = 0.25 T_{i-1}^n + 0.5 T_i^n + 0.25 T_{i+1}^n$
2. $r = 0.50$: $\quad T_i^{n+1} = 0.5(T_{i-1}^n + T_{i+1}^n)$

Table 5.2 shows a comparison of finite difference solutions with the exact results for the cases with $r = 0.5$ and $r = 0.25$. Clearly, the smaller value of r produces more accurate results.

TABLE 5.2

A Comparison of the Finite Difference Solution with Exact Results at the Location x = 0.3 for Two Different Values of r

T(0.3.t) r	Time	Finite Difference	Exact	% Error
0.5	0.0050	7.6942	7.8070	−1.444
	0.0150	5.0359	5.2605	−4.269
	0.0250	3.2960	3.5447	−7.014
	0.0350	2.1573	2.3884	−9.679
	0.0450	1.4120	1.6094	−12.268
0.25	0.0025	8.6024	8.6167	−0.167
	0.0075	7.0379	7.0732	−0.499
	0.0125	5.7580	5.8062	−0.830
	0.0175	4.7108	4.7661	−1.160
	0.0225	3.8541	3.9112	−1.489

Example 5.3

Consider the following transient heat conduction problem with energy generation

$$\alpha \frac{\partial^2 T}{\partial x^2} + \frac{\alpha}{k}g = \frac{\partial T}{\partial t} \quad 0 < x < L, \quad t > 0$$

$$-k\frac{\partial T}{\partial x} + hT = hT_\infty \quad x = 0 \quad\quad t > 0$$

$$T = 0 \quad\quad x = L, \quad t > 0$$

$$T = F(x) \quad\quad t = 0, \quad 0 \leq x \leq L$$

By dividing the region $0 \leq x \leq L$ into M equal parts, develop the second-order accurate finite difference approximation for this problem by using the explicit method.

Solution

The finite difference approximation for this problem is immediately obtained from equation (5.4a) by including the contribution of the generation term.

$$\frac{T_i^{n+1} - T_i^n}{\Delta t} = \alpha \frac{T_{i-1}^n - 2T_i^n + T_{i+1}^n}{(\Delta x)^2} + \frac{\alpha}{k}g_i^n$$

Solving for T_i^{n+1} we obtain

$$T_i^{n+1} = rT_{i-1}^n + (1-2r)T_i^n + rT_{i+1}^n + \frac{\alpha \Delta t}{k}g_i^n \quad \text{for } i = 1, 2, \ldots, M-1 \quad\quad (a)$$

A second-order accurate finite difference approximation for the convection boundary condition at x = 0 is obtained from equation (5.11a) by properly adding the contribution of the energy generation term. We find

$$T_0^{n+1} = (1 - 2r\beta_0)T_0^n + 2rT_1^n + 2r\gamma_0 + \frac{\alpha \Delta t}{k}g_0^n \quad \text{for } i = 0 \quad\quad (b)$$

where
$$\beta_0 = 1 + \frac{\Delta x\, h}{k}, \quad \gamma_0 = \frac{\Delta x\, h}{k} T_\infty, \quad r = \frac{\alpha \Delta t}{(\Delta x)^2}$$

We also have $T_M^n = 0$ for $i = M$, and the initial condition is given by
$$T_i^0 = F(i\Delta x) \quad \text{for} \quad i = 0, 1, 2, \ldots, M.$$

The complete finite difference approximation of this problem is summarized as

$$T_0^{n+1} = (1 - 2r\beta_0)T_0^n + 2rT_1^n + 2r\gamma_0 + \frac{\alpha \Delta t}{k} g_0^n, \qquad i = 0$$

$$T_i^{n+1} = rT_{i-1}^n + (1 - 2r)T_i^n + rT_{i+1}^n + \frac{\alpha \Delta t}{k} g_i^n, \qquad i = 1, 2, \ldots, M-1$$

$$T_M^{n+1} = 0, \qquad i = M$$

and the initial conditions
$$T_i^0 = F(i\Delta x), \quad i = 0, 1, 2, \ldots, M$$

Here, β_0 and γ_0 are defined previously. The stability criteria are obtained from equation (5.20a) as

$$0 < r \leq \frac{1}{2 + 2\frac{\Delta x\, h_0}{k}}$$

Example 5.4

Show that in Example 5.3 the finite difference approximation of the convection boundary condition at $x = 0$, given by equation (b), is also obtainable by the control volume approach utilizing the conservation principle for a control volume of thickness $\Delta x/2$ about the node $i = 0$.

Solution

The accompanying figure shows the control volume of thickness $\Delta x/2$ and area A about the boundary node $i = 0$. The conservation principle for this control volume can be stated as

$$\underbrace{\begin{pmatrix} \text{Rate of energy} \\ \text{entering through} \\ \text{boundaries} \end{pmatrix}}_{I} + \underbrace{\begin{pmatrix} \text{Rate of energy} \\ \text{generation} \end{pmatrix}}_{II} = \underbrace{\begin{pmatrix} \text{Rate of increase} \\ \text{of internal} \\ \text{energy} \end{pmatrix}}_{III}$$

Introducing the mathematical expressions for each of these terms, we obtain

$$\underbrace{Ah(T_\infty - T_0^n) + Ak\frac{T_1^n - T_0^n}{\Delta x}}_{I} + \underbrace{A\frac{\Delta x}{2}g_0}_{II} = \underbrace{A\frac{\Delta x}{2}\rho C_p \frac{T_0^{n+1} - T_0^n}{\Delta t}}_{III}$$

where g_0 is the volumetric energy generation rate in the control volume next to the surface at $x = 0$. After cancellation and rearrangement, we obtain

$$T_0^{n+1} = \left[1 - 2\frac{\alpha \Delta t}{(\Delta x)^2}\left(1 + \frac{h\Delta x}{k}\right)\right]T_0^n + 2\frac{\alpha \Delta t}{(\Delta x)^2}T_1^n$$
$$+ 2\frac{\alpha \Delta t}{(\Delta x)^2}\frac{h\Delta x T_\infty}{k} + \frac{\alpha \Delta t}{k}g_0$$

This expression is written as

$$T_0^{n+1} = (1 - 2r\beta_0)T_0^n + 2rT_1^n + 2r\gamma_0 + \frac{\alpha \Delta t}{k}g_0^n$$

where

$$\alpha = \frac{k}{\rho C_p}, \quad \beta_0 = 1 + \frac{h\Delta x}{k}, \quad \gamma_0 = \frac{h\Delta x}{k}T_\infty, \quad r = \frac{\alpha \Delta t}{(\Delta x)^2}$$

Clearly, this result is the same as that given by equation (b) in Example 5.3.

5.1.2 Simple Implicit Method

The simple explicit method discussed previously is very simple computationally, but the maximum size of the time step is restricted by stability considerations. If calculations are to be performed over a large period of time, the number of steps—hence the number of calculations—needed may become prohibitively large. To alleviate this difficulty, finite difference schemes that are not restrictive to the size of the time step Δt have been developed. One such method is the simple implicit method. We consider the one-dimensional diffusion equation

$$\frac{\partial T}{\partial t} = \alpha \frac{\partial^2 T}{\partial x^2} \tag{5.34}$$

Differently from the simple explicit method [see equation (5.4a)], the space derivative is now discretized at the n + 1 time level as

$$\left.\frac{\partial^2 T}{\partial x^2}\right|_{i,n+1} = \frac{T_{i-1}^{n+1} - 2T_i^{n+1} + T_{i+1}^{n+1}}{(\Delta x)^2} + 0[(\Delta x)^2] \tag{5.35a}$$

One-Dimensional Transient Systems

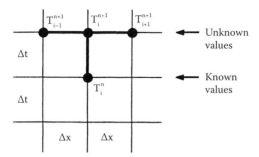

FIGURE 5.4
The finite difference molecules for the simple implicit scheme.

while the time derivative is discretized as a backward first order approximation from the n+1 time level, that is,

$$\left.\frac{\partial T}{\partial t}\right|_{i,n+1} = \frac{T_i^{n+1} - T_i^n}{\Delta t} + 0(\Delta t) \tag{5.35b}$$

Introducing equations (5.35a,b) into equation (5.34), we obtain the simple implicit finite difference approximation for the diffusion equation as

$$\frac{T_i^{n+1} - T_i^n}{\Delta t} = \alpha \frac{T_{i-1}^{n+1} - 2T_i^{n+1} + T_{i+1}^{n+1}}{(\Delta x)^2} \tag{5.36}$$

which is accurate to $0[(\Delta x)^2, \Delta t]$ and unconditionally stable. This is an implicit scheme because at each time level equations are to be solved simultaneously in order to determine the nodal temperatures.

Figure 5.4 illustrates the expansion point (i, n+1) and the implicit finite difference molecule. If the problem involves M unknown node temperatures, the simultaneous solution of M equations at each time level is more involved than that of the explicit method; but the method has the advantage that a larger time step Δt can be used than in the explicit method.

5.1.2.1 Stability Analysis

We apply the Fourier method of stability analysis to demonstrate that the simple implicit scheme is unconditionally stable.

As discussed previously, the numerical solution T_N is the sum of the exact solution, T_E, of the problem, plus an error term ε, given in the form

$$T_N = T_E + \varepsilon \tag{5.37}$$

We introduce equation (5.37) into equation (5.36) and note that T_E should also satisfy the difference equation. We obtain

$$\frac{\varepsilon_j^{n+1} - \varepsilon_j^n}{\Delta t} = \alpha \frac{\varepsilon_{j-1}^{n+1} - 2\varepsilon_{j+1}^{n+1} + \varepsilon_{j+1}^{n+1}}{(\Delta x)^2} \tag{5.38}$$

where we replaced the space variable index i by j. The error terms ε_j^n are represented as given by equations (5.25a–c). Introducing ε_j's from equations (5.25a–c) into equation (5.38) and after cancellations and some rearrangement, we obtain

$$\xi - 1 = \frac{2\alpha \Delta t}{(\Delta x)^2} \xi \left(\frac{e^{i\beta_m \Delta x} + e^{-i\beta_m \Delta x}}{2} - 1 \right) \tag{5.39}$$

where $i = \sqrt{-1}$. Noting that

$$\cos(\beta_m \Delta x) = \frac{e^{i\beta_m \Delta x} + e^{-i\beta_m \Delta x}}{2}, \tag{5.40}$$

Equation (5.39) is written as

$$\xi - 1 = 2r\xi(\cos\beta_m \Delta x - 1) \tag{5.41}$$

or

$$\xi - 1 = -4r\xi \sin^2\left(\frac{\beta_m \Delta x}{2}\right) \tag{5.42a}$$

where

$$r = \frac{\alpha \Delta t}{(\Delta x)^2} \tag{5.42b}$$

Equation (5.42a) is solved for ξ

$$\xi = \left[1 + 4r \sin^2\left(\frac{\beta_m \Delta x}{2}\right) \right]^{-1} \tag{5.43}$$

For stability, we need $|\xi| \leq 1$, and this condition is satisfied for all positive values of r. Therefore, the simple implicit finite difference approximation is stable for all values of the time step Δt. However, Δt must be kept reasonably small to obtain results sufficiently close to the exact solution of the partial differential equation because the method is $0(\Delta t)$.

5.1.3 Crank–Nicolson Method

The basic idea in the use of the implicit scheme has been further developed by numerous investigators with the objective of finding efficient schemes that are more accurate and that also have no restriction on the size of the time step. One such scheme, called the Crank–Nicolson method, has been successful in achieving such an objective. This alternative implicit differencing scheme proposed by Crank and Nicolson (1947) retains the left-hand side of the implicit equation (5.36), but modifies the right-hand side by taking the arithmetic average of the right-hand sides of the explicit equation (5.4a) and the implicit equation (5.36). Then, the finite difference approximation of the diffusion equation (5.34) with the Crank–Nicolson method becomes

$$\frac{T_i^{n+1} - T_i^n}{\Delta t} = \frac{\alpha}{2}\left[\frac{T_{i-1}^{n+1} - 2T_i^{n+1} + T_{i+1}^{n+1}}{(\Delta x)^2} + \frac{T_{i-1}^n - 2T_i^n + T_{i+1}^n}{(\Delta x)^2}\right] \quad (5.44)$$

By Taylor series expansion about the node (i,n), it can be shown that the Crank–Nicolson method (Fox 1962; Smith 1978) is second-order accurate in both time and space, that is, $0[(\Delta t)^2,(\Delta x)^2]$. This is an advantage over the simple implicit scheme, which is accurate only to $0[\Delta t,(\Delta x)^2]$. In addition, as in the case of the simple implicit scheme, it has no restriction on the size of the time step Δt for computations.

A qualitative discussion of the significance of the Crank–Nicolson differencing scheme is instructive. The right-hand side of equation (5.44) is the arithmetic average of the central difference expressions for the second derivative $\partial^2 T/\partial x^2$ about the node i at time steps n + 1 and n. Such an averaging scheme can be regarded as an estimate of the second derivative at the node i about the time n + 1/2, illustrated as the node A in Figure 5.5. Then, the left-hand side of equation (5.44) can be regarded as the central difference representation of $\partial T/\partial t$ about the point A. We conclude that with the finite differencing of derivatives with respect to both the space and time derivatives being central differences, the truncation errors are expected to be second order in both Δt and Δx. Another interpretation is that the Crank–Nicolson scheme is constructed from the summation of the explicit and implicit discretizations of the diffusion equation, which results in the first order error term cancellation from the forward (explicit scheme) and backward (implicit scheme) representations of the first derivative in time, of opposite signs.

Equation (5.44) is now rearranged in the form

$$-rT_{i-1}^{n+1} + (2+2r)T_i^{n+1} - rT_{i+1}^{n+1} = rT_{i-1}^n + (2-2r)T_i^n + rT_{i+1}^n \quad (5.45)$$

where

$$r = \frac{\alpha \Delta t}{(\Delta x)^2}$$

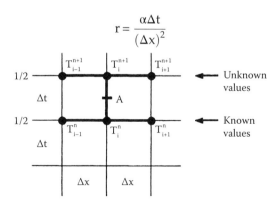

FIGURE 5.5
The finite difference molecules for the Crank–Nicolson scheme.

This equation is unconditionally stable with no restriction on the value of the parameter r. The only restriction on r is that, for a given α and Δx, the resulting value of the time step Δt should not be large to impair the accuracy.

If the temperatures are prescribed at the boundary surfaces $x = 0$ and $x = L$, then the node temperatures T_0 and T_M are known. Thus, at each time level n, equation (5.45) provides M−1 simultaneous algebraic equations for the determination of M−1 unknown internal node temperatures for the next time level n + 1.

When the boundary conditions at $x = 0$ and $x = L$ are convection or prescribed heat flux, then, the node temperatures T_0 and T_M at the boundaries are unknown. Therefore, two additional relationships are needed for such cases, as described next (see also Chapter 2).

A second-order accurate differencing of the boundary conditions is obtainable by using the central difference formula about the boundary nodes $i = 0$ and $i = M$. We consider the convection boundary conditions at both boundaries, $x = 0$ and $x = L$, given by

$$-k\frac{\partial T}{\partial x} + h_0 T = h_0 T_{\infty,0} \quad \text{at} \quad x = 0 \qquad (5.46)$$

$$k\frac{\partial T}{\partial x} + h_L T = h_L T_{\infty,L} \quad \text{at} \quad x = L \qquad (5.47)$$

The finite difference approximation of these boundary conditions about the nodes $i = 0$ and $i = M$, with the second-order accurate central difference formula, gives

$$-k\frac{T_1^n - T_{-1}^n}{2\Delta x} + h_0 T_0^n = h_0 T_{\infty,0} \qquad (5.48)$$

$$k\frac{T_{M+1}^n - T_{M-1}^n}{2\Delta x} + h_L T_M^n = h_L T_{\infty,L} \qquad (5.49)$$

where T_{-1}^n and T_{M+1}^n are the fictitious temperatures at the fictitious nodes, as illustrated in Figure 5.2. Equations (5.48) and (5.49) can also be written for the time level (n + 1) by merely replacing the superscript n by (n + 1). To eliminate the fictitious temperatures, two additional relations are obtained from equation (5.45) by evaluating it for $i = 0$ and $i = M$, yielding, respectively

$$-rT_{-1}^{n+1} + (2+2r)T_0^{n+1} - rT_1^{n+1} = rT_{-1}^n + (2-2r)T_0^n + rT_1^n \qquad (5.50)$$

$$-rT_{M-1}^{n+1} + (2+2r)T_M^{n+1} - rT_{M+1}^{n+1} = rT_{M-1}^n + (2-2r)T_M^n + rT_{M+1}^n \qquad (5.51)$$

The fictitious temperatures T_{-1}^n and T_{-1}^{n+1} are eliminated from equation (5.50) by utilizing equation (5.48) for time levels n and n + 1. Similarly, T_{M+1}^n and T_{M+1}^{n+1} are eliminated from equation (5.51) by means of equation (5.49).

One-Dimensional Transient Systems

Then, the following two expressions are obtained for the finite difference approximation of the convection boundary conditions at the boundary nodes $i = 0$ and $i = M$, respectively.

$$(2+2r\beta_0)T_0^{n+1} - 2rT_1^{n+1} = (2-2r\beta_0)T_0^n + 2rT_1^n + 4r\gamma_0 \quad \text{for} \quad i=0 \quad (5.52)$$

$$-2rT_{M-1}^{n+1} + (2+2r\beta_L)T_M^{n+1} = 2rT_{M-1}^n + (2-2r\beta_L)T_M^n + 4r\gamma_L \quad \text{for} \quad i=M \quad (5.53)$$

where

$$\begin{aligned}
\beta_0 &= 1 + \frac{\Delta x h_0}{k}, & \gamma_0 &= \frac{\Delta x h_0}{k} T_{\infty,0} \\
\beta_L &= 1 + \frac{\Delta x h_L}{k}, & \gamma_L &= \frac{\Delta x h_L}{k} T_{\infty,L} \\
r &= \alpha \Delta t / (\Delta x)^2
\end{aligned} \quad (5.54)$$

Thus, equations (5.52) and (5.53), together with equation (5.45) for $i = 1,2,\ldots,M-1$, are the complete finite difference approximation (with the Crank–Nicolson method) of the diffusion equation in a slab subjected to convection at both boundary surfaces. They provide $M + 1$ simultaneous algebraic equations for the $M + 1$ unknown node temperatures. The system is unconditionally stable.

5.1.3.1 Prescribed Heat Flux Boundary Condition

We consider prescribed heat flux boundary conditions at the surfaces $x = 0$ and $x = L$ given by

$$-k\frac{\partial T}{\partial x} = q_0 \quad \text{at} \quad x=0 \quad (5.55a)$$

$$k\frac{\partial T}{\partial x} = q_L \quad \text{at} \quad x=L \quad (5.55b)$$

where q_0 and q_L are the prescribed heat fluxes supplied to the surfaces at $x = 0$ and $x = L$, respectively. The finite difference equations for prescribed heat flux boundary conditions are readily obtainable from equations (5.52) and (5.53), respectively, by making

$$\beta_0 = \beta_L = 1, \quad \gamma_0 = \frac{\Delta x}{k}q_0, \quad \gamma_L = \frac{\Delta x}{k}q_L \quad (5.56a\text{-}c)$$

We obtain

$$(2+2r)T_0^{n+1} - 2rT_1^{n+1} = (2-2r)T_0^n + 2rT_1^n + 4r\frac{\Delta x q_0}{k} \quad \text{for} \quad i=0 \quad (5.57a)$$

$$-2rT_{M-1}^{n+1} + (2+2r)T_M^{n+1} = 2rT_{M-1}^n + (2-2r)T_M^n + 4r\frac{\Delta x q_L}{k} \quad \text{for} \quad i = M$$

(5.57b)

where

$$r = \frac{\alpha \Delta t}{(\Delta x)^2} \quad (5.58)$$

Thus, equations (5.57a,b) together with equation (5.45) for $i = 1,2,\ldots,M-1$ provide a complete finite difference representation with the Crank–Nicolson method of heat diffusion in a slab subjected to prescribed heat flux at both boundaries. They provide $M + 1$ simultaneous algebraic equations for the determination of $M + 1$ unknown node temperatures, and the system is unconditionally stable. Next, proof of the stability of the Crank–Nicolson method will be given in connection with the combined method.

5.1.4 Combined Method

We recall that the Crank–Nicolson method was developed by taking the arithmetic average of the right-hand sides of the explicit equation (5.4a) and the implicit equation (5.36), while retaining the left-hand side of the implicit equation (5.36). This idea can be generalized such that, instead of taking the arithmetic average, one can take a weighted average of the right-hand sides of equations (5.4a) and (5.36), while retaining the left-hand side of equation (5.36). The resulting algorithm for the finite difference representation of the one-dimensional diffusion equation (5.2) is called the combined method and is given by

$$\frac{T_i^{n+1} - T_i^n}{\Delta t} = \alpha \left[\theta \frac{T_{i-1}^{n+1} - 2T_i^{n+1} + T_{i+1}^{n+1}}{(\Delta x)^2} + (1-\theta) \frac{T_{i-1}^n - 2T_i^n + T_{i+1}^n}{(\Delta x)^2} \right] \quad (5.59)$$

where the constant θ $(0 \leq \theta \leq 1)$ is the weight factor that represents the degree of implicitness. That is, equation (5.59) reduces to the simple explicit form for $\theta = 0$, to the Crank–Nicolson method for $\theta = \frac{1}{2}$, and to the simple implicit form for $\theta = 1$.

To establish the order of accuracy of the combined method given by equation (5.59) for different values of the weight factor θ, one needs to examine the leading truncation error term associated with various θ values. The truncation error is determined by examining the *modified equation* associated with equation (5.59), which is the partial differential equation that is

One-Dimensional Transient Systems

actually solved when a finite difference method is applied. The modified equation for the weighted method is given by Pletcher et al. (2012) in the form

$$\frac{\partial T}{\partial t} - \alpha \frac{\partial^2 T}{\partial x^2} = \left[\left(\theta - \frac{1}{2}\right)\alpha^2 \Delta t + \frac{\alpha(\Delta x)^2}{12} \right] \frac{\partial^4 T}{\partial x^4}$$
$$+ \left[\left(\theta^2 - \theta + \frac{1}{3}\right)\alpha^3 (\Delta t)^2 + \frac{1}{6}\left(\theta - \frac{1}{2}\right)\alpha^2 \Delta t (\Delta x)^2 + \frac{1}{360}\alpha(\Delta x)^4 \right] \frac{\partial^6 T}{\partial x^6} + \cdots \quad (5.60)$$

which is determined by substituting the Taylor series expansion for the terms T_i^{n+1}, T_{i-1}^n, and so on about the point (i,n) and then replacing all time derivatives with the spatial derivatives.

The order of accuracy of various difference schemes corresponding to specific values of θ is obtainable from the modified equation (5.60) by examining the leading truncation error terms. For a given value of θ, the lowest-order term on the right-hand side of the modified equation (5.60) gives the order of the method. We obtain:

1. $\theta = 0$, the explicit method: $0[\Delta t, (\Delta x)^2]$
2. $\theta = 1$, the fully implicit method: $0[\Delta t, (\Delta x)^2]$
3. $\theta = \frac{1}{2}$, the Crank–Nicolson method: $0[(\Delta t)^2, (\Delta x)^2]$
4. $\theta = \frac{1}{2} - \frac{(\Delta x)^2}{12\alpha \Delta t} : 0[(\Delta t)^2, (\Delta x)^4]$

Clearly, finite difference schemes with varying degrees of accuracy are obtainable from the combined method by properly choosing the value of the weight factor θ. Figure 5.6 shows the finite difference molecules for the combined method.

To solve equation (5.59), all the unknown temperatures T^{n+1} are moved on one side and all the known temperatures T^n are moved on the other side. We obtain

$$-r\theta T_{i-1}^{n+1} + (1+2r\theta)T_i^{n+1} - r\theta T_{i+1}^{n+1} = r(1-\theta)T_{i-1}^n + [1-2r(1-\theta)]T_i^n + r(1-\theta)T_{i+1}^n \quad (5.61)$$

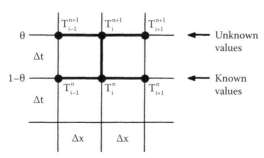

FIGURE 5.6
The finite difference molecules for the combined scheme.

where $r = (\alpha \Delta t)/(\Delta x)^2$. The resulting system of equations (5.61) has a tridiagonal linear coefficient matrix. Hence, it can be solved with the Thomas algorithm discussed in Chapter 3.

When temperatures are prescribed at all boundaries, the system [equation (5.61)] provides a complete set of algebraic equations for the determination of all the unknown internal node temperatures. With convection or prescribed heat flux boundary conditions, the temperatures at the boundary nodes are not known. Additional equations are obtained by either discretizing the boundary condition directly about the boundary node or by the application of a conservation principle for a control volume about the boundary node.

5.1.4.1 Stability of Combined Method

We consider the one-dimensional diffusion equation (5.2) approximated in finite differences by using the combined method as given by equation (5.59). The error terms $\varepsilon_j^n, \varepsilon_{j+1}^n$, and so on defined by equations (5.25a–c) should also satisfy this equation. Then, introducing the error terms given by equations (5.25a–c) into equation (5.59), we obtain

$$e^{i\beta_m j \Delta x}(\xi^{n+1} - \xi^n) = r[\theta e^{i\beta_m j \Delta x} \xi^{n+1}(e^{i\beta_m \Delta x} + e^{-i\beta_m \Delta x} - 2) \\ + (1-\theta)e^{i\beta_m \Delta x} \xi^n (e^{i\beta_m \Delta x} + e^{-i\beta_m \Delta x} - 2)] \quad (5.62)$$

After cancellations and utilizing the identity $\cos(\beta_m \Delta x) = \dfrac{1}{2}(e^{i\beta_m \Delta x} + e^{-i\beta_m \Delta x})$, equation (5.62) takes the form

$$\xi - 1 = -2r\theta\xi(1 - \cos\beta_m \Delta x) - 2r(1 - \theta)(1 - \cos\beta_m \Delta x) \quad (5.63)$$

An alternative form of this equation is obtained by utilizing the trigonometric identity

$$1 - \cos(\beta_m \Delta x) = 2\sin^2\left(\frac{\beta_m \Delta x}{2}\right) \quad (5.64)$$

Then equation (5.63) becomes

$$\xi - 1 = -4r\theta\xi\sin^2\left(\frac{\beta_m \Delta x}{2}\right) - 4r(1-\theta)\sin^2\left(\frac{\beta_m \Delta x}{2}\right) \quad (5.65)$$

This result is solved for ξ

$$\xi = \frac{1 - 4r(1-\theta)\sin^2\left(\dfrac{\beta_m \Delta x}{2}\right)}{1 + 4r\theta\sin^2\left(\dfrac{\beta_m \Delta x}{2}\right)} \quad (5.66)$$

For stability, we require

$$-1 \leq \xi \leq 1 \quad (5.67)$$

Introducing equation (5.66) into equation (5.67), we obtain

$$-1 - 4r\theta\sin^2\left(\frac{\beta_m \Delta x}{2}\right) \leq 1 - 4r(1-\theta)\sin^2\left(\frac{\beta_m \Delta x}{2}\right) \leq 1 + 4r\theta\sin^2\left(\frac{\beta_m \Delta x}{2}\right) \quad (5.68)$$

which must be satisfied for all values of β_m. The right-hand side of this inequality is satisfied for all β_m. Therefore, we need to consider only the left-hand side.

$$-1 - 4r\theta\sin^2\left(\frac{\beta_m \Delta x}{2}\right) \leq 1 - 4r(1-\theta)\sin^2\left(\frac{\beta_m \Delta x}{2}\right)$$

or

$$2r(1-2\theta)\sin^2\left(\frac{\beta_m \Delta x}{2}\right) \leq 1 \quad (5.69)$$

To satisfy this under the most strict condition, we must have $\sin^2\left(\frac{\beta_m \Delta x}{2}\right) = 1$; hence, equation (5.69) reduces to

$$2r(1-2\theta) \leq 1 \quad (5.70)$$

1. When $1 - 2\theta \leq 0$ or $\frac{1}{2} \leq \theta \leq 1$, the combined method is unconditionally stable for all values of r.

2. When $1 - 2\theta > 0$ or $0 \leq \theta < \frac{1}{2}$, the combined method is stable for

$$0 \leq r \leq \frac{1}{2 - 4\theta} \quad (5.71)$$

In this section, we considered the simple explicit, simple implicit, Crank–Nicolson, and combined schemes for the discretization of the one-dimensional diffusion equation (5.1). Table 5.3 summarizes the finite difference forms, the stability criterion, and the order of accuracy for each of these methods.

5.1.5 Cylindrical and Spherical Symmetry

In the previous sections, we presented various finite difference schemes for use with one-dimensional parabolic systems by choosing the one-dimensional diffusion equation in the rectangular coordinate system as the model equation. In this section, we illustrate the application of these finite difference schemes in problems with cylindrical and spherical symmetry by choosing the one-dimensional diffusion equation with a source term as the model equation given by

$$\frac{1}{\alpha}\frac{\partial T(x,t)}{\partial t} = \frac{\partial^2 T}{\partial R^2} + \frac{p}{R}\frac{\partial T}{\partial R} + \frac{1}{k}g(R,t) \quad \text{for } R \neq 0 \quad (5.72)$$

where R denotes the radial variable and $p = \begin{cases} 1 & \text{cylinder} \\ 2 & \text{sphere} \end{cases}$

TABLE 5.3

A Summary of Various Finite Difference Schemes for a Discretizing Diffusion Equation (5.1)

Scheme	Finite Difference Discretization[a]	Stability Criteria[b]	Order of Accuracy
Simple explicit	$\dfrac{T_i^{n+1} - T_i^n}{\Delta t} = \alpha \Delta_{xx} T_i^n$	Stable for $r \leq \tfrac{1}{2}$	$0[(\Delta x)^2, \Delta t]$
Simple implicit	$\dfrac{T_i^{n+1} - T_i^n}{\Delta t} = \alpha \Delta_{xx} T_i^{n+1}$	Always stable	$0[(\Delta x)^2, \Delta t]$
Crank–Nicolson	$\dfrac{T_i^{n+1} - T_i^n}{\Delta t} = \dfrac{\alpha}{2}[\Delta_{xx} T_i^{n+1} + \Delta_{xx} T_i^n]$	Always stable	$0[(\Delta x)^2, (\Delta t)^2]$
Combined	$\dfrac{T_i^{n+1} - T_i^n}{\Delta t} = \alpha[\theta \Delta_{xx} T_i^{n+1} + (1-\theta)\Delta_{xx} T_i^n]$	$\theta = 0$: Simple explicit $\theta = 1$: Simple implicit $\theta = 1/2$: Crank–Nicolson	

[a] $\Delta_{xx} T_i \equiv \dfrac{T_{i-1} - 2T_i + T_{i+1}}{(\Delta x)^2}$.

[b] $r = \dfrac{\alpha \Delta t}{(\Delta x)^2}$.

When the solution domain includes the origin $R = 0$, as in the case of a solid cylinder and sphere, the apparent singularity at $R = 0$ is avoided if equation (5.72) is replaced by (see Chapter 4):

$$\frac{1}{\alpha}\frac{\partial T(R,t)}{\partial t} = (1+p)\frac{\partial^2 T}{\partial R^2} + \frac{1}{k}g(R,t) \quad \text{for} \quad R=0 \qquad (5.73)$$

We now examine the finite difference approximation of these equations by using the finite differencing schemes discussed previously.

5.1.6 Application of Simple Explicit Method

The simple explicit method is now applied to discretize the one-dimensional diffusion equation with a source term in cylindrical and spherical symmetry. The cases of a solid cylinder and sphere and a hollow cylinder and sphere are considered separately because the former includes the origin in the solution domains; hence, it requires the finite difference approximation of equation (5.73) in addition to equation (5.72).

5.1.6.1 *Solid Cylinder and Sphere*

Consider a solid cylinder or sphere of radius $R = b$. The solution domain $0 \leq R \leq b$ is divided into M layers each of thickness

$$\delta = \frac{b}{M} \qquad (5.74)$$

as illustrated in Figure 5.7. The finite difference equations for the internal nodes i = 1,2,...,M–1 are developed by discretizing the differential equation (5.72) and, for the center node R = 0, by discretizing equation (5.73). We obtain

$$\frac{T_i^{n+1} - T_i^n}{\alpha \Delta t} = \frac{T_{i-1}^n - 2T_i^n + T_{i+1}^n}{\delta^2} + \frac{p}{i\delta} \frac{T_{i+1}^n - T_{i-1}^n}{2\delta} + \frac{g_i^n}{k} \quad (5.75)$$
$$\text{for } i = 1, 2 \ldots, M - 1$$

and

$$\frac{T_0^{n+1} - T_0^n}{\alpha \Delta t} = \frac{1}{\delta^2}\left[2(1+p)(T_1^n - T_0^n) + \frac{\delta^2 g_0^n}{k}\right] \quad \text{for} \quad i = 0 \quad (5.76)$$

since $\left.\dfrac{\partial^2 T}{\partial R^2}\right|_{R=0} = \dfrac{T_1 - 2T_0 + T_{-1}}{\delta^2}$ and, with the symmetry condition, we can write $\left.\dfrac{\partial^2 T}{\partial R^2}\right|_{R=0} = \dfrac{2(T_1 - T_0)}{\delta^2}$.

Equations (5.75) and (5.76) are rearranged, respectively, as

$$T_i^{n+1} = r\left(1 - \frac{p}{2i}\right)T_{i-1}^n + (1 - 2r)T_i^n + r\left(1 + \frac{p}{2i}\right)T_{i+1}^n + r\frac{\delta^2 g_i^n}{k} \quad (5.77)$$
$$\text{for } i = 1, 2, \ldots, M - 1$$

and

$$T_0^{n+1} = [1 - 2r(1+p)]T_0^n + 2r(1+p)T_1^n + r\frac{\delta^2 g_0^n}{k} \quad \text{for } i = 0 \quad (5.78)$$

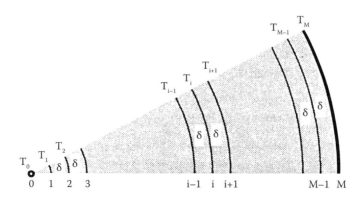

FIGURE 5.7
Nomenclature for finite difference representation for cylindrical and spherical symmetry.

where

$$r = \frac{\alpha \Delta t}{\delta^2}, \quad \delta = \frac{b}{M}, \quad \text{and} \quad p = \begin{cases} 1 & \text{cylinder} \\ 2 & \text{sphere} \end{cases} \quad (5.79)$$

which are accurate to $O[\delta^2, \Delta t]$.

Equations (5.77) and (5.78) provide M relations but they contain M + 1 unknown node potentials T_i^{n+1}, $i = 0,1,2,\ldots,M$ for the time level n + 1. An additional relationship is developed by utilizing the boundary condition at the outer surface R = b. The boundary condition can be a prescribed surface temperature, heat flux, or convection. We examine each of these three cases separately.

1. Prescribed temperature T_b at the boundary R = b: For such a case, we have

$$T_M = T_b \quad \text{for } i = M \quad (5.80)$$

   for all time levels. Then, equations (5.77) and (5.78), together with equation (5.80), are sufficient to calculate the M unknown internal node temperatures T_i^{n+1} (i = 0,1,2,...,M−1) from the knowledge of the node temperatures at the time level n.

2. Convection at the boundary R = b: For this case, the boundary condition at R = b is given by

$$k \frac{\partial T}{\partial R} + h_b T = h_b T_{\infty,b} \quad \text{at } R = b \quad (5.81)$$

   where $T_{\infty,b}$ is the ambient temperature and h_b the heat transfer coefficient. To discretize the derivative term in this equation with a second-order accurate central difference formula, we consider a fictitious node M + 1 at a fictitious temperature T_{M+1} located at a distance δ to the right of the node M, as illustrated in Figure 5.8. Then equation (5.81) can be discretized by using the central difference formula as

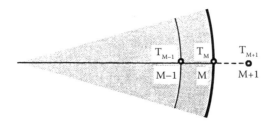

FIGURE 5.8
Fictitious node M + 1.

One-Dimensional Transient Systems

$$k \frac{T_{M+1}^n - T_{M-1}^n}{2\delta} + h_b T_M^n = h_b T_{\infty,b} \tag{5.82}$$

To eliminate the fictitious temperature T_{M+1}^n from this expression, an additional relationship is obtained by evaluating equation (5.77) for $i = M$, to yield

$$T_M^{n+1} = r\left(1 - \frac{p}{2M}\right) T_{M-1}^n + (1 - 2r) T_M^n + r\left(1 + \frac{p}{2M}\right) T_{M+1}^n + r \frac{\delta^2 g_M^n}{k},$$
$$\text{for } i = M \tag{5.83}$$

Eliminating T_{M+1}^n between equations (5.82) and (5.83), we obtain

$$T_M^{n+1} = 2r T_{M-1}^n + (1 - 2r\beta_M) T_M^n + 2r\gamma_M + r G_M^n \quad \text{for } i = M \tag{5.84}$$

where

$$\beta_M = 1 + \left(1 + \frac{p}{2M}\right) \frac{\delta h_b}{k} \tag{5.85a}$$

$$\gamma_M = \left(1 + \frac{p}{2M}\right) \frac{\delta}{k} h_b T_{\infty,b} \tag{5.85b}$$

$$G_M = \frac{\delta^2 g_M^n}{k} \tag{5.85c}$$

$$p = \begin{cases} 1 & \text{cylinder} \\ 2 & \text{sphere} \end{cases} \tag{5.85d}$$

Then, equations (5.77) and (5.78), together with equation (5.84), provide $M + 1$ relations for the determination of $M + 1$ unknown node temperatures T_i^{n+1}, $i = 0,1,2,\ldots,M$ when the node temperatures T_i^n at the previous time level are available. The calculations are started with the initial condition.

3. **Prescribed heat flux at the boundary $R = b$:** For this case, the boundary condition is given by

$$k \frac{\partial T}{\partial R} = q_b \quad \text{at } R = b \tag{5.86}$$

Clearly, this boundary condition is obtainable from the convection boundary condition [equation (5.81)] as a special case. Then the finite difference form of equation (5.86) is immediately obtained from equation (5.84) by making

$$\beta_M = 1, \gamma_M = \left(1 + \frac{p}{2M}\right)\frac{\delta}{k} q_b \qquad (5.87a,b)$$

We obtain

$$T_M^{n+1} = 2rT_{M-1}^n + (1-2r)T_M^n + 2r\left(1 + \frac{p}{2M}\right)\frac{\delta}{k} q_b + \frac{r\delta^2 g_M^n}{k}, \quad \text{for } i = M \qquad (5.88)$$

Then, equations (5.77) and (5.78), together with equation (5.88), provide M + 1 relations for the determination of M + 1 unknown node temperatures i = 0,1,...,M at the time level n + 1 from the knowledge of the node temperatures T_i^{n+1}, i = 0,1,2,...,M, at the time level n + 1, from the knowledge of the node temperature at the time level n.

The limitations of the solutions obtained with prescribed heat flux boundary conditions should also be recognized. When a solid cylinder or sphere is subjected to a uniform heat supply at its boundary surfaces, the temperature of the solid will continuously rise because there is no way for heat to escape from the medium. Therefore, under such conditions, no steady-state solution is obtainable for the problem. If the sum of the heat rate generated within the medium with the heat rate supplied at the boundaries (in the form of an applied heat flux) is zero, a steady-state solution exists, but the solution for temperature will be unique only to within an additive constant.

5.1.6.2 Stability of Solution

With the finite difference scheme considered here being explicit, some restriction should be imposed on the maximum permissible value of the parameter $r = \frac{\alpha \Delta t}{\delta^2}$. To illustrate this matter, we consider the finite difference equation (5.77) for the internal nodes i = 1,2,...,M−1 and examine the stability criterion associated with it by following a physical argument similar to that used previously for determining the stability of finite
difference equations in the rectangular coordinate system. Suppose at any time level n the node temperatures T_{i-1}^n and T_{i+1}^n are equal and that there is no energy generation in the medium. Then equation (5.77) reduces to

$$T_i^{n+1} = 2rT_{i-1}^n + (1-2r)T_i^n \qquad (5.89)$$

For illustration purposes, let $T_i^n = 100°C$ and the temperatures T_{i-1}^n and T_{i+1}^n of the two neighboring nodes be $0°C$. Then equation (5.89) reduces to

$$T_i^{n+1} = (1 - 2r)100 \tag{5.90}$$

For the specific case considered here, the physical situation requires that T_i^{n+1} cannot go below the temperature of the two neighboring nodes, $T_{i-1}^n = T_{i+1}^n = 0°C$. Therefore, if equation (5.90) is used to determine T_i^{n+1}, a negative value of $(1-2r)$ will violate such a requirement. To obtain meaningful results from the solution of the finite difference equation (5.77), the following stability criterion should be satisfied

$$1 - 2r \geq 0 \quad \text{or} \quad r \equiv \frac{\alpha \Delta t}{\delta^2} \leq \frac{1}{2} \tag{5.91}$$

Clearly, this stability criterion is the same as that determined for the rectangular coordinate system.

In the case of equation (5.78) for the center node $i = 0$, the application of a physical argument similar to the one discussed earlier or the use of a Fourier method of stability analysis leads to a restriction on r. A discussion of the stability of a finite difference equation for the center node $i = 0$ of a solid sphere using the matrix method of analysis is given by Smith (1978).

In the case of a convection boundary condition at the outer surface $R = b$, by following a similar physical argument, one finds the following stability criterion

$$r \equiv \frac{\alpha \Delta t}{\delta^2} \leq \frac{1}{2\beta_m} \tag{5.92a}$$

where

$$\beta_m = 1 + \left(1 + \frac{p}{2M}\right)\frac{\delta h_b}{k} \tag{5.92b}$$

$$p = \begin{cases} 1 & \text{cylinder} \\ 2 & \text{sphere} \end{cases} \tag{5.92c}$$

and h_b is the heat transfer coefficient. For all practical purposes, M is sufficiently large so that the term $p/2M$ can be neglected; then, the stability criterion given by equations (5.92a–c) reduces to that given by equation (5.20b) for the rectangular coordinate system.

Clearly, the most restrictive stability criterion should be used for the value of r in numerical calculations.

5.1.6.3 Hollow Cylinder and Sphere

In the case of a hollow cylinder or sphere of inner radius a and outer radius b, the solution domain $a \leq R \leq b$ is divided into M equal layers, each of thickness

$$\delta = \frac{b-a}{M} \tag{5.93}$$

as illustrated in Figure 5.9. Then the governing differential equation is given by

$$\frac{1}{\alpha}\frac{\partial T(R,t)}{\partial t} = \frac{\partial^2 T}{\partial R^2} + \frac{p}{R}\frac{\partial T}{\partial R} + \frac{1}{k}g(R,t), \text{ in } a < R < b \tag{5.94}$$

This equation is discretized by using the simple explicit scheme. The resulting finite difference equations for the internal nodes are given by

$$\frac{T_i^{n+1} - T_i^n}{\alpha \Delta t} = \frac{T_{i-1}^{n+1} - 2T_i^n + T_{i+1}^n}{\delta^2} + \frac{p}{a+i\delta}\frac{T_{i+1}^n - T_{i-1}^n}{2\delta} + \frac{1}{k}g_i^n \tag{5.95}$$

$$\text{for } i = 1, 2, \ldots, M-1$$

Equation (5.95) is solved for T_i^{n+1}:

$$T_i^{n+1} = r\left[1 - \frac{p}{2\left(\frac{a}{\delta}+i\right)}\right]T_{i-1}^n + (1-2r)T_i^n + r\left[1 + \frac{p}{2\left(\frac{a}{\delta}+i\right)}\right]T_{i+1}^n + r\frac{\delta^2 g_i^n}{k}$$

$$\text{for } i = 1, 2, \ldots, M-1$$

(5.96a)

where

$$\delta = \frac{b-a}{M}, \quad r = \frac{\alpha \Delta t}{\delta^2}, \quad p = \begin{cases} 1 & \text{cylinder} \\ 2 & \text{sphere} \end{cases} \tag{5.96b}$$

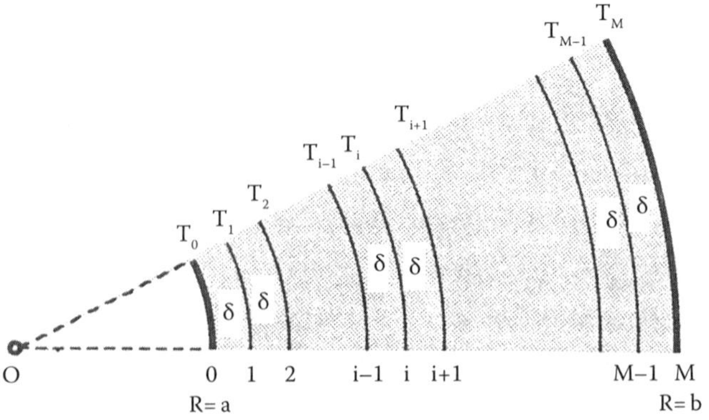

FIGURE 5.9
Nomenclature for finite difference representation for a hollow cylinder or sphere.

Note that, for a = 0, equation (5.96a) reduces to equation (5.77) for a solid cylinder or solid sphere.

Equations (5.96a,b) provide M−1 relations, but the problem involves M + 1 unknown node temperatures $T_i^{n+1}, i = 0, 1, 2, \ldots, M$ at each time level. Two additional relations needed to make the number of equations equal to the number of unknowns are obtained by utilizing the boundary conditions at the surfaces R = a and R = b. If the temperatures are prescribed at both boundaries, then equations (5.96a,b) are sufficient to solve for the unknown node temperatures, $T_i^{n+1}, i = 1, 2, \ldots, M-1$.

In the case of convection or prescribed heat flux at the boundaries, the boundary temperatures are unknown. Two additional relations are developed by discretizing the boundary conditions. In order to obtain a second-order accurate result, central differencing should be used by considering fictitious nodes T_{-1} and T_{M+1} to the left of the boundary R = a and to the right of the boundary R = b, respectively. The fictitious temperatures T_{-1} and T_{M+1} appearing in the resulting equations are then eliminated by means of the expressions obtained by evaluating equation (5.96a) for i = 0 and i = M, respectively, in a similar manner as discussed in Chapter 4.

Example 5.5

Consider the following transient radial heat conduction in a solid cylinder, $0 \leq \eta \leq 1$, given in the dimensionless form as

$$\frac{\partial^2 T}{\partial \eta^2} + \frac{1}{\eta}\frac{\partial T}{\partial \eta} = \frac{\partial T}{\partial t} \quad \text{in} \quad 0 < \eta < 1, \quad t > 0$$

$$\frac{\partial T}{\partial \eta} = 0 \quad \text{at} \quad \eta = 0, \quad t > 0$$

$$T = 0 \quad \text{at} \quad \eta = 1, \quad t > 0$$

$$T = 100 J_0(\beta_1 \eta) \quad \text{for} \quad t = 0, \quad 0 \leq \eta \leq 1$$

where $J_0(z)$ is the zero-order Bessel function of the first kind and β_1 is the first root of $J_0(z) = 0$. The exact analytic solution of this problem is given by

$$T(\eta, t) = 100 J_0(\beta_1 \eta) e^{-\beta_1^2 t}$$

By dividing the solution domain into five equal parts and using the simple explicit scheme, solve this problem with finite differences and compare the center temperature T(0,t) with the exact solution at dimensionless times t = 0.2, 0.4, 0.6,..., and 1.6.

Solution
The finite difference equations are obtained from equations (5.77) and (5.78) by setting p = 1 (i.e., cylindrical coordinate) and $g_i^n = 0$ (i.e., no source).

TABLE 5.4

A Comparison of the Finite Difference Solution and the Exact Results for Example 5.5

	T(0,t), Center Temperature	
Time t	Exact	Explicit r = 0.2, δ = 0.2
0.2	31.4550	31.4028
0.4	9.8942	9.8693
0.6	3.1122	3.1017
0.8	0.9789	0.9748
1.0	0.3079	0.3064
1.2	0.0969	0.0963
1.4	0.0305	0.0303
1.6	0.0096	0.0095

Equations (5.77) and (5.78), respectively, give

$$T_i^{n+1} = r\left(1 - \frac{1}{2i}\right)T_{i-1}^n + (1 - 2r)T_i^n + r\left(1 + \frac{1}{2i}\right)T_{i+1}^n \quad \text{for} \quad i = 1, 2, 3, 4$$

and

$$T_0^{n+1} = (1 - 4r)T_0^n + 4rT_1^n \quad \text{for} \quad i = 0$$

These equations are solved and the results for the center temperature T(0,t) are listed in Table 5.4.

5.1.7 Application of Simple Implicit Scheme

We now illustrate the application of the simple implicit method of finite differencing in cylindrical and spherical coordinates with symmetry by taking the one-dimensional diffusion equation with the source term as the model equation. The cases of a solid cylinder and sphere and a hollow cylinder and sphere are considered separately because in the former, the solution domain includes the origin R = 0; hence, equation (5.73) needs to be considered in addition to equation (5.72).

5.1.7.1 Solid Cylinder and Sphere

Consider a solid cylinder or sphere of radius R = b. To develop the finite difference approximation, the solution domain $0 \leq R \leq b$ is divided into M layers each of thickness $\delta = \frac{b}{M}$, as illustrated in Figure 5.7. The finite difference approximation of equations (5.72) and (5.73), with the simple implicit scheme, respectively, gives

$$\frac{T_i^{n+1} - T_i^n}{\alpha \Delta t} = \frac{1}{\delta^2}\left\{\left[1 - \frac{p}{2i}\right]T_{i-1}^{n+1} - 2T_i^{n+1} + \left[1 + \frac{p}{2i}\right]T_{i+1}^{n+1} + \frac{\delta^2 g_i^{n+1}}{k}\right\}$$

$$\text{for the internal nodes} \quad i = 1, 2, \ldots, M - 1 \quad (5.97)$$

One-Dimensional Transient Systems

and

$$\frac{T_0^{n+1} - T_0^n}{\alpha \Delta t} = \frac{1}{\delta^2}\left[2(1+p)(T_1^{n+1} - T_0^{n+1}) + \frac{\delta^2 g_0^{n+1}}{k}\right] \text{ for the center node } i = 0$$

(5.98a)

where

$$\delta = \frac{b}{M}$$ (5.98b)

These equations are unconditionally stable, accurate to $0(\delta^2, \Delta t)$, and provide M relations, but they contain M + 1 unknown node temperatures. An additional relation is obtained by utilizing the boundary condition at R = b. If the boundary condition is a prescribed temperature, equations (5.97) and (5.98) are sufficient to determine the M unknown internal node temperatures at a time level n + 1 from the knowledge of temperatures at the time level n, starting with the initial temperature distribution. In the case of convection or prescribed heat flux at the boundary R = b, an additional relation is obtained by discretizing the boundary condition. A central difference approximation should be used to discretize the boundary condition by following the procedures described previously, in order to obtain a second-order accurate finite difference approximation.

5.1.7.2 Hollow Cylinder and Sphere

In the case of a hollow cylinder or sphere of inner radius a and outer radius b, the solution domain $a \leq R \leq b$ is divided into M equal layers, each of thickness $\delta = \frac{b-a}{M}$, as illustrated in Figure 5.9, and the governing differential equation (5.72) is discretized by using the simple implicit scheme to obtain

$$\frac{T_i^{n+1} - T_i^n}{\alpha \Delta t} = \frac{1}{\delta^2}\left\{\left[1 - \frac{p}{2\left(\frac{a}{\delta} + i\right)}\right]T_{i-1}^{n+1} - 2T_i^{n+1} + \left[1 + \frac{p}{2\left(\frac{a}{\delta} + i\right)}\right]T_{i+1}^{n+1} + \frac{\delta^2 g_i^{n+1}}{k}\right\}$$

for the internal nodes $i = 1, 2, \ldots, M-1$, (5.99a)

where

$$\delta = \frac{b-a}{M}$$ (5.99b)

These equations are unconditionally stable, accurate to $0(\delta^2, \Delta t)$, and provide M−1 relations, but the problem contains M + 1 unknown node temperatures. Two additional relations are obtained from the boundary conditions at R = a and R = b. If temperatures are prescribed at the boundaries, equations (5.99a,b) are sufficient to determine the unknown internal node temperatures starting with the initial condition. If the boundary conditions are convection or prescribed heat flux, two additional relations are obtained by discretizing these boundary conditions. A central difference approximation should be used to discretize these boundary conditions in order to obtain a second-order accurate finite difference approximation for the boundary conditions.

5.1.8 Application of Crank–Nicolson Method

The application of the Crank–Nicolson method for finite differencing in cylindrical and spherical coordinates with symmetry is now illustrated. A one-dimensional diffusion equation with a source term is taken as the model equation, and only the case of a solid cylinder and sphere of radius $R = b$ is considered. To develop the finite difference approximation, the solution domain $0 \leq R \leq b$ is divided into M layers, each of thickness $\delta = \dfrac{b}{M}$, as illustrated in Figure 5.7. The finite difference equations applicable for the internal nodes $i = 1, 2, \ldots, M-1$ are developed by discretizing equation (5.72) with the Crank–Nicolson method. We obtain

$$\frac{T_i^{n+1} - T_i^n}{\alpha \Delta t} = \frac{1}{2\delta^2} \left\{ \left[1 - \frac{p}{2i}\right] T_{i-1}^{n+1} - 2T_i^{n+1} + \left[1 + \frac{p}{2i}\right] T_{i+1}^{n+1} \right.$$
$$\left. + \left[1 - \frac{p}{2i}\right] T_{i-1}^n - 2T_i^n + \left[1 + \frac{p}{2i}\right] T_{i+1}^n + \frac{\delta^2}{k}(g_i^{n+1} + g_i^n) \right\} \quad (5.100)$$

for the internal nodes $i = 1, 2, \ldots, M-1$.

The finite difference equation for the center node $i = 0$ is developed by discretizing equation (5.73) with the Crank–Nicolson method to give

$$\frac{T_0^{n+1} - T_0^n}{\alpha \Delta t} = \frac{1}{2\delta^2} \left\{ 2(1+p)(T_1^{n+1} - T_0^{n+1}) + 2(1+p)(T_1^n - T_0^n) + \frac{\delta^2}{k}(g_0^{n+1} + g_0^n) \right\}$$

for the internal nodes $i = 0$ \quad (5.101)

Equations (5.100) and (5.101) are rearranged, respectively, in the form

$$-r\left[1 - \frac{p}{2i}\right] T_{i-1}^{n+1} + (2 + 2r) T_i^{n+1} - r\left[1 + \frac{p}{2i}\right] T_{i+1}^{n+1} = r\left[1 - \frac{p}{2i}\right] T_{i-1}^n + (2 - 2r) T_i^n$$
$$+ r\left[1 + \frac{p}{2i}\right] T_{i-1}^n + \frac{r\delta^2}{k}(g_i^{n+1} + g_i^n) \quad \text{for} \quad i = 1, 2, \ldots, M-1$$

(5.102)

and

$$[2 + 2r(1+p)] T_0^{n+1} - 2r(1+p) T_1^{n+1} = [2 - 2r(1+p)] T_0^n + 2r(1+p) T_1^n$$
$$+ \frac{r\delta^2}{k}(g_0^{n+1} + g_0^n) \quad \text{for} \quad i = 0$$

(5.103)

where

$$\delta = \frac{b}{M} \quad \text{and} \quad r = \frac{\alpha \Delta t}{\delta^2}$$

Equations (5.102) and (5.103) are unconditionally stable, accurate to $0[\delta^2, (\Delta t)^2]$, provide M simultaneous algebraic equations, but contain M + 1 unknown node temperatures. An additional relation is obtained from the boundary condition at R = b. We consider the following three different situations at the boundary surface R = b.

1. Prescribed temperature T_b at the boundary R = b: For such a case, we have

$$T_M = T_b \quad \text{for} \quad i = M \tag{5.104}$$

for all time levels. Then equations (5.102) and (5.103) are sufficient to calculate the M unknown internal node temperatures, T_i^{n+1}, $i = 0, 1, 2, \ldots, M - 1$, at the time level n + 1 from the knowledge of the T_i^n at the previous time level.

2. Convection at the boundary R = b: For this case, the boundary condition at R = b is given by

$$k \frac{\partial T}{\partial R} + h_b T = h_b T_{\infty,b} = \text{known at } R = b \tag{5.105}$$

This expression should be discretized with a second-order accurate central difference formula by considering a fictitious node M + 1 at a fictitious temperature T_{M+1}, located at a distance δ to the right of the node M, as illustrated in Figure 5.8. The discretization of equation (5.105) for the time levels n and n + 1, respectively, gives

$$k \frac{T_{M+1}^n - T_{M-1}^n}{2\delta} + h_b T_M^n = h_b T_{\infty,b} \tag{5.106a}$$

and

$$k \frac{T_{M+1}^{n+1} - T_{M-1}^{n+1}}{2\delta} + h_b T_M^{n+1} = h_b T_{\infty,b} \tag{5.106b}$$

Then, we evaluate the finite difference equation (5.102) for i = M and obtain

$$-r\left(1 - \frac{P}{2M}\right) T_{M-1}^{n+1} + (2 + 2r) T_M^{n+1} - r\left(1 + \frac{P}{2M}\right) T_{M+1}^{n+1} = r\left(1 - \frac{P}{2M}\right) T_{M-1}^n$$
$$+ (2 - 2r) T_M^n + r\left(1 + \frac{P}{2M}\right) T_{M-1}^n + \frac{r\delta^2}{k}(g_M^{n+1} + g_M^n) \quad \text{for} \quad i = M$$

$$\tag{5.107}$$

where $r = \dfrac{\alpha \Delta t}{\delta^2}$.

The fictitious temperatures T_{M+1}^n and T_{M+1}^{n+1} appearing in this expression are eliminated by utilizing equations (5.106a) and (5.106b), respectively. Then, the second-order accurate finite difference approximation of the boundary condition [equation (5.105)] becomes

$$-2rT_{M-1}^{n+1} + (2 + 2r\beta_M)T_M^{n+1} = 2rT_{M-1}^n + (2 - 2r\beta_M)T_M^n + 4r\gamma_M + rG_M^n$$
$$\text{for } i = M \tag{5.108}$$

where

$$\beta_M = 1 + \left(1 + \frac{p}{2M}\right)\frac{\delta h_b}{k} \tag{5.109a}$$

$$\gamma_M = \left(1 + \frac{p}{2M}\right)\frac{\delta}{k}h_b T_{\infty,b} \tag{5.109b}$$

$$G_M^n = \frac{\delta^2}{k}(g_M^{n+1} + g_M^n), \quad r = \frac{\alpha \Delta t}{\delta^2} \tag{5.109c,d}$$

Summarizing, Equations (5.102), (5.103), and (5.108) provide $M + 1$ simultaneous algebraic equations for the determination of $M + 1$ unknown node temperatures, $T_i^{n+1}, i = 0, 1, 2, \ldots, M$, at the time level $n + 1$ from the knowledge of the node temperatures at the previous time level n, starting with the knowledge of the initial distribution.

3. Prescribed heat flux at the boundary $R = b$: In this case, we have

$$k\frac{\partial T}{\partial R} = q_b \quad \text{at} \quad R = b \tag{5.110}$$

A comparison of this heat flux boundary condition with the convection boundary condition, equation (5.105), reveals that the second-order accurate finite difference approximation to the prescribed heat flux boundary condition, equation (5.110), is readily obtained from equations (5.108) by making

$$\beta_M = 1, \quad \gamma_M = \left(1 + \frac{p}{2M}\right)\frac{\delta}{k}q_b \tag{5.111a,b}$$

We find

$$-2rT_{M-1}^{n+1} + (2 + 2r)T_M^{n+1} = 2rT_{M-1}^n + (2 - 2r)T_M^n + 4r\left[1 + \frac{p}{2M}\right]\frac{\delta}{k}q_b + rG_M^n,$$
$$\text{for } i = M \tag{5.112}$$

The Crank–Nicolson method of finite differencing given here is unconditionally stable; hence, there is no restriction on the maximum value of the parameter r.

5.2 Advective–Diffusive Systems

After the purely diffusive models considered in the previous section, we now address cases where fluid flow influences the transport of the conserved quantity in transient one-dimensional problems. Although the Taylor series and the finite control volume approaches for the discretization of the governing equations are applicable to the cases examined in this section, care must be exercised with respect to the discretization of advective terms, which actually gave rise to several discretization schemes in the past, some of which are discussed here. Both the purely advective equation and the advective–diffusive equations are examined in this section.

5.2.1 Purely Advective (Wave) Equation

The purely one-dimensional linear advective equation is hyperbolic and is given in the form

$$\underbrace{\frac{\partial T}{\partial t}}_{\text{unsteady term}} + \underbrace{u \frac{\partial T}{\partial x}}_{\text{advective term}} = 0 \qquad (5.113)$$

which is generally called the first-order wave equation. For u constant and positive, it describes a wave motion in the positive x direction with a velocity u.

The exact solution of this equation for an initial condition $T(x,0) = F(x)$ in the domain $-\infty < x < \infty$, is given by

$$T(x, t) = F(x - ut) \qquad (5.114)$$

which implies that the initial distribution is translated along the x-axis with a velocity u, without a change in shape.

Although equation (5.113) characterizes a very simple, idealized situation, it serves well as a model equation to illustrate the behavior of more complicated hyperbolic equations when they are to be solved numerically. Several different finite difference schemes have been proposed for the numerical solution of hyperbolic equations. Here, as applied to the first-order wave equation, we consider the standard upwind differencing, MacCormack's method, and Warming and Beam's second-order upwind scheme. These schemes are important from a historical perspective and permit a discussion regarding the influence of the truncation error on the behavior of the numerical solution. More recent and accurate schemes are presented in Chapter 9, as applied to quasi-one-dimensional compressible flow equations and to Euler's two-dimensional equations.

5.2.1.1 Upwind Method

Upwind differencing was discussed previously. It is a one-step, explicit, finite differencing scheme that uses forward differencing in time and one-sided differencing in space, such that backward differencing in space is used for $u > 0$ and forward differencing for $u < 0$. We adopt the notation

$$T(x,t) = T(j\Delta x, n\Delta t) \equiv T_j^n \quad (5.115)$$

Then the finite difference approximation of equation (5.113) with the upwind scheme becomes

$$\frac{T_j^{n+1} - T_j^n}{\Delta t} = -u \frac{T_j^n - T_{j-1}^n}{\Delta x} \quad \text{backward differencing for } u > 0 \quad (5.116a)$$

and

$$\frac{T_j^{n+1} - T_j^n}{\Delta t} = -u \frac{T_{j+1}^n - T_j^n}{\Delta x} \quad \text{forward differencing for } u < 0 \quad (5.116b)$$

Solving these equations for T_j^{n+1}, we obtain

$$T_j^{n+1} = T_j^n - c(T_j^n - T_{j-1}^n) \quad \text{for } u > 0 \quad (5.117a)$$

and

$$T_j^{n+1} = T_j^n - c(T_{j+1}^n - T_j^n) \quad \text{for } u < 0 \quad (5.117b)$$

where c is the *Courant number*

$$c = \frac{u\Delta t}{\Delta x} \quad (5.118)$$

By Taylor series expansion, it can be shown that the method is accurate to $0(\Delta t, \Delta x)$.

5.2.1.1.1 Stability Criteria

The method being explicit, the stable solution of the aforementioned finite difference equations is restricted to certain values of the Courant number. It can be shown by Fourier stability analysis that the stability criterion is given by

$$0 < c \leq 1 \quad (5.119)$$

5.2.1.1.2 Modified Equation

It is instructive to examine the modified equation associated with the wave equation (5.113). It is given by Pletcher et al. (2012):

$$\frac{\partial T}{\partial t} + u\frac{\partial T}{\partial x} = \frac{u\Delta x}{2}(1-c)\frac{\partial^2 T}{\partial x^2} - \frac{u(\Delta x)^2}{6}(2c^2 - 3c + 1)\frac{\partial^3 T}{\partial x^3}$$

$$+ 0[(\Delta x)^3, (\Delta x)^2 \Delta t, \Delta x(\Delta t)^2, (\Delta t)^3] \quad (5.120)$$

According to the *heuristic* stability analysis, the necessary condition for stability can be obtained from the coefficient of the lowest-order even derivative on the right-hand side of equation (5.120) as

$$\frac{u\Delta x}{2}(1-c) \geq 0 \text{ or } c \leq 1 \quad (5.121\text{a,b})$$

which is the same as that given by equation (5.119), developed by Fourier stability analysis. Equation (5.121b) is called the Courant–Friedrichs–Lewy condition.

The right-hand side of the modified equation (5.120) represents the difference between the original partial differential equation and its finite difference approximation. For $c = 1$, the right-hand side of the modified equation becomes zero; hence, the equation is solved exactly. Furthermore, the upstream difference equations (5.117a,b) reduce to

$$T_j^{n+1} = T_{j-1}^n \text{ for } u > 0 \quad (5.122\text{a})$$

$$T_j^{n+1} = T_{j+1}^n \text{ for } u < 0 \quad (5.122\text{b})$$

which are equivalent to solving the problem exactly.

The modified Equation (5.120) can also provide information on the dispersion and dissipation errors associated with the numerical scheme. The dissipation error is the result of the lowest-order term in the modified equation that contains an even derivative. The dissipation error tends to spread a sharp wave front. The dispersion error is the result of the lowest-order term in the modified equation that contains an odd derivative and tends to distort a sharp wave front by producing wiggles immediately before and/or after the front. Dissipation is characteristic of first-order methods, such as the upwind scheme of equations (5.117a,b), while dispersion is characteristic of second-order methods, such as the MacCormack and the Warming and Beam methods presented as follows. The effects of dissipation and dispersion will be apparent with two examples, after the presentation of the MacCormack and the Warming and Beam methods.

Example 5.6

By applying the Fourier stability analysis, show that the stability criterion given by equation (5.119) should be satisfied for stable solution of the finite difference equation given by equation (5.117).

Solution
As discussed previously, the error term is taken in the form

$$\varepsilon_j^n = \xi^n \, e^{i\beta_m j \Delta x} \tag{a}$$

where $i = \sqrt{-1}$ and ξ is the amplification factor. Equation (a) is introduced into equation (5.117a), and cancellations are made. We obtain

$$\xi = 1 - c(1 - e^{-i\beta_m \Delta x}) \tag{b}$$

which is written in the form

$$\xi = (1 - c + \cos\theta) - i\, c \, \sin\theta \tag{c}$$

where

$$\theta = \beta_m \Delta x \tag{d}$$

The amplification factor is complex. For stability, the modulus of the amplification factor should be less than unity, that is, $|\xi| \leq 1$. The modulus of ξ is determined from Equation (c) as

$$|\xi|^2 = \xi \cdot \xi^* = (1 - c + c\cos\theta)^2 + c^2(1 - \cos^2\theta) \tag{e}$$

where ξ^* is the complex conjugate. To determine the stability criterion, $|\xi|$ needs to be plotted for several different values of c for all values of θ from 0 to π. Such a plot will reveal that the stability condition $|\xi| \leq 1$ is satisfied if the Courant number remains less than or equal to unity.

5.2.1.2 MacCormack's Method

This explicit finite difference scheme proposed by MacCormack (1969) involves a predictor and corrector procedure. The method is particularly useful for solving nonlinear partial differential equations involving a propagating front. The main advantage of this method over other numerical schemes is that it allows high resolution of discontinuities when the finite differencing of the predictor is in the direction of a propagating front. The finite differencing of equation (5.113) with the MacCormack method leads to the following steps for the determination of T_j^{n+1}:

$$\text{Predictor: } \overline{T_j^{n+1}} = T_j^n - u \frac{\Delta t}{\Delta x}(T_{j+1}^n - T_j^n) \tag{5.123a}$$

One-Dimensional Transient Systems

$$\text{Corrector}: T_j^{n+1} = \frac{1}{2}\left[T_j^n + \overline{T_j^{n+1}} - u\frac{\Delta t}{\Delta x}(\overline{T_j^{n+1}} - \overline{T_{j-1}^{n+1}})\right] \quad (5.123b)$$

with a truncation error $0[(\Delta x)^2,(\Delta t)^2]$.

Here the predictor, denoted by $\overline{T_j^{n+1}}$, gives a temporary approximate value of T_j at an intermediate time level $\overline{n+1}$, while the corrector, denoted by T_j^{n+1}, gives the final value of T_j at time level n + 1 at the grid point j. We note that, in the predictor, a forward difference in space is used, while in the corrector, a backward difference in space is used.

This explicit scheme is second-order accurate, has a truncation error $0[(\Delta x)^2,(\Delta t)^2]$, and is stable for

$$|c| \leq 1 \quad (5.124)$$

where $c = \dfrac{u\Delta t}{\Delta x} =$ Courant number.

The two-step MacCormack method given here can be expressed as a one-step algorithm by introducing equation (5.123a) into equation (5.123b) in order to eliminate the predictor step. We obtain

$$T_j^{n+1} = T_j^n - \frac{1}{2}c\,(T_{j+1}^n - T_{j-1}^n) + \frac{1}{2}c^2(T_{j+1}^n - 2T_j^n + T_{j-1}^n) \quad (5.125)$$

The MacCormack scheme belongs to the family of the so-called Lax–Wendroff methods. The modified equation for the MacCormack method is given by

$$\frac{\partial T}{\partial t} + u\frac{\partial T}{\partial x} = -u\frac{(\Delta x)^2}{6}(1-c^2)\frac{\partial^3 T}{\partial x^3} - u\frac{(\Delta x)^3}{8}c(1-c^2)\frac{\partial^4 T}{\partial x^4} + \ldots \quad (5.126)$$

Clearly, for c = 1, the right-hand side of the modified equation (5.126) vanishes; hence, the equation is solved exactly.

5.2.1.3 Warming and Beam's Method

The second-order upwind method proposed by Warming and Beam (1975) also involves a predictor–corrector procedure. With this scheme, the finite difference representation of equation (5.113) is given by

$$\text{Predictor}: \overline{T_j^{n+1}} = T_j^n - \frac{u\Delta t}{\Delta x}(T_j^n - T_{j-1}^n) \quad (5.127a)$$

$$\text{Corrector}: T_j^{n+1} = \frac{1}{2}\left[T_j^n + \overline{T_j^{n+1}} - \frac{u\Delta t}{\Delta x}(\overline{T_j^{n+1}} - \overline{T_{j-1}^{n+1}}) - \frac{u\Delta t}{\Delta x}(T_j^n - 2T_{j-1}^n + T_{j-2}^n)\right]$$

$$(5.127b)$$

The stability analysis shows that the solution is stable for $0 \leq c \leq 2$, and the scheme is second-order accurate with a truncation error $O[(\Delta t)^2, (\Delta t)(\Delta x), (\Delta x)^2]$.

The method is a variation of the MacCormack method. It uses backward differences both in the predictor and corrector steps for $u > 0$; in addition, there is the second backward difference in the corrector step.

This two-step scheme can be expressed as a one-step algorithm by introducing the predictor given by equation (5.127a) into equation (5.127b). We obtain

$$T_j^{n+1} = T_j^n - c(T_j^n - T_{j-1}^n) + \frac{1}{2}c(c-1)(T_j^n - 2T_{j-1}^n + T_{j-2}^n) \tag{5.128}$$

where $c = \dfrac{u\Delta t}{\Delta x}$.

The modified equation for the second-order upwind method is given by

$$\frac{\partial T}{\partial t} + u\frac{\partial T}{\partial x} = u\frac{(\Delta x)^2}{6}(1-c)(2-c)\frac{\partial^3 T}{\partial x^3} - \frac{(\Delta x)^4}{8\Delta t}c(1-c)^2(2-c)\frac{\partial^4 T}{\partial x^4} + \ldots \tag{5.129}$$

The right-hand side of this equation vanishes for $c = 1$ or $c = 2$. Hence, for such situations the equation is solved exactly.

Example 5.7

Apply the upwind, MacCormack, and Warming and Beam methods to the first-order wave equation in $0 < x < L$, where $L = 1$, for an initial condition given by

$$T(x,0) = \begin{cases} 1, & 0 < x \leq 0.2 \\ 0.2, & 0.2 < x < 1 \end{cases}$$

The wave speed is $u = 0.1$ and the boundary conditions are reflective, that is, $\dfrac{\partial T}{\partial x} = 0$ at $x = 0$ and at $x = 1$. Compare the numerical solutions at the final time of 5, for the Courant numbers 1, 0.6, and 0.3.

Solution

The three schemes were applied in a discretized domain containing 51 grid points ($\Delta x = 0.02$) and a time step that satisfies the Courant number given by equation (5.119). The results obtained for $c = 1$ are presented in Figure 5.10a. This figure shows that the three schemes exactly solve the problem. Such a result was actually expected from the analysis of the modified equations (5.120), (5.126), and (5.129), which have a null right-hand side (truncation error) for $c = 1$.

The numerical solutions for $c = 0.6$ are presented in Figure 5.10b. We notice in this figure that the sharp wave front is smeared out in the upwind solution due to the second derivative of the term of highest order in the truncation error—the right-hand side of the modified equation (5.120).

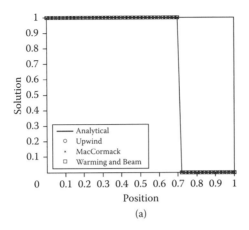
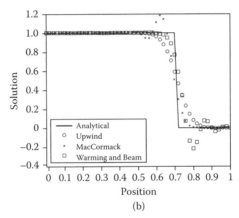
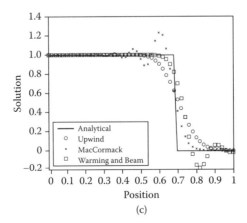

FIGURE 5.10
(a) Solution for $c = 1$; (b) solution for $c = 0.6$; and (c) solution for $c = 0.3$.

The coefficient of the second derivative is positive for c = 0.6, so that it behaves like a viscosity that induces dissipation of the solution, which is typical of first-order schemes. Notice that a value of c > 1 would result in a negative coefficient (viscosity) that would make the upwind scheme unstable. Figure (5.10b) also shows that the second-order schemes of MacCormack and Beam and Warming result in oscillations (dispersion) in the neighborhood of the sharp wave front. The dispersion tends to distort the phase relations among the infinite waves that compound the step wave. Dispersion results from the derivative of odd order in the highest-order term of the truncation error; see equations (5.126) and (5.129). Note that, for c = 0.6, the coefficient of such a term is negative for the MacCormack method, although it is positive for the Warming and Beam method. As a result, the oscillations in the solution obtained with MacCormack's method appear behind the wave front, while those in the solution with Warming and Beam's method appear ahead of the wave front; that is, these solutions lag and lead, respectively, in comparison to the exact solution. Dissipation and dispersion effects are amplified for smaller values of c, as depicted in Figure 5.10c, which presents the numerical solutions for c = 0.3.

Example 5.8

Apply the upwind, MacCormack, and Warming and Beam methods to the first-order wave equation in $0 < x < L$, where $L = 40$, for an initial condition given by $T(x,0) = \sin\left(\dfrac{6\pi x}{L}\right)$ and with periodic boundary conditions. The wave speed is 1. Compare the numerical solutions at the final time of 19.2 for the Courant numbers of 1, 0.6, and 0.3.

Solution

The upwind, MacCormack, and Warming and Beam schemes were applied in a discretized domain containing 51 grid points ($\Delta x = 0.8$) and a time step that satisfies the Courant number given by equation (5.119). This example is illustrative in further demonstrating the dissipation and dispersion effects of these three schemes, when the solution is continuous but periodic. Figure 5.11a presents the solutions obtained for c = 1. As is the case for the step wave in Example 5.7, the three numerical solutions perfectly recover the analytical solution because the right-hand side of the associated modified equations is null for c = 1. On the other hand, the dissipation effect of the first-order upwind scheme strongly manifests itself for c = 0.6, as demonstrated by Figure 5.11b, by damping the amplitude of the sinusoidal wave. The dispersion effects of the second-order MacCormack, and Warming and Beam schemes can also be noticed in Figure 5.11b. MacCormack's solution lags the exact solution due to the negative coefficient of the third-order derivative in the modified equation (5.126), while Warming and Beam's solution leads the exact solution because such coefficient is positive for its modified equation. The dissipation and dispersion effects can be

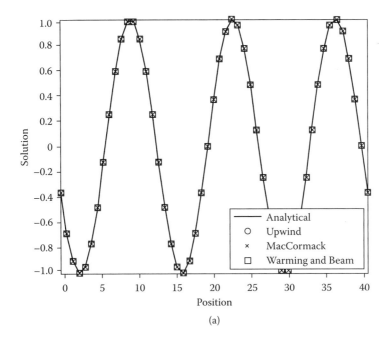

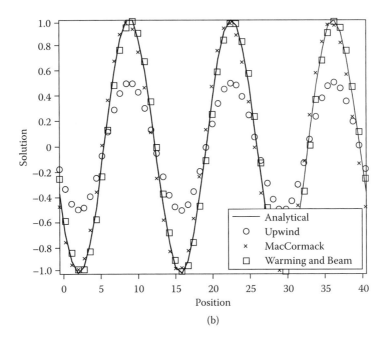

FIGURE 5.11
(a) Solution for c = 1; and (b) solution for c = 0.6. (*Continued*)

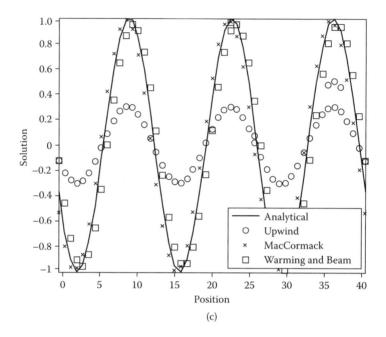

FIGURE 5.11 (CONTINUED)
(c) Solution for c = 0.3.

more clearly noticed for the case with c = 0.3, as shown by the results in Figure 5.11c.

The results in Examples 5.7 and 5.8 show that the time step must be taken as close to that for c = 1 as the problem allows, in order to avoid dispersion and dissipation in the numerical solution for the wave equation. Although the solution could be computed with a time step that satisfies c = 1 in the previous examples, such is generally not the case, like for nonlinear problems. The results in Examples 5.7 and 5.8 also show that, for c < 1, although second-order methods have a better agreement with smooth solutions, they exhibit oscillations near discontinuities. Such oscillations may increase without bounds and the method becomes unstable, thus imposing limitations in the size of the time step for nonlinear problems. On the other hand, first-order methods are stable near solution discontinuities, although sharp jumps in the solution are spread throughout the neighborhood of the discontinuity and the amplitudes of continuous solutions are damped. After this discussion, the reader might be thinking: Why not use adaptive methods that are second order when the solution is smooth and become first-order near solution discontinuities? Many such methods were actually developed, and they receive different designations in the literature, such as high-resolution, total variation diminishing, hybrid, adaptive, self-adjusting, essentially nonoscillatory, and so on (Laney 1998). A flux average method of this kind will be presented in Chapter 9, as applied to compressible flow models.

One-Dimensional Transient Systems

5.2.2 Advection–Diffusion Equation

5.2.2.1 Simple Explicit Scheme

We now consider the one-dimensional transient advection–diffusion equation given by

$$\frac{\partial T}{\partial t} + u\frac{\partial T}{\partial x} = \alpha\frac{\partial^2 T}{\partial x^2} \tag{5.130}$$

where the flow velocity u and the diffusion coefficient α are constants. For energy conservation, this parabolic equation allows for diffusion and convection in the streamwise direction but none in the direction perpendicular to the flow; hence, it characterizes transient temperature problems for flow in wide channels. To discretize this equation, we introduce the notation

$$T(x,t) = T(j\Delta x, n\Delta t) \equiv T_j^n \tag{5.131}$$

and apply a forward time and centered space finite difference scheme to obtain

$$\frac{T_j^{n+1} - T_j^n}{\Delta t} + u\frac{T_{j+1}^n - T_{j-1}^n}{2\Delta x} = \alpha\frac{T_{j+1}^n - 2T_j^n + T_{j-1}^n}{(\Delta x)^2} \tag{5.132}$$

with a truncation error $0[\Delta t, (\Delta x)^2]$. This equation is rearranged as

$$T_j^{n+1} = T_j^n - \frac{c}{2}(T_{j+1}^n - T_{j-1}^n) + r(T_{j+1}^n - 2T_j^n + T_{j-1}^n) \tag{5.133}$$

where

$$r = \frac{\alpha \Delta t}{(\Delta x)^2}, \quad c = \frac{u\Delta t}{\Delta x} = \text{Courant number} \tag{5.134}$$

The method being explicit, the time step Δt is restricted by stability considerations as discussed in the following section.

5.2.2.1.1 Stability

The Fourier stability analysis is now applied to establish the stability criterion. We consider the error term ε_j^n defined as

$$\varepsilon_j^n = \xi^n e^{i\beta_m j \Delta x} \tag{5.135}$$

where $i = \sqrt{-1}$ and ξ is the amplification factor. For stable solution of the finite difference equations, the requirement $|\xi| \leq 1$ should be satisfied for all values of β_m.

To develop an expression for the amplification factor, equation (5.135) is substituted into the finite difference equation (5.133). After cancellations, we obtain

$$\xi = 1 - 2r[1 - \cos(\beta_m \Delta x)] - i\, c\, \sin(\beta_m \Delta x) \tag{5.136}$$

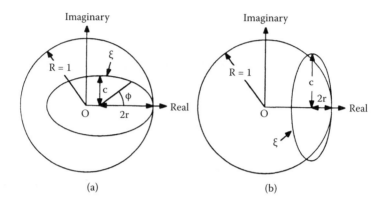

FIGURE 5.12
A plot of the amplification factor ξ.

which is a complex quantity as contrasted to the real value of the amplification factor for the case of pure diffusion; see equation (5.30a). For a no-flow condition, we have u = 0; hence, c = 0 and the complex amplification factor ξ given by equation (5.136) reduces to the real amplification factor for the diffusion equation; see equation (5.30a). To examine the stability requirement, equation (5.136) is rearranged in the form

$$\xi = [1 - 2r(1 - \cos\phi)] - i(c \sin\phi) \qquad (5.137)$$

where $\phi = \beta_m \Delta x$. Figure 5.12a shows a plot of ξ for a given c and r. It is an ellipse centered on the positive real axis at (1–2r) and has a semi-major and semi-minor axis 2r and c, respectively. Included in this figure is a unit circle. We note that the ellipse is tangent to this unit circle at the point where the circle intersects the real positive axis. The stability condition $|\xi| \leq 1$ is satisfied if the ellipse remains entirely within the unit circle. This requirement leads to the following constraints on the lengths of the semi-minor and semi-major axes of the ellipse:

$$c = \frac{u\Delta t}{\Delta x} \leq 1, \quad r = \frac{\alpha \Delta t}{(\Delta x)^2} \leq \frac{1}{2} \qquad (5.138a,b)$$

We now refer to Figure 5.12b, which illustrates that even though the constraints given by equation (5.138) are satisfied, the solution can still be unstable. A thorough examination of the stability constraints can be performed by computing the modulus of the complex amplification factor and requiring that it remains less than unity. Such an analysis leads to the stability limitations given by Pletcher et al. (2012) as

$$c^2 \leq 2r \quad \text{or} \quad \Delta t \leq \frac{2\alpha}{u^2} \qquad (5.139a)$$

One-Dimensional Transient Systems

$$\text{and } r \leq \frac{1}{2} \quad \text{or} \quad \Delta t \leq \frac{(\Delta x)^2}{2\alpha} \tag{5.139b}$$

We now define the mesh *Peclet number*, $Pe_{\Delta x}$, as

$$Pe_{\Delta x} = \frac{u\Delta x}{\alpha} = \frac{u\Delta t (\Delta x)^2}{\Delta x \, \alpha \Delta t} = \frac{c}{r} \tag{5.140}$$

Then the stability constraints given by equations (5.139a,b) are written in the form

$$2c \leq Pe_{\Delta x} \leq \frac{2}{c} \tag{5.141}$$

where the left-hand side is equivalent to equation (5.139b) and the right-hand side to equation (5.139a). If the chosen time step Δt is too large to violate the aforementioned stability criterion for a fixed value of Δx, the numerical results exhibit oscillations that grow rapidly and, after a few steps, the amplitude becomes so large that the computer may register an "overflow." If the results given using equation (5.141) are combined, this leads to the inequality

$$c \leq 1 \tag{5.142}$$

The stability criteria given by equation (5.141) arise from the discretization of the convection and diffusion terms. In flow situations involving high velocities, the constraint arising from the convection term imposes severe restrictions on the time step Δt because it is inversely proportional to the square of the velocity. Therefore, it is difficult to apply the scheme when α/u^2 is too small, and it is impossible to use for the case of a pure convection equation when $\alpha = 0$. To alleviate such difficulties, the non-centered *upwind scheme* can be used to discretize the convection term in equation (5.130) as described as follows.

5.2.2.1.2 Upwind Scheme

The space derivative $\partial T/\partial x$ contained in the convection term in equation (5.130) is discretized with *backward differencing* for $u > 0$ and *forward differencing* for $u < 0$. The scheme is called the upwind method. The reason the sign of u governs the choice of backward or forward differencing has been discussed by Peyret and Taylor (1983).

Suppose $\alpha = 0$ in equation (5.130). The resulting equation, which represents pure advection with no diffusion, is a hyperbolic equation with a general solution of the form $T(x,t) = F(x-ut)$, where F is an arbitrary function, as discussed in Section 5.2.1. The form of this exact solution implies that F remains constant along the characteristic line $x-ut = $ constant. Then, if $u > 0$, the information propagates in the direction $x > 0$; hence, a backward differencing should be used. Conversely, if $u < 0$, the information propagates in the direction $x < 0$, and a forward differencing should be used. With these considerations, the finite difference approximation of equation (5.130) with the simple explicit method, using the upwind scheme for discretizing the convection term, is given by

$$\frac{T_j^{n+1} - T_j^n}{\Delta t} + \frac{u}{2\Delta x}[(1-\varepsilon)(T_{j+1}^n - T_j^n) + (1+\varepsilon)(T_j^n - T_{j-1}^n)]$$
$$= \alpha \frac{T_{j+1}^n - 2T_j^n + T_{j-1}^n}{(\Delta x)^2} \qquad (5.143a)$$

where ε denotes the sign of u, that is

$$\varepsilon \equiv \text{sign}(u) = \begin{cases} +1 & \text{for} \quad u > 0 \\ -1 & \text{for} \quad u < 0 \end{cases} \qquad (5.143b)$$

Clearly, the finite difference representation of the derivative $\partial T/\partial x$ is backward differencing for $u > 0$ and forward differencing for $u < 0$. The stability criterion for equation (5.143a) is given by Peyret and Taylor (1983) as

$$\Delta t \leq \frac{(\Delta x)^2}{|u|\Delta x + 2\alpha} = \left\{ \frac{|u|}{\Delta x} + \frac{2\alpha}{(\Delta x)^2} \right\}^{-1} \qquad (5.144a)$$

When $\alpha = 0$ (i.e., no diffusion), equation (5.144a) reduces to

$$c \equiv \frac{|u|\Delta t}{\Delta x} \leq 1 \qquad (5.144b)$$

The stability criterion given by equation (5.144a) can also be written in terms of the parameters $Pe_{\Delta x}$ and r as

$$r \leq \frac{1}{2 + |Pe_{\Delta x}|} \qquad (5.144c)$$

where $Pe_{\Delta x}$ is the mesh Peclet number and r is the stability parameter defined previously [see equation (5.138b)].

5.2.2.2 Implicit Finite Volume Method

In this section, we will apply the finite volume method described earlier to discretize the advective–diffusive equation (5.130). The method will be further detailed when applied to the transient Navier-Stokes equation in Chapter 7. Let us consider equation (5.130) for energy conservation, which is rewritten in the following form by taking into account that the thermal diffusivity $\alpha = k/(\rho\, C_p)$:

$$\rho \frac{\partial T}{\partial t} + \rho u \frac{\partial T}{\partial x} = \frac{k}{C_p} \frac{\partial^2 T}{\partial x^2} \qquad (5.145)$$

We can integrate equation (5.145) in the control volume around a point P represented in Figure 4.7, where the centers of the volumes located in the west and east directions are denoted by W and E, and their corresponding

One-Dimensional Transient Systems

interfaces are represented by w and e, respectively. Integrating equation (5.145) from time t to t + Δt in the control volume around the node P gives

$$\int_{t}^{t+\Delta t}\int_{w}^{e}\left[\rho\frac{\partial T}{\partial t}+\rho u\frac{\partial T}{\partial x}\right]dxdt = \int_{t}^{t+\Delta t}\int_{w}^{e}\left[\frac{k}{C_p}\frac{\partial^2 T}{\partial x^2}\right]dxdt \qquad (5.146)$$

Considering constant physical properties and the integrand as a representative average value within the control volume, the following equation can be obtained

$$\rho\{[T_P]_{t+\Delta t}-[T_P]_t\}\Delta x + \rho u\{[T]_e - [T]_w\}\Delta t = \frac{k}{C_p}\left\{\left[\frac{\partial T}{\partial x}\right]_e - \left[\frac{\partial T}{\partial x}\right]_w\right\}\Delta t \qquad (5.147)$$

We can now define the following variables, which represent the mass within the control volume P

$$M_P = \rho \Delta x \qquad (5.148)$$

and the mass flow rate through the two boundaries of the control volume

$$\dot{M}_e = [\rho u]_e \qquad (5.149)$$

$$\dot{M}_w = [\rho u]_w \qquad (5.150)$$

Notice that since ρ and u are constants, equations (5.149) and (5.150) are identical (in Chapter 7 we will consider a nonuniform velocity u).

We can also define the following coefficient, which will be part of the diffusive terms of the conservation equations

$$D_{11} = \frac{k}{C_p} \qquad (5.151)$$

By placing equations (5.148)–(5.151) into equation (5.147), we obtain:

$$M_P\frac{\{[T_P]_{t+\Delta t}-[T_P]_t\}}{\Delta t} + \{[\dot{M}T]_e - [\dot{M}T]_w\} = \left\{\left[D_{11}\frac{\partial T}{\partial x}\right]_e - \left[D_{11}\frac{\partial T}{\partial x}\right]_w\right\} \qquad (5.152)$$

In equation (5.152), we can choose to evaluate the second term on the left-hand side and the term on the right-hand side at time t (explicit scheme), at time t + Δt (implicit scheme), or alternatively by making a weighted average between times t and t + Δt (combined scheme). Here, we use the implicit scheme and adopt the notations $T_{t+\Delta t} = T^{n+1}$ and $T_t = T^n$, so that equation (5.152) becomes

$$\frac{M_P}{\Delta t}(T_P^{n+1} - T_P^n) + \dot{M}(T_e^{n+1} - T_w^{n+1}) = D_{11}\left(\left.\frac{\partial T^{n+1}}{\partial x}\right|_e - \left.\frac{\partial T^{n+1}}{\partial x}\right|_w\right) \qquad (5.153)$$

Central differences can be applied for the derivative terms resulting from diffusion and the upwind scheme for the advective terms in equation (5.153), that is,

$$\begin{cases} \left[\dfrac{\partial T^{n+1}}{\partial x}\right]_e = \dfrac{T_E^{n+1} - T_P^{n+1}}{\Delta x} \\ \left[\dfrac{\partial T^{n+1}}{\partial x}\right]_w = \dfrac{T_P^{n+1} - T_W^{n+1}}{\Delta x} \\ T_e^{n+1} = \begin{cases} T_P^{n+1} & \text{for } u > 0 \\ T_E^{n+1} & \text{for } u < 0 \end{cases} \quad T_w^{n+1} = \begin{cases} T_W^{n+1} & \text{for } u > 0 \\ T_P^{n+1} & \text{for } u < 0 \end{cases} \end{cases} \quad (5.154\text{a–d})$$

Considering only the case with u > 0, we can place equations (5.154a–d) into equation (5.153) and obtain

$$\frac{M_P}{\Delta t}(T_P^{n+1} - T_P^n) + \dot{M}(T_P^{n+1} - T_W^{n+1}) = D_{11}\left(\frac{T_E^{n+1} - T_P^{n+1}}{\Delta x} - \frac{T_P^{n+1} - T_W^{n+1}}{\Delta x}\right) \quad (5.155)$$

or

$$T_E^{n+1}\left(-\frac{D_{11}}{\Delta x}\right) + T_P^{n+1}\left(\frac{M_P}{\Delta t} + \dot{M} + \frac{2D_{11}}{\Delta x}\right) + T_W^{n+1}\left(-\dot{M} - \frac{D_{11}}{\Delta x}\right) = T_P^n\left(\frac{M_P}{\Delta t}\right) \quad (5.156)$$

We can then define the following coefficients

$$A_W = \left(\dot{M} + \frac{D_{11}}{\Delta x}\right) \quad (5.157)$$

$$A_P = \left(\frac{M_P}{\Delta t} + \dot{M} + \frac{2D_{11}}{\Delta x}\right) \quad (5.158)$$

$$A_E = \left(\frac{D_{11}}{\Delta x}\right) \quad (5.159)$$

$$B_P = \left(\frac{M_P}{\Delta t}\right) \quad (5.160)$$

and then write equation (5.156) as

$$A_P T_P^{n+1} = A_E T_E^{n+1} + A_W T_W^{n+1} + B_P T_P^n \quad (5.161)$$

Equation (5.161), written for all internal volumes of the domain, with the application of proper discretization of the boundary conditions by using the method described in Chapter 2, results in a system that can be solved by the techniques presented in Chapter 3. We note that equation (5.156) reduces to the discretization of the diffusion equation examined in Section 5.1 by making $\dot{M}=0$.

5.3 Hyperbolic Heat Conduction Equation

The classical heat conduction theory based on the Fourier heat flux model leads to an infinite speed of propagation of thermal waves within the solids. Despite the fact that such an assumption is physically unrealistic, the Fourier heat conduction model serves very well for the analysis of heat conduction encountered in most practical applications. However, the classical heat conduction equation breaks down at temperatures near absolute zero, at moderate temperatures when the elapsed time during a transient is very small (i.e., of the order of about a pico second), and under extremely large heat fluxes. In such situations, the wave nature of thermal transport becomes dominant, that is, a thermal disturbance travels in a medium with a finite speed of propagation.

A heat flux equation that accommodates the finite speed of propagation of thermal waves has been proposed by Cattaneo (1958), Vernotte (1958), Chester (1963), and Weymann (1967). However, the original idea dates back to Maxwell (1867). The so-called Cattaneo–Vernotte's constitutive equation in one dimension is given in the form

$$\Gamma \frac{\partial q(x,t)}{\partial t} + q(x,t) = -k \frac{\partial T(x,t)}{\partial x} \qquad (5.162)$$

where

$q(x,t)$ = heat flux, W/m^2
k = thermal conductivity, $W/m°C$
$\Gamma = \dfrac{\alpha}{u^2}$ = the relaxation time, s
α = thermal diffusivity, m^2/s
u = wave speed (i.e., speed of propagation of thermal waves within the solid)

The values of the relaxation parameter Γ for homogeneous materials have been quoted by Sieniutycz (1977) to range from 10^{-8} to 10^{-10} s for gases and from 10^{-10} to 10^{-12} s for liquids and dielectric solids. Clearly, for an infinite propagation speed (i.e., $u \rightarrow \infty$), the relaxation time Γ vanishes, and equation (5.162) reduces to the classical Fourier law

$$q(x,t) = -k \frac{\partial T(x,t)}{\partial x} \qquad (5.163)$$

Cattaneo–Vernotte's constitutive equation is a special case of the more general dual phase dual lagging model, where the constitutive equation contains relaxation times for the heat flux and for the temperature gradient (Qiu and Tien 1992, 1993, 1994; Qiu et al. 1994; Özişik and Tzou 1994; Tzou et al. 1994; Orlande et al. 1995; Tzou 1996; Quaresma et al. 2010; Wang et al. 2010; Nóbrega et al. 2011).

The one-dimensional energy conservation equation for transient heat conduction in an isotropic solid is given by

$$-\frac{\partial q(x,t)}{\partial x} + g(x,t) = \rho C_p \frac{\partial T(x,t)}{\partial t} \tag{5.164}$$

where

$g(x,t)$ = volumetric energy generation rate, W/m^3
C_p = specific heat, $J/(kg°C)$
ρ = density, kg/m^3

The elimination of heat flux between equations (5.162) and (5.164) for a medium with constant k leads to the following *hyperbolic heat conduction equation*

$$\frac{\partial^2 T}{\partial x^2} + \frac{1}{k}\left[g(x,t) + \Gamma \frac{\partial g(x,t)}{\partial t}\right] = \frac{1}{\alpha}\left(\Gamma \frac{\partial^2 T}{\partial t^2} + \frac{\partial T}{\partial t}\right) \tag{5.165}$$

For the special case of $\Gamma = 0$, equation (5.165) reduces to the classical heat conduction equation.

We have two alternative forms of the hyperbolic heat conduction equation: (i) Two first-order partial differential equations, equations (5.162) and (5.164), in the two variables $q(x,t)$ and $T(x,t)$, respectively, and (ii) a single second-order partial differential equation (5.165) in the variable $T(x,t)$.

5.3.1 Finite Difference Representation of Hyperbolic Heat Conduction Equation

The solution of wave propagation problems generally involves a sharp discontinuity at the wave front. MacCormack's explicit predictor–corrector scheme has been applied successfully for the solution of the hyperbolic heat conduction equation under situations leading to the formation of a very sharp moving front. We consider the hyperbolic heat conduction equation expressed as two coupled first-order partial differential equations given by equations (5.162) and (5.164). These equations can be written in the dimensionless form as (Glass et al. 1985a)

$$\frac{\partial \Theta}{\partial \xi} + \frac{\partial Q}{\partial \eta} - 2S = 0 \tag{5.166a}$$

$$\frac{\partial Q}{\partial \xi} + \frac{\partial \Theta}{\partial \eta} + 2Q = 0 \tag{5.166b}$$

where the dimensionless time ξ and space variable η are defined by

$$\xi = \frac{u^2 t}{2\alpha}, \quad \eta = \frac{ux}{2\alpha} \tag{5.167a,b}$$

The definition of dimensionless temperature Θ and flux Q depends on whether the thermal disturbance is initiated by an applied surface temperature, heat flux, or by a volumetric energy source. The definitions of Θ and Q for a number of other specific situations are listed by Glass et al. (1986). For example, for a plate of thickness L, initially at constant temperature T_0, and for times $t > 0$, the boundary surfaces at $x = 0$ and $x = L$ are kept at temperatures T_w and 0°C, respectively, the dimensionless temperature Θ and heat flux Q are defined as

$$\Theta = \frac{T(x,t) - T_0}{T_w}, \quad Q = \frac{q}{(T_w k u)/\alpha} \tag{5.168a,b}$$

while the dimensionless heat source is given by

$$S = \frac{\alpha^2 g}{T_w k u^2} \tag{5.168c}$$

Then, the dimensionless thickness, η_L, of the plate becomes $\eta_L = uL/2\alpha$. Equations (5.166a,b) are expressed in the vector form as

$$\frac{\partial E}{\partial \xi} + \frac{\partial F}{\partial \eta} + H = 0 \tag{5.169}$$

where

$$E = \begin{bmatrix} \Theta \\ Q \end{bmatrix}, \quad F = \begin{bmatrix} Q \\ \Theta \end{bmatrix}, \quad H = \begin{bmatrix} -2S \\ 2Q \end{bmatrix} \tag{5.170a–c}$$

We adopt the notation

$$F(\xi, \eta) = F(n\Delta\xi, i\Delta\eta) \equiv F_i^n \tag{5.171}$$

Then, the finite difference representation of the vector equation (5.169), using MacCormack's explicit predictor–corrector scheme, becomes

$$\text{Predictor}: \overline{E_i^{n+1}} = E_i^n - \frac{\Delta\xi}{\Delta\eta}(F_{i+1}^n - F_i^n) - \Delta\xi H_i^n \tag{5.172a}$$

Corrector : $E_i^{n+1} = \dfrac{1}{2}\left[E_i^n + \overline{E_i^{n+1}} - \dfrac{\Delta\xi}{\Delta\eta}(\overline{F_i^{n+1}} - \overline{F_{i-1}^{n+1}}) - \Delta\xi\overline{H_i^{n+1}}\right]$ (5.172b)

Here the predicted quantities $\overline{E_i^{n+1}}$, $\overline{F_i^{n+1}}$, and so on give temporary approximate values of E_i, F_i, and so on at an intermediate time level $\overline{n+1}$, while the corrected quantities E_i^{n+1}, F_i^{n+1}, and so on give the final values of E_i, F_i, and so on at time level n + 1. Note that the predictor step utilizes forward difference in space while the corrector step uses backward difference in space. In equations (5.172), the ratio $\Delta\xi/\Delta\eta$ is the Courant number, that is

$$\dfrac{\Delta\xi}{\Delta\eta} = \dfrac{u^2\Delta t}{2\alpha}\dfrac{2\alpha}{u\Delta x} = \dfrac{u\Delta t}{\Delta x} = \text{Courant number} \quad (5.173)$$

A comparison of the vector equations (5.172a,b) with equations (5.123a,b) for the first-order wave equation reveals that they are of the identical form, except the former is in the vector form and contains a source term.

Example 5.9

A plate of thickness L is initially at temperature T_0. For times t > 0, the boundary surface at x = 0 is kept at temperature T_w and the boundary surface at x = L at temperature T_0. There is no energy generation within the medium. Formulate the finite difference approximation of the hyperbolic heat conduction model for this problem and determine the temperature profile as a function of position at different times.

Solution
The mathematical formulation of this hyperbolic heat conduction problem in vector form is given by

$$\dfrac{\partial E}{\partial \xi} + \dfrac{\partial F}{\partial \eta} = 0 \quad \text{in} \quad 0 < \eta < 1 \quad (a)$$

where

$$E = \begin{bmatrix}\Theta \\ Q\end{bmatrix}, \quad F = \begin{bmatrix}Q \\ \Theta\end{bmatrix} \quad (b,c)$$

$$\Theta = \dfrac{T(x,t) - T_0}{T_w - T_0}, \quad Q = \dfrac{q}{(T_w ku/\alpha)} \quad (d,e)$$

The initial conditions are given by

$$\Theta = 0 \quad \text{and} \quad Q = 0 \quad \text{for } \xi = 0 \text{ in } 0 < \eta < 1 \quad (f,g)$$

One-Dimensional Transient Systems

and the boundary conditions by

$$\Theta = 1 \quad \text{and} \quad \left(\frac{\partial Q}{\partial \eta} = 0\right) \quad \eta = 0, \quad \xi > 0 \quad \text{(h,i)}$$

$$\Theta = 0 \quad \text{and} \quad \left(\frac{\partial Q}{\partial \eta} = 0\right) \quad \eta = \eta_L = 1, \quad \xi > 0 \quad \text{(j,k)}$$

where the dimensionless thickness of the plate is taken as $\eta_L = 1$. The additional conditions for Q, shown in parenthesis, are obtained from equation (5.166a). This problem has been numerically solved by Glass et al. (1986) by using MacCormack's explicit predictor–corrector scheme with 1000 mesh intervals and a Courant number c = 0.98. Exact analytic solution of the problem is given by Carey and Tsai (1982).

Figure 5.13 shows the temperature profile as a function of position η at two dimensionless times, $\xi = 0.5978$ (before reflection of the wave front from the wall at $\eta_L = 1$) and at $\xi = 1.2005$ (after reflection of the wave front from the wall at $\eta_L = 1$). Included in this figure is the exact analytic solution of the problem. The agreement between the numerical and analytic solutions is excellent. Each curve shows a slight numerical oscillation just prior to the wave front caused by numerical dispersion.

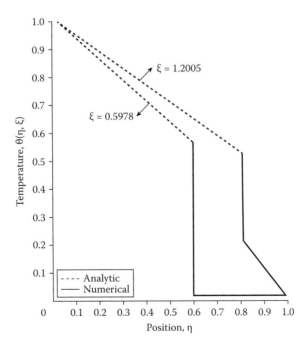

FIGURE 5.13
Comparison of numerical and exact analytic solutions.

PROBLEMS

5.1. Consider the following diffusion problem

$$\frac{\partial^2 T}{\partial x^2} + \frac{1}{k}g = \frac{1}{\alpha}\frac{\partial T}{\partial t} \quad \text{in} \quad 0 < x < L, \quad t > 0$$

$$-k\frac{\partial T}{\partial x} + h_0 T = h_0 T_\infty \quad \text{at} \quad x = 0, \quad t > 0$$

$$k\frac{\partial T}{\partial x} = q_L \quad \text{at} \quad x = L, \quad t > 0$$

$$T = F(x) \quad \text{for} \quad t = 0, \quad 0 \leq x \leq L$$

Develop the finite difference form of this problem by using an explicit scheme for the differential equation of heat conduction and the fictitious node approach for the discretization of the boundary conditions. Discuss the order of accuracy of the differential equation.

5.2. Repeat Problem 5.1 by using the implicit scheme and the combined method for the differential equation.

5.3. Consider the following diffusion problem for a semi-infinite region, $x \geq 0$:

$$\frac{\partial^2 T}{\partial x^2} = \frac{1}{\alpha}\frac{\partial T}{\partial t} \quad x > 0, \quad t > 0$$

$$T = T_0 \quad x = 0, \quad t > 0$$

$$T \to T_b \quad \text{as} \quad x \to \infty, \quad t > 0$$

$$T = T_b \quad \text{for} \quad t = 0, \quad \text{in} \quad x \geq 0$$

Using the explicit scheme, develop a finite difference formulation of this time-dependent heat condition problem and discuss the stability criterion.

5.4. Consider the following diffusion problem for a semiregion, $x \geq 0$:

$$\frac{\partial^2 T}{\partial x^2} = \frac{1}{\alpha}\frac{\partial T}{\partial t} \quad x > 0, \quad t > 0$$

$$-k\frac{\partial T}{\partial x} + h_0 T = h_0 T_{\infty,0} \quad \text{at} \quad x = 0, \quad t > 0$$

$$T \to T_b \quad \text{at} \quad x \to \infty, \quad t > 0$$

$$T = T_b \quad \text{for} \quad t = 0, \quad x \geq 0$$

Develop the finite difference form of this heat conduction problem by using an explicit differencing for the differential equation and central difference formula for the boundary condition.

5.5. Consider the following diffusion problem

$$\frac{\partial^2 T}{\partial x^2} = \frac{1}{\alpha}\frac{\partial T}{\partial t} \qquad 0 < x < L, \quad t > 0$$

$$T = 0 \qquad \text{at} \quad x = 0, \qquad t > 0$$

$$k\frac{\partial T}{\partial x} = q_L \qquad \text{at} \quad x = L, \qquad t > 0$$

$$T = F(x) \qquad t = 0, \qquad 0 \leq x \leq L$$

Develop the finite difference form of this problem using:

a. The explicit method for the differential equation and fictitious node concept for the boundary condition at x = L
b. The fully implicit method for the differential equation and fictitious node concept for the boundary condition at x = L
c. The Crank–Nicolson method for the differential equation and fictitious node concept for the boundary condition at x = L

Discuss the order of accuracy for each of these cases.

5.6. Consider the following transient heat-conduction problem given in the dimensionless form as

$$\frac{\partial^2 T}{\partial x^2} = \frac{\partial T}{\partial t} \qquad 0 < x < 1, \quad t > 0$$

$$\frac{\partial T}{\partial x} = 0 \qquad x = 0, \qquad t > 0$$

$$T = 0 \qquad x = 1, \qquad t > 0$$

$$T = 100 \qquad t = 0, \qquad 0 \leq x \leq 1$$

Develop the finite difference formulation of this problem using an explicit scheme for the differential equation and the fictitious node concept for the boundary condition, by taking $\Delta x = 0.2$. Solve the resulting difference equations and calculate the temperature of the insulated surface at x = 0 at each time step Δt up to dimensionless time t = 0.4 by using the following two different time steps:

a. $\Delta t = 0.02$, which satisfies the stability criterion
b. $\Delta t = 0.04$, which violates the stability criterion

Plot the temperature at x = 0 as a function of time.

5.7. Consider the following diffusion problem given in the dimensionless form as

$$\frac{\partial^2 T}{\partial x^2} = \frac{\partial T}{\partial t} \quad 0 < x < \frac{1}{2}, \quad t > 0$$

$$T = 0 \quad x = 0, \quad t > 0$$

$$\frac{\partial T}{\partial x} = 0 \quad x = \frac{1}{2}, \quad t > 0$$

$$T = x \quad t = 0, \quad 0 \le x \le \frac{1}{2}$$

Develop the finite difference form of this problem by using an explicit scheme for the differential equation and central difference formula for the derivative boundary condition. Solve the resulting difference equations and calculate the temperature of the insulated surface $x = \frac{1}{2}$ by taking:

(i) $\Delta x = 0.1$, $r = 0.5$; (ii) $\Delta x = 0.1$, $r = 0.1$

and compare the results with the following exact solution for the problem.

t, time	T(0,5,t), exact
0.005	0.4202
0.010	0.3871
0.020	0.3404
0.100	0.1510

5.8. Consider Problem 5.7

$$\frac{\partial^2 T}{\partial x^2} = \frac{\partial T}{\partial t} \quad 0 < x < \frac{1}{2}, \quad t > 0$$

$T = 0 \qquad x = 0, \qquad t > 0$

$\dfrac{\partial T}{\partial x} = 0 \qquad x = \dfrac{1}{2}, \qquad t > 0$

$T = x \qquad t = 0, \qquad 0 \le x \le \dfrac{1}{2}$

Develop the finite difference form of this problem using the Crank–Nicolson method for the differential equation and the fictitious node concept for the boundary condition at $x = \frac{1}{2}$. Solve the resulting difference equations and calculate the temperature at the surface $x = \frac{1}{2}$ by setting $r = 1$ and compare the results with the following exact solution for the problem.

t, time	T(0,5,t), exact
0.005	0.4202
0.010	0.3871
0.020	0.3404
0.100	0.1510

5.9. Consider the diffusion problem given in the dimensionless form as

$$\frac{\partial^2 T}{\partial x^2} = \frac{\partial T}{\partial t} \quad 0 < x < 1, \quad t > 0$$

$$T = 0 \quad x = 0, \quad t > 0$$
$$T = 0 \quad x = 1, \quad t > 0$$
$$T = \sin \pi x \quad t = 0, \quad 0 \le x \le 1$$

The exact solution of this problem is given by

$$T(x, t) = e^{-\pi^2 t} \sin \pi x$$

Write the finite difference formulation of this problem by using the explicit method. Solve the resulting difference equations with a step size $\Delta x = 0.1$ and calculate the center temperature (i.e., $x = 0.5$) by taking

a. $r = \dfrac{1}{6}$

b. $r = 0.5$

c. $r = 0.6$

and compare the results with the exact solution at times $t = 0.005, 0.01, 0.02, 0.05,$ and 0.10.

5.10. Consider the following diffusion problem given in the dimensionless form as

$$\frac{\partial^2 T}{\partial x^2} = \frac{\partial T}{\partial t} \quad 0 < x < \frac{1}{2}, \quad t > 0$$

$$\frac{\partial T}{\partial x} = 0 \quad x = 0, \quad t > 0$$

$$\frac{\partial T}{\partial x} = 10 \quad x = \frac{1}{2}, \quad t > 0$$

$$T = 0 \quad t = 0, \quad 0 \leq x \leq \frac{1}{2}$$

Solve this problem with finite differences by taking the step size $\Delta x = 0.1$ and using

a. Explicit differencing for the differential equation and the fictitious node concept for the boundary conditions

b. The Crank–Nicolson method for the differential equation and the fictitious node concept for the boundary conditions

Then compare your results for the location $x = 0.3$ with the following exact solution of the problem.

t, time	T(0.3,t), exact
0.01	0.101
0.03	0.573
0.05	1.023
0.10	2.061
0.50	10.067
1.00	20.067

Note that this problem has no steady-state solution.

5.11. Consider the following diffusion problem given in the dimensionless form as

$$\frac{\partial^2 T}{\partial x^2} = \frac{\partial T}{\partial t} \quad 0 < x < 1, \quad t > 0$$

$$\frac{\partial T}{\partial x} = 10 \quad x = 0, \quad t > 0$$

$$T = 0 \quad x = 1, \quad t > 0$$

$$T = 0 \quad t = 0, \quad 0 \leq x \leq 1$$

Solve this problem with finite differences by taking a step size $\Delta x = 0.2$ and using
a. Explicit differencing of the differential equation and the fictitious node concept for the boundary condition at $x = 0$
b. The Crank–Nicolson method for the differential equation and the fictitious node concept for the boundary condition at $x = 0$

and list your results for the location $x = 0$ for dimensionless times $t = 0.01, 0.02, 0.1$, and 0.2.

5.12. Consider the following diffusion problem for a slab of thickness L:

$$\frac{\partial^2 T}{\partial x^2} + \frac{1}{k} g(x, t) = \frac{1}{\alpha} \frac{\partial T}{\partial t} \quad 0 < x < L, \quad t > 0$$

$$-k_0 \frac{\partial T}{\partial x} + h_0 T = h_0 T_{\infty,0} \equiv f_0 \quad x = 0, \quad t > 0$$

$$k_L \frac{\partial T}{\partial x} + h_L T = h_L T_{\infty,L} \equiv f_L \quad x = L, \quad t > 0$$

$$T = F(x) \quad t = 0, \quad 0 \leq x \leq L$$

where k_0, h_0, k_L, and h_L are treated as coefficients. By choosing these coefficients appropriately, nine different combinations of boundary conditions are obtainable. Write the finite difference form of this problem by:
a. Explicit method
b. Fully implicit method
c. Crank–Nicolson method

Use the fictitious node concept formula for finite difference approximation of the boundary conditions.

5.13. A thick concrete wall ($\alpha = 8 \times 10^{-7}$ m^2/s, k = 0.8 W/[m°C]) is initially at a uniform temperature $T_i = 60°C$. Suddenly, it is exposed to a cool airstream at $T_\infty = 5°C$. The heat transfer coefficient between the airstream and the surface is h = 10 W/(m^2°C). The wall can be regarded as a semi-infinite medium confined to the region $x \geq 0$ with the surface $x = 0$ subjected to convection. By using an explicit finite difference scheme, calculate the wall temperature at a depth $x = 10$ cm from the surface at $t = 10$ min and $t = 1$ h after the start of cooling, by taking $\Delta x = 2$ cm.

5.14. A slab of thickness L = 12 cm and thermal diffusivity $\alpha = 2 \times 10^{-5}$ m^2/s has an initial temperature distribution $T_i = 100 \sin(\pi x/L)$. For $t > 0$, both boundaries are kept at zero temperature. By using an explicit finite difference scheme and a mesh size $\Delta x = 2$ cm, calculate the center temperature at $t = 5$ min and $t = 15$ min after the start of cooling.

5.15. A slab of thickness L = 6 cm and thermal diffusivity $\alpha = 8 \times 10^{-5}$ m^2/s has an initial temperature distribution $T_i = 100 \sin([\pi/2][x/L])$. For $t > 0$, the boundary surface at $x = 0$ is kept at zero temperature and that at $x = 6$ cm is kept insulated. By using an explicit finite difference scheme and a mesh size $\Delta x = 1$ cm, calculate the temperature of the insulated boundary $t = 1$ min and $t = 5$ min after the start of cooling.

5.16. An 8-cm-thick chrome-steel plate, $\alpha = 1.6 \times 10^{-5}$ m^2/s, k = 61 W/(m°C), initially at a uniform temperature, $T_i = 325°C$, is suddenly exposed to a cool airstream at $T_\infty = 25°C$ at both of its surfaces. The heat transfer coefficient between the air and the surface is h = 400 W/(m^2°C). By using an explicit finite difference scheme and a mesh size $\Delta x = 1$ cm, determine the center plane temperature $t = 5$ min and $t = 15$ min after the start of cooling.

5.17. A large and very thick brick wall ($\alpha = 5 \times 10^{-7}$ m^2/s) that is initially at a uniform temperature $T_i = 125°C$ is suddenly exposed to cooling by maintaining its surface at $x = 0$ at $T_0 = 25°C$. To calculate the temperature transients at depths small in comparison to the thickness, the wall can be regarded as a semi-infinite medium confined to the region $x \geq 0$. By using an explicit scheme and a mesh size $\Delta x = 0.3$ cm, calculate the temperature at $x = 1.2$ cm from the surface at $t = 1$ min and $t = 5$ min after the exposure.

5.18. Repeat Problem 5.17 with the combined method of discretization. Using the Fourier stability analysis, show that this explicit scheme is unconditionally stable.

5.19. Consider the following diffusion problem for a hollow cylinder or sphere

$$\frac{\partial^2 T}{\partial R^2} + \frac{p}{R}\frac{\partial T}{\partial R} = \frac{1}{\alpha}\frac{\partial T}{\partial t} \quad \text{in} \quad a < R < b, \quad t > 0$$

$$-k\frac{\partial T}{\partial R} + h_a T = h_a T_{\infty,a} \quad \text{at} \quad R = a, \quad t > 0$$

$$k\frac{\partial T}{\partial R} + h_b T = h_b T_{\infty,b} \quad \text{at} \quad R = b, \quad t > 0$$

$$T = F(R) \quad \text{for} \quad t = 0$$

Develop the finite difference approximation for this problem by using the simple explicit method to discretize the differential equation and the fictitious node approach to discretize the boundary conditions.

5.20. Repeat Problem 5.19 using the Crank–Nicolson method to discretize the differential equation and the fictitious node approach to discretize the boundary conditions.

5.21. Consider the following diffusion problem for a solid cylinder or sphere

$$\frac{\partial^2 T}{\partial R^2} + \frac{p}{R}\frac{\partial T}{\partial R} + \frac{1}{k}g(R,t) = \frac{1}{\alpha}\frac{\partial T}{\partial t} \quad \text{in} \quad 0 < R < b, \quad t > 0$$

$$\frac{\partial T}{\partial R} = 0 \quad \text{at} \quad R = 0, \quad t > 0$$

$$k\frac{\partial T}{\partial R} + h_b T = h_b T_{\infty,b} \quad \text{at} \quad R = b, \quad t > 0$$

$$T = F(R) \quad \text{for} \quad t = 0$$

Develop the finite difference approximation for this problem by using the simple implicit method to discretize the differential equation and the fictitious node approach to discretize the boundary condition at $R = b$.

5.22. Consider the following transient heat conduction problem for a solid cylinder given in the dimensionless form as

$$\frac{\partial^2 T}{\partial R^2} + \frac{1}{R}\frac{\partial T}{\partial R} = \frac{\partial T}{\partial t} \quad 0 \leq R < 1, \quad t > 0$$

$$\frac{\partial T}{\partial R} = 0 \quad R = 0, \quad t > 0$$

$$\frac{\partial T}{\partial R} + HT = 0 \quad R = 1, \quad t > 0$$

$$T = F(R) \quad t = 0, \quad 0 \leq R \leq 1$$

Develop the explicit finite difference form of this heat conduction problem by using the fictitious node approach for the boundary conditions.

5.23. Consider the following transient heat conduction problem for a solid sphere given in the dimensionless form as

$$\frac{\partial^2 T}{\partial R^2} + \frac{2}{R}\frac{\partial T}{\partial R} + G = \frac{\partial T}{\partial t} \quad 0 < R < 1, \quad t > 0$$

$$\frac{\partial T}{\partial R} = 0 \quad R = 0, \quad t > 0$$

$$\frac{\partial T}{\partial R} + HT = 0 \quad R = 1, \quad t > 0$$

$$T = F(R) \quad t = 0, \quad 0 \leq R \leq 1$$

Develop an explicit finite difference form of this heat conduction problem by using the fictitious node approach for the boundary conditions.

5.24. Repeat Problem 5.23 by using the Crank–Nicolson method of a finite differencing scheme.

5.25. By using the Crank–Nicolson method, develop the finite difference representation of the following dimensionless transient heat conduction problem for a solid cylinder.

$$\frac{\partial^2 T}{\partial R^2} + \frac{1}{R}\frac{\partial T}{\partial R} = \frac{\partial T}{\partial t} \quad 0 < R < \quad t > 0$$

$$\frac{\partial T}{\partial R} = 0 \quad R = 0, \quad t > 0$$

$$T = 0 \quad R = 1, \quad t > 0$$

$$T = 100 \quad t = 0, \quad 0 \leq R \leq 1$$

5.26. Consider the following dimensionless transient heat conduction for a hollow cylinder

$$\frac{\partial^2 T}{\partial R^2} + \frac{1}{R}\frac{\partial T}{\partial R} = \frac{\partial T}{\partial t} \quad 1 < R < 2, \quad t > 0$$

$$\frac{\partial T}{\partial R} = 0 \quad R = 1, \quad t > 0$$

$$T = 0 \quad R = 2, \quad t > 0$$

$$T = 100 \quad t = 0, \quad 1 \leq R \leq 2$$

Solve this transient heat conduction problem by using an explicit finite difference scheme. Calculate the temperature of the insulated surface $R = 1$ at dimensionless times $t = 0.2, 0.4, 0.6, 0.8, 1.0, 1.2,$ and 1.4.

5.27. Repeat Problem 5.26 for a hollow sphere in $1 \leq R \leq 2$.

5.28. Consider the following transient heat conduction problem for a hollow cylinder

$$\frac{\partial^2 T}{\partial R^2} + \frac{1}{R}\frac{\partial T}{\partial R} + \frac{1}{k}g(R) = \frac{\partial T}{\partial t} \quad 1 < R < 2, \quad t > 0$$

$$-\frac{\partial T}{\partial R} + H_1 T = 0 \quad R = 1, \quad t > 0$$

$$\frac{\partial T}{\partial R} + H_2 T = 0 \quad R = 2, \quad t > 0$$

$$T = F(R) \quad t = 0, \quad 1 \leq R \leq 2$$

Using the explicit scheme, develop the finite difference form of this problem.

5.29. Consider the following transient heat conduction problem for a hollow sphere.

$$\frac{\partial^2 T}{\partial R^2} + \frac{2}{R}\frac{\partial T}{\partial R} = \frac{\partial T}{\partial t} \quad 1 < R < 2, \quad t > 0$$

$$-\frac{\partial T}{\partial R} = q_0 \quad R = 1, \quad t > 0$$

$$\frac{\partial T}{\partial R} + HT = 0 \quad R = 2, \quad t > 0$$

$$T = F(R) \quad t = 0, \quad 1 \leq R \leq 2$$

Using the implicit scheme, develop the finite difference form of this problem.

5.30. Using the simple implicit scheme with upwind discretization, develop a finite difference approximation for the following one-dimensional, transient, convection–diffusion equation

$$\frac{\partial T}{\partial t} + u\frac{\partial T}{\partial x} = \alpha \frac{\partial^2 T}{\partial x^2}$$

where the flow velocity u and the thermal diffusivity α are considered constants.

5.31. Repeat Problem 5.30 using the Crank–Nicolson method.

5.32. Consider the first-order wave equation given by

$$\frac{\partial T}{\partial t} + u\frac{\partial T}{\partial x} = 0, \quad u > 0$$

The finite difference approximation of this equation, with the Euler explicit method, is given by

$$\frac{T_j^{n+1} - T_j^n}{\Delta t} + u\frac{T_{j+1}^n - T_{j-1}^n}{2\Delta x} = 0$$

Using Fourier stability analysis, show that this scheme is unconditionally unstable and hence has no utility for solving the wave equation.

5.33. Consider the first-order wave equation given by

$$\frac{\partial T}{\partial t} + u\frac{\partial T}{\partial x} = 0$$

Write the finite difference approximation of this equation using MacCormack's method.

5.34. The finite difference approximation of the first-order wave equation with the upwind method is given by

$$\frac{T_j^{n+1} - T_j^n}{\Delta t} + u\frac{T_j^n - T_{j-1}^n}{\Delta x} = 0 \quad \text{for} \quad u > 0.$$

Using the Fourier stability analysis, show that the scheme is stable for $0 < c \leq 1$ where $c \equiv \dfrac{u\Delta t}{\Delta x}$ is the Courant number.

5.35. Consider the first-order wave equation given by
$$\frac{\partial T}{\partial t} + u\frac{\partial T}{\partial x} = 0 \quad \text{in } 0 < x < L, \, t > 0$$
subjected to reflective boundary conditions, that is, $\frac{\partial T}{\partial x} = 0$ at $x = 0$ and at $x = 1$.

Apply the upwind, MacCormack, and Warming and Beam methods for the discretization of this equation and use the fictitious node approach for the discretization of the boundary conditions.

5.36. Repeat Problem 5.35 but for periodic boundary conditions, that is, $T_{M+1} = T_1$, $T_{-1} = T_{M-1}$, and $T_0 = T_M$.

5.37. Develop a manufactured solution for the heat conduction problem given by Problem 5.1.

5.38. Develop a manufactured solution for the heat conduction problem given by Problem 5.9. Apply the code and solution verification techniques presented in Chapter 2 for the finite difference solution obtained in Problem 5.9.

5.39. Develop a manufactured solution for the heat conduction problem given by Problem 5.14. Apply the code and solution verification techniques presented in Chapter 2 for the finite difference solution obtained in Problem 5.14.

5.40. Using the finite control volume approach, develop implicit discretized equations for Problem 5.1.

5.41. Using the finite control volume approach, develop implicit discretized equations for Problem 5.5.

5.42. Using the finite control volume approach, develop implicit discretized equations for the following advection–diffusion problem:
$$\rho\frac{\partial T}{\partial t} + \rho u\frac{\partial T}{\partial x} = \frac{k}{C_p}\frac{\partial^2 T}{\partial x^2} \quad \text{in } 0 < x < L, \, t > 0$$
$$T = T_0 \quad \text{at } x = 0, \, t > 0$$
$$\frac{\partial T}{\partial x} = 0 \quad \text{at } x = L, \, t > 0$$
$$T = T_i \quad \text{in } 0 < x < L, \, t = 0$$

One-Dimensional Transient Systems

5.43. Consider the second-order undamped wave equation given in the form

$$\frac{\partial^2 T}{\partial t^2} = u^2 \frac{\partial^2 T}{\partial x^2}$$

where the wave propagation speed u is constant. Let

$$T_1 = \frac{\partial T}{\partial t}, \quad T_2 = u\frac{\partial T}{\partial x}$$

and transform this second-order equation into two first-order equations; express the resulting equations in the vector form.

5.44. Consider the nonlinear hyperbolic heat conduction equation with temperature-dependent thermal conductivity given in the form

$$\frac{\partial \Theta}{\partial t} + \frac{\partial Q}{\partial X} - 2S = 0$$

$$\frac{\partial Q}{\partial t} + \frac{\partial}{\partial X}\left[\Theta(1+\gamma\Theta)\right] + 2Q = 0$$

where γ is a constant. Write this equation in the quasilinear form given by equation (1.29) and determine the Jacobian matrix.

5.45. To develop the finite difference approximation of the first-order wave equation with the Lax–Wendroff scheme, one considers a Taylor series expansion in the time variable in the form

$$T_j^{n+1} = T_j^n + \Delta t \frac{\partial T}{\partial t} + \frac{1}{2}(\Delta t)^2 \frac{\partial^2 T}{\partial t^2} + 0[(\Delta x)^2]$$

Then, one relates the time derivatives $\partial T/\partial t$ and $\partial^2 T/\partial t^2$ to the space derivatives $\partial T/\partial x$ and $\partial^2 T/\partial x^2$ by

$$\frac{\partial T}{\partial t} = -u\frac{\partial T}{\partial x}, \quad \frac{\partial^2 T}{\partial t^2} = u^2 \frac{\partial^2 T}{\partial x^2}$$

and substitute in this equation. Show that, for the linear problem considered here, the finite difference approximation of the first-order wave equation with the Lax–Wendroff scheme is identical to the MacCormack method given by equation (5.125).

5.46. Consider the dimensionless hyperbolic heat conduction equation given by equation (5.166), that is,

$$\frac{\partial \Theta}{\partial t} + \frac{\partial Q}{\partial X} - 2S = 0 \tag{1}$$

$$\frac{\partial Q}{\partial t} + \frac{\partial \Theta}{\partial X} + 2Q = 0 \tag{2}$$

If these linear equations are expressed in the quasilinear vector form as defined by equation (1.29), show that the Jacobian matrix **A** is given by

$$\mathbf{A} = \begin{bmatrix} 0 & 1 \\ 1 & 0 \end{bmatrix}$$

5.47. Consider the hyperbolic heat conduction equation given in the form

$$\frac{\Gamma}{\alpha}\frac{\partial^2 T}{\partial t^2} + \frac{1}{\alpha}\frac{\partial T}{\partial t} = \frac{\partial^2 T}{\partial x^2}$$

Write the finite difference approximation of this equation by applying centered time and centered space discretization and determine the truncation error.

5.48. Consider the hyperbolic heat conduction problem in the dimensionless form given by (see Example 5.9):

$$\frac{\partial \mathbf{E}}{\partial \xi} + \frac{\partial \mathbf{F}}{\partial \eta} = 0 \quad \text{in} \quad 0 < \eta < 1$$

where

$$\mathbf{E} = \begin{bmatrix} \Theta \\ Q \end{bmatrix} \qquad \mathbf{F} = \begin{bmatrix} Q \\ \Theta \end{bmatrix}$$

subject to the initial conditions

$$\Theta = 0 \quad \text{and} \quad Q = 0 \quad \text{for } \xi = 0$$

and the boundary conditions

$$\Theta = 1 \quad \text{and} \quad \left(\frac{\partial Q}{\partial \eta} = 0\right) \quad \text{for } \eta = 0, \ \xi > 0$$

$$\Theta = 0 \quad \text{and} \quad \left(\frac{\partial Q}{\partial \eta} = 0\right) \quad \text{for } \eta = 1, \ \xi > 0$$

Solve this problem with the MacCormack scheme and plot Θ versus η at the time $\xi = 0.5978$, and compare your results with those shown in Example 5.9.

5.49. Consider the modulus of the amplification factor $|\xi|$ given by Example 5.6. By plotting $|\xi|$ for $c = 0.5$, 0.75, and 1.25, for different values of θ from 0 to π, show that the stability criteria $|\xi| \leq 1$ are satisfied if the value of c is less than or equal to one.

6
Transient Multidimensional Systems

When temperature gradients within a body become important in more than one direction, as in the case of finite-sized bodies and spatially varying boundary conditions, multidimensional effects need to be included in the analysis. In this chapter, we examine the numerical solution of transient multidimensional parabolic systems by finite difference methods. There is a large variety of heat or mass transfer problems that are parabolic in nature. For example, problems of transient heat or mass diffusion in solids and advection–diffusion type transient problems can be categorized under parabolic systems.

The discretization schemes examined earlier in this book can be readily extended for multidimensional problems. The simple explicit method, for example, can be adopted for the solution of multidimensional parabolic problems; but the restriction imposed on the allowable marching step, as well as the increase in the number of grid points for multiple dimensions, increases the computational time enormously. The Crank–Nicolson or the fully implicit methods can also be adopted for the solution of two- and three-dimensional problems, but the resulting system of algebraic equations is no longer tridiagonal. To alleviate such difficulties, various alternative approaches have been proposed for the finite difference solution of multidimensional problems. For example, alternating direction implicit (ADI) methods have been proposed by Douglas (1955), Peaceman and Rachford (1955), and Douglas and Gun (1964), and closely related methods are described by Yanenko (1971). Alternating direction explicit (ADE) methods have been proposed by Saul'yev (1957), Barakat and Clark (1966), Larkin (1964), and Allada and Quon (1966). Keller's box method (1970) has been used for solving two-dimensional parabolic problems. Several other alternative schemes have also been proposed by researchers such as Yuen and Wong (1980), Sommeljer et al. (1981), and Evans and Avdelas (1978). In this chapter, we present the ADI and ADE methods as well as the use of explicit and combined methods for finite difference representation of two- and three-dimensional model problems. Purely diffusive systems and advective–diffusive systems with known velocity field are examined as follows.

6.1 Simple Explicit Method

The use of the simple explicit method and the stability constraints associated with it are presented here for the finite difference representation of typical parabolic problems including: (i) two-dimensional heat diffusion in solids, (ii) two-dimensional steady, boundary layer flow, and (iii) transient convection–diffusion.

6.1.1 Two-Dimensional Diffusion

We consider two-dimensional, linear heat diffusion in isotropic solids as the model problem. The temperature fields in the medium are governed by the partial differential equation

$$\frac{\partial T}{\partial t} = \alpha \left(\frac{\partial^2 T}{\partial x^2} + \frac{\partial^2 T}{\partial y^2} + \frac{1}{k} g \right) \qquad (6.1)$$

subject to some specified boundary and initial conditions, where $T \equiv T(x,y,t)$ and $g \equiv g(x,y,t)$. To discretize this differential equation, we introduce the notation

$$T(x, y, t) = T(i\Delta x, j\Delta y, n\Delta t) \equiv T_{i,j}^n \qquad (6.2)$$

Then, the finite difference approximation of the differential equation (6.1) at a grid point (x,y) by the simple explicit method using forward time and central space (FTCS) discretization gives

$$\frac{T_{i,j}^{n+1} - T_{i,j}^n}{\Delta t} = \alpha \left[\frac{T_{i-1,j}^n - 2T_{i,j}^n + T_{i+1,j}^n}{(\Delta x)^2} + \frac{T_{i,j-1}^n - 2T_{i,j}^n + T_{i,j+1}^n}{(\Delta y)^2} + \frac{1}{k} g_{i,j}^n \right] \qquad (6.3)$$

This expression is rearranged in the form

$$T_{i,j}^{n+1} = T_{i,j}^n + r_x(T_{i-1,j}^n - 2T_{i,j}^n + T_{i+1,j}^n) + r_y(T_{i,j-1}^n - 2T_{i,j}^n + T_{i,j+1}^n) + \frac{\alpha \Delta t}{k} g_{i,j}^n \qquad (6.4a)$$

where

$$r_x \equiv \frac{\alpha \Delta t}{(\Delta x)^2} \qquad r_y \equiv \frac{\alpha \Delta t}{(\Delta y)^2} \qquad (6.4b)$$

For a square mesh $\Delta x = \Delta y = \delta$, equation (6.4a) reduces to

$$T_{i,j}^{n+1} = r(T_{i-1,j}^n + T_{i+1,j}^n + T_{i,j-1}^n + T_{i,j+1}^n) + (1 - 4r)T_{i,j}^n + rG_{ij}^n \qquad (6.5a)$$

where

$$r \equiv \frac{\alpha \Delta t}{\delta^2} \qquad G_{ij}^n \equiv \frac{\delta^2 g_{i,j}^n}{k} \qquad (6.5b)$$

Equation (6.4a) or (6.5a) provides an explicit expression for the determination of $T_{i,j}^{n+1}$ at the time level n + 1 from the knowledge of grid-point

Transient Multidimensional Systems

temperatures at the previous time level n. If temperature is prescribed at all boundaries, the number of equations is equal to the number of unknown temperatures at the internal grid points; hence, the problem is solvable.

For derivative boundary conditions, such as convection or prescribed heat flux, the temperatures at the boundary nodes are not known. For such cases, additional relations are developed by discretizing the boundary conditions, such as presented in Chapter 2, by using fictitious nodes.

The control volume approach can also be used to develop the discretized form of equation (6.1) by applying an extension of the procedure presented in Chapter 2 to multidimensional problems. In this case, the development of the discretized equations for the boundary nodes presented in Chapter 2 is equally extended for multidimensional problems.

Stability: To obtain meaningful results from the solution of the difference equation (6.4a), the stability criterion associated with it should be established. We rewrite equation (6.4a) in the form

$$T_{j,k}^{n+1} = T_{j,k}^{n} + r_x(T_{j-1,k}^{n} - 2T_{j,k}^{n} + T_{j+1,k}^{n}) + r_y(T_{j,k-1}^{n} - 2T_{j,k}^{n} + T_{j,k+1}^{n}) \quad (6.6)$$

where the heat generation term is not taken into account for the analysis of the propagation of errors, and the subscript i is replaced by j, in order to distinguish the subscript from $i = \sqrt{-1}$, which will appear in the analysis.

The Fourier stability analysis described previously is now generalized for the two-dimensional case considered here by choosing the error term in the form

$$\varepsilon_{j,k}^{n} = \xi^{n} e^{i\beta_m j\Delta x}\, e^{i\eta_p k\Delta y} \quad \text{where} \quad \xi \equiv e^{\gamma \Delta t} \quad (6.7)$$

$i = \sqrt{-1}$, β_m and η_p are the Fourier modes. In view of the definition of ξ, the error term $\varepsilon_{j,k}^{n}$ will not increase without bounds as t increases, provided that

$$|\xi| \leq 1 \quad (6.8)$$

The error term should also satisfy the finite difference equation (6.6). Therefore, equation (6.7) is substituted into equation (6.6) and after cancellations we obtain

$$\xi = 1 + r_x(e^{-i\beta_m \Delta x} + e^{i\beta_m \Delta x} - 2) + r_y(e^{-i\eta_p \Delta y} + e^{i\eta_p \Delta y} - 2) \quad (6.9)$$

which can be written as

$$\xi = 1 - 2r_x(1 - \cos \beta_m \Delta x) - 2r_y(1 - \cos \eta_p \Delta y) \quad (6.10)$$

since

$$\cos z = \frac{1}{2}(e^{-iz} + e^{iz})$$

The application of the stability criterion defined by equations (6.8)–(6.10) yields

$$-1 \leq [1 - 2r_x(1 - \cos \beta_m \Delta x) - 2r_y(1 - \cos \eta_p \Delta y)] \leq 1$$

which must be satisfied for all values of β_m and η_p. The right-hand side is always satisfied. To satisfy the left-hand side under most strict conditions, we must have $1 - \cos \beta_m \Delta x = 2$ and $1 - \cos \eta_p \Delta y = 2$, yielding

$$-1 \leq [1 - 4r_x - 4r_y] \tag{6.11a}$$

that is,

$$(r_x + r_y) \leq \frac{1}{2} \tag{6.11b}$$

or

$$\left[\frac{\alpha \Delta t}{(\Delta x)^2} + \frac{\alpha \Delta t}{(\Delta y)^2}\right] \leq \frac{1}{2} \tag{6.11c}$$

For the case $\Delta x = \Delta y = \delta$, the stability criterion becomes

$$r = \frac{\alpha \Delta t}{\delta^2} \leq \frac{1}{4} \tag{6.12}$$

which is twice as restrictive as the one-dimensional constraint $r \leq \frac{1}{2}$.

Example 6.1

Develop the stability criterion for the finite difference approximation by the simple explicit method of the three-dimensional linear transient diffusion equation in the x,y,z rectangular coordinates.

Solution

The finite difference equation (6.6) and the corresponding error term [equation (6.7)] are readily generalized to the three-dimensional case. The error term is substituted into the finite difference equation, and a procedure similar to that described previously is followed. The following stability criterion results.

$$(r_x + r_y + r_z) \leq \frac{1}{2}$$

or

$$\left[\frac{\alpha \Delta t}{(\Delta x)^2} + \frac{\alpha \Delta t}{(\Delta y)^2} + \frac{\alpha \Delta t}{(\Delta z)^2}\right] \leq \frac{1}{2}$$

For the case $\Delta x = \Delta y = \Delta z = \delta$, the stability criteria becomes

$$r = \frac{\alpha \Delta t}{\delta^2} \leq \frac{1}{6}$$

which is thrice as restrictive as the one-dimensional constraint $r \leq \frac{1}{2}$.

Example 6.2

Using the simple explicit method, write the finite difference form of the following two-dimensional diffusion equation in cylindrical coordinates.

$$\frac{1}{\alpha}\frac{\partial T}{\partial t} = \frac{\partial^2 T}{\partial R^2} + \frac{1}{R}\frac{\partial T}{\partial R} + \frac{1}{R^2}\frac{\partial^2 T}{\partial \phi^2} + \frac{g}{k}$$

where $T \equiv T(R,\phi,t)$, $g \equiv g(R,\phi,t)$.

Solution

We adopt the notation

$$T(R,\phi,t) = T(i\Delta R, j\Delta\phi, n\Delta t) \equiv T_{i,j}^n$$

Using FTCS discretization, we obtain

$$\frac{T_{i,j}^{n+1} - T_{i,j}^n}{\alpha \Delta t} = \frac{T_{i-1,j}^n - 2T_{i,j}^n + T_{i+1,j}^n}{(\Delta R)^2} + \frac{1}{i\Delta R}\frac{T_{i+1,j}^n - T_{i-1,j}^n}{2\Delta R}$$
$$+ \frac{1}{(i\Delta R)^2}\frac{T_{i,j-1}^n - 2T_{i,j}^n + T_{i,j+1}^n}{(\Delta\phi)^2} + \frac{1}{k}g_{i,j}^n$$

This equation is now rearranged in the form

$$T_{i,j}^{n+1} = r_\xi\left(1 - \frac{1}{2i}\right)T_{i-1,j}^n + r_\xi\left(1 + \frac{1}{2i}\right)T_{i+1,j}^n + \frac{1}{i^2}r_\phi T_{i,j-1}^n + \frac{1}{i^2}r_\phi T_{i,j+1}^n$$
$$+ \left(1 - 2r_\xi - \frac{1}{i^2}2r_\phi\right)T_{i,j}^n + r_\xi\frac{(\Delta R)^2}{k}g_{i,j}^n \qquad (a)$$

where

$$r_\xi = \frac{\alpha \Delta t}{(\Delta R)^2} \qquad r_\phi = \frac{\alpha \Delta t}{(\Delta\phi \Delta R)^2}$$

Equation (d) is applicable at all the internal nodes

$$i = 1, 2, \ldots, M-1 \text{ and } j = 0, 1, 2, \ldots$$

except at the origin $i = 0$, which is examined in the next example. Furthermore, boundary conditions are applied for the nodes at the surface of the cylinder, which correspond to $i = M$, in a fashion similar to that described in Chapter 5.

Example 6.3

Using the simple explicit method, write the finite difference approximation of the following diffusion equation

$$\frac{1}{\alpha}\frac{\partial T}{\partial t} = \frac{\partial^2 T}{\partial R^2} + \frac{1}{R}\frac{\partial T}{\partial R} + \frac{1}{R^2}\frac{\partial^2 T}{\partial \phi^2} + \frac{g}{k}$$

at the origin $R = 0$.

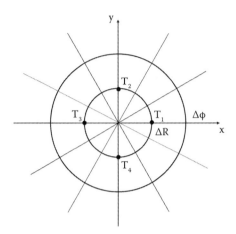

FIGURE 6.1
Approximation of the Laplacian at R = 0.

Solution

The Laplacian term has an apparent singularity at R = 0. Therefore, in order to develop a finite difference approximation at R = 0, the Laplacian in cylindrical coordinates is replaced by its Cartesian equivalent as

$$\frac{1}{\alpha}\frac{\partial T}{\partial t} = \nabla^2 T + \frac{g}{k}$$

where

$$\nabla^2 T = \frac{\partial^2 T}{\partial x^2} + \frac{\partial^2 T}{\partial y^2} \quad \text{at} \quad R = 0$$

To construct the finite difference form of $\nabla^2 T$ at the origin, the values of T at the circle with radius ΔR centered at R = 0 are used (see Figure 6.1). This circle intersects the x and y axes at the nodal points 1, 2, 3, and 4. Let T_0 be the temperature at the center R = 0 and T_1, T_2, T_3, T_4 the temperatures at these four points. Then, by using these four terms, the finite difference approximation to the Laplacian operator $\nabla^2 T|_{R=0}$ becomes

$$\nabla^2 T|_{R=0} = \frac{T_1 + T_2 + T_3 + T_4 - 4T_0}{(\Delta R)^2} + 0[(\Delta R)^2]$$

The rotation of the axes by the angle increment $\Delta\phi$ leads to similar results. Repetition of this rotation and the addition of such results leads to

$$\nabla^2 T|_{R=0} = \frac{4(\overline{T} - T_0)}{(\Delta R)^2} + 0[(\Delta R)^2]$$

where $\overline{T}$ is the mean value of T at all nodes over the circle of radius ΔR.

6.1.2 Two-Dimensional Transient Convection–Diffusion

The one-dimensional transient convection–diffusion problem considered in Chapter 5 is now extended to the two-dimensional energy conservation equation given by

$$\frac{\partial T}{\partial t} = -u\frac{\partial T}{\partial x} - v\frac{\partial T}{\partial y} + \alpha\left(\frac{\partial^2 T}{\partial x^2} + \frac{\partial^2 T}{\partial y^2}\right) + \frac{1}{\rho C_p}g \qquad (6.13)$$

where $T \equiv T(x,y,t)$, and u and v are the temperature and the velocity components along the x and y directions, respectively, in an incompressible flow of a fluid with constant physical properties. Clearly, this equation allows for convection and diffusion in both the x and y directions.

To discretize this equation, we adopt the notation

$$T(x,y,t) = T(i\Delta x, j\Delta y, n\Delta t) \equiv T_{i,j}^n \qquad (6.14)$$

6.1.2.1 FTCS Differencing

We apply the FTCS explicit scheme and obtain its finite difference approximation in the form

$$\begin{aligned}\frac{T_{i,j}^{n+1} - T_{i,j}^n}{\Delta t} &= -u_{i,j}\frac{T_{i+1,j}^n - T_{i-1,j}^n}{2\Delta x} - v_{i,j}\frac{T_{i,j+1}^n - T_{i,j-1}^n}{2\Delta y} \\ &+ \alpha\left[\frac{T_{i+1,j}^n - 2T_{i,j}^n + T_{i-1,j}^n}{(\Delta x)^2} + \frac{T_{i,j+1}^n - 2T_{i,j}^n + T_{i,j-1}^n}{(\Delta y)^2}\right] \\ &+ (1/\rho C_p)g_{i,j}^n\end{aligned} \qquad (6.15)$$

with a truncation error $0[\Delta t, (\Delta x)^2, (\Delta y)^2]$.

The finite difference scheme being explicit, some restriction is imposed on the time step Δt. The stability constraint is given by Peyret and Taylor (1983) as

$$\Delta t \leq \frac{4\alpha}{(|u_{i,j}| + |v_{i,j}|)^2} \qquad (6.16a)$$

and

$$\Delta t \leq \frac{1}{2\alpha}\left[\frac{1}{(\Delta x)^2} + \frac{1}{(\Delta y)^2}\right]^{-1} \qquad (6.16b)$$

where the constraint given by equation (6.16a) arises from the convection terms and that given by equation (6.16b) arises from the diffusion terms in equation (6.13.) For the special case of $\Delta x = \Delta y$, equation (6.16b) reduces to

$$r \equiv \frac{\alpha \Delta t}{(\Delta x)^2} \leq \frac{1}{4}$$

which is the criterion given by equation (6.12). When the constraint of equation (6.16a) from the convection term becomes too restrictive, the application of this finite difference scheme might be impractical. To alleviate the difficulty in such situations, the upwind differencing scheme can be used to discretize the convection terms as described in the following section.

6.1.2.2 Upwind Differencing

The use of central differencing, although it is second-order accurate, gives rise to a severe stability constraint on the time step Δt when applied to discretize the convection terms, as discussed in Chapter 5. The stability constraint resulting from the flow velocity can be removed if the upwind differencing scheme is used to discretize the first derivatives in the convection terms in equation (6.13). That is, backward differencing is used when the velocity is positive and forward differencing when the velocity is negative.

Thus, when $u_{i,j}^n$ and $v_{i,j}^n$ are positive, the finite difference approximation of equation (6.13) with the upwind scheme gives

$$\frac{T_{i,j}^{n+1} - T_{i,j}^n}{\Delta t} = -u_{i,j}^n \frac{T_{i,j}^n - T_{i-1,j}^n}{\Delta x} - v_{i,j}^n \frac{T_{i,j}^n - T_{i,j-1}^n}{\Delta y}$$

$$+ \alpha \left[\frac{T_{i+1,j}^n - 2T_{i,j}^n + T_{i-1,j}^n}{(\Delta x)^2} + \frac{T_{i,j+1}^n - 2T_{i,j}^n + T_{i,j-1}^n}{(\Delta y)^2} \right]$$

$$+ \frac{1}{\rho C_p} g_{i,j}^n \quad \text{for} \quad u_{i,j}^n > 0, \quad v_{i,j}^n > 0 \quad (6.17)$$

with the truncation error $O(\Delta t, \Delta x, \Delta y)$. The stability criterion for equation (6.17) is given by Roache (1976) as

$$\Delta t \leq \frac{1}{2\alpha \left(\frac{1}{(\Delta x)^2} + \frac{1}{(\Delta y)^2} \right) + \frac{|u|}{\Delta x} + \frac{|v|}{\Delta y}} \quad (6.18a)$$

which reduces to

$$\Delta t \leq \frac{(\Delta x)^2}{4\alpha + (|u| + |v|)\Delta x} \quad \text{for} \quad \Delta x = \Delta y \quad (6.18b)$$

If the velocity components $u_{i,j}$ and $v_{i,j}$ are both negative at the location i,j, the finite difference representation of the convection terms in equation (6.17) should be replaced by the forward differences given by

$$-u_{i,j}^n \frac{T_{i+1,j}^n - T_{i,j}^n}{\Delta x} - v_{i,j}^n \frac{T_{i,j+1}^n - T_{i,j}^n}{\Delta y}$$

6.1.2.3 Control Volume Approach

The control volume approach, due to its conservative nature, provides a consistent and systematic discretization of the transient convective-diffusive equation (6.13). For the application of the control volume discretization, equation (6.13) is rewritten in conservative form as:

$$\frac{\partial(\rho T)}{\partial t} = -\frac{\partial(\rho u T)}{\partial x} - \frac{\partial(\rho v T)}{\partial y} + \frac{\partial}{\partial x}\left(\frac{k}{C_p}\frac{\partial T}{\partial x}\right) + \frac{\partial}{\partial y}\left(\frac{k}{C_p}\frac{\partial T}{\partial y}\right) + \frac{g}{C_p} \quad (6.19)$$

Equation (6.19) is integrated in the time interval $[t, t + \Delta t]$ and in the finite volume around the point $(i\Delta x, j\Delta y)$ illustrated by Figure 6.2. The usual notation of the finite control volume discretization is utilized here, where the subscript P refers to the point $(i\Delta x, j\Delta y)$, while the subscripts W and E refer to the points at $[(i-1)\Delta x, j\Delta y]$ and $[(i+1)\Delta x, j\Delta y]$, respectively. Similarly, S and N refer to the points at $[i\Delta x, (j-1)\Delta y]$ and $[i\Delta x, (j+1)\Delta y]$, respectively. In the finite volume notation, the lowercase subscripts w, e, s, and n designate the surfaces of the control volume, as depicted in Figure 6.2. The reader should recognize that the superscript n, widely used in this book, denotes the number of time steps [see, for example, equation (6.17)] while the subscript n designates the control volume surface between points P and N. The integration of equation (6.19) in time and in the control volume gives

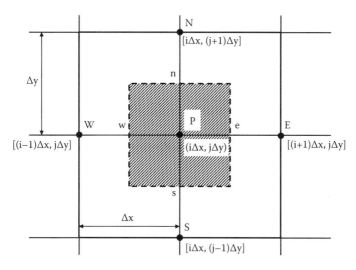

FIGURE 6.2
Finite control volume discretization for a two-dimensional domain.

$$\int_t^{t+\Delta t}\int_{(i-1/2)\Delta x}^{(i+1/2)\Delta x}\int_{(j-1/2)\Delta y}^{(j+1/2)\Delta y}\frac{\partial(\rho T)}{\partial t}dydxdt+\int_t^{t+\Delta t}\int_{(i-1/2)\Delta x}^{(i+1/2)\Delta x}\int_{(j-1/2)\Delta y}^{(j+1/2)\Delta y}\frac{\partial(\rho uT)}{\partial x}dydxdt$$

$$+\int_t^{t+\Delta t}\int_{(i-1/2)\Delta x}^{(i+1/2)\Delta x}\int_{(j-1/2)\Delta y}^{(j+1/2)\Delta y}\frac{\partial(\rho vT)}{\partial y}dydxdt=\int_t^{t+\Delta t}\int_{(i-1/2)\Delta x}^{(i+1/2)\Delta x}\int_{(j-1/2)\Delta y}^{(j+1/2)\Delta y}\frac{\partial}{\partial x}\left(\frac{k}{C_p}\frac{\partial T}{\partial x}\right)dydxdt$$

$$+\int_t^{t+\Delta t}\int_{(i-1/2)\Delta x}^{(i+1/2)\Delta x}\int_{(j-1/2)\Delta y}^{(j+1/2)\Delta y}\frac{\partial}{\partial y}\left(\frac{k}{C_p}\frac{\partial T}{\partial y}\right)dydxdt+\int_t^{t+\Delta t}\int_{(i-1/2)\Delta x}^{(i+1/2)\Delta x}\int_{(j-1/2)\Delta y}^{(j+1/2)\Delta y}\frac{g}{C_p}dydxdt$$

(6.20)

The transient term can be integrated by considering the average value of the integrand in the control volume. The heat source term is also integrated by considering the average value of the integrand in the control volume, while the remaining terms are integrated by considering average values of the integrand over the surfaces of the control volume, thus resulting in

$$[(\rho T)_P^{t+\Delta t}-(\rho T)_P^t]\Delta y\Delta x+[(\rho uT)_e^t-(\rho uT)_w^t]\Delta y\Delta t+[(\rho vT)_n^t-(\rho vT)_s^t]\Delta x\Delta t$$

$$=\left[\left(\frac{k}{C_p}\frac{\partial T}{\partial x}\right)_e^t-\left(\frac{k}{C_p}\frac{\partial T}{\partial x}\right)_w^t\right]\Delta y\Delta t+\left[\left(\frac{k}{C_p}\frac{\partial T}{\partial y}\right)_n^t-\left(\frac{k}{C_p}\frac{\partial T}{\partial y}\right)_s^t\right]\Delta x\Delta t+\left(\frac{g}{C_p}\right)_P^t\Delta y\Delta x\Delta t$$

(6.21)

where an explicit discretization was used in time.

The mass of fluid inside the finite control volume is given by

$$M=\rho\Delta y\Delta x \qquad (6.22a)$$

Similarly, the mass fluxes through the surfaces of the control volume can be defined as

$$\dot{M}_e=(\rho u)_e\Delta y, \quad \dot{M}_w=(\rho u)_w\Delta y \qquad (6.22b,c)$$

$$\dot{M}_n=(\rho v)_n\Delta x, \quad \dot{M}_s=(\rho v)_s\Delta x \qquad (6.22d,e)$$

The following diffusion coefficients are defined

$$D_{11}=\frac{k}{C_p}\Delta y, \quad D_{22}=\frac{k}{C_p}\Delta x \qquad (6.22f,g)$$

and the source term of the discretized equation is given by

$$S_P=\left(\frac{g}{C_p}\right)_P \qquad (6.22h)$$

By using equations (6.22a–h), equation (6.21) is rewritten as

$$\frac{(MT)_P^{t+\Delta t} - (MT)_P^t}{\Delta t} + \dot{M}_e^t T_e^t - \dot{M}_w^t T_w^t + \dot{M}_n^t T_n^t - \dot{M}_s^t T_s^t$$
$$= \left(D_{11}\frac{\partial T}{\partial x}\right)_e^t - \left(D_{11}\frac{\partial T}{\partial x}\right)_w^t + \left(D_{22}\frac{\partial T}{\partial y}\right)_n^t - \left(D_{22}\frac{\partial T}{\partial y}\right)_s^t + S_P^t \Delta y\, \Delta x \quad (6.23)$$

The temperatures and the temperature gradients at the surfaces of the control volume are now respectively approximated in terms of the temperatures at the centers of the neighboring control volumes in the forms

$$T_e = \left(\frac{1}{2} + \alpha_e\right) T_P + \left(\frac{1}{2} - \alpha_e\right) T_E \quad (6.24a)$$

$$T_w = \left(\frac{1}{2} + \alpha_w\right) T_W + \left(\frac{1}{2} - \alpha_w\right) T_P \quad (6.24b)$$

$$T_n = \left(\frac{1}{2} + \alpha_n\right) T_P + \left(\frac{1}{2} - \alpha_n\right) T_N \quad (6.24c)$$

$$T_s = \left(\frac{1}{2} + \alpha_s\right) T_S + \left(\frac{1}{2} - \alpha_s\right) T_P \quad (6.24d)$$

and

$$\left(\frac{\partial T}{\partial x}\right)_e = \beta_e \left(\frac{T_E - T_P}{\Delta x}\right) \quad (6.25a)$$

$$\left(\frac{\partial T}{\partial x}\right)_w = \beta_w \left(\frac{T_P - T_W}{\Delta x}\right) \quad (6.25b)$$

$$\left(\frac{\partial T}{\partial y}\right)_n = \beta_n \left(\frac{T_N - T_P}{\Delta y}\right) \quad (6.25c)$$

$$\left(\frac{\partial T}{\partial y}\right)_s = \beta_s \left(\frac{T_P - T_S}{\Delta y}\right) \quad (6.25d)$$

where the coefficients α and β are obtained from the one-dimensional weighted upstream differencing scheme (WUDS) interpolation scheme presented in Chapter 4. For the surfaces e and w, the velocity u in the x direction is used for the computation of the grid Peclet number, while the velocity v is used for the surfaces n and s in the y direction. Thus,

$$Pe_e = \left(\frac{\rho C_p}{k} u\right)_e \Delta x, \quad Pe_w = \left(\frac{\rho C_p}{k} u\right)_w \Delta x \qquad (6.26a,b)$$

$$Pe_n = \left(\frac{\rho C_p}{k} v\right)_n \Delta y, \quad Pe_s = \left(\frac{\rho C_p}{k} v\right)_s \Delta y \qquad (6.26c,d)$$

and

$$\alpha = \frac{Pe^2}{10 + 2Pe^2}\,\text{sign}(Pe), \quad \beta = \frac{1 + 0.005Pe^2}{1 + 0.05Pe^2} \qquad (6.27a,b)$$

In equations (6.26a–d), the properties at the surfaces of the control volume are approximated by harmonic means, such as given by equation (2.33b).

Equations (6.24) and (6.25) are now substituted into equation (6.23) to obtain:

$$T_P^{t+\Delta t} = \frac{\Delta t}{M_P^t}(A_P^t T_P^t + A_e^t T_E^t + A_w^t T_W^t + A_n^t T_N^t + A_s^t T_S^t + S_P^t \Delta x \Delta y) \qquad (6.28)$$

where

$$A_P^t = \frac{M_P^t}{\Delta t} - \dot{M}_e^t\left(\frac{1}{2}+\alpha_e^t\right) + \dot{M}_w^t\left(\frac{1}{2}-\alpha_w^t\right) - \dot{M}_n^t\left(\frac{1}{2}+\alpha_n^t\right) + \dot{M}_s^t\left(\frac{1}{2}-\alpha_s^t\right)$$
$$- \frac{D_{11,e}^t \beta_e^t}{\Delta x} - \frac{D_{11,w}^t \beta_w^t}{\Delta x} - \frac{D_{22,n}^t \beta_n^t}{\Delta y} - \frac{D_{22,s}^t \beta_s^t}{\Delta y} \qquad (6.29a)$$

$$A_e^t = -\dot{M}_e^t\left(\frac{1}{2}-\alpha_e^t\right) + \frac{D_{11,e}^t \beta_e^t}{\Delta x} \qquad (6.29b)$$

$$A_w^t = \dot{M}_w^t\left(\frac{1}{2}+\alpha_w^t\right) + \frac{D_{11,w}^t \beta_w^t}{\Delta x} \qquad (6.29c)$$

$$A_n^t = -\dot{M}_n^t\left(\frac{1}{2}-\alpha_n^t\right) + \frac{D_{22,n}^t \beta_n^t}{\Delta y} \qquad (6.29d)$$

$$A_s^t = \dot{M}_s^t\left(\frac{1}{2}+\alpha_s^t\right) + \frac{D_{22,s}^t \beta_s^t}{\Delta y} \qquad (6.29e)$$

The stability criterion for this scheme is obtained from the requirement that all coefficients given by equations (6.29a–e) must be positive (Patankar 1980). Therefore, the maximum Δt for a stable solution is obtained from the condition that $A_P^t > 0$.

For well-posedness of the mathematical formulation of the problem, initial and boundary conditions need to be prescribed. Discretization of boundary conditions readily follows the schemes presented in Chapter 2, depending on whether the finite difference formulae or the control volume approach is used for the discretization.

6.2 Combined Method

The application of the combined method for finite difference approximation of multidimensional parabolic systems will now be illustrated for a three-dimensional diffusion problem. We will consider a three-dimensional linear diffusion problem in an isotropic solid governed by the partial differential equation

$$\frac{\partial T}{\partial t} = \alpha \left(\frac{\partial^2 T}{\partial x^2} + \frac{\partial^2 T}{\partial y^2} + \frac{\partial^2 T}{\partial z^2} \right) \tag{6.30}$$

with appropriate boundary and initial conditions. To discretize this equation, we introduce the notation

$$T(x, y, z, t) = T(i\Delta x, j\Delta y, k\Delta z, n\Delta t) \equiv T_{i,j,k}^n \tag{6.31}$$

Then, the finite difference approximation of the differential equation (6.30) with the combined method becomes

$$\frac{T_{i,j,k}^{n+1} - T_{i,j,k}^n}{\alpha \Delta t} = \theta [\Delta_{xx} T_{i,j,k}^{n+1} + \Delta_{yy} T_{i,j,k}^{n+1} + \Delta_{zz} T_{i,j,k}^{n+1}] \\ + (1-\theta)[\Delta_{xx} T_{i,j,k}^n + \Delta_{yy} T_{i,j,k}^n + \Delta_{zz} T_{i,j,k}^n] \tag{6.32}$$

where the weight factor θ assumes values $0 \leq \theta \leq 1$, and the finite difference operators Δ_{xx}, Δ_{yy}, and Δ_{zz} are defined as

$$\Delta_{xx} T_{i,j,k}^n \equiv \frac{1}{(\Delta x)^2} [T_{i+1,j,k}^n - 2T_{i,j,k}^n + T_{i-1,j,k}^n] \tag{6.33}$$

$$\Delta_{yy} T_{i,j,k}^n \equiv \frac{1}{(\Delta y)^2} [T_{i,j+1,k}^n - 2T_{i,j,k}^n + T_{i,j-1,k}^n] \tag{6.34}$$

$$\Delta_{zz} T_{i,j,k}^n \equiv \frac{1}{(\Delta z)^2} [T_{i,j,k+1}^n - 2T_{i,j,k}^n + T_{i,j,k-1}^n] \tag{6.35}$$

Clearly, depending on the value chosen for the weight factor θ, the simple explicit, the simple implicit, and the Crank–Nicolson methods are readily obtained as special cases, that is, for

1. θ = 0: The simple explicit scheme is obtained, where the truncation error is $0[\Delta t, (\Delta x)^2, (\Delta y)^2, (\Delta z)^2]$ and the stability constraint on the time step Δt is

$$\left[\frac{\alpha \Delta t}{(\Delta x)^2} + \frac{\alpha \Delta t}{(\Delta y)^2} + \frac{\alpha \Delta t}{(\Delta z)^2}\right] \leq \frac{1}{2} \quad (6.36)$$

2. $\theta = \frac{1}{2}$: The Crank–Nicolson scheme is obtained, where the truncation error is $0[(\Delta t)^2, (\Delta x)^2, (\Delta y)^2, (\Delta z)^2]$.
3. θ = 1: The simple implicit scheme is obtained, where the truncation error is $0[\Delta t, (\Delta x)^2, (\Delta y)^2, (\Delta z)^2]$.

For $0.5 \leq \theta \leq 1$, the combined scheme is unconditionally stable.

6.3 ADI Method

The Crank–Nicolson and the simple implicit methods discussed previously have the advantage that they are unconditionally stable, but the computational problems become enormous for two- and three-dimensional situations. For example, a three-dimensional problem with N interior nodes in each direction involves a total of N^3 interior points. Therefore, a system of $N^3 \times N^3$ equations must be solved for each time level. To alleviate such difficulties, Peaceman and Rachford (1955) and Douglas (1955) developed the ADI method. In such methods, the discretization is performed so that more than one reduced system is solved at each time level, instead of solving one single large system of equations. For example, for a three-dimensional problem containing N interior nodes in each direction, the ADI method reduces the system of N^3 equations to three systems of N equations for each time level. Furthermore, the ADI method requires minimal computer storage and is quite accurate.

The ADI method originally proposed by Peaceman and Rachford is now presented for the case of two-dimensional diffusion chosen as the model problem. We consider the diffusion equation

$$\frac{1}{\alpha}\frac{\partial T}{\partial t} = \frac{\partial^2 T}{\partial x^2} + \frac{\partial^2 T}{\partial y^2} + \frac{1}{k}g(x,y,t) \quad (6.37)$$

Transient Multidimensional Systems

subject to appropriate boundary and initial conditions, and introduce the notation

$$T(x, y, t) = T(i\Delta x, j\Delta y, n\Delta t) \equiv T_{i,j}^n \qquad (6.38)$$

The finite difference approximation of the differential equation (6.37) with the ADI method is based on the following concepts. Suppose the computations are to be advanced from the (n)th time level to the (n + 1)th time level. The simple implicit method is used for one of the directions, say x, and the simple explicit method is used for the other direction (i.e., y) for computing the solution in an intermediate time $(n + 1/2)\Delta t$. Then, the advancement from the (n + 1/2)th level to the (n + 1)th level is done by reversing the directions of the implicit and explicit methods. The computational procedure is thus continued by alternately changing the directions of the explicit and implicit methods.

We now illustrate the application of the ADI method for the discretization of equation (6.37). The implicit scheme is used in the x direction and the explicit scheme in the y direction to advance from the nth to the (n + 1/2)th time level. The finite difference approximation of equation (6.37) is given by

$$\frac{T_{i,j}^{n+1/2} - T_{i,j}^n}{\alpha \frac{\Delta t}{2}} = \frac{T_{i-1,j}^{n+1/2} - 2T_{i,j}^{n+1/2} + T_{i+1,j}^{n+1/2}}{(\Delta x)^2} + \frac{T_{i,j-1}^n - 2T_{i,j}^n + T_{i,j+1}^n}{(\Delta y)^2} + \frac{1}{k} g_{i,j}^{n+1/2} \qquad (6.39)$$

An explicit formulation is now used for the x direction and an implicit formulation for the y direction. Then, the finite difference approximation for equation (6.37) from the intermediate (n + 1/2)th to the (n + 1)th time step becomes

$$\frac{T_{i,j}^{n+1} - T_{i,j}^{n+1/2}}{\alpha \frac{\Delta t}{2}} = \frac{T_{i-1,j}^{n+1/2} - 2T_{i,j}^{n+1/2} + T_{i+1,j}^{n+1/2}}{(\Delta x)^2} + \frac{T_{i,j-1}^{n+1} - 2T_{i,j}^{n+1} + T_{i,j+1}^{n+1}}{(\Delta y)^2} + \frac{1}{k} g_{i,j}^{n+1} \qquad (6.40)$$

For computational purposes, it is convenient to rearrange equations (6.39) and (6.40) as

$$-r_x T_{i-1,j}^{n+1/2} + (1 + 2r_x) T_{i,j}^{n+1/2} - r_x T_{i+1,j}^{n+1/2}$$
$$= r_y T_{i,j-1}^n + (1 - 2r_y) T_{i,j}^n + r_y T_{i,j+1}^n + \frac{\alpha \Delta t}{2k} g_{i,j}^{n+1/2} \qquad (6.41a)$$

for the time level n + 1/2 and

$$-r_y T_{i,j-1}^{n+1} + (1 + 2r_y) T_{i,j}^{n+1} - r_y T_{i,j+1}^{n+1}$$
$$= r_x T_{i-1,j}^{n+1/2} + (1 - 2r_x) T_{i,j}^{n+1/2} + r_x T_{i+1,j}^{n+1/2} + \frac{\alpha \Delta t}{2k} g_{i,j}^{n+1} \qquad (6.41b)$$

for the time level n + 1, where

$$r_x = \frac{\alpha \Delta t}{2(\Delta x)^2} \quad \text{and} \quad r_y = \frac{\alpha \Delta t}{2(\Delta y)^2} \qquad (6.42a,b)$$

The advantage of this approach over the fully implicit or Crank–Nicolson methods is that each equation, although implicit, is only tridiagonal and can be efficiently solved with the Thomas algorithm presented in Chapter 3. That is, equation (6.41a) contains implicit unknowns $T_{i,j}^{n+1/2}, T_{i-1,j}^{n+1/2}$, and $T_{i+1,j}^{n+1/2}$, while equation (6.41b) contains implicit unknowns $T_{i,j}^{n+1}, T_{i,j-1}^{n+1}$, and $T_{i,j+1}^{n+1}$. After application of the boundary conditions of the problem by using the discretization procedures outlined in Chapter 2, equations (6.41a,b) result in tridiagonal systems of the form:

$$\begin{bmatrix} b_{1,j}^x & c_{1,j}^x & & & \\ a_{2,j}^x & b_{2,j}^x & c_{2,j}^x & & \\ & \ddots & \ddots & \ddots & \\ & & a_{I-1,j}^x & b_{I-1,j}^x & c_{I-1,j}^x \\ & & & a_{I,j}^x & b_{I,j}^x \end{bmatrix} \begin{bmatrix} T_{1,j}^{n+1/2} \\ T_{2,j}^{n+1/2} \\ \vdots \\ T_{I-1,j}^{n+1/2} \\ T_{I,j}^{n+1/2} \end{bmatrix} = \begin{bmatrix} d_{1,j}^x \\ d_{2,j}^x \\ \vdots \\ d_{I-1,j}^x \\ d_{I,j}^x \end{bmatrix} \quad (6.43)$$

which is obtained from the implicit discretization for the x direction and is solved for $j = 1,\ldots, J$, while

$$\begin{bmatrix} b_{i,1}^y & c_{i,1}^y & & & \\ a_{i,2}^y & b_{i,2}^y & c_{i,2}^y & & \\ & \ddots & \ddots & \ddots & \\ & & a_{i,J-1}^y & b_{i,J-1}^y & c_{i,J-1}^y \\ & & & a_{i,J}^y & b_{i,J}^y \end{bmatrix} \begin{bmatrix} T_{i,1}^{n+1} \\ T_{i,2}^{n+1} \\ \vdots \\ T_{i,J-1}^{n+1} \\ T_{i,J}^{n+1} \end{bmatrix} = \begin{bmatrix} d_{i,1}^y \\ d_{i,2}^y \\ \vdots \\ d_{i,J-1}^y \\ d_{i,J}^y \end{bmatrix} \quad (6.44)$$

is obtained from the implicit discretization for the y direction are solved for $i = 1,\ldots, I$, where the index i is supposed to vary from 1 to I and the index j from 1 to J.

The tridiagonal system [equation (6.43)] can be solved independently for each $j = 1,\ldots, J$ because vector dependencies in j were removed through the explicit discretization along the y direction. Therefore, one can benefit from vector computations in modern computers by simultaneously solving as many systems [equation (6.43)] as the vector capabilities of the computer allow. Similarly, the system [equation (6.44)] can be solved simultaneously for several indexes, i. In order to take advantage of vector computations and speed up the solutions of the systems resulting from the ADI discretization, the Thomas algorithm is rewritten in the form that follows.

For the system [equation (6.43)] resulting from the implicit discretization in the x direction, we have:

Forward Sweep
  For j = 1,...,J
  $$Z^x_{1,j} = b^x_{1,j}$$
  $$S^x_{1,j} = d^x_{1,j}$$
  end
  For i = 2,...,I
    For j = 1,...,J
    $$Z^x_{i,j} = b^x_{i,j} - \frac{a^x_{i,j}}{Z^x_{i-1,j}} c^x_{i-1,j}$$
    $$S^x_{i,j} = d^x_{i,j} - \frac{a^x_{i,j}}{Z^x_{i-1,j}} S^x_{i-1,j}$$
    end
  end

Backward Sweep
  For j = 1,...,J
  $$T^{n+1/2}_{I,j} = \frac{S^x_{I,j}}{Z^x_{I,j}}$$
  end
  For i = I − 1,...,1
    For j = 1,...,J
    $$T^{n+1/2}_{i,j} = \frac{1}{Z^x_{i,j}} (S^x_{i,j} - c^x_{i,j} T^{n+1/2}_{i+1,j})$$
    end
  end

Similarly, for the system [equation (6.44)] resulting from the implicit discretization in the y direction, we have:

Forward Sweep
  For i = 1,...,I
  $$Z^y_{i,1} = b^y_{i,1}$$
  $$S^y_{i,1} = d^y_{i,1}$$
  end

For j = 2,...,J
   For i = 1,...,I

$$Z_{i,j}^y = b_{i,j}^y - \frac{a_{i,j}^y}{Z_{i,j-1}^y} c_{i,j-1}^y$$

$$S_{i,j}^y = d_{i,j}^y - \frac{a_{i,j}^y}{Z_{i,j-1}^y} S_{i,j-1}^y$$

   end
end

Backward Sweep
For i = 1,...,I

$$T_{i,J}^{n+1} = \frac{S_{i,J}^y}{Z_{i,J}^y}$$

end
For j = J–1,...,1
   For i = 1,...,I

$$T_{i,j}^{n+1} = \frac{1}{Z_{i,j}^y} (S_{i,j}^y - c_{i,j}^y T_{i,j+1}^{n+1})$$

   end
end

The ADI method illustrated previously for a two-dimensional case involving a diffusion equation can be straightforwardly extended for a three-dimensional case as well as be used with the control volume discretization approach.

6.4 ADE Method

ADE methods not only provide computational simplicity but also possess the advantages of the implicit methods in that no severe limitation is imposed on the time step. Saul'yev (1957) was the first to propose a two-step scheme, which is unconditionally stable and has a truncation error of order $0[(\Delta t)^2, (\Delta x)^2, (\Delta t/\Delta x)^2]$. However, in practice, the scheme is first-order accurate due to the presence of the term $(\Delta t/\Delta x)^2$ in the truncation error.

Later on, Larkin (1964) and Barakat and Clark (1966) proposed alternative ADE schemes. Numerical tests appear to indicate that the Barakat and Clark scheme is more accurate than the Larkin scheme. Therefore, we present the

ADE method originally proposed by Barakat and Clark. To illustrate the basic concepts of this method, we choose the one-dimensional diffusion as the model problem before considering the two-dimensional situation.

Consider one-dimensional heat diffusion in a slab of thickness L governed by the differential equation

$$\frac{1}{\alpha}\frac{\partial T}{\partial t} = \frac{\partial^2 T}{\partial x^2} + \frac{1}{k}g(x,t) \quad \text{in } 0 < x < L,\ t > 0 \tag{6.45}$$

subject to prescribed temperatures at both boundaries and an initial condition. For discretization purposes, we introduce the notation

$$T(x,t) = T(j\Delta x, n\Delta t) \equiv T_j^n \tag{6.46}$$

Let U_j^n and V_j^n be the solutions of the following two finite difference equations, which are multilevel finite difference representations of the differential equation (6.45):

$$\frac{U_j^{n+1} - U_j^n}{\Delta t} = \alpha \frac{U_{j-1}^{n+1} - U_j^{n+1} - U_j^n + U_{j+1}^n}{(\Delta x)^2} + \frac{\alpha}{2k}(g_j^{n+1} + g_j^n) \tag{6.47}$$

$$\frac{V_j^{n+1} - V_j^n}{\Delta t} = \alpha \frac{V_{j-1}^n - V_j^n - V_j^{n+1} + V_{j+1}^{n+1}}{(\Delta x)^2} + \frac{\alpha}{2k}(g_j^{n+1} + g_j^n)$$

$$\text{for } j = 1, 2, \ldots, J-1. \tag{6.48}$$

Equations (6.47) and (6.48) are rearranged in order to obtain explicit expressions for U_j^{n+1} and V_j^{n+1}. They become

$$U_j^{n+1} = aU_j^n + b(U_{j-1}^{n+1} + U_{j+1}^n) + bG_j^* \tag{6.49}$$

$$V_j^{n+1} = aV_j^n + b(V_{j-1}^n + V_{j+1}^{n+1}) + bG_j^* \tag{6.50}$$

respectively, where

$$j = 1, 2, \ldots, M-1, \quad a = \frac{1-r}{1+r}, \quad b = \frac{r}{1+r} \tag{6.51a–c}$$

$$G_j^* = \frac{(\Delta x)^2}{2k}(g_j^{n+1} + g_j^n), \quad r = \frac{\alpha \Delta t}{(\Delta x)^2} \tag{6.51d,e}$$

The computational procedure for calculating U_j^{n+1} and V_j^{n+1} from equations (6.47) and (6.48) follows.

Equation (6.47) marches the solution from left to right by starting at the node i = 1, with U_0^{n+1} being always available from the prescribed temperature at the left boundary. Similarly, equation (6.48) marches the solution from right to left by starting at the node j = J − 1, while V_M^{n+1} is available from the prescribed temperature at the right boundary. The two solutions are performed simultaneously. Once U_j^{n+1} and V_j^{n+1} are determined from these calculations, the temperatures T_j^{n+1} at the time level n + 1 at the interior nodes j are computed from the arithmetic average of U_j^{n+1} and V_j^{n+1} as

$$T_j^{n+1} = \frac{1}{2}(U_j^{n+1} + V_j^{n+1}) \tag{6.52}$$

The advantage of this method is twofold. First, it is unconditionally stable, and second, the truncation error is approximately $0[(\Delta t)^2, (\Delta x)^2]$ because the averaging of the two marching solutions as given by equation (6.52) tends to cancel the error terms of opposite signs.

Stability: To illustrate that the ADE method is unconditionally stable, we apply the Fourier stability analysis to the finite difference equation using either equation (6.47) or (6.48).

Consider, for example, equation (6.47), without the energy generation term, written in the form

$$U_j^{n+1} - U_j^n = r(U_{j-1}^{n+1} - U_j^{n+1} - U_j^n + U_{j+1}^n) \tag{6.53}$$

where

$$r = \alpha \Delta t / (\Delta x)^2 \tag{6.54}$$

A typical error term, as discussed previously, is taken in the form

$$\varepsilon_j^n = \xi^n e^{i\beta_m j \Delta x} \tag{6.55}$$

where $i = \sqrt{-1}$ and β_m is the Fourier mode. This error term should satisfy the difference equation (6.53). Introducing equation (6.55) into equation (6.53) and after cancellations, we obtain

$$\xi - 1 = r(\xi e^{-i\beta_m \Delta x} - \xi - 1 + e^{-i\beta_m \Delta x}) \tag{6.56}$$

which is solved for ξ as

$$\xi = \frac{1 - r(1 - e^{i\beta_m \Delta x})}{1 + r(1 + e^{i\beta_m \Delta x})} \tag{6.57}$$

The errors will not increase without bound as t increases if

$$|\xi| \leq 1 \tag{6.58}$$

Examination of equation (6.57) shows that inequality [equation (6.58)] is satisfied for all values of r, which indicates unconditional stability.

The foregoing analysis is based on the assumption that temperatures are prescribed at all boundaries; hence, temperatures at the boundary nodes are known. In the case of convection or prescribed heat flux boundary conditions, it has been shown by the stability analysis that no constraints are needed on the size of the time step either.

The ADE method is now generalized for the two-dimensional case by considering diffusion as the model problem. The governing differential equation is taken as

$$\frac{1}{\alpha}\frac{\partial T}{\partial t} = \frac{\partial^2 T}{\partial x^2} + \frac{\partial^2 T}{\partial y^2} + \frac{g}{k}, \quad 0 < x < a, \quad 0 < y < b, \quad t > 0 \tag{6.59}$$

subject to prescribed temperatures at all boundaries and the initial condition. For discretization purposes, we introduce the notation

$$T(x, y, t) = T(i\Delta x, j\Delta y, n\Delta t) \equiv T_{i,j}^n \tag{6.60}$$

Let $U_{i,j}^n$ and $V_{i,j}^n$ be the solutions of the following two finite difference equations, which are multilevel finite difference representations of the differential equation (6.59).

$$\frac{U_{i,j}^{n+1} - U_{i,j}^n}{\alpha \Delta t} = \frac{U_{i-1,j}^{n+1} - U_{i,j}^{n+1} - U_{i,j}^n + U_{i+1,j}^n}{(\Delta x)^2}$$
$$+ \frac{U_{i,j-1}^{n+1} - U_{i,j}^{n+1} - U_{i,j}^n + U_{i,j+1}^n}{(\Delta y)^2} + \frac{1}{2k}(g_{i,j}^{n+1} + g_{i,j}^n) \tag{6.61}$$

and

$$\frac{V_{i,j}^{n+1} - V_{i,j}^n}{\alpha \Delta t} = \frac{V_{i-1,j}^n - V_{i,j}^n - V_{i,j}^{n+1} + V_{i+1,j}^{n+1}}{(\Delta x)^2}$$
$$+ \frac{V_{i,j-1}^n - V_{i,j}^n - V_{i,j}^{n+1} + V_{i,j+1}^{n+1}}{(\Delta y)^2} + \frac{1}{2k}(g_{i,j}^{n+1} + g_{i,j}^n) \tag{6.62}$$

Equations (6.61) and (6.62) are rearranged in order to obtain explicit expressions for $U_{i,j}^{n+1}$ and $V_{i,j}^{n+1}$. They become

$$U_{i,j}^{n+1} = AU_{i,j}^n + B(U_{i-1,j}^{n+1} + U_{i+1,j}^n) + C(U_{i,j-1}^{n+1} + U_{i,j+1}^n) + G_{i,j}^* \tag{6.63}$$

$$V_{i,j}^{n+1} = AV_{i,j}^n + B(V_{i-1,j}^n + V_{i+1,j}^{n+1}) + C(V_{i,j-1}^n + V_{i,j+1}^{n+1}) + G_{i,j}^* \tag{6.64}$$

respectively, where $i = 1, 2, \ldots, I - 1$ and $j = 1, 2, \ldots, J - 1$

$$A = \frac{1-r_x-r_y}{1+r_x+r_y}, \quad B = \frac{r_x}{1+r_x+r_y}, \quad C = \frac{r_y}{1+r_x+r_y} \quad (6.65\text{a–c})$$

$$G_{i,j}^* = \frac{\alpha \Delta t}{2(1+r_x+r_y)k}(g_{i,j}^{n+1} + g_{i,j}^n) \quad (6.65\text{d})$$

$$r_x = \frac{\alpha \Delta t}{(\Delta x)^2} \quad r_y = \frac{\alpha \Delta t}{(\Delta y)^2} \quad (6.65\text{e,f})$$

The computational procedure for calculating $U_{i,j}^{n+1}$ and $V_{i,j}^{n+1}$ from equations (6.63) and (6.64) is as follows: Consider, for example, the computation of $U_{i,j}^{n+1}$ from equation (6.63). The calculation starts from the grid point nearest to the boundaries x = 0 and y = 0 (i.e., i = 1, j = 1) and is carried out in a sequence of increasing i,j, while $U_{0,1}^{n+1}$ and $U_{1,0}^{n+1}$ are always available from the boundary conditions.

Similarly, the $V_{i,j}^{n+1}$ are computed from equation (6.64) by starting calculations from the node nearest to the boundaries x = a and y = b and are carried out by marching in a sequence of decreasing i, j, while $V_{I,J-1}^{n+1}$ and $V_{I-1,J}^{n+1}$ are always available from the boundary conditions.

Once $U_{i,j}^{n+1}$ and $V_{i,j}^{n+1}$ are computed, the temperatures $T_{i,j}^{n+1}$ at the interior nodes i, j are determined from

$$T_{i,j}^{n+1} = \frac{1}{2}(U_{i,j}^{n+1} + V_{i,j}^{n+1}) \quad (6.66)$$

The truncation error of the two-dimensional ADE scheme is approximately $0[(\Delta t)^2, (\Delta x)^2, (\Delta y)^2]$ and the Fourier stability analysis shows that it is unconditionally stable.

6.5 An Application Related to the Hyperthermia Treatment of Cancer

Hyperthermia designates the temperature increase of body tissues (Cho and Krishnan 2013). There are records on the use of hyperthermia for therapeutic purposes by Hippocrates. However, the first report on its use is from Dr. William Coley in 1891. He recognized the effect of fever on tumors and invented a cocktail of bacteria that was administered to his patients to increase their body temperatures, which was also lately recognized as the precursor of immunotherapy treatment (Cho and Krishnan 2013).

Due to their subcellular size, nanoparticles have been devised for drug delivery and to target specific cancer cells. In special, noble metal nanoparticles exhibit surface plasmon resonance, which increases absorption and scattering of light. Therefore, these nanoparticles can be specifically designed—in terms of geometry, size, and materials—to be resonant at specific wavelengths, such as in the near-infrared (NIR) range (700–1400 nm), where hemoglobin and water absorption are minimum (Cho and Krishnan 2013). Nanoparticles injected into tumors and heated by lasers can then increase the temperature in the tumor cells without significantly affecting the healthy cells surrounding the cancerous tissues.

Phantoms are materials designed in a way that their properties simulate those of human tissues or even human organs. Materials such as intralipid, polyacrylamide gel, agar gel, and polyvinyl chloride-plastisol (PVC-P) have been used for the preparation of phantoms of soft tissues, which are commonly used for the evaluation of new technologies prior to experiments with animals or human tissues (Eibner et al. 2014; Jaime et al. 2013; Spirou et al. 2005; Tanaka 1981; Xu et al. 2003).

This section follows the work by Lamien et al. (2014, 2016a, 2016b, 2016c) and deals with the numerical simulation of a cylindrical phantom, supposedly containing plasmonic nanoparticles in a specified region that is aimed to represent a tumor. Related works were also performed by Varon et al. (2015, 2016), where the heating was imposed by radiofrequency waves and the tumor contained magnetic nanoparticles.

The physical problem involves the heating of a cylindrical phantom with an external collimated Gaussian laser beam. The phantom is assumed to be made of PVC-P (Eibner et al. 2014; Jaime et al. 2013) and to contain a disk inclusion coaxial with the cylinder. This inclusion simulates the tumor and is supposed to be made of PVC-P loaded with gold nanorods, as illustrated by Figure 6.3. The dimensions of the phantom are also presented in this figure.

The laser propagation in the phantom is modeled with a linear diffusion approximation. The laser beam is assumed to be coaxial with the cylindrical medium so that the problem can be formulated as two-dimensional with axial symmetry. The incident laser is assumed to be partially reflected (specular reflection) at the external surface, with a reflection coefficient R_{sc}. The internal surface of the irradiated boundary is assumed to partially and diffusively reflect the incident radiation, with reflectivity characterized by Fresnel's coefficient A_1, while opacity is assumed for the remaining boundaries. The refractive indexes of the different materials are assumed as constant and homogeneous.

The diffuse component of the fluence rate, which is a radiant flux, $\Phi_s(r,z)$, is computed from the solution of the following boundary value problem:

$$\nabla \cdot \left[-D(r,z)\nabla\Phi_s(r,z) + \frac{\sigma'_s(r,z)g'(r,z)}{\beta_{tr}(r,z)}\Phi_p(r,z)\hat{s}_c\right] + \kappa(r,z)\Phi_s(r,z)$$
$$= \sigma'_s(r,z)\Phi_p(r,z) \quad \text{in } 0 < r < L_r \text{ and } 0 < z < L_z \quad (6.67a)$$

$$-D(r,z)\nabla\Phi_s(r,z)\cdot\mathbf{n} + \frac{1}{2A_1}\Phi_s(r,z) = -\frac{\sigma'_s(r,z)g'(r,z)}{\beta_{tr}(r,z)}\Phi_p(r,z)$$
$$\text{at } z=0, 0 < r < L_r \quad (6.67b)$$

$$\Phi_s(r,z) = 0 \quad \text{at } z = L_z, \; 0 < r < L_r \quad (6.67c)$$

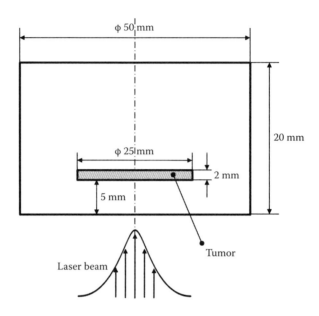

FIGURE 6.3
Sketch of the phantom containing the tumor (Lamien et al., 2016a).

$$\nabla \Phi_s(r,z) \cdot n = 0 \quad \text{at } r = 0, \ 0 < z < L_z \quad (6.67d)$$

$$\Phi_s(r,z) = 0 \quad \text{at } r = L_r, \ 0 < z < L_z \quad (6.67e)$$

where

$$D = \frac{1}{3\beta_{tr}}; \quad \sigma'_s = (1-g^2)\sigma_s; \quad g' = \frac{g}{1+g};$$
$$A_1 = (1+R_2)/(1-R_1); \quad \beta_{tr} = \kappa + \sigma_s(1-g) \quad (6.68\text{a–e})$$

with g being the anisotropy factor of scattering, κ the absorption coefficient, and σ_s the scattering coefficient, while R_1 and R_2 are the first and second moments of Fresnel's reflection coefficient, respectively. In equation (6.67a), $\hat{s}_c$ is a unit vector in the direction of propagation of the collimated laser beam.

The collimated component of the fluence rate follows the generalized Beer–Lambert's law and is given by:

$$\Phi_p(r,z) = \Phi_{0,i}(r,z) = \Phi_{0,i-1}\left(r, d_{i-1}(r)\right) \exp[-\beta'_i(z-z_i)] \quad (6.69a)$$

with

$$\beta' = \kappa + \sigma'_s \quad (6.69b)$$

where the subscript i refers to the layer i, and d_i is the thickness of each layer, while z_i and $\Phi_{0,i-1}$ are the axial position at which the collimated light enters layer i and the collimated fluence rate at this position, respectively. For i = 1, we have

$$\Phi_{0,1}(r,z) = (1-R_{sc})E_0 \exp(-2r^2/r_0^2)\exp(\beta'_1 z) \quad (6.69c)$$

where r_0 is the Gaussian beam radius, that is, the radial location where the irradiance falls to $1/e^2$ of the maximum irradiance and is related to the full width half maximum (FWHM) by

$$r_0 = \frac{\text{FWHM}}{\sqrt{2\ln 2}} \qquad (6.69d)$$

The total fluence rate is obtained by adding both diffuse and collimated components, that is,

$$\Phi(r,z) = \Phi_p(r,z) + \Phi_s(r,z) \qquad (6.70)$$

The light propagation problem given earlier is coupled to a two-dimensional transient heat conduction problem—given in this work in cylindrical coordinates with axial symmetry. Both surfaces at $z = 0$ and at $z = L_z$ exchange heat with the surrounding media by convection and linearized radiation. Heat transfer is neglected through the lateral surfaces of the phantom. The heat conduction problem is then formulated by using position-dependent properties as:

$$\rho(r,z)C_p(r,z)\frac{\partial T(r,z,t)}{\partial t} = \nabla \cdot [k(r,z)\nabla T(r,z,t)] + \kappa(r,z)\Phi(r,z),$$
$$0 < z < L_z, \quad 0 \le r < L_r \quad t > 0 \qquad (6.71a)$$

$$-k\frac{\partial T}{\partial z} + h_1 T = h_1 T_1, \quad z = 0, \quad 0 \le r < L_r \quad t > 0 \qquad (6.71b)$$

$$k\frac{\partial T}{\partial z} + h_2 T = h_2 T_2, \quad z = L_z, \quad 0 \le r < L_r \quad t > 0 \qquad (6.71c)$$

$$\frac{\partial T}{\partial r} = 0, \quad r = L_r, \quad 0 < z < L_z \quad t > 0 \qquad (6.71d)$$

$$T = T_s, \quad 0 < z < L_z, \quad 0 \le r < L_r, \quad t = 0 \qquad (6.71e)$$

where the last term on the right-hand side of equation (6.71a) represents the volumetric heat source given by the laser absorption within the medium.

The thermophysical and optical properties of PVC-P used in the simulations are summarized in Table 6.1. The disk inclusion that simulates the tumor was also assumed to be made of PVC-P but loaded with gold nanorods with different nanoparticle concentrations. The nanorods were supposed to have a radius of 11.43 nm and an aspect ratio of 3.9 because they exhibit plasmonic resonance in the NIR range at the wavelength of 797 nm, with absorption and scattering cross sections given by $C_{abs} = 2.2128 \times 10^{-14} m^2$

TABLE 6.1

Thermophysical and Optical Properties of PVC-P

Density ρ (kg/m^3)	Specific Heat C_p (kJ/kg K)	Thermal Conductivity k (W/m K)	Absorption Coefficient κ (m^{-1})	Scattering Coefficient σ_s (m^{-1})	Scattering Anisotropy Factor g
995.1	1.79	0.15	2	12000	0.9

TABLE 6.2

Optical Properties of the Region Containing Gold Nanorods

Concentration (Nanoparticles/m^3)	κ (m^{-1})	σ_s (m^{-1})
2×10^{15}	44.26	3.46
4×10^{15}	88.51	6.91
8×10^{15}	177.02	13.83

and $C_{sca} = 1.7286 \times 10^{-15}$ m^2, respectively. The absorption and the scattering coefficients of the region containing the nanorods were computed as

$$\kappa_{tumor} = \kappa + C_{abs} f_v \quad \sigma_{s,tumor} = \sigma_s + C_{sca} f_v \quad (6.72a,b)$$

where f_v is the concentration of nanoparticles. Table 6.2 presents the values of these properties for the three different concentrations of nanoparticles examined here, namely: 2×10^{15} nanoparticles/m^3, 4×10^{15} nanoparticles/m^3, and 8×10^{15} nanoparticles/m^3. The other properties of the region containing the nanoparticles were taken as those of the base material.

The dimensions of the phantom were considered as 20 mm of thickness and 25 mm of radius, with an inclusion of 2 mm of thickness and 12.5 mm of radius. The disk inclusion was assumed to be located 5 mm below the heated surface at $z = 0$ (see Figure 6.3). The phantom was assumed to be initially in thermal equilibrium with the ambient, that is, $T_s = T_1 = T_2 = 25°C$. The heat transfer coefficients were taken as $h_1 = 10$ W/(m K) at the irradiated boundary ($z = 0$) and as $h_2 = 1000$ W/(m K) at the nonheated boundary ($z = L_z$). The laser exposure time was taken as 3 min for all the results presented as follows.

For the solution of both radiation and heat conduction problems, given by equations (6.67) and (6.71), respectively, a finite volume code was developed based on the ADI method. Code verification was performed by independently comparing the solutions for the radiation and the conduction problems with the analytical solutions for limiting cases.

Figure 6.4 shows the effects on the fluence rate and temperature fields of the addition of nanoparticles in the disk inclusion that simulates a tumor. These simulations were carried out by assuming a laser output power of

0.35 W with a Gaussian-shaped beam of FWHM of 5 mm and a concentration of nanoparticles of $f_v = 8 \times 10^{15}$ m^{-3}. Fluence rate and temperature fields (at t = 180 s) are presented by Figure 6.4a and c, respectively, for the case without nanoparticles, and by Figure 6.4b and d for the case with nanoparticles. By comparing Figure 6.4a and c, it can be noticed that nanoparticles do not allow photons to penetrate deep into the phantom, thus reducing

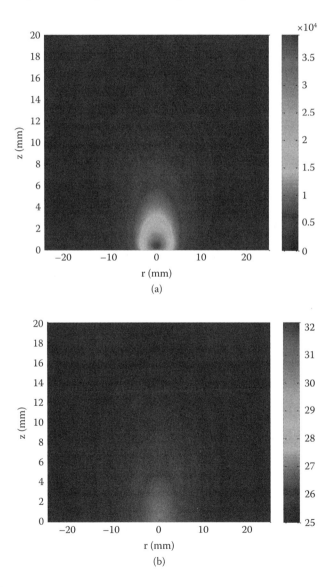

FIGURE 6.4
Phantom without nanoparticles: (a) fluence rate distribution (W/m^2) and (b) temperature distribution (°C) at t = 180 s. *(Continued)*

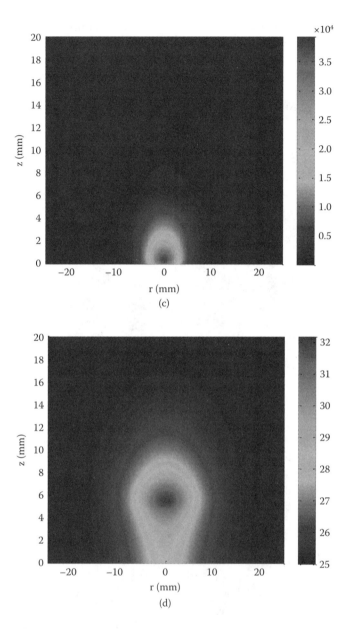

FIGURE 6.4 (CONTINUED)
Phantom containing nanoparticles: (c) fluence rate distribution (W/m^2) and (d) temperature distribution (°C) at t = 180 s.

the fluence rate beyond the disk inclusion. On the other hand, the increase of the absorption coefficient due to the presence of the nanoparticles in the disk inclusion locally increased the heat source. Such an effect can be clearly seen by comparing Figure 6.4b and d. It can be noticed in Figure 6.4b (for the case without nanoparticles) that after 180 s of irradiation, the maximum temperature was only of 26°C at the irradiated surface. On the other hand, when nanoparticles were added to the disk inclusion in the phantom, the maximum temperature achieved was 32°C (see Figure 6.4d). Figure 6.4d also shows that the higher temperatures were in the tumor region that contained the nanoparticles, although the temperatures around this region also increased due to heat diffusion.

The collimated and diffuse components of the fluence rate for the case with nanoparticles in the disk inclusion are shown by Figure 6.5a and b, respectively. We notice in Figure 6.5a the fast decay of the collimated beam—a result of absorption and scattering within the medium. On the other hand, a comparison of Figures 6.4c and 6.5b reveals the large contribution of the diffusive component to the total fluence rate.

A parametric study was performed in order to investigate the effects of the laser output power, the beam radius, and nanoparticle concentration on the fluence rate and temperature fields. We first consider the variation of the concentration of nanoparticles, with a laser output power of 0.35 W and FWHM = 5 mm. Figure 6.6a and b presents the fluence rate and the temperature distributions (at the end of heating period, t = 180 s) along the centerline of the phantom, while Figure 6.7a and b presents these same quantities along the diameter of the phantom at the depth z = 6 mm, which corresponds to a line crossing the mid-height of the disk inclusion. Figures 6.6a and 6.7a show that the increase in nanoparticle concentration decreased the magnitude of the fluence rate within the medium. On the other hand, Figures 6.6b and 6.7b reveal that the temperature increased when the nanoparticle concentration was increased. Furthermore, it can be noticed in these figures that the maximum temperatures take place in the region loaded with nanoparticles. In addition, in Figures 6.6b and 6.7b, one can clearly see the temperature increase beyond the limits of the disk inclusion (z < 5 mm, z > 7 mm, r > 12.5 mm) due to heat diffusion.

We now consider the nanoparticle concentration of 4×10^{15} m^{-3} and the laser output power of 0.35 W, while the laser beam FWHM is assumed to be 5 mm or 7 mm. The fluence rate and the temperature (at t = 180 s) fields are presented, respectively, by Figure 6.8a and b for FHWM = 5 mm and by Figure 6.8c and d for FHWM = 7 mm. By comparing Figure 6.8a and c, it can be noticed that the fluence rate is increased when the beam size is increased. Furthermore, it can be noticed that light penetrates deeper in the phantom for the FWHM = 7 mm laser beam; the penetration depth is around 13 mm for FWHM = 7 mm, whereas it is about 10 mm for FWHM = 5 mm. As a consequence of the difference among the fluence rate distributions in these cases, the heating patterns and temperature fields are different

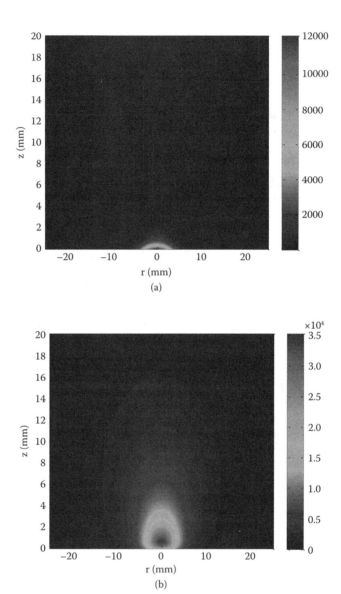

FIGURE 6.5
(a) Collimated fluence rate distribution (W/m^2) and (b) diffusive fluence rate distribution (W/m^2).

(see Figure 6.8b and d). An analysis of Figure 6.8b and d shows that the maximum temperature reached in the region of the inclusion is around 31°C for FWHM = 5 mm, while it is about 34°C for FWHM = 7 mm.

Finally, by keeping the nanoparticle concentration as 4×10^{15} m^{-3} and the laser beam size as FWHM = 5 mm, the effects of the laser output power on the temperature distribution were investigated. The temperature distributions

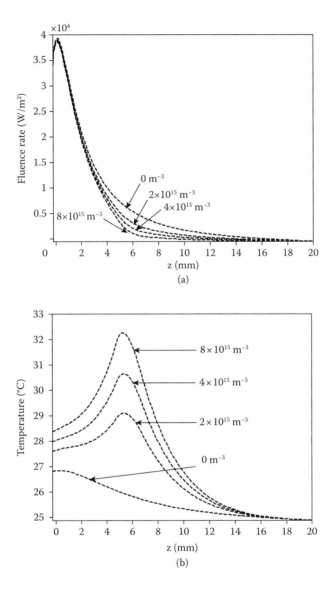

FIGURE 6.6
(a) Fluence rate distribution (W/m²) and (b) temperature distribution (°C) at t = 180 s, along the centerline of the phantom.

along the centerline and along the line z = 6 mm are presented by Figure 6.9a and b, respectively, for different laser powers. It can be noticed in these figures that the higher the laser output power, the larger is the temperature increase in the medium. However, it is interesting to note that the temperature increase in the regions surrounding the inclusion also increase significantly as the laser

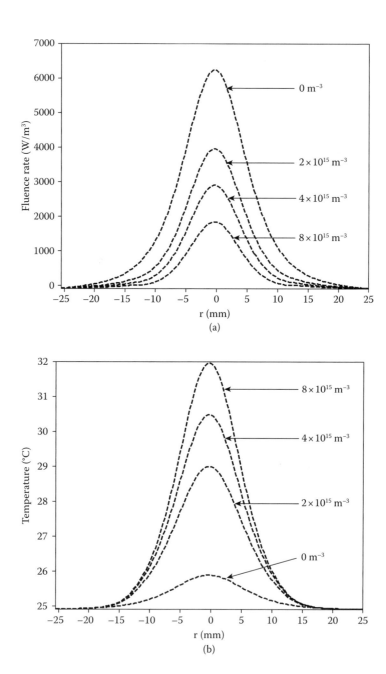

FIGURE 6.7
(a) Fluence rate distribution (W/m^2) and (b) temperature distribution (°C) at t = 180 s, along the line at z = 6 mm.

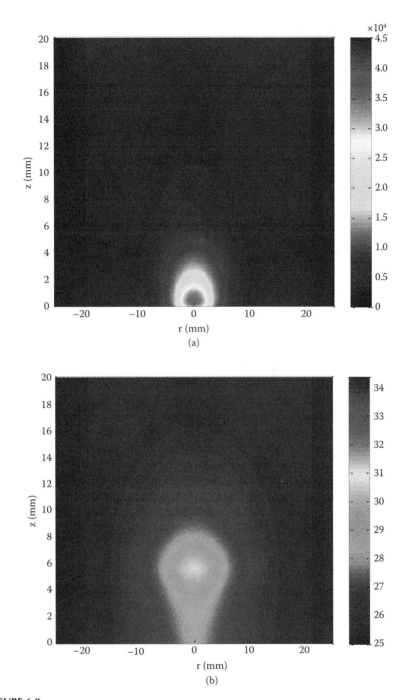

FIGURE 6.8
FWHM = 5 mm: (a) fluence rate distribution (W/m^2) and (b) temperature distribution (°C) at t = 180 s. *(Continued)*

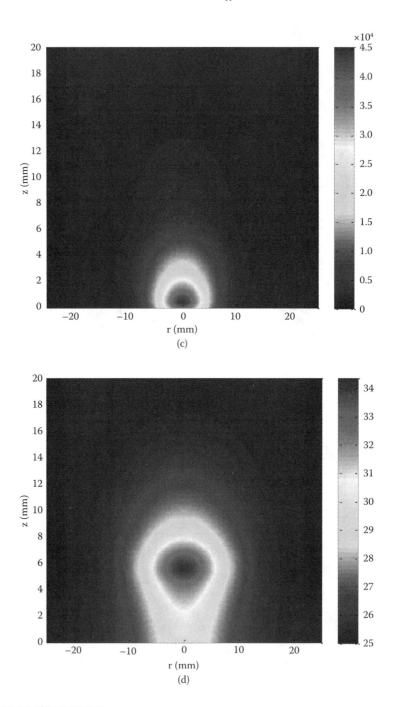

FIGURE 6.8 (CONTINUED)
FWHM = 7 mm: (c) fluence rate distribution (W/m^2) and (d) temperature distribution (°C) at t = 180 s.

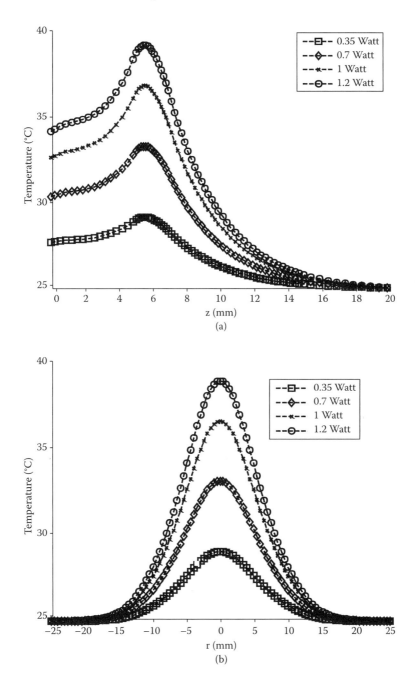

FIGURE 6.9
Temperature distribution at t = 180 s for different laser powers (a) along the centerline of the phantom and (b) along the radial direction at z = 6 mm.

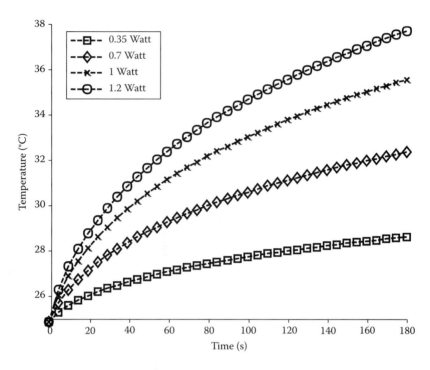

FIGURE 6.10
Effect of laser output power on the transient temperature at r = 0 mm, z = 6 mm.

output power is increased. This is of special interest when dealing with hyperthermia therapy because the objective is to locally increase the temperature inside the tumor without harming the surrounding healthy tissues. In Figure 6.10, one can also notice the transient temperature for a point inside the inclusion (r = 0 mm, z = 6 mm), which increases with larger laser exposure time and with larger laser output power.

PROBLEMS

6.1. Consider the following transient heat conduction problem in a rectangular region:

$$\frac{1}{\alpha}\frac{\partial T}{\partial t} = \frac{\partial^2 T}{\partial x^2} + \frac{\partial^2 T}{\partial y^2} \quad 0 < x < a, \quad 0 < x < b, \quad t > 0$$

$T = 0$ at all boundaries

$T = F(x, y)$ for $t = 0$

Using a square mesh $\Delta x = \Delta y = l$ and taking $a = Ml$, $b = Nl$, write the finite difference representation of this heat conduction problem with:
   a. Explicit scheme
   b. Fully implicit scheme
   c. Crank–Nicolson method

6.2. Using Fourier stability analysis, examine the stability criterion for the finite difference approximation of the heat conduction equation

$$\frac{1}{\alpha}\frac{\partial T}{\partial t} = \frac{\partial^2 T}{\partial x^2} + \frac{\partial^2 T}{\partial y^2}$$

by using
   a. Explicit method
   b. Fully implicit method

6.3. Repeat Problem 6.2 using the Crank–Nicolson method.

6.4. Consider the following transient heat conduction equation for a solid cylinder

$$\frac{1}{\alpha}\frac{\partial T}{\partial t} = \frac{\partial^2 T}{\partial r^2} + \frac{1}{r}\frac{\partial T}{\partial r} + \frac{1}{r^2}\frac{\partial^2 T}{\partial \phi^2} + \frac{1}{k}g(r, \phi, t)$$

in $0 \leq r < b$, $0 \leq \phi \leq 2\pi$, $t > 0$

Write a finite difference representation of this equation using: (i) explicit method, (ii) fully implicit method, and (iii) Crank–Nicolson method.

6.5. Consider the following transient heat conduction problem in a rectangular region $0 < x < a$, $0 < y < b$:

$$\frac{1}{\alpha}\frac{\partial T}{\partial t} = \frac{\partial^2 T}{\partial x^2} + \frac{\partial^2 T}{\partial y^2} + \frac{g}{k} \quad \text{in} \quad 0 < x < a, \quad 0 < y < b, \; t > 0$$

$$-k\frac{\partial T}{\partial x} = q_0 \qquad \text{at} \quad x = 0, \qquad t > 0$$

$$k\frac{\partial T}{\partial x} + hT = f_a \qquad \text{at} \quad x = a, \qquad t > 0$$

$$\frac{\partial T}{\partial y} = 0 \qquad \text{at} \quad y = a, \qquad t > 0$$

$$k\frac{\partial T}{\partial y} + hT = f_b \qquad \text{at} \quad y = b, \qquad t > 0$$

$$T = T_i \qquad \text{for} \quad t = 0$$

Write a finite difference representation of this equation using: (i) explicit method and (ii) fully implicit method, with a square mesh of size $\Delta x = \Delta y = \delta$ and for $a = M\delta$ and $b = N\delta$.

6.6. Consider the following transient heat conduction problem

$$\frac{1}{\alpha}\frac{\partial T}{\partial t} = \frac{\partial^2 T}{\partial x^2} + \frac{\partial^2 T}{\partial y^2} \quad \text{in} \quad 0 < x < L, \quad 0 < y < L, \quad t > 0$$

$T = 0$	$x = 0$
$T = 100$	$x = L$
$T = 0$	$y = 0$
$T = 100$	$y = L$
$T = 0$	$t = 0$

Solve this heat conduction problem with finite differences by using: (i) explicit scheme and (ii) Crank–Nicolson method by taking $L = 10$ cm, $\alpha = 10^{-5}$ m^2/s, and a square mesh of size 2 cm × 2 cm. Calculate the center temperature as a function of time.

6.7. Consider the following transient heat conduction problem in a rectangular region $0 < x < a,\ 0 < y < b$:

$$\frac{1}{\alpha}\frac{\partial T}{\partial t} = \frac{\partial^2 T}{\partial x^2} + \frac{\partial^2 T}{\partial y^2} + \frac{1}{k}g, \qquad 0 < x < a, \quad 0 < y < b, t > 0$$

$T = 0$ at all boundaries
$T = T_0$ for $t = 0$

Using a square mesh $\Delta x = \Delta y = \delta$, write the finite difference representation of this heat conduction equation with: (i) explicit scheme and (ii) fully implicit scheme.

6.8. Consider the following two-dimensional transient heat conduction in a solid cylinder of radius b and height c:

$$\frac{1}{\alpha}\frac{\partial T}{\partial t} = \frac{\partial^2 T}{\partial R^2} + \frac{1}{R}\frac{\partial T}{\partial R} + \frac{\partial^2 T}{\partial z^2} + \frac{1}{k}g(R, z, t) \quad \text{in } 0 \leq R < b,\ 0 < z < c$$

$T = 0$ at all boundaries surfaces
$T = F(R, z)$ for $t = 0$

Write the finite difference representation of this heat conduction problem using:

a. Explicit scheme
b. Fully implicit scheme
c. Crank–Nicolson method

6.9. Consider the following transient heat conduction problem in a solid cylinder of radius b and height c:

$$\frac{1}{\alpha}\frac{\partial T}{\partial t} = \frac{\partial^2 T}{\partial R^2} + \frac{1}{R}\frac{\partial T}{\partial R} + \frac{\partial^2 T}{\partial z^2}, \qquad 0 \leq R < b, \quad 0 < z < c, \quad t > 0$$

$\dfrac{\partial T}{\partial R} = 0$ $R = 0$
$T = 0$ $R = b$
$T = 100$ $z = 0$
$T = 0$ $z = c$
$T = 0$ for $t = 0$

Taking $b = 5$ cm, $c = 5$ cm, and $\alpha = 10^{-5}$ m^2/s, solve this problem with finite differences by using the explicit scheme and calculate the center temperature as a function of time.

6.10. Repeat Problem 6.9 by using the fully implicit scheme and determine the temperature at the center of the cylinder.

6.11. Consider the following transient heat conduction problem in a rectangular region $0 < x < a$, $0 < y < b$, $t > 0$

$$\frac{1}{\alpha}\frac{\partial T}{\partial t} = \frac{\partial^2 T}{\partial x^2} + \frac{\partial^2 T}{\partial y^2} + \frac{1}{k}g(x,y,t), \quad 0 < x < a, \quad 0 < y < b, \quad t > 0$$

$T = 0$ at all boundaries

$T = F(x, y)$ for $t = 0$

Using a square mesh $\Delta x = \Delta y = l$ and $a = Ml$ and $b = Nl$, write the finite difference representation of this heat conduction problem with:

i. ADI method

ii. ADE method

6.12. Consider the following two-dimensional transient heat conduction problem in a rectangular region $0 < x < a$, $0 < y < b$:

$$\frac{1}{\alpha}\frac{\partial T}{\partial t} = \frac{\partial^2 T}{\partial x^2} + \frac{\partial^2 T}{\partial y^2} \quad 0 < x < a, \quad 0 < y < b, \quad t > 0$$

$T = 0$ $x = 0$

$T = 0$ $x = a$

$T = T_0$ $y = 0$

$T = 0$ $y = b$

$T = 0$ $t = 0$

Solve this transient heat conduction problem with an explicit finite difference scheme for an iron bar $\alpha = 10^{-5}$ m²/s, $a = b = 5$ cm, $T_0 = 100°C$ using a square mesh 1 cm × 1 cm in size and taking $r = \dfrac{\alpha \Delta t}{(\Delta x)^2} = \dfrac{1}{4}$. Calculate the temperature at the center of the bar as a function of time and compare the steady-state temperature with the following exact analytic solution for the problem.

$$T(x,y) = \frac{4T_0}{\pi}\sum_{n=0}^{\infty}\frac{1}{(2n+1)}\sin\left[\frac{(2n+1)\pi x}{a}\right]\frac{\sinh\left[\frac{(b-y)(2n+1)\pi}{a}\right]}{\sinh\left[\frac{(2n+1)\pi b}{a}\right]}$$

6.13. Repeat Problem 6.5 with the Crank–Nicolson method.

6.14. Repeat Problem 6.5 using the ADI method.

6.15. Consider two-dimensional transient heat conduction in a rectangular region $-a < x < a$, $-b < y < b$ given by

$$\frac{1}{\alpha}\frac{\partial T}{\partial t} = \frac{\partial^2 T}{\partial x^2} + \frac{\partial^2 T}{\partial y^2} + \frac{1}{k}g, \quad -a < x < a, \quad -b < y < b, \quad t > 0$$

$T = 0$ at all boundary surfaces

$T = 0$ at $t = 0$

Solve this transient heat conduction problem with an explicit finite difference scheme for $\alpha = 10^{-5}$ m²/s, $a = b = 5$ cm, $g = 10^8$ W/m³, and $k = 40$ W/m·°C by using 1 cm × 1 cm square mesh and taking $r = \dfrac{\alpha \Delta t}{(\Delta x)^2} = \dfrac{1}{4}$. Calculate the center temperature as a function of time and compare the steady-state temperature with the following exact analytic solution.

$$T(x,y) = \frac{g}{k}\left(\frac{a^2 - x^2}{2}\right) - 2a^2\frac{g}{k}\sum_{n=0}^{\infty}\frac{(-1)^n \cosh\left(\beta_n \frac{y}{b}\right)\cos\left(\beta_n \frac{x}{a}\right)}{\beta_n^3 \cosh\left(\beta_n \frac{b}{a}\right)}$$

where $\beta_n = \dfrac{(2n+1)\pi}{2}$

6.16. Consider the following transient heat conduction problem for a solid cylinder of finite height:

$$\frac{1}{\alpha}\frac{\partial T}{\partial t} = \frac{\partial^2 T}{\partial R^2} + \frac{1}{R}\frac{\partial T}{\partial R} + \frac{\partial^2 T}{\partial z^2} \quad \text{in} \quad 0 \leq R < b, \quad 0 < z < c, \quad t > 0$$

$T = 0$ at all surfaces

$T = T_0$ initial temperature

Taking $b = 5$ cm, $c = 5$ cm, $T_0 = 500$ °C, and $\alpha = 1 \times 10^{-5}$ m²/s, solve this transient heat conduction problem with an explicit finite difference scheme. Calculate the center temperature as a function of time.

6.17. Using the simple implicit scheme, with upwind and central space discretizations for the convective and diffusive terms, respectively, develop finite difference approximation for the following transient, convection–diffusion equation

$$\frac{\partial T}{\partial t} + u\frac{\partial T}{\partial x} + v\frac{\partial T}{\partial y} = \alpha\frac{\partial^2 T}{\partial x^2}$$

where the flow velocity components, u and v, and the thermal diffusivity α are considered constants.

6.18. Repeat Problem 6.17 using the Crank–Nicolson method.

6.19. Consider the diffusion equation in the cylindrical coordinates given in the form

$$\frac{1}{\alpha}\frac{\partial T}{\partial t} = \frac{\partial^2 T}{\partial R^2} + \frac{1}{R}\frac{\partial T}{\partial R} + \frac{1}{R^2}\frac{\partial^2 T}{\partial \phi^2} + \frac{\partial^2 T}{\partial z^2} + \frac{1}{k}g(R,\phi,z,t)$$

where $T \equiv T(R,\phi,z,t)$, $0 \leq \phi \leq 2\pi$, $0 \leq R < b$, $0 < z < L$.

Using the notation

$$T(R,\phi,z,t) = T(i\Delta R, j\Delta\phi, k\Delta z, n\Delta t) \equiv T_{i,j,k}^n$$

Write the finite difference approximation of this equation for the internal nodes by using the simple explicit scheme for (a) $R \neq 0$ and (b) $R = 0$.

6.20. Consider the three-dimensional transient heat conduction equation in the spherical coordinate system given in the form

$$\frac{1}{\alpha}\frac{\partial T}{\partial t} = \frac{\partial^2 T}{\partial R^2} + \frac{1}{R}\frac{\partial T}{\partial R} + \frac{1}{R^2 \sin\theta}\frac{\partial}{\partial \theta}\left(\sin\theta \frac{\partial T}{\partial \theta}\right) + \frac{1}{R^2 \sin^2\theta}\frac{\partial^2 T}{\partial \phi^2} + \frac{1}{k}g(R,\theta,\phi,t)$$

This equation is rearranged as

$$\frac{1}{\alpha}\frac{\partial T}{\partial t} = \frac{\partial^2 T}{\partial R^2} + \frac{2}{R}\frac{\partial T}{\partial R} + \frac{2}{R^2}\frac{\partial^2 T}{\partial \theta^2} + \frac{\cot\theta}{R^2}\frac{\partial T}{\partial \theta} + \frac{1}{R^2 \sin^2\theta}\frac{\partial^2 T}{\partial \phi^2} + \frac{1}{k}g(R,\theta,\phi,t)$$

where $T \equiv T(R,\theta,\phi,t)$, $0 \leq \theta \leq \pi$, $0 \leq \phi \leq 2\pi$, $0 \leq R < b$.

Using the simple explicit method, write the finite difference approximation of this equation for (a) $R \neq 0$ and (b) $R = 0$.

6.21. Consider the energy equation for transient-forced convection inside ducts under a slug flow assumption given in the dimensionless form as

$$\frac{\partial \theta}{\partial \tau} + w\frac{\partial \theta}{\partial z} = \frac{\partial^2 \theta}{\partial y^2}$$

where w = constant. Develop the finite difference approximation for this equation with the simple explicit scheme using upwind differencing. Use the Fourier stability analysis to establish the stability criterion for the finite difference equation.

6.22. Consider the energy equation

$$\frac{\partial \theta}{\partial t} + u\frac{\partial \theta}{\partial x} + v\frac{\partial \theta}{\partial y} = \alpha\left(\frac{\partial^2 \theta}{\partial x^2} + \frac{\partial^2 \theta}{\partial y^2}\right)$$

Develop the finite difference representation of this differential equation with the simple explicit scheme by employing backward difference for the first-order space derivatives and central difference for the second-order space derivatives.

6.23. Repeat Problem 6.17 by applying the control volume approach of discretization, with the WUDS interpolation scheme.

6.24. Repeat Problem 6.21 by applying the control volume approach of discretization, with the WUDS interpolation scheme.

6.25. Repeat Problem 6.22 by applying the control volume approach of discretization, with the WUDS interpolation scheme. Make $u = v = 0$ and present the discretized equation for the purely diffusive problem.

6.26. Apply code verification and solution verification techniques to Problem 6.12.

6.27. Apply code verification and solution verification techniques to Problem 6.15.

6.28. Develop a manufactured solution for Problem 6.16 and apply code verification and solution verification techniques using the manufactured solution.

6.29. Consider a two-dimensional steady-state heat conduction problem for a solid cylinder given by

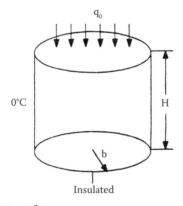

$$\frac{\partial^2 T}{\partial R^2} + \frac{1}{R}\frac{\partial T}{\partial R} + \frac{\partial^2 T}{\partial z^2} + \frac{g}{k} = 0 \quad \text{in} \quad 0 < R < b, \quad 0 < z < H$$

$$\frac{\partial T}{\partial R} = 0 \qquad R = 0$$
$$T = 0 \qquad R = b$$
$$\frac{\partial T}{\partial z} = 0 \qquad z = 0$$
$$k\frac{\partial T}{\partial z} = q_0 \qquad z = H$$

where q_0 is heat supply in W/m², and g is the volumetric energy generation rate in W/m³. The accompanying figure illustrates the geometry and the boundary conditions. By dividing the height H and radius b into five equal parts, write the finite difference representation of this heat conduction problem.

6.30. Consider two-dimensional steady-state heat conduction with no energy generation in a square region subject to the boundary conditions, as illustrated in the accompanying figure. By dividing the region into six equal parts in the x and y directions, develop the finite difference approximation of this heat conduction problem.

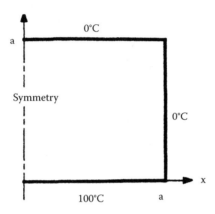

6.31. Consider a two-dimensional steady-state heat conduction problem in a square region given in the dimensionless form as

$$\frac{\partial^2 T}{\partial x^2} + \frac{\partial^2 T}{\partial y^2} + G = 0 \quad \text{in} \quad -1 < x < 1, \quad -1 < y < 1,$$
$$T = 0 \quad \text{at} \quad x = \pm 1, \quad y = \pm 1$$

The exact solution of this problem for temperature at $x = 0, y = 0$ is given by

$$T(0,0) = \left[\frac{1}{2} - 2 \sum_{n=0}^{\infty} \frac{(-1)^n}{\beta_n^3 \cosh \beta_n} \right] G = 0.293 G$$

where $\beta_n = \dfrac{(2n+1)\pi}{2}$.

Solve this heat conduction problem with finite differences and compare the center temperature with the exact solution given earlier for $G = 100$.

NOTES

Steady-State Diffusion

Physical processes such as steady-state heat or mass diffusion, with or without sources within the medium, irrotational flow of incompressible fluid, the slow motion of incompressible viscous fluid, and many others are modeled with elliptic partial differential equations. For example, Laplace's equation,

$$\nabla^2 T = 0 \tag{a}$$

is well-known for modeling the steady-state diffusion without sources in the medium or the subsonic irrotational flow of a fluid. Poisson's equation,

$$\nabla^2 T + f(\mathbf{r}) = 0 \tag{b}$$

is used to model steady-state diffusion with sources within the medium.

Commonly used linear boundary conditions for these equations include the specification of the value of the function (e.g., prescribed temperature) or its derivative (e.g., prescribed heat flux) or a linear combination of the value and its derivative (e.g., convection). For a heat conduction problem, if the solution domain is finite and the derivative of the function is specified at every point of the boundary, such that the sum of heat entering and generated equals to that leaving the body through its boundaries, then the solution of the problem is unique only to within an additive constant. That is, if T is a solution, "T + constant" is also a solution. If the sum of heat entering and generated is not equal to heat leaving the medium, the problem has no steady-state solution.

The two-dimensional steady-state heat diffusion in a constant property medium without sources in Cartesian coordinates is given by

$$\frac{\partial^2 T}{\partial x^2} + \frac{\partial^2 T}{\partial y^2} = 0 \tag{c}$$

subject to appropriate thermal boundary conditions.

To discretize this equation, we adopt the notation

$$T(x,y) = T(i\Delta x, j\Delta y) \equiv T_{i,j} \tag{d}$$

Using the second-order accurate standard finite differencing scheme, the approximation of equation (c) for the internal nodes is given by

$$\frac{T_{i+1,j} - 2T_{i,j} + T_{i-1,j}}{(\Delta x)^2} + \frac{T_{i,j+1} - 2T_{i,j} + T_{i,j-1}}{(\Delta y)^2} = 0 \tag{e}$$

For prescribed temperature at all boundaries, the number of equations provided by the system is equal to the number of unknown internal node temperatures; hence, the equations can be solved. For boundary conditions of the second or third kinds, the temperatures at the boundary nodes are not known. Additional relations can be developed by the application of the techniques presented in Chapter 2 and widely used earlier in this book.

Finite difference approximations of Laplace's or Poisson's equations are therefore a straightforward matter. The procedure leads to a system that contains a large number of linear algebraic equations, and the choice of a proper solution algorithm is important. In Chapter 3, the methods of solving systems of linear algebraic equations by direct and iterative methods were discussed. If convergence requirements are satisfied and the coefficient matrix is sparse, iterative methods are preferred. The systems of algebraic equations resulting from the finite difference approximation of Laplace's or Poisson's equations satisfy such conditions; hence, iterative methods such as Gauss–Seidel iteration, successive overrelaxation, or the conjugate gradient method are well suited for their solutions. Multidimensional steady-state problems may also be handled as a false transient formulation, just by adding a transient term to the formulation and proposing a guessed initial condition. Then, all of the numerical schemes discussed within this chapter can be directly applicable. The false transient solution is then carried on until the steady-state is reached, which is in fact the only solution of interest in this case. The closer the guess of the artificial initial condition, the faster will be the convergence to the final steady state. The false transient approach will also be discussed in the next chapter, in the context of nonlinear steady-state problems.

7
Nonlinear Diffusion

In principle, there is no difficulty in applying finite difference methods to nonlinear parabolic systems. Diffusion-type problems become nonlinear due to the nonlinearity of the governing differential equation or the boundary condition or both. Consider, for example, the heat conduction equation

$$\nabla \cdot [k(T)\nabla T] = \rho C_p(T)\frac{\partial T}{\partial t} \quad \text{in region R,} \quad t > 0 \tag{7.1a}$$

or the mass diffusion equation

$$\nabla \cdot [D(C)\nabla C] = \frac{\partial C}{\partial t} \quad \text{in region R,} \quad t > 0 \tag{7.1b}$$

They are nonlinear because the properties depend on temperature (or concentration). Consider the boundary condition given in the form

$$\frac{\partial T}{\partial \mathbf{n}} = f(T) \quad \text{on boundary,} \quad t > 0 \tag{7.2}$$

where $\frac{\partial}{\partial \mathbf{n}}$ is the derivative along the outward drawn normal to the boundary. This boundary condition becomes nonlinear if the function f(T) involves a power of T, as in the case of a radiation boundary condition

$$k\frac{\partial T}{\partial \mathbf{n}} = \varepsilon\,\sigma(T^4 - T_\infty^4) \tag{7.3a}$$

or a natural convection boundary condition

$$k\frac{\partial T}{\partial \mathbf{n}} = c|T - T_\infty|^{1/4}(T - T_\infty) \tag{7.3b}$$

where ε is the emissivity, σ is the Stefan–Boltzmann constant, T_∞ is the ambient temperature in which radiation or free convection takes place, and c is a constant that depends on the geometry and flow regime.

Coupled conduction and radiation in participating media are also among this important class of problems with many practical applications, such as in heat transfer within semitransparent solids at high temperatures or in combustion processes.

Various schemes are available for finite difference approximation of nonlinear diffusion problems as a system of linear algebraic equations. They include, among others, the lagging of temperature or concentration-dependent properties by one time step, the use of three-time-level finite differencing, and linearization procedures. In this chapter, we examine the application of such approaches. We also describe the solution of nonlinear, steady-state diffusion problems by the false transient method. In this approach, the steady-state problem is replaced by the relevant time-dependent parabolic system, which is solved by any one of the standard finite difference methods until the solution ceases to change with time; that is, the steady-state condition is reached. The method readily yields the steady-state solution if the steady state exists and is unique. The basic idea in the false transient technique is simple, the algorithm is straightforward, and for sufficiently large times the transients die out and the steady-state condition is approached.

7.1 Lagging Properties by One Time Step

We consider the nonlinear diffusion equation given in the form

$$\rho C_p(T) \frac{\partial T}{\partial t} = \frac{\partial}{\partial x}\left[k(T)\frac{\partial T}{\partial x}\right] \qquad (7.4)$$

where the specific heat $C_p(T)$ and the thermal conductivity $k(T)$ vary with temperature. This equation can be discretized by using any one of the finite difference schemes described previously. Here we prefer to use the combined method because of its versatility in yielding the simple explicit, simple implicit, Crank–Nicolson, and other methods merely by the adjustment of a coefficient. With the combined method, the finite difference representation of equation (7.4) is given by

$$\Delta x (\rho C_p)_i \frac{T_i^{n+1} - T_i^n}{\Delta t} = \theta \left[k_{i-1/2} \frac{T_{i-1}^{n+1} - T_i^{n+1}}{\Delta x} + k_{i+1/2} \frac{T_{i+1}^{n+1} - T_i^{n+1}}{\Delta x} \right]$$
$$+ (1-\theta) \left[k_{i-1/2} \frac{T_{i-1}^n - T_i^n}{\Delta x} + k_{i+1/2} \frac{T_{i+1}^n - T_i^n}{\Delta x} \right] \qquad (7.5)$$

where the constant $\theta (0 \leq \theta \leq 1)$ is the weight factor that represents the degree of implicitness. The values $\theta = 0, \frac{1}{2}$, and 1 correspond to the explicit, Crank–Nicolson, and simple implicit schemes, respectively. The thermal conductivities at $i - 1/2$ or $i + 1/2$ can be computed with equations (2.33a,b).

We note that the thermal properties $(\rho C_p)_i$ and $k_{i \pm 1/2}$ depend on temperature, but at this stage of the analysis, it is not yet specified how they will be computed. This matter will be discussed later on.

Nonlinear Diffusion

Equation (7.5) can be written more compactly in the form

$$T_i^{n+1} - T_i^n = \theta[A_i T_{i-1}^{n+1} - 2B_i T_i^{n+1} + D_i T_{i+1}^{n+1}] \\ + (1-\theta)[A_i T_{i-1}^n - 2B_i T_i^n + D_i T_{i+1}^n] \quad (7.6)$$

where

$$A_i = \frac{k_{i-1/2}}{(\rho C_p)_i} \frac{\Delta t}{(\Delta x)^2} \quad (7.7a)$$

$$D_i = \frac{k_{i+1/2}}{(\rho C_p)_i} \frac{\Delta t}{(\Delta x)^2} \quad (7.7b)$$

$$B_i = \frac{1}{2}(A_i + D_i) = \frac{1}{2}\frac{k_{i-1/2} + k_{i+1/2}}{(\rho C_p)_i} \frac{\Delta t}{(\Delta x)^2} \quad (7.7c)$$

equation (7.6) is now rearranged so that all unknown temperatures (i.e., those at the time level n + 1) appear on one side and all the known temperatures (i.e., those at the time level n) on the other side.

$$-\theta A_i T_{i-1}^{n+1} + (1 + 2\theta B_i) T_i^{n+1} - \theta D_i T_{i+1}^{n+1} \\ = (1-\theta) A_i T_{i-1}^n + [1 - 2(1-\theta) B_i] T_i^n + (1-\theta) D_i T_{i+1}^n \quad (7.8)$$

We note that, for the case of constant thermal properties, we have

$$A_i = B_i = D_i = \frac{k}{\rho C_p} \frac{\Delta t}{(\Delta x)^2} = \frac{\alpha \Delta t}{(\Delta x)^2} \equiv r$$

and equation (7.8) reduces to the linear case given by equation (5.61). Assuming that the coefficients A_i, D_i, and B_i are available, the system [equation (7.8)] provides a complete set of equations for the determination of the unknown internal node temperatures when the temperatures at the boundary surfaces are prescribed. For the case of prescribed heat flux or convection boundary conditions, the temperatures at the boundaries are unknown; additional relations are developed by discretizing such boundary conditions as described in Chapter 2 and as exhaustively discussed in previous chapters. Because equation (7.8) has a tridiagonal coefficient matrix, any one of the algorithms discussed in Chapter 3 can be used for the solution provided that the coefficients A_i, B_i, and D_i are known at the time level n + 1.

The simplest but least accurate method for computing these coefficients is to lag the evaluation of the temperature-dependent properties by one time step. To perform the computations at the time level n + 1, the coefficients are evaluated at the previous time level n; that is,

$$A_i \equiv A_i^n, \quad B_i \equiv B_i^n \quad \text{and} \quad D_i \equiv D_i^n \quad (7.9)$$

A more accurate approach for calculating the temperature-dependent properties is the use of an extrapolation scheme. Consider, for example, the thermal conductivity k^{n+1} at the time level $n + 1$ expanded in terms of k at the time level n in the form

$$k^{n+1} \cong k^n + \left(\frac{\partial k}{\partial t}\right)^n \Delta t$$
$$\cong k^n + \left(\frac{\partial k}{\partial T}\right)^n \left(\frac{\partial T}{\partial t}\right)^n \Delta t \qquad (7.10a)$$

The time derivative of temperature is approximated by

$$\left(\frac{\partial T}{\partial t}\right)^n \cong \frac{T^n - T^{n-1}}{\Delta t} \qquad (7.10b)$$

Introducing equation (7.10b) into equation (7.10a), the following expression is obtained for the determination of thermal conductivity at the time level $n + 1$ from the knowledge of k^n

$$k^{n+1} = k^n + \left(\frac{\partial k}{\partial T}\right)^n (T^n - T^{n-1}) \qquad (7.11)$$

A similar expression can be written for the specific heat

$$C_p^{n+1} = C_p^n + \left(\frac{\partial C_p}{\partial T}\right)^n (T^n - T^{n-1}) \qquad (7.12)$$

With equations (7.11) and (7.12), the coefficients A_i, D_i, and B_i are then approximately calculated at time level $n + 1$. Clearly, if the second terms on the right-hand sides of equations (7.11) and (7.12) are neglected, the result is equivalent to the lagging of the coefficients.

7.2 Use of Three-Time-Level Implicit Scheme

The finite difference approximation of parabolic equations is customarily done using only two-time-level schemes. It is possible to use time discretization other than the two-time-level schemes, but a three- (or more) time-level scheme is used only to achieve some advantage over the two-time-level scheme, such as a smaller local truncation error, greater stability, and to transform a nonlinear problem to a linear one. Here, we present the three-time-level implicit Dupont-II scheme (Dupont et al. 1974; Hogge 1981). We consider the following one-dimensional nonlinear energy equation with temperature-dependent thermal properties and energy generation

$$\frac{\partial T}{\partial t} = w(T) \frac{1}{R^p} \frac{\partial}{\partial R} \left[R^p \, k(T) \frac{\partial T}{\partial R} \right] + G(T) \qquad (7.13)$$

where

$$w(T) = 1/[\rho C_p(T)] \tag{7.14a}$$

$$G(T) = g(T)/[\rho C_p(T)] \tag{7.14b}$$

$$g(T) = \text{volumetric energy generation rate} \tag{7.14c}$$

$$p = \begin{cases} 0 & \text{slab} \\ 1 & \text{cylinder} \\ 2 & \text{sphere} \end{cases} \tag{7.14d}$$

For mass diffusion, $w(T) = 1$, $k(T)$ is the diffusion coefficient, T represents the concentration, and $g(T)$ represents mass production rate.

The advantages of the three-time-level Dupont-II scheme, as pointed out by Hogge (1981), include high accuracy even for large time steps and the damping out of oscillations. On the other hand, it requires results at the two previous time levels; hence, a two-time-level scheme, such as the Crank-Nicolson method, is needed to start the computations—generally with a very small Δt.

We now describe the finite difference approximation of the nonlinear differential equation (7.13) for the internal nodes and the discretization of boundary conditions for the boundary nodes.

7.2.1 Internal Nodes

Hogge (1981), after critically examining the accuracy of various finite difference schemes for the solution of the nonlinear heat conduction equation, concluded that the implicit Dupont-II scheme, with the parameter α of the Dupont et al. (1974) paper taken as $\alpha = \frac{1}{4}$, produces excellent results for the finite difference solution of the nonlinear heat conduction equation.

The finite difference approximation of equation (7.13) with the implicit Dupont-II scheme, for the parameter $\alpha = \frac{1}{4}$, is given by

$$\frac{T_i^{n+1} - T_i^n}{\Delta t} = w_i^* \frac{1}{R_i^p} \frac{1}{(\Delta R)^2} \left[\frac{3}{4} F_i^{n+1} + \frac{1}{4} F_i^{n-1} \right] + G_i^* \tag{7.15}$$

where

$$F_i^{n+1} = R_{i+1/2}^p k_{i+1/2}^* (T_{i+1}^{n+1} - T_i^{n+1}) - R_{i-1/2}^p k_{i-1/2}^* (T_i^{n+1} - T_{i-1}^{n+1}) \tag{7.16a}$$

$$F_i^{n-1} = R_{i+1/2}^p k_{i+1/2}^* (T_{i+1}^{n-1} - T_i^{n-1}) - R_{i-1/2}^p k_{i-1/2}^* (T_i^{n-1} - T_{i-1}^{n-1}) \tag{7.16b}$$

where the quantities with asterisks refer to averaging on time as

$$w_i^* = \frac{3}{2}w_i^n - \frac{1}{2}w_i^{n-1} \tag{7.17a}$$

$$G_i^* = \frac{3}{2}G_i^n - \frac{1}{2}G_i^{n-1} \tag{7.17b}$$

$$k^* = \frac{3}{2}k^n - \frac{1}{2}k^{n-1} \tag{7.17c}$$

The finite difference scheme given by equation (7.15) makes use of central differencing for derivatives in the space variable and forward differencing for the derivative in the time variable. The functions F_i^{n+1} and F_i^{n-1} are weighted in proportions 3/4 and 1/4 at time levels $n+1$ and $n-1$, respectively. According to equations (7.17a–c), the temperature-dependent quantities are computed as weighted averages of the values of these functions at time levels n and $n-1$. The scheme requires no iteration; however, being a three-time-level scheme, it requires that the calculations should be started with a two-time-level scheme and, after the first step, the computations should be switched to this scheme.

Hogge (1981) found that the Dupont-II implicit scheme quickly damps out the initial errors and gives excellent results for short-time responses, while the long-time response also remains very accurate.

7.2.2 Limiting Case R = 0 for Cylinder and Sphere

For cylindrical and spherical symmetry, when the solution domain includes the origin $R = 0$, the finite difference equation (7.15) is not applicable at $R = 0$ because it has an apparent singularity at this point. An equivalent expression applicable at $R = 0$ can be developed by writing the original differential equation (7.13) in the form

$$\frac{\partial T}{\partial t} = w(T)\frac{\partial}{\partial R}\left[k(T)\frac{\partial T}{\partial R}\right] + w(T)\frac{p}{R}k(T)\frac{\partial T}{\partial R} + G(T) \tag{7.18}$$

When the first term on the right-hand side is differentiated, this equation becomes

$$\frac{\partial T}{\partial t} = w(T)\left[\frac{\partial k(T)}{\partial R}\frac{\partial T}{\partial R} + k(T)\frac{\partial^2 T}{\partial R^2}\right] + w(T)k(T)\frac{p}{R}\frac{\partial T}{\partial R} + G(T) \tag{7.19}$$

In this equation, the term $\partial T/\partial R$ inside the bracket vanishes at $R = 0$ because of symmetry. The term $(1/R)(\partial T/\partial R)$ takes an undetermined form, 0/0, at $R = 0$; however, its limiting value at $R = 0$ is determined by the application of L'Hospital's rule to give

Nonlinear Diffusion

$$\lim_{R \to 0} \left(\frac{1}{R} \frac{\partial T}{\partial R} \right) = \frac{\partial^2 T}{\partial R^2} \qquad (7.20)$$

Then, equation (7.19) at the origin takes the form

$$\frac{\partial T}{\partial t} = w(T) k(T) (1+p) \frac{\partial^2 T}{\partial R^2} + G(T) \quad \text{at} \quad R = 0 \qquad (7.21)$$

where

$$p = \begin{cases} 1 & \text{cylinder} \\ 2 & \text{sphere} \end{cases}$$

The finite difference approximation of equation (7.21) at $R = 0$ with the implicit Dupont-II scheme becomes

$$\frac{T_0^{n+1} - T_0^n}{\Delta t} = w_0^* k_0^* \frac{p+1}{(\Delta R)^2} \left[\frac{3}{4} F_0^{n+1} + \frac{1}{4} F_0^{n-1} \right] + G_0^* \qquad (7.22)$$

where

$$F_0^{n+1} = 2(T_1^{n+1} - T_0^{n+1}) \qquad (7.23a)$$

$$F_0^{n-1} = 2(T_1^{n-1} - T_0^{n-1}) \qquad (7.23b)$$

$$w_0^* = \frac{3}{2} w_0^n - \frac{1}{2} w_0^{n-1} \qquad (7.23c)$$

$$G_0^* = \frac{3}{2} G_0^n - \frac{1}{2} G_0^{n-1} \qquad (7.23d)$$

$$k_0^* = \frac{3}{2} k_0^n - \frac{1}{2} k_0^{n-1} \qquad (7.23e)$$

and the subscript 0 refers to the center $R = 0$.

7.2.3 Boundary Nodes

For a prescribed flux or convection boundary condition of the form

$$k \frac{\partial T}{\partial R} = f_b(T) \quad \text{on boundary} \quad R = b \qquad (7.24)$$

the temperature at the boundary is unknown; however, the additional relation can be determined by discretizing the derivative term with the Dupont-II scheme by following a procedure described in the following three steps.

1. We consider the following Taylor series expansion

$$\left(k\frac{\partial T}{\partial R}\right)_{b-\frac{\Delta R}{2}} = \left(k\frac{\partial T}{\partial R}\right)_b - \frac{\Delta R}{2}\left[\frac{\partial}{\partial R}\left(k\frac{\partial T}{\partial R}\right)\right]_b \quad (7.25a)$$

which is rearranged as

$$\left[\frac{\partial}{\partial R}\left(k\frac{\partial T}{\partial R}\right)\right]_b = \frac{2}{\Delta R}\left(k\frac{\partial T}{\partial R}\right)_b - \frac{2}{\Delta R}\left(k\frac{\partial T}{\partial R}\right)_{b-\frac{\Delta R}{2}} \quad (7.25b)$$

2. Equation (7.13) is written in the form

$$\frac{\partial T}{\partial t} = w(T)\frac{\partial}{\partial R}\left[k(T)\frac{\partial T}{\partial R}\right] + w(T)\frac{p}{R}\left[k(T)\frac{\partial T}{\partial R}\right] + G(T) \quad (7.26a)$$

which is evaluated at $R = b$, and equation (7.25b) is utilized

$$\left.\frac{\partial T}{\partial t}\right|_b = -w_b(T)\frac{2}{\Delta R}\left(k\frac{\partial T}{\partial R}\right)_{b-\frac{\Delta R}{2}} + w_b(T)\left(\frac{2}{\Delta R}+\frac{p}{b}\right)\left[k\frac{\partial T}{\partial R}\right]_b + G_b(T)$$

$$(7.26b)$$

The term $\left(k\frac{\partial T}{\partial R}\right)_b$ is eliminated from equation (7.26b) by utilizing the boundary condition [equation (7.24)]

$$\left.\frac{\partial T}{\partial t}\right|_b = -w_b(T)\frac{2}{\Delta R}\left(k\frac{\partial T}{\partial R}\right)_{b-\frac{\Delta R}{2}} + w_b(T)\left(\frac{2}{\Delta R}+\frac{p}{b}\right)f_b(T) + G_b(T)$$

$$(7.27)$$

3. The Dupont-II scheme is applied to discretize equation (7.27)

$$\frac{T_b^{n+1} - T_b^n}{\Delta t} = -w_b^* \frac{2}{\Delta R}\frac{1}{\Delta R}\left(\frac{3}{4}F_b^{n+1} + \frac{1}{4}F_b^{n-1}\right) + w_b^*\left(\frac{2}{\Delta R}+\frac{p}{b}\right)f_b^* + G_b^*$$

$$(7.28)$$

which is rearranged as

$$T_b^{n+1} - T_b^n = -2w_b^*\frac{\Delta t}{(\Delta R)^2}\left[\frac{3}{4}F_b^{n+1} + \frac{1}{4}F_b^{n-1}\right] + w_b^*\Delta t\left(\frac{2}{\Delta R}+\frac{p}{b}\right)f_b^* + \Delta t G_b^*$$

$$(7.29)$$

where various quantities are defined as

$$F_b^{n+1} = k_{b-\frac{\Delta R}{2}}^*(T_b^{n+1} - T_{b-\Delta R}^{n+1}) \quad (7.30a)$$

Nonlinear Diffusion

$$F_b^{n-1} = k^*_{b-\frac{\Delta R}{2}}(T_b^{n-1} - T_{b-\Delta R}^{n-1}) \tag{7.30b}$$

$$w_b^* = \frac{3}{2}w_b^n - \frac{1}{2}w_b^{n-1} \tag{7.30c}$$

$$f_b^* = \frac{3}{2}f_b^n - \frac{1}{2}f_b^{n-1} \tag{7.30d}$$

$$G_b^* = \frac{3}{2}G_b^n - \frac{1}{2}G_b^{n-1} \tag{7.30e}$$

$$k^* = \frac{3}{2}k^n - \frac{1}{2}k^{n-1} \tag{7.30f}$$

Thus, equation (7.29) is the finite difference approximation of the boundary condition equation (7.24) at the boundary R = b.

7.3 Linearization

We consider a nonlinear partial differential equation in the form

$$\frac{\partial T}{\partial t} = \frac{\partial^2}{\partial x^2}(T^m) \tag{7.31}$$

where m is a positive integer and $m \geq 2$. This equation can also be written in the form

$$\frac{\partial T}{\partial t} = \frac{\partial}{\partial x}\left[(mT^{m-1})\frac{\partial T}{\partial x}\right] \tag{7.32}$$

which represents a diffusion equation in which the diffusion coefficient is proportional to the power $(m-1)$ of T.

The finite difference approximation of equation (7.31) using the combined method is given by

$$\frac{(T)_i^{n+1} - (T)_i^n}{\Delta t} = \theta \Delta_{xx}(T^m)_i^{n+1} + (1-\theta)\Delta_{xx}(T^m)_i^n \tag{7.33a}$$

where the operator Δ_{xx} is defined by

$$\Delta_{xx} F_i \equiv \frac{1}{(\Delta x)^2}(F_{i-1} - 2F_i + F_{i+1}) \tag{7.33b}$$

and the weight factor θ ($0 \leq \theta \leq 1$) controls the degree of implicitness. The finite difference approximation given by equation (7.33a) is not convenient for computational purposes because the resulting system of algebraic equations is highly nonlinear; hence, it is difficult to solve. To alleviate this difficulty, the unknown $(T^m)^{n+1}$ is linearized by following a procedure discussed by Richtmyer and Morton (1967) and Smith (1982).

A Taylor's series expansion of $(T^m)_i^{n+1}$ about n gives

$$(T^m)_i^{n+1} = (T^m)_i^n + \frac{d(T^m)_i^n}{dt}\Delta t + \cdots$$

$$= (T^m)_i^n + \frac{d(T^m)_i^n}{d(T)_i^n}\frac{d(T)_i^n}{dt}\Delta t + \cdots \qquad (7.34a)$$

The time derivative $d(T)_i^n/dt$ is discretized forward in time and the term $d(T^m)_i^n/d(T)_i^n$ is differentiated.

$$(T^m)_i^{n+1} = (T^m)_i^n + m(T^{m-1})_i^n \frac{(T)_i^{n+1}-(T)_i^n}{\Delta t}\Delta t + \cdots$$

$$= (T^m)_i^n + m(T^{m-1})_i^n[(T)_i^{n+1}-(T)_i^n] + \cdots \qquad (7.34b)$$

A new dependent variable U_i is introduced as

$$(T)_i^{n+1} - (T)_i^n = U_i \qquad (7.35)$$

Then equation (7.34b) becomes

$$(T^m)_i^{n+1} \cong (T^m)_i^n + m(T^{m-1})_i^n U_i \qquad (7.36)$$

Equations (7.35) and (7.36) are introduced into equation (7.33a)

$$\frac{U_i}{\Delta t} = \theta \Delta_{xx}[(T^m)_i^n + m(T^{m-1})_i^n U_i] + (1-\theta)\Delta_{xx}(T^m)_i^n \qquad (7.37a)$$

And, after cancellations, equation (7.37a) becomes

$$\frac{U_i}{\Delta t} = m\theta \Delta_{xx}[(T^{m-1})_i^n U_i] + \Delta_{xx}(T^m)_i^n \qquad (7.37b)$$

Equation (7.37b) is written in the expanded form as

$$\frac{U_i}{\Delta t} = \theta \frac{m}{(\Delta x)^2}[(T^{m-1})_{i+1}^n U_{i+1} - 2(T^{m-1})_i^n U_i + (T^{m-1})_{i-1}^n U_{i-1}]$$

$$+ \frac{1}{(\Delta x)^2}[(T^m)_{i+1}^n - 2(T^m)_i^n + (T^m)_{i-1}^n] \qquad (7.38)$$

We define

Nonlinear Diffusion

$$r = \frac{\Delta t}{(\Delta x)^2} \tag{7.39}$$

and rearrange equation (7.38) in the form

$$-\theta mr(T^{m-1})_{i-1}^n U_{i-1} + [1 + 2\theta mr(T^{m-1})_i^n]U_i - \theta mr(T^{m-1})_{i+1}^n U_{i+1}$$
$$= r(T^m)_{i-1}^n - 2r(T^m)_i^n + r(T^m)_{i+1}^n \tag{7.40}$$

which is valid for all internal nodes i = 1,2,...,M−1. The system [equation (7.40)] provides a complete set of equations for the determination of U_i at the time level n + 1 when T is prescribed at the boundaries. Once U_i are available, $(T)_i^{n+1}$ at the time level n + 1 for all internal nodes are determined from equation (7.35), that is

$$(T)_i^{n+1} = U_i + (T)_i^n \tag{7.41}$$

because $(T)_i^n$ are known.

The system [equation (7.40)] can readily be solved by any one of the algorithms discussed in Chapter 3. For prescribed flux or convection boundary conditions, T is unknown at the boundary and additional relationships are developed by discretizing such boundary conditions, as illustrated in Chapter 2.

The system of equations (7.40) is unconditionally stable for $\frac{1}{2} \leq \theta \leq 1$. For $0 \leq \theta < \frac{1}{2}$, the time step Δt must satisfy the stability constraint, which will be discussed next.

7.3.1 Stability Criterion

We consider the differential equation (7.32) written in the form

$$\frac{\partial T}{\partial t} = \frac{\partial}{\partial x}\left[D(T)\frac{\partial T}{\partial x}\right] \tag{7.42a}$$

where D(T) is an effective diffusion coefficient given by

$$D(T) = mT^{m-1} \tag{7.42b}$$

Richtmyer and Morton (1967) used a heuristic approach to answer the stability question by considering equations (7.42a,b) for the case $D(T) = 5T^4$. They concluded that the method would be stable up to time $t = t_0$ if and only if

$$\frac{5T^4 \Delta t}{(\Delta x)^2} < \frac{1}{2 - 4\theta} \quad \text{for} \quad 0 \leq \theta < \frac{1}{2} \tag{7.43}$$

for all x in $0 \leq x \leq L$ and for all t in $0 \leq t \leq t_0$, with no restriction on Δt for $\frac{1}{2} \leq \theta \leq 1$. Richtmyer and Morton (1967) also pointed out that, "For nonlinear problems, stability depends not only on the structure of the finite difference system but also generally on the solution being obtained; and for a given solution, the system may be stable for some values of t and not for others." Therefore, for the solution of nonlinear systems with an explicit scheme such as that discussed in this section, it is desirable to keep a constant check for stability by making tests for Δt with an inequality such as that given by equation (7.43).

Example 7.1

Consider the finite difference approximation of the nonlinear diffusion equation (7.31) for m = 5 with the linearization procedure described earlier. Suppose the weight factor θ and the parameter r are chosen as

$$\theta = 0.4, \quad r = \frac{\Delta t}{(\Delta x)^2} = 0.001$$

Estimate the value of T above which instability is expected in the numerical solution.

Solution
We consider the stability criterion given by equation (7.43) and set $\theta = 0.4$, $[\Delta t/(\Delta x)^2] = 0.001$, to obtain

$$5\, T^4(0.001) < \frac{1}{2 - (4 \times 0.4)}$$
$$T^4 < 500$$

or

$$T < 4.7$$

Thus, it is expected that the calculations will remain stable until a time t_0 when T reaches a value $T \simeq 4.7$ and beyond, at which the instability is expected to occur. Richtmyer and Morton (1967) numerically examined such a case and indeed observed that instability occurred soon after such a value of T was reached.

7.4 False Transient

The method of *false transient* is an alternative and powerful technique for solving nonlinear, steady-state diffusion problems. In this approach, instead of solving the steady-state problem directly, the relevant transient problem is solved until the solution no longer varies with time. Because the interest is in the final steady-state solution, the exact shape of the transient profile is immaterial, and it does not matter how one approaches the steady-state condition.

The false transient method is effective for solving steady-state nonlinear problems if a transient equation relevant to the steady-state problem can be constructed. For diffusion-type problems, this requirement is readily satisfied. For diffusion-type nonlinear problems, the false transient method has the following distinct advantages (Kubicek and Hlavacek 1983):

1. Stiff and parametrically sensitive boundary value problems can readily be handled.
2. Strongly nonlinear problems can be handled without calculating derivatives.
3. The sensitivity to a poor initial guess is very low, and a stable profile convergence is usually assured.

The method also has a shortcoming. For example, if the problem is highly nonlinear and has multiple solutions, it is difficult to locate such solutions because of the instability of some of the steady-state solutions.

To illustrate the basic idea in the false transient method, we consider the following nonlinear, steady-state mass diffusion problem with concentration-dependent volumetric sources.

$$\frac{d^2 C(x)}{dx^2} - S\left(x, C, \frac{dC}{dx}\right) = 0 \quad \text{in } 0 < x < 1 \tag{7.44a}$$

subject to the boundary conditions

$$C = C_0 \quad \text{at } x = 0 \tag{7.44b}$$

$$C = C_1 \quad \text{at } x = 1 \tag{7.44c}$$

Here, the source term $S\left(x, C, \frac{dC}{dx}\right)$ may depend on the position, concentration, and on the gradient of the concentration.

To construct the false transient form of this problem, a fictitious transient term is added to the differential equation and an arbitrary but physically consistent initial condition is chosen. We obtain

$$w \frac{\partial C}{\partial t} = \frac{\partial^2 C}{\partial x^2} - S\left(x, C, \frac{dC}{dx}\right) \quad \text{in } 0 < x < 1, \quad t > 0 \tag{7.45a}$$

$$C = C_0 \quad \text{at } x = 0 \tag{7.45b}$$

$$C = C_1 \quad \text{at } x = 1 \tag{7.45c}$$

and the initial condition

$$C = F(x) \quad \text{for} \quad t = 0 \quad \text{in} \quad 0 < x < 1 \tag{7.45d}$$

Here, w is a weight parameter for the false transient term, with values varying from w = 1 to 5. On the other hand, there is no general simple rule for choosing the optimal value of the weight parameter, and numerical experimentation is necessary for each particular problem.

Finite difference schemes such as the simple explicit, simple implicit, or combined method can be used for the finite differencing of the transient non-linear diffusion equation (7.45a). Suppose the region $0 \leq x \leq 1$ is divided into M equal parts, so that $\Delta x = 1/M$, and the following finite difference notation is adopted

$$C(x,t) = C(i\Delta x, n\Delta t) \equiv C_i^n \tag{7.46}$$

The time-dependent, nonlinear differential equation (7.45a) can be represented in the finite difference form by using the simple explicit and the simple implicit methods as described in the following sections.

7.4.1 Simple Explicit Scheme

The finite difference representation of equation (7.45a) with the simple explicit scheme is given by

$$w \frac{C_i^{n+1} - C_i^n}{\Delta t} = \frac{C_{i-1}^n - 2C_i^n + C_{i+1}^n}{(\Delta x)^2} - S\left[x_i, C_i^n, \frac{C_{i+1}^n - C_{i-1}^n}{2\Delta x}\right] \tag{7.47}$$

for i = 1,2,...,M–1 and n = 0,1,2,... .

These finite difference equations, together with the boundary conditions [equations (7.45b,c)] and the initial condition [equation (7.45d)], are sufficient to evaluate C_i^{n+1} at the time level n + 1 from the knowledge of the C_i^n values at the previous time level n. The scheme being explicit, the time step Δt should satisfy the stability criteria

$$\frac{\Delta t}{w(\Delta x)^2} \leq \frac{1}{2} \tag{7.48}$$

If the steady-state distribution C(x) must be calculated with high precision, small space steps Δx must be used, which in turn requires the use of very small time step Δt. For such situations, a large number of calculations is needed to reach the steady state; hence, one must consider the use of the implicit method.

7.4.2 Simple Implicit Scheme

Finite difference representation of equation (7.45a) with the simple implicit scheme is given by

$$w \frac{C_i^{n+1} - C_i^n}{\Delta t} = \frac{C_{i-1}^{n+1} - 2C_i^{n+1} + C_{i+1}^{n+1}}{(\Delta x)^2} - S\left[x_i, C_i^n, \frac{C_{i+1}^n - C_{i-1}^n}{2\Delta x}\right] \quad (7.49)$$

for $i = 1,2,\ldots,M-1$ and $n = 0,1,2,\ldots$.

We note that the evaluation of the source term S is lagging by one time step. The scheme being implicit, a set of linear algebraic equations must be solved simultaneously at each time step to evaluate C_i^{n+1} at the time level $n + 1$ from the knowledge of C_i^n values at the previous time step. The calculation is simple because the resulting linear system of equations has a tridiagonal coefficient matrix, which can be solved with the Thomas algorithm.

Initial calculations can be performed with only a few spatial nodes in order to bring the initial condition close to the steady-state solution. The number of spatial mesh points is then gradually increased as the steady-state solution is approached.

7.4.3 A Set of Diffusion Equations

In most applications, the formulation of simultaneous heat and mass diffusion problems involves more than one diffusion equation, and the system of equations may be coupled. Consider, for example, simultaneous heat and mass diffusion accompanied by an exothermic chemical reaction taking place in a porous catalyst. The mathematical formulation of the problem involves two nonlinear diffusion-type ordinary differential equations given by

$$\frac{d^2C}{dx^2} - \frac{\delta}{\gamma\beta} C^n \exp\left(\frac{T}{1 + \frac{T}{\gamma}}\right) = 0 \quad \text{in } 0 < x < 1 \quad (7.50)$$

$$\frac{d^2T}{dx^2} + \delta C^n \exp\left(\frac{T}{1 + \frac{T}{\gamma}}\right) = 0 \quad \text{in } 0 < x < 1 \quad (7.51)$$

subject to appropriate boundary conditions, taken in this example as

$$\frac{dC}{dx} = 0, \quad \frac{dT}{dx} = 0 \quad \text{at } x = 0 \quad (7.52a,b)$$

$$C = 1, \quad T = 0 \quad \text{at } x = 1 \quad (7.52c,d)$$

where δ, γ, β, and n are parameters related to the chemical reaction.

This nonlinear system of equations cannot be solved analytically, but a numerical solution is possible. The *false transient* formulation of this problem is given by

$$w_1 \frac{\partial C}{\partial t} = \frac{\partial^2 C}{\partial x^2} - \frac{\delta}{\gamma \beta} C^n \exp\left(\frac{T}{1 + \frac{T}{\gamma}}\right) \quad \text{in } 0 < x < 1, \ t > 0 \quad (7.53)$$

$$w_2 \frac{\partial T}{\partial t} = \frac{\partial^2 T}{\partial x^2} + \delta C^n \exp\left(\frac{T}{1 + \frac{T}{\gamma}}\right) \quad \text{in } 0 < x < 1, \ t > 0 \quad (7.54)$$

subject to the boundary conditions

$$\frac{\partial C}{\partial x} = 0, \quad \frac{\partial T}{\partial x} = 0 \quad \text{at } x = 0, \ t > 0 \quad (7.55a,b)$$

$$C = 1, \quad T = 0 \quad \text{at } x = 1, \ t > 0 \quad (7.55c,d)$$

and the initial conditions chosen as

$$C = F(x) \quad \text{and} \quad T = H(x) \quad \text{for } t = 0 \quad (7.55e,f)$$

For the dimensionless parameters $\gamma = 20$, $\beta = 0.2$, $\delta = 2.56$, and $n = 1$, the problem has only one steady-state solution, which can be calculated by the false transient method for $w_1 = w_2 = 1$ (Kubicek and Hlavacek 1983) by using the simple explicit or the simple implicit schemes described previously.

7.5 Applications in Coupled Conduction and Radiation in Participating Media

7.5.1 One-Dimensional Problem with Diffusion Approximation

When heat conduction takes place through semitransparent materials such as glass and quartz at high temperatures, the radiation flux may become of the same order of magnitude as the conduction heat flux. In heat transfer through porous insulating materials such as fibers, powders, and foam, thermal radiation may become equally important as conduction. A treatment of simultaneous conduction and radiation in a medium that absorbs, emits, and scatters radiation can be found in the texts by Özişik (1973), Sparrow and Cess (1978), Siegel and Howell (2002), and Modest (2013). When the conduction and radiation are of comparable magnitude, a separate calculation of conductive and radiative fluxes without any consideration of the interaction between them may introduce error in the heat transfer results.

Nonlinear Diffusion

The energy equation for simultaneous conduction and radiation in a participating medium is given by (Özişik 1973)

$$-\nabla \cdot (\mathbf{q}^c + \mathbf{q}^r) + g(\mathbf{r},t) = \rho C_p \frac{\partial T(\mathbf{r},t)}{\partial t} \qquad (7.56)$$

where $\mathbf{q}^c$ and $\mathbf{q}^r$ are the conduction and radiation flux vectors, respectively, $g(\mathbf{r},t)$ is the volumetric energy source per unit volume per unit time in the medium, ρ is the density, C_p is the specific heat, and $\mathbf{r}$ is the position vector.

The conduction heat flux vector $\mathbf{q}^c$ is given by Fourier's law as

$$\mathbf{q}^c = -k\nabla T(\mathbf{r},t) \qquad (7.57)$$

and the radiation flux vector is determined from the solution of the equation of radiative transfer for the participating medium, subject to the appropriate radiation boundary conditions. The numerical solution of the equation of radiative transfer for a participating medium is a very complicated matter and will be the subject of the next section in this chapter. However, to illustrate the handling of simultaneous conduction and radiation problems, here we consider a very simple but least accurate diffusion approximation to the radiation heat flux vector given in the form

$$\mathbf{q}^r = -\frac{4n^2\bar{\sigma}}{3\beta_R} \nabla T^4 \qquad (7.58a)$$

which is written in the alternative form as

$$\mathbf{q}^r = -\frac{16n^2\bar{\sigma}\,T^3}{3\beta_R} \nabla T \qquad (7.58b)$$

where n is the refractive index of the medium, $\bar{\sigma}$ is the Stefan–Boltzmann constant, and β_R is the *Roseland mean extinction coefficient*.

Introducing equations (7.57) and (7.58b) into the energy equation (7.56), we obtain

$$\nabla \cdot [k_e(T)\nabla T] + g(\mathbf{r},t) = \rho C_p \frac{\partial T(\mathbf{r},t)}{\partial t} \qquad (7.59a)$$

where

$$k_e(T) \equiv k + \frac{16\bar{\sigma} n^2 T^3}{3\beta_R} \qquad (7.59b)$$

Clearly, the energy equation (7.59) is just like a nonlinear transient heat conduction equation with temperature-dependent thermal conductivity, $k_e(T)$. Such a nonlinear equation can be solved with finite differences using the solution techniques described previously.

Consider, for example, the one-dimensional form of equation (7.59a) in the rectangular coordinates without the energy generation term given by

$$\rho C_p \frac{\partial T}{\partial t} = \frac{\partial}{\partial y}\left[k_e(T)\frac{\partial T}{\partial y}\right] \tag{7.60}$$

This equation is similar to equation (7.4); hence, the finite difference scheme described for the discretization of equation (7.4) becomes applicable for the finite difference representation of equation (7.60).

The limitations of the diffusion approximation for the determination of radiation heat flux should be recognized. This approximation is valid only for the optically thick medium,* that is,

$$\beta_R L \equiv \tau_o \gg 1 \tag{7.61}$$

where L is the physical thickness, β_R is the extinction coefficient, and τ_o is the optical thickness of the medium. Furthermore, the diffusion approximation is not accurate near the boundaries because it does not accommodate the effects of the boundary conditions. Despite its shortcomings, the diffusion approximation may be helpful in estimating the effects of radiation on temperature distribution in the interior of an optically thick region for coupled conduction and radiation.

Example 7.2

Consider simultaneous steady-state conduction and radiation in a plate of thickness L with a extinction coefficient β_R. Boundaries at $y = 0$ and $y = L$ are kept at constant temperatures T_1 and T_2, respectively. Develop the mathematical formulation of this simultaneous conduction and radiation problem by using the diffusion approximation described earlier.

Solution

The governing differential equation is obtained from equation (7.60) as

$$\frac{d}{dy}\left[k_e(T)\frac{dT}{dy}\right] = 0 \quad \text{in} \quad 0 < y < L \tag{a}$$

$$T(y) = T_1 \quad \text{at} \quad y = 0 \tag{b}$$

$$T(y) = T_2 \quad \text{at} \quad y = L \tag{c}$$

where

$$k_e(T) = k\left(1 + \frac{16n^2 \bar{\sigma} T^3}{3k\beta_R}\right) \tag{d}$$

---

* In radiation problems, the optical variable τ is defined by

$$d\tau = \beta \, dy$$

where β is the extinction coefficient and y is the space variable. Thus, for constant β, the optical variable becomes $\tau = \beta y$ and the optical thickness τ_0 of a layer L is given by $\tau_0 = \beta L$.

These equations are expressed in the dimensionless form as (Özişik 1973)

$$\frac{d}{d\tau}\left[k_e(\theta)\frac{d\theta}{d\tau}\right] = 0 \quad \text{in} \quad 0 < \tau < \tau_o \tag{e}$$

$$\theta(\tau) = \frac{T_1}{T_2} \equiv \theta_1 \quad \text{at} \quad \tau = 0 \tag{f}$$

$$\theta(\tau) = \frac{T_2}{T_2} = 1 \quad \text{at} \quad \tau = \tau_o \tag{g}$$

where

$$k_e(\theta) = 1 + \frac{4}{3N}\theta^3 \tag{h}$$

$$N = \frac{k\beta_R}{4n^2\overline{\sigma}T_2^3} \tag{i}$$

$$\theta(\tau) = \frac{T(y)}{T_2} \tag{j}$$

$$\tau = \beta_R y, \quad \tau_o = \beta_R L \tag{k}$$

Here, the parameter N is called the conduction-to-radiation parameter. A small value of N characterizes strong radiation (i.e., a radiation-dominant case) and large values indicate weak radiation (i.e., a conduction-dominant case).

Figure 7.1 shows a plot of the dimensionless temperature θ as a function of position in the plate for $N = 0.01$ and $\theta_1 = 0.5$. In this figure, the solid line represents the solution obtained with the diffusion approximation, and the broken lines correspond to the exact solution of the problem for two different optical thicknesses: $\tau_o = 1$ and $\tau_o = 10$. Clearly, the diffusion approximation is closer to the exact solution for $\tau_o = 10$, which represents an optically thicker case than that for $\tau_o = 1$. Furthermore, in the vicinity of the boundaries, the slope of the curve obtained with the diffusion approximation is inaccurate. Because the heat flux is proportional to the temperature gradient, the heat transfer results also become inaccurate.

In order to obtain a more accurate analysis of simultaneous conduction and radiation, the radiation problem should be formulated in terms of the equation of radiative transfer, and the resulting divergence of the radiation flux vector $\nabla \cdot \mathbf{q}^r$, or $\partial q^r/\partial y$ for a one-dimensional case should be introduced into the energy equation, which is then solved numerically. For example, for the one-dimensional steady-state problem considered earlier, the governing energy equation can be written in the form

$$k\frac{d^2T}{dy^2} - \frac{dq^r}{dy} = 0 \tag{7.62}$$

However, the radiation problem needs the temperature distribution in the medium for its solution, whereas the energy equation (7.62) needs the

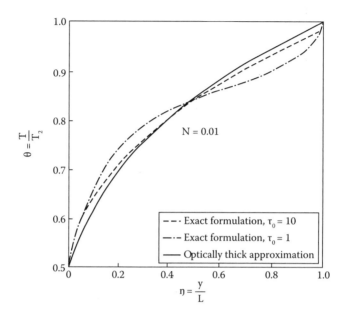

FIGURE 7.1
A comparison of diffusion approximation with exact results.

dq^r/dy term for its solution. Therefore, the solution must be performed by iteration, and the solution of the radiation problem needs to be obtained by numerical methods such as finite volumes, as presented in the next section.

7.5.2 Solution of the Three-Dimensional Equation of Radiative Transfer

In this section, the previously performed analysis is extended to a three-dimensional case, where the coupled conduction and radiation problems are numerically solved by finite volumes. The radiation problem is now formulated in terms of the equation of radiative transfer (Özişik 1973) instead of the diffusion approximation used in the previous section. The equation of radiative transfer represents the conservation of radiant energy and is formulated in terms of the radiation intensity as the dependent variable. The monochromatic radiation intensity is defined as the amount of radiative energy streaming through a unit area perpendicular to a particular direction of propagation, per unit solid angle around this direction, per unit frequency (or wavelength) that characterizes the spectral dependence of radiation, and per unit of time (Özişik 1973). The radiation intensity is a function of the spatial position within the body, the direction of propagation, time, and wavelength. Typically, the transient variation of the radiation intensity can be neglected for most engineering applications because of the large speeds of propagation. The equation of radiative transfer can be further simplified for the so-called gray media, where the propagation of radiation is supposed to be independent of the wavelength.

Different techniques can be found in the literature for the numerical solution of the equation of radiative transfer, such as, for example, the discrete ordinates method (Chandrasekhar 1960; Özişik 1973; Barichello and Siewert 1999, 2002; Siegel and Howell 2002; Barichello 2011; Modest 2013) and the finite volume method (Raithby and Chui 1990; Chui and Raithby 1993; Raithby 1999; Kim and Huh 2000). In this section, the finite volume method is applied to the physical problem examined by Wellele et al. (2006), related to the heat transfer processes taking place in the application of the flash method at high temperatures (André and Degiovanni 1995).

The flash method is a standard technique for the measurement of thermal diffusivity (ASTM 2001). An environmental chamber is used to heat a thin sample to high temperatures and, after thermal equilibrium is reached, a xenon lamp or a laser pulse heats one of the sample's surfaces. During the experiment, an optical detector, such as an infrared camera, measures the temperature variation of the sample's opposite surface. The environmental control chamber can be gas or vacuum tight if operation in a protective atmosphere is desired.

For the case examined here, the sample is considered to be a parallelepiped with sides 2a*, 2b*, and c*, along the x*, y*, and z* directions, respectively, where the superscript "*" denotes dimensional variables. As traditionally used in the flash method, the sample is assumed to be coated with graphite in order to increase the energy absorbed, so that the boundaries are opaque. The laser heating is supposed to be imposed at the center of the sample, with a function symmetric with respect to the sample's mid-planes. The sample, initially at the uniform temperature $T^* = T_0^*$, is heated for $t^* > 0$ by the laser through its top surface. The sample loses heat by linearized radiation and convection through all surfaces with a combined heat transfer coefficient $h^{rad*} = h^* + 4\varepsilon\sigma T_0^{*3}$ (Özişik 1985), where h^* is the convection heat transfer coefficient, ε is the emissivity of the graphite paint that coats the sample, and σ (=5.67 × 10^{-8} W/m^{2}K^4) is the Stefan–Boltzmann constant. The solid is assumed to be orthotropic (Özişik 1993) with thermal conductivities k_x^*, k_y^*, and k_z^*, along the x*, y*, and z* directions, respectively. In addition, the sample is assumed to be a gray body, that is, its radiative properties do not vary with the wavelength. Figures 7.2a and b illustrates the sample and the computational domain, which takes into account the symmetry with respect to the planes at x* = 0 and at y* = 0.

In order to write the mathematical formulation for the present physical problem in dimensionless form, we define the following dimensionless groups:

$$k_x = \frac{k_x^*}{k_{ref}^*}, \quad k_y = \frac{k_y^*}{k_{ref}^*}, \quad k_z = \frac{k_z^*}{k_{ref}^*} \qquad (7.63a\text{--}c)$$

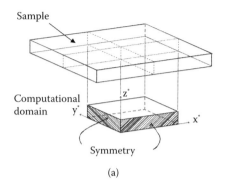

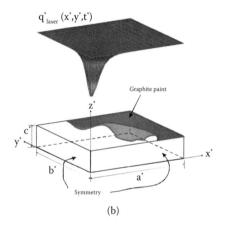

FIGURE 7.2
(a) Sample's geometry and (b) Computational domain (Wellele et al., 2006).

$$x = \frac{x^*}{d_{ref}^*}, \quad y = \frac{y^*}{d_{ref}^*}, \quad z = \frac{z^*}{d_{ref}^*} \qquad (7.63\text{d--f})$$

$$a = \frac{a^*}{d_{ref}^*}, \quad b = \frac{b^*}{d_{ref}^*}, \quad c = \frac{c^*}{d_{ref}^*} \qquad (7.63\text{g--i})$$

$$C = \frac{\rho^* C_p^*}{C_{ref}^*}, \quad q = \frac{q^*}{q_{ref}^*}, \quad T = \frac{T^* - T_0^*}{\Delta T_{ref}^*} \qquad (7.63\text{j--l})$$

$$t = \frac{k_{ref}^* t^*}{C_{ref}^* d_{ref}^{*2}}, \quad \nabla \cdot \mathbf{q}^{rad} = \frac{d_{ref}^{*2}}{k_{ref}^* \Delta T_{ref}^*} \nabla \cdot \mathbf{q}^{rad*} \qquad (7.63\text{m,n})$$

$$Bi^{rad} = \frac{h^{rad*} d_{ref}^*}{k_{ref}^*} \qquad (7.63\text{o})$$

where k_{ref}^*, C_{ref}^*, and q_{ref}^* are reference values for thermal conductivity, volumetric heat capacity, and heat flux, respectively, and d_{ref}^* is a characteristic length of the sample. The characteristic temperature difference for the problem is taken as

$$\Delta T_{ref}^* = \frac{q_{ref}^* d_{ref}^*}{k_{ref}^*} \quad (7.64)$$

In addition to equations (7.63a–o), the following dimensionless groups are required to write the equation of radiative transfer in dimensionless form:

$$I = \frac{I^*}{4\sigma T_0^{*4}}, \quad \tau_0 = \beta^* d_{ref}^*, \quad \kappa_a = \kappa_a^* d_{ref}^* \quad (7.65a\text{--}c)$$

$$\sigma_s = \sigma_s^* d_{ref}^*, \quad N_{pl} = \frac{\beta^* k_{ref}^* \Delta T_{ref}^*}{4\sigma T_0^{*4}} \quad (7.65d,e)$$

In equation (7.65a), $I^* \equiv I^*(x^*, y^*, z^*, \vec{s})$ is the radiation intensity propagating in the direction $\vec{s}$ at a position (x^*, y^*, z^*). Other quantities of interest for the analysis are β^*, κ_a^*, and σ_s^*, which are the extinction, absorption, and scattering coefficients of the medium, respectively. These coefficients have the units of m^{-1}. The absorption coefficient, κ_a^*, represents the fraction of the incident radiation intensity that is absorbed by the matter per unity length along the path of the beam. Similarly, the scattering coefficient, σ_s^*, represents the fraction of the incident radiation intensity that is scattered by the matter in all directions per unity length along the path of the beam. The extinction coefficient jointly represents the effects of absorption and scattering, that is, $\beta^* = \kappa_a^* + \sigma_s^*$. In equation (7.65b), τ_0 is denoted as the optical thickness of the medium (Özişik 1973).

With the aforementioned dimensionless variables defined, the *equation of radiative transfer* can be written as (Özişik 1973)

$$\vec{s} \cdot \nabla I = -(\kappa_a + \sigma_s)I + S \quad \text{in } 0 < x < a, \ 0 < y < b, \ 0 < z < c \quad (7.66a)$$

where

$$S = \kappa_a n_r^2 I_b(T) + \frac{\sigma_s}{4\pi} \int_{\Omega' = 4\pi} I p(\vec{s}' \to \vec{s}) d\Omega' \quad (7.66b)$$

In equation (7.66a), the first term on the right-hand side gives the loss of radiation intensity propagating in the direction $\vec{s}$ due to absorption and scattering, while the source function, S, represents the gain of radiant energy along this direction. The first and second terms in the source function, given by equation (7.66b), represent emission and in-scattering radiation, respectively, where $I_b(T) = T^4/4\pi$ is the black-body dimensionless intensity and n_r is the refractive index. The scattering phase function, $p(\vec{s}' \to \vec{s})$, is the probability that the incident radiation at the direction $\vec{s}'$ be scattered into an element of a solid angle about the direction $\vec{s}$.

Equation (7.66a) is subjected to the following boundary conditions for this problem:

$$I(\xi,\eta,\mu) = I(-\xi,\eta,\mu) \quad \text{for} \quad \xi > 0 \quad \text{at} \quad \Gamma_1:(x=0,\ 0<y<b,\ 0<z<c)$$
(7.66c)

$$I(-\xi,\eta,\mu) = \varepsilon n_r^2 I_b + \frac{1-\varepsilon}{\pi}\int_{\xi'>0} I(\xi',\eta',\mu')\xi'd\Omega'$$
(7.66d)

for $\xi < 0$ at $\Gamma_2:(x=a,\ 0<y<b,\ 0<z<c)$

$$I(\xi,\eta,\mu) = I(\xi,-\eta,\mu) \quad \text{for} \quad \eta > 0 \quad \text{at} \quad \Gamma_3:(0<x<a, y=0, 0<z<c)$$
(7.66e)

$$I(\xi,-\eta,\mu) = \varepsilon n_r^2 I_b + \frac{1-\varepsilon}{\pi}\int_{\eta'>0} I(\xi',\eta',\mu')\eta'd\Omega'$$
(7.66f)

for $\eta < 0$ at $\Gamma_4:(0<x<a,\ y=b,\ 0<z<c)$

$$I(\xi,\eta,\mu) = \varepsilon n_r^2 I_b + \frac{1-\varepsilon}{\pi}\int_{\mu'<0} I(\xi',\eta',\mu')\mu'd\Omega'$$
(7.66g)

for $\mu > 0$ at $\Gamma_5:(0<x<a, 0<y<b, z=0)$

$$I(\xi,\eta,-\mu) = \varepsilon\, n_r^2 I_b + \frac{1-\varepsilon}{\pi}\int_{\mu'>0} I(\xi',\eta',\mu')\mu'd\Omega'$$
(7.66h)

for $\mu < 0$ at $\Gamma_6:(0<x<a,\ 0<y<b,\ z=c)$

where ξ, η, and μ are the cosines of direction $\vec{s}$ along the x, y, and z directions, respectively. Equations (7.66c) and (7.66e) provide the symmetry conditions at $x=0$ and $y=0$, respectively. In equations (7.66d), (7.66f), (7.66g), and (7.66h), the first term on the right-hand side characterizes the radiation emitted, while the second term represents the radiation reflected by the boundary surfaces that are opaque.

With the dimensionless groups described earlier, the *energy conservation equation* can be written as (Özişik 1973):

$$C\frac{\partial T}{\partial t} = \nabla\cdot(k\nabla T) - \nabla\cdot\mathbf{q}^{rad} \quad \text{in } 0<x<a,\ 0<y<b,\ 0<z<c \text{ for } t>0$$
(7.67a)

where the divergence of the radiative flux is given by

$$\nabla\cdot\mathbf{q}^{rad} = \frac{\kappa_a \tau_0}{N_{pl}}\left[4\pi n_r^2 I_b - \int_{\Omega=4\pi} I d\Omega\right]$$
(7.67b)

Equation (7.67a) is subjected to the following boundary conditions, which take into account the radiative heat transfer at the body surfaces and the flux imposed by the laser at z = c:

$$\frac{\partial T}{\partial x} = 0 \quad \Gamma_1:(x=0,\ 0<y<b,\ 0<z<c) \quad \text{for } t>0 \tag{7.67c}$$

$$k_x \frac{\partial T}{\partial x} + Bi^{rad}\, T = \frac{\varepsilon \tau_0}{N_{pl}} \left[\int_{\xi>0} I\xi d\Omega - n_r^2 \pi I_b \right] \tag{7.67d}$$
$$\text{at} \quad \Gamma_2:(x=a, 0<y<b, 0<z<c), \quad \text{for } t>0$$

$$\frac{\partial T}{\partial y} = 0 \quad \text{at} \quad \Gamma_3:(0<x<a, y=0, 0<z<c), \quad \text{for } t>0 \tag{7.67e}$$

$$k_y \frac{\partial T}{\partial y} + Bi^{rad}\, T = \frac{\varepsilon \tau_0}{N_{pl}} \left[\int_{\eta>0} I\eta d\Omega - n_r^2 \pi I_b \right] \tag{7.67f}$$
$$\text{at} \quad \Gamma_4:(0<x<a,\ y=b,\ 0<z<c), \quad \text{for } t>0$$

$$-k_z \frac{\partial T}{\partial z} + Bi^{rad}\, T = \frac{\varepsilon \tau_0}{N_{pl}} \left[\int_{\mu<0} I\mu d\Omega - n_r^2 \pi I_b \right] \tag{7.67g}$$
$$\text{at} \quad \Gamma_5:(0<x<a, 0<y<b, z=0), \quad \text{for } t>0$$

$$k_z \frac{\partial T}{\partial z} + Bi^{rad}\, T = \frac{\varepsilon \tau_0}{N_{pl}} \left[\int_{\mu>0} I\mu d\Omega - n_r^2 \pi I_b \right] + (1 - \rho_{graphite,\lambda})\, q_{laser}(x,y,t)$$
$$\text{at} \quad \Gamma_6:(0<x<a,\ 0<y<b,\ z=c), \quad \text{for } t>0 \tag{7.67h}$$

with initial condition

$$T=0 \quad \text{in } 0<x<a, 0<y<b, 0<z<c \quad \text{for } t>0 \tag{7.67i}$$

where $\rho_{graphite,\lambda}$ is the graphite reflectivity at the laser wavelength and $q_{laser}(x,y,t)$ is the heat flux imposed by the laser. On the right-hand sides of equations (7.67d), (7.67f), (7.67g), and (7.67h), the first terms represent the internal radiant flux absorbed by the boundaries, while the second terms represent the radiation emitted by the boundary surfaces toward the interior of the body.

The radiation and energy conservation problems, given by equations (7.66a–h) and (7.67a-i), respectively, are coupled through the temperature distribution required to compute the black-body intensity, appearing in the equation of radiative transfer and in its boundary conditions, as well as through the

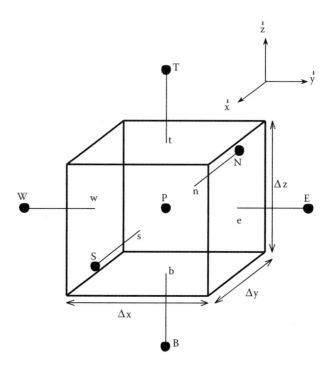

FIGURE 7.3
Control volume used for the spatial discretization (Wellele et al., 2006).

divergence of the radiative heat flux, appearing in the energy conservation equation. Therefore, these problems need to be solved simultaneously.

For the solution of the coupled conduction–radiation problem given by equations (7.66) and (7.67), we use the finite volume method (Raithby and Chui 1990; Chui and Raithby 1993; Raithby 1999; Kim and Huh 2000). This method requires the discretization of the whole medium with nonoverlapping control volumes. There is one node within each control volume for which the dependent variables are computed. The governing equations for the problem are integrated over each control volume, and piecewise profiles, expressing the variation of the conserved variables between the grid points, are used to evaluate the desired integrals. The simplest profile, which is generally used for problems such as the one under analysis, is that the dependent variables are uniform over the control volume. Similarly, each flux entering/leaving the control volume is assumed to be uniform over each surface that composes its boundary (surfaces e, w, s, n, t, and b). Such fluxes are approximated with algebraic equations that involve geometric parameters as well as the dependent variables at node P and its neighbors (nodes E, W, S, N, T, and B; see Figure 7.3), as thoroughly discussed earlier in this book.

The radiative intensity for a gray medium depends not only on the spatial variables but also on the direction of propagation. Hence, in addition to the traditional spatial discretization needed for the energy conservation equation (7.67a), the equation of radiative transfer [equation (7.66a)] requires the discretization along the intensity directions. In this case, the directional domain is subdivided into N_l solid angles Ω^l, so that, $\sum_{l=1}^{N_l} \Omega^l = 4\pi$.

We summarize the basic steps for the discretization of the energy conservation and radiative transfer equations as follows.

By integrating equation (7.67a) in the general elementary control volume ΔV and from time t to t + Δt, we obtain:

$$\int_t^{t+\Delta t} \int_{\Delta V} C \frac{\partial T}{\partial t} dVdt = \int_t^{t+\Delta t} \int_{\Delta V} \nabla \cdot (k\nabla T) dVdt - \int_t^{t+\Delta t} \int_{\Delta V} \nabla \cdot \mathbf{q}^{rad} dVdt \quad (7.68)$$

For the term on the left-hand side of equation (7.68), the temperature and the volumetric heat capacities are assumed to be uniform over the elementary volume. Then, by approximating the time derivative with forward finite differences, such an integral term becomes:

$$\int_t^{t+\Delta t} \int_{\Delta V} C \frac{\partial T}{\partial t} dVdt = C_P[T_P(t+\Delta t) - T_P(t)] \Delta V \quad (7.69a)$$

The second term on the right-hand side of equation (7.68), involving the divergence of the radiative flux, is approximated considering the integrand as a representative average value within the control volume. The following expression results:

$$\int_t^{t+\Delta t} \int_{\Delta V} \nabla \cdot \mathbf{q}^{rad} dVdt = [\nabla \cdot \mathbf{q}^{rad}]_P \Delta V \Delta t \quad (7.69b)$$

In the first term on the right-hand side of equation (7.68), the volume integral is replaced by an area integral by using Gauss's divergence theorem. The conduction heat flux at the control volume boundary is then approximated as being uniform over each of its surfaces. Therefore, the first term on the right-hand side of equation (9.68) becomes:

$$\int_t^{t+\Delta t} \int_{\Delta V} \nabla \cdot (k\nabla T) dVdt = \sum_{i=e,w,n,s,t,b} A_i (k\nabla T)_i \Delta t \quad (7.69c)$$

where A_i is the area of each of the boundary surfaces of the elementary control volume, i = e, w, n, s, t, and b. The gradient at each of such surfaces is approximated by finite differences by using the temperature values at node P and its neighboring nodes E, W, S, N, T, and B. Equations (7.69a–c) are then substituted into equation (7.68) in order to obtain the discretized form of the energy conservation equation.

The equation of radiative transfer [equation (7.66a)] is integrated in the elementary volume ΔV, as well as in the solid-angle Ω^l around the direction $\vec{s}^l$ (see Figure 7.4), in order to derive its finite volume approximation, that is,

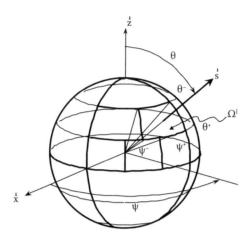

FIGURE 7.4
Solid angle used for the directional discretization (Wellele et al., 2006).

$$\int_{\Omega^l}\int_{\Delta V} \vec{s}\cdot\nabla I dV d\Omega = \int_{\Omega^l}\int_{\Delta V} -(\kappa_a+\sigma_s)I dV d\Omega + \int_{\Omega^l}\int_{\Delta V} S dV d\Omega \quad (7.70a)$$

where

$$\int_{\Omega^l}\int_{\Delta V} S dV d\Omega = \int_{\Omega^l}\int_{\Delta V} \kappa_a n_r^2 I_b dV d\Omega + \int_{\Omega^l}\int_{\Delta V} \frac{\sigma_s}{4\pi}\int_{\Omega'=4\pi} I p(\vec{s}'\rightarrow\vec{s}) d\Omega' dV d\Omega$$
$$(7.70b)$$

The integral term on the left-hand side of equation (7.70a) is approximated by using Gauss's divergence theorem and by assuming that the intensity at each boundary surface can be represented in terms of an average value over the surface. By defining

$$D_i^l = \int_{\Omega^l} \vec{n}_{A_i}\cdot\vec{s}\, d\Omega \quad (7.71)$$

such a term can be written as

$$\int_{\Omega^l}\int_{\Delta V} \vec{s}\cdot\nabla I dV d\Omega = \sum_{i=e,w,n,s,t,b} A_i D_i^l I_i^l \quad (7.72)$$

where the superscript l refers to the direction $\vec{s}^{\,l}$ (see Figure 7.4).

The first term on the right-hand side of equation (7.70a) is approximated by considering an average value for the intensity within each elementary volume in the direction $\vec{s}^{\,l}$ as well as average values for the absorption and scattering coefficients. We then write:

$$\int_{\Omega^l}\int_{\Delta V} -(\kappa_a+\sigma_s)I dV d\Omega = -(\kappa_{a,P}+\sigma_{s,P})I_P^l \Omega^l \Delta V \quad (7.73)$$

Similarly, the contribution of emission in the source term [equation (7.70b)] is approximated as:

$$\int_{\Omega^l}\int_{\Delta V} \kappa_a n_r^2 I_b dV d\Omega = \kappa_{a,P} n_{r,P}^2 I_{b,P} \Omega^l \Delta V \tag{7.74}$$

The second term on the right-hand side of equation (7.70b), containing the double integral with respect to the solid angle, which gives the in-scattering contribution of the source term, is approximated as:

$$\int_{\Omega^l}\int_{\Delta V} \frac{\sigma_s}{4\pi} \int_{\Omega'=4\pi} I p(\vec{s}' \to \vec{s}) d\Omega' dV d\Omega = \Omega^l \Delta V \frac{\sigma_{s,P}}{4\pi} \sum_{l'=1}^{N_l} I_P^{l'} \overline{p_P^{l'l}} \Omega^{l'} \tag{7.75}$$

where

$$\overline{p_P^{l'l}} = \frac{1}{\Omega^l \Omega^{l'}} \int_{\Omega^{l'}}\int_{\Omega^l} p(\vec{s}' \to \vec{s}) d\Omega d\Omega' \tag{7.76}$$

is the discrete phase function.

Equations (7.71)–(7.76) are then substituted into equations (7.70a) and (7.70b) in order to obtain the discretized form of the equation of radiative transfer, which is solved simultaneously with the discretized form of the energy conservation equation.

With the solution of such nonlinear systems of equations, the temperatures T_P at each node P and the intensity I_P^l for each direction $\vec{s}^l$ at each node P are computed for each time step. For the results presented as follows, the solution was marched in time with the alternating direction implicit (ADI) scheme described in Chapter 6.

We consider, as an example, the heating of a sample with dimensions $2a^* = 2b^* = 0.01$ m and $c^* = 0.001$ m by a laser with a Gaussian distribution, as depicted in Figure 7.2b. For the heat flux imposed by the laser, 99% of its power is assumed to be delivered within a circle with a radius of 2 mm centered at the sample. The temperature variations are presented here for points A, B, and C, located at the positions (0,0,0) m, (0.002,0,0) m, and (0,0.002,0) m, respectively (see Figure 7.5). The physical properties are given by $\rho^* C^* = 2.5 \times 10^6$ J/m³K, $k_x^* = k_y^* = k_z^* = 5$ W/mK, $\kappa_a^* = 10$ m^{-1}, and $\sigma_s^* = 1000$ m^{-1}, which are typical of ceramics such as alumina. The Henyey–Greenstein phase function (Siegel and Howell 2002)

$$p(\vec{s}' \to \vec{s}) = \frac{(1-g^2)}{(1+g^2-2g\mu_0)^{3/2}} \tag{7.77}$$

with a $g = 0.7$ asymmetry factor was chosen in order to describe the highly forward scattering character of such porous ceramics, where μ_0 is the cosine of the scattering angle. The convection heat transfer coefficient was taken as

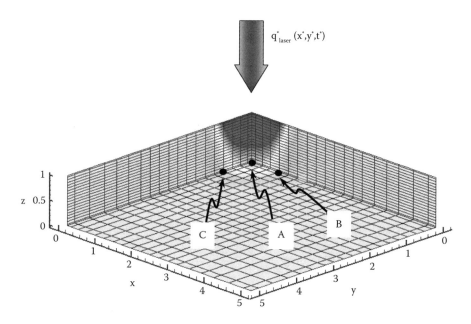

FIGURE 7.5
Heated sample (Wellele et al., 2006).

TABLE 7.1

Effects of Radiation Heat Transfer on the Maximum Temperature Rise of Point A

$T_0^*[K]$	N_{pl}	$h^{rad\,*}$ [W/m²K]	Q_{laser}^* [W]	$T_{max}^* - T_0^*[K]$ Conduction	$T_{max}^* - T_0^*[K]$ Conduction–Radiation	Δ (%)
300	824.7	11	0.7	5.0	5.0	0.0
800	43.5	121	7.3	5.0	4.9	−1.3
1300	10.1	503	15	5.0	4.7	−6.7
1800	3.8	1328	25	5.0	4.3	−13.3
2300	1.8	2764	42	5.0	4.1	−18.1

$h^* = 5\,W/m^2K$, and the duration of the laser pulse as 5 s. The graphite coating was assumed as a black surface ($\varepsilon = 1$).

Table 7.1 presents the effects of radiation heat transfer due to the increase in the initial temperature in the medium. The initial temperature of the sample is given in the first column of Table 7.1. The second and third columns of this table give the conduction-to-radiation parameter [see equation (7.65e)] and the combined radiation–convection heat transfer coefficient, respectively, computed with the corresponding initial temperature. The fourth column of Table 7.1 presents the laser power required to induce a maximum temperature rise of 5 K on Point A (see the fifth column of Table 7.1) if heat transfer within the sample takes place by conduction only. The maximum temperature rise of Point A with the coupled conduction–radiation model is presented in the sixth column of Table 7.1. The percent difference between the maximum temperature rise of Point A, with the conduction and with the coupled conduction-radiation models, is given in the seventh column of this table. In Table 7.1, we note an increase in the laser power required to induce 5 K of temperature rise on Point A with the conduction model, when the initial temperature is increased. Such is the case because of the larger heat losses for larger values of the initial temperature due to the increase of the combined convection-radiation heat transfer coefficient. Table 7.1 also shows that radiation within the sample is negligible for small initial temperatures, so that the maximum temperature rises obtained with the conduction and with the coupled conduction–radiation models are identical. As the initial temperature is increased, the coupling between conduction and radiation becomes more significant and the maximum temperature rise of Point A is reduced. In fact, for an initial temperature of 1800 K, which is typical of modern flash method apparatuses, the use of a model that does not take into account the radiation within the sample results in an error of 10% for such a ceramic material.

Figure 7.6 presents a comparison of the temperature rise for Points A, B, and C, obtained with the purely conductive (C) and with the coupled conductive-radiative (C + R) models. The experimental conditions involve an initial temperature of $T_0^* = 1800$ K. The sample is assumed to be heated with a 25-W powered laser and a pulse duration of 5 s, such as in the fourth test case presented by Table 7.1. The variation of the laser power is also presented in Figure 7.6. We note that the temperature rises of Points B and C are identical because the medium is isotropic. Similar to Table 7.1, Figure 7.6 shows a temperature reduction when the radiative effects are taken into account. The net radiative flux at the surface $z^* = 0$, defined as the difference between the

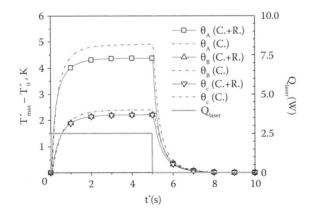

FIGURE 7.6
Temperature rise for Points A, B, and C for an orthotropic medium (Wellele et al., 2006).

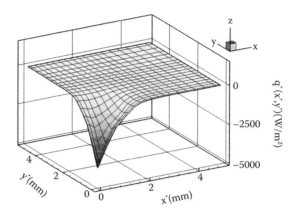

FIGURE 7.7
Net radiative flux at the boundary $z^* = 0$ (Wellele et al., 2006).

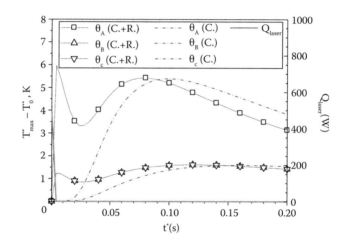

FIGURE 7.8
Temperature rise for Points A, B, and C for an orthotropic medium under a fast laser pulse (Wellele et al., 2006).

radiative flux reaching the surface from the interior of the body and that leaving the surface to the interior of the body, at the end of the heating period, is shown in Figure 7.7. We can notice in this figure larger magnitudes of the net radiative heat flux below the region where the laser heating is imposed. Because the net radiative flux is negative, energy is reradiated into the medium from the surface at $z^* = 0$, resulting in the temperature reduction observed in Figure 7.6.

In order to examine the effects of the coupling between radiation and conduction for cases involving fast laser pulses, we considered the same conditions described previously but with a laser power of 2500 W and a pulse duration of 0.005 s. For this case, Figure 7.8 illustrates the temperature rises for Points A, B, and C, which were obtained with the purely conductive (C) and with the coupled conductive-radiative (C + R) models. We can notice in Figure 7.8 that the temperature rises computed with the two models are significantly different in such case. In special, the coupled model results in faster temperature responses due to the radiative heat transfer between the top and bottom surfaces of the body, which takes place immediately after the top surface is heated by the laser. As a result, the temperature rises for small times are larger for the coupled conduction–radiation model. However, for large times, the temperature rises obtained with this model are smaller than those obtained with the purely conductive model. This behavior is due to the energy reradiated into the medium by the bottom surface as discussed earlier for the case involving a long laser pulse.

PROBLEMS

7.1. Consider the nonlinear mass diffusion equation

$$\frac{\partial}{\partial x}\left[D(C)\frac{\partial}{\partial x}\right] + g(C) = \frac{\partial C}{\partial t}$$

Write the finite difference approximation of this equation for the internal nodes by using the Crank–Nicolson method.

7.2. Write the finite difference approximation of the nonlinear diffusion equation given in Problem 7.1 by using the three-time-level implicit Dupont-II scheme.

7.3. Consider the following nonlinear heat conduction problem

$$\rho C_p(T)\frac{\partial T}{\partial t} = \frac{1}{R^2}\frac{\partial}{\partial R}\left[R^2 k(T)\frac{\partial T}{\partial R}\right] + G(T) \quad \text{in } 0 \leq R < b, \ t > 0$$

$$\frac{\partial T}{\partial R} = 0 \quad \text{at } R = 0, \ t > 0$$

$$k\frac{\partial T}{\partial R} + hT = hT_\infty \quad \text{at } R = b, \ t > 0$$

$$T = F(R) \quad \text{for } t = 0, \text{ in } 0 \leq R < b$$

Use the combined method to develop a finite difference approximation for this differential equation at the internal nodes and for its boundary condition at the boundary node R = b.

7.4. Repeat Problem 7.3 by using the implicit Dupont-II three-level-time scheme.

7.5. Consider the following nonlinear diffusion equation

$$\frac{\partial T}{\partial t} = \frac{1}{R}\frac{\partial}{\partial R}\left[RD(T)\frac{\partial T}{\partial R}\right] + G(T) \quad \text{in} \quad 0 \leq R < b$$

Develop finite difference representation of this equation using:
a. Combined method
b. Implicit Dupont-II three-time-level scheme

7.6. Consider the following nonlinear diffusion equation

$$\frac{\partial T}{\partial t} = \frac{\partial^2}{\partial x^2}(T^3) + G(T)$$

Develop finite difference approximation for this equation using the linearization procedure of Section 7.3.

7.7. Consider the mass transfer in a porous catalyst governed by the following equation

$$\frac{d^2 C}{dx^2} = \phi^2\, C \exp\left[\frac{\gamma\beta(1-C)}{1+\beta(1-C)}\right] \quad \text{in} \quad 0 < x < 1$$

subject to the boundary conditions

$$\frac{dC}{dx} = 0 \quad \text{at} \quad x = 0$$
$$C = 1 \quad \text{at} \quad x = 1$$

where ϕ^2, γ, and β are parameters of the problem. Write the false transient form of this system and discretize the false transient equations using both the simple explicit and the simple implicit finite difference schemes.

Solve the resulting finite difference equations for the cases:
a. $\gamma = 20$, $\gamma\beta = 2$, and $\phi = 1$
b. $\gamma = 20$, $\gamma\beta = 4$, and $\phi = 2$
and determine the value of C at $x = 0$.

7.8. Consider the problem of axial mixing in a tubular reactor governed by the differential equation

$$\frac{1}{Pe}\frac{d^2C}{dx^2} - \frac{dC}{dx} - RC^2 = 0 \quad \text{in} \quad 0 < x < 1$$

subject to the boundary conditions

$$-\frac{1}{Pe}\frac{dC}{dx} + C = 1 \quad \text{at} \quad x = 0$$

$$\frac{dC}{dx} = 0 \quad \text{at} \quad x = 1$$

Write the false transient form of this system and discretize the resulting false transient equation using the simple implicit scheme.

Solve the resulting finite difference equations for Pe = 6 and R = 2 and compute C at the locations x = 0, 0.2, 0.4, 0.6, 0.8, and 1.0.

7.9. Dimensionless steady-state temperature distribution in a slab of reacting explosive material is governed by the differential equation

$$\frac{d^2T}{dx^2} + \delta e^T = 0 \quad \text{in} \quad 0 < x < 1$$

subject to the boundary conditions

T = 0 at x = 0

T = 0 at x = 1

where δ is the system parameter.

Write the false transient form of this problem and discretize the resulting equation using the simple implicit scheme. For $\delta = 1$, calculate the steady-state temperature at the locations x = 0.2, 0.4, 0.6, and 0.8.

7.10. Consider Problem 7.9 for a solid infinite cylinder governed by the nonlinear ordinary differential equation

$$\frac{d^2T}{dR^2} + \frac{1}{R}\frac{dT}{dR} = -\delta e^T \quad \text{in} \quad 0 < R < 1$$

subject to the boundary conditions

$$\frac{dT}{dR} = 0 \quad \text{at} \quad R = 0$$
$$T = 0 \quad \text{at} \quad R = 1$$

where δ is the system parameter.

The analytic solution of this problem can be expressed in the form (Kubicek and Hlavacek 1983):

$$T = \ln\left[\frac{8B/\delta}{(BR^2+1)^2}\right]$$

where the integration constant B is given by

$$\frac{8B/\delta}{(B+1)^2} = 1$$

Clearly, for δ in the range $0 < \delta < 2$, B has two distinct real values and thus the problem has two solutions. For $\delta = 2$, B has only one value (i.e., $B = 1$) and the problem has only one solution. For $\delta > 2$, the problem possesses no solution.

Write the false transient form of this problem, discretize the resulting false transient equation using the simple implicit scheme, solve the finite difference equations for the case $\delta = 2$, and compare the results with the aforementioned exact analytic solution.

7.11. Heat and mass transfer in a porous spherical catalyst is governed by the following two nonlinear ordinary differential equations.

$$\frac{d^2C}{dR^2} + \frac{2}{R}\frac{dC}{dR} = \phi^2 C \exp\left(\frac{T}{1+\frac{T}{\gamma}}\right) \quad \text{in } 0 < R < 1$$

$$\frac{d^2T}{dR^2} + \frac{2}{R}\frac{dT}{dR} = -\gamma_\beta \phi^2 C \exp\left(\frac{T}{1+\frac{T}{\gamma}}\right) \quad \text{in } 0 < R < 1$$

subject to the boundary conditions

$$\frac{dC}{dR} = \frac{dT}{dR} = 0 \quad \text{at } R = 0$$

$$C = 1, T = 0 \quad \text{at } R = 1$$

Write the false transient form of this system and express the resulting equations in the finite difference form using: (i) simple explicit scheme and (ii) simple implicit scheme.

8

Multidimensional Incompressible Laminar Flow

The two-dimensional incompressible Navier–Stokes equations for constant property flow without body forces are a mixed set of elliptic–parabolic equations. By using the vorticity-stream function approach, these equations are transformed into a *vorticity transport equation* that is parabolic in time and a *stream function equation* that is an elliptic Poisson's equation. However, for three-dimensional problems, this approach is no longer simple, and the equations usually are solved for the primitive variables, that is, for the velocity components and pressure.

For an incompressible flow with constant properties and no body forces, the velocity and the temperature problems are uncoupled. Therefore, once the velocity components are determined from the solution of the equations of motion, they are used as input to the energy equation, which can be solved subject to appropriate thermal boundary and initial conditions.

In this chapter, we examine the finite difference representation of the two-dimensional Navier–Stokes equations for incompressible and laminar flow of a constant property fluid, including the energy conservation as well as the boundary layer equations.

The reader should consult standard references, such as Schlichting (1979), Kays and Crawford (1980), and Pletcher et al. (2012) for the development of boundary layer equations from the simplification of the Navier–Stokes equations, which are elliptic in the steady state. For simple geometries and flow conditions, similarity solutions are possible for the problems of velocity and temperature distribution in boundary layers; but a numerical method of solution is necessary when dealing with the complicated situations. For computational purposes, the advantage of parabolic boundary layer equations is the fact that the behavior of flow at any location is influenced only by the conditions on the upstream side of that location. As a result, computations are started from a known upstream condition and marched to successive downstream locations by calculating the conditions at one location at a time; but the model is not applicable in the region of flow reversal, such as separated flow.

8.1 Vorticity-Stream Function Formulation

We consider a laminar two-dimensional incompressible flow field for a constant property fluid, without considering the boundary layer simplifications.

Typical applications include isothermal flow in enclosures, flow near the leading and trailing ends of a flat plate, and the flow in a wake where no boundary layer simplifications are applicable. To compute the velocity and pressure distribution in the flow in such situations, the full Navier–Stokes equations are to be solved. The governing two-dimensional flow equations include the continuity and the momentum equations; in the rectangular coordinates, they are given by

$$\text{continuity: } \frac{\partial u}{\partial x} + \frac{\partial v}{\partial y} = 0 \tag{8.1}$$

$$\text{x-momentum: } \frac{\partial u}{\partial t} + u\frac{\partial u}{\partial x} + v\frac{\partial u}{\partial y} = -\frac{1}{\rho}\frac{\partial p}{\partial x} + \upsilon\left(\frac{\partial^2 u}{\partial x^2} + \frac{\partial^2 u}{\partial y^2}\right) \tag{8.2}$$

$$\text{y-momentum: } \frac{\partial v}{\partial t} + u\frac{\partial v}{\partial x} + v\frac{\partial v}{\partial y} = -\frac{1}{\rho}\frac{\partial p}{\partial y} + \upsilon\left(\frac{\partial^2 v}{\partial x^2} + \frac{\partial^2 v}{\partial y^2}\right) \tag{8.3}$$

where u and v are the velocity components in the x and y directions, respectively; υ is the kinematic viscosity, ρ is the density, and p is the pressure. For the constant property, incompressible flow with no body forces, the equations of motion are uncoupled from the energy equation. Therefore, they can be solved separately for the unknowns u, v, and p.

For two-dimensional cases, the most successful numerical technique for solving such a system is based on the vorticity-stream function formulation that will be discussed next. However, a single-stream function does not exist for three-dimensional flow problems. For these cases, the numerical solution of the previous equations, subject to appropriate boundary conditions, is more adequate. This case will be discussed after the vorticity-stream function method.

8.1.1 Vorticity and Stream Function

This approach, commonly used for the solution of two-dimensional, constant property Navier–Stokes equations, is based on the transformation of the dependent variables from (u,v) to (ω,ψ), where ω is the *vorticity* and ψ is the *stream function*.

The vorticity ω is defined as

$$\omega = \nabla \times \mathbf{v} \tag{8.4}$$

where **v** is the velocity vector. For the two-dimensional x,y Cartesian coordinate system considered here, the magnitude of the vorticity vector is given by

$$\omega = \frac{\partial v}{\partial x} - \frac{\partial u}{\partial y} \tag{8.5}$$

and the stream function ψ is defined by

$$\frac{\partial \psi}{\partial y} = u, \quad \frac{\partial \psi}{\partial x} = -v \tag{8.6a,b}$$

with this definition of the stream function, the continuity equation (8.1) is identically satisfied. The transformation of the dependent variables from (u,v) to (ω,ψ) is achieved by eliminating the pressure term between the momentum equations (8.2) and (8.3). That is, equation (8.2) is differentiated with respect to y, equation (8.3) with respect to x, the results are subtracted, and the definitions of ω and ψ are applied to obtain the following equation for the vorticity, ω,

$$\frac{\partial \omega}{\partial t} + u\frac{\partial \omega}{\partial x} + v\frac{\partial \omega}{\partial y} = \upsilon\left(\frac{\partial^2 \omega}{\partial x^2} + \frac{\partial^2 \omega}{\partial y^2}\right) \tag{8.7a}$$

which is written in the conservative form as

$$\frac{\partial \omega}{\partial t} + \frac{\partial}{\partial x}(u\omega) + \frac{\partial}{\partial y}(v\omega) = \upsilon\left(\frac{\partial^2 \omega}{\partial x^2} + \frac{\partial^2 \omega}{\partial y^2}\right) \tag{8.7b}$$

The equivalence of these two equations becomes apparent if, in equation (8.7b), the convective terms are expanded and the continuity equation is utilized. Equation (8.7a) or (8.7b) is called the *vorticity transport equation*, which is parabolic in time.

An additional relationship is obtained by introducing equations (8.6a,b) into equation (8.5). We find

$$\frac{\partial^2 \psi}{\partial x^2} + \frac{\partial^2 \psi}{\partial y^2} = -\omega \tag{8.8}$$

which is an *elliptic Poisson's equation* for the stream function.

Thus, the two momentum equations (8.2) and (8.3) in the (u,v) variables are transformed to equations (8.7) and (8.8) in the (ω,ψ) variables. We note that the velocity components u and v in equations (8.7a,b) are related to the stream function ψ by equations (8.6a,b).

Thus, using the definition of vorticity and stream function, we transformed the mixed elliptic–parabolic two-dimensional incompressible constant property Navier–Stokes equations in the (u,v) variables to the vorticity transport equation, which is parabolic in time, and to an elliptic Poisson's equation in the (ω,ψ) variables. An additional equation for the pressure in the flow field is determined by manipulating the Navier–Stokes equations (8.1)–(8.3), as described next.

The differential equation for pressure is determined as follows. The x-momentum equation (8.2) is differentiated with respect to x, the y-momentum equation (8.3) with respect to y, the results are added, and the continuity equation (8.1) is utilized to obtain

$$\frac{\partial^2 p}{\partial x^2} + \frac{\partial^2 p}{\partial y^2} = 2\rho\left(\frac{\partial u}{\partial x}\frac{\partial v}{\partial y} - \frac{\partial u}{\partial y}\frac{\partial v}{\partial x}\right) \qquad (8.9)$$

This result is expressed in terms of the stream function as

$$\frac{\partial^2 p}{\partial x^2} + \frac{\partial^2 p}{\partial y^2} = s \qquad (8.10a)$$

where

$$s = 2\rho\left[\left(\frac{\partial^2 \psi}{\partial x^2}\right)\left(\frac{\partial^2 \psi}{\partial y^2}\right) - \left(\frac{\partial^2 \psi}{\partial x \partial y}\right)^2\right] \qquad (8.10b)$$

Thus, equations (8.10a,b) provides the relation for determining the pressure in the flow field when the stream function is known.

We now summarize the governing equations for the vorticity-stream function formulation. The vorticity transport equation for ω is given by

$$\frac{\partial \omega}{\partial t} + u\frac{\partial \omega}{\partial x} + v\frac{\partial \omega}{\partial y} = \upsilon\left(\frac{\partial^2 \omega}{\partial x^2} + \frac{\partial^2 \omega}{\partial y^2}\right) \qquad (8.11)$$

Poisson's equation for ψ by

$$\frac{\partial^2 \psi}{\partial x^2} + \frac{\partial^2 \psi}{\partial y^2} = -\omega \qquad (8.12)$$

and Poisson's equation for pressure p by

$$\frac{\partial^2 p}{\partial x^2} + \frac{\partial^2 p}{\partial y^2} = s \qquad (8.13a)$$

where

$$s = 2\rho\left[\left(\frac{\partial^2 \psi}{\partial x^2}\right)\left(\frac{\partial^2 \psi}{\partial y^2}\right) - \left(\frac{\partial^2 \psi}{\partial x \partial y}\right)^2\right] \qquad (8.13b)$$

These equations can be expressed in dimensionless form as

$$\frac{\partial \Omega}{\partial \tau} + U\frac{\partial \Omega}{\partial X} + V\frac{\partial \Omega}{\partial Y} = \frac{1}{Re}\left(\frac{\partial^2 \Omega}{\partial X^2} + \frac{\partial^2 \Omega}{\partial Y^2}\right) \qquad (8.14)$$

$$\frac{\partial^2 \Psi}{\partial X^2} + \frac{\partial^2 \Psi}{\partial Y^2} = -\Omega \qquad (8.15)$$

$$\frac{\partial^2 P}{\partial X^2} + \frac{\partial^2 P}{\partial Y^2} = S \qquad (8.16a)$$

where

$$S = 2\left[\left(\frac{\partial^2 \Psi}{\partial X^2}\right)\left(\frac{\partial^2 \Psi}{\partial Y^2}\right) - \left(\frac{\partial^2 \Psi}{\partial X \partial Y}\right)^2\right] \quad (8.16b)$$

and various dimensionless quantities are defined as

$$U = \frac{u}{u_o} \quad X = \frac{x}{L} \quad P = \frac{p}{\rho u_o^2}$$
$$V = \frac{v}{u_o} \quad Y = \frac{y}{L} \quad \tau = \frac{\mu_o t}{L} \quad (8.16\text{c–k})$$
$$\Omega = \frac{\omega L}{u_o} \quad \psi = \frac{\Psi}{u_o L} \quad \text{Re} = \frac{u_o L}{\upsilon}$$

where u_o is a reference velocity, L is a reference length, and $\text{Re} = u_o L/\upsilon$ is the Reynolds number.

The conservative form of the vorticity transport equation is given by

$$\frac{\partial \Omega}{\partial \tau} + \frac{\partial}{\partial X}(U\Omega) + \frac{\partial}{\partial Y}(V\Omega) = \frac{1}{\text{Re}}\left(\frac{\partial^2 \Omega}{\partial X^2} + \frac{\partial^2 \Omega}{\partial Y^2}\right) \quad (8.17)$$

We note that in the vorticity transport equation, the velocity components u and v can be expressed in terms of the stream function according to the definition of the stream function given by equations (8.6a,b).

8.1.2 Finite Difference Representation of Vorticity-Stream Function Formulation

The vorticity-stream function formulation described previously consists of three different partial differential equations for the dependent variables ω, ψ, and p. The vorticity transport equation for ω is parabolic in time, whereas equation (8.12) for ψ and equation (8.13) for p are Poisson's equations in which time enters merely as a parameter. To obtain the steady-state solutions, which are usually of practical interest, the equation for ω, which is parabolic in time, is solved together with the equation for ψ until the asymptotic state is reached. Once the steady-state values of ω and ψ are available, the pressure equation is solved.

For finite difference approximation of these equations, we adopt the following notation

$$F(x, y, t) = F(i\Delta x, j\Delta y, n\Delta t) \equiv F_{i,j}^n \quad (8.18)$$

where

$$F \equiv \omega, \psi, \text{ or } p \quad (8.19)$$

and present in the next section the discretization of the vorticity transport equation and Poisson's equation for the stream function and pressure.

8.1.2.1 Vorticity Transport Equation

Because this equation is parabolic in time and contains convective terms, several possibilities exist for its discretization. For example, explicit or implicit methods can be applied; central differencing or upwind differencing can be used to discretize the convection terms. For simplicity, we consider only the simple explicit scheme with forward differencing in time. The upwind differencing discretizes the convective terms and the central differencing discretizes the second derivatives with respect to the space variables. Then, the finite difference approximation of the vorticity transport equation (8.11) becomes

$$\frac{\omega_{i,j}^{n+1} - \omega_{i,j}^n}{\Delta t} = -u_{i,j}^n \left(\frac{\partial \omega}{\partial x}\right)_{i,j}^n - v_{i,j}^n \left(\frac{\partial \omega}{\partial y}\right)_{i,j}^n$$
$$+ \upsilon \left[\frac{\omega_{i+1,j}^n - 2\omega_{i,j}^n + \omega_{i-1,j}^n}{(\Delta x)^2} + \frac{\omega_{i,j+1}^n - 2\omega_{i,j}^n + \omega_{i,j-1}^n}{(\Delta y)^2}\right] \quad (8.20)$$

where the first derivatives in the convective terms are discretized with the upwind scheme as follows

$$\left(\frac{\partial \omega}{\partial x}\right)_{i,j}^n = \frac{\omega_{i,j}^n - \omega_{i-1,j}^n}{\Delta x} \quad \text{for } u_{i,j} > 0 \quad (8.21a)$$

$$\left(\frac{\partial \omega}{\partial x}\right)_{i,j}^n = \frac{\omega_{i+1,j}^n - \omega_{i,j}^n}{\Delta x} \quad \text{for } u_{i,j} < 0 \quad (8.21b)$$

and

$$\left(\frac{\partial \omega}{\partial y}\right)_{i,j}^n = \frac{\omega_{i,j}^n - \omega_{i,j-1}^n}{\Delta y} \quad \text{for } v_{i,j} > 0 \quad (8.21c)$$

$$\left(\frac{\partial \omega}{\partial y}\right)_{i,j}^n = \frac{\omega_{i,j+1}^n - \omega_{i,j}^n}{\Delta y} \quad \text{for } v_{i,j} < 0 \quad (8.21d)$$

The scheme has a truncation error $0(\Delta x, \Delta y, \Delta t)$. The velocities $u_{i,j}$ and $v_{i,j}$ are related to the stream function ψ by equations (8.6a,b), and in the finite difference form, they are given by

$$u_{i,j}^n = \frac{\psi_{i,j+1}^n - \psi_{i,j-1}^n}{2\Delta y}, \quad 0[(\Delta y)^2] \quad (8.22a)$$

$$-v_{i,j}^n = \frac{\psi_{i+1,j}^n - \psi_{i-1,j}^n}{2\Delta x}, \quad 0[(\Delta x)^2] \quad (8.22b)$$

If the dimensionless vorticity transport equation (8.14) was discretized, the resulting finite difference equation would be similar to that given by equation (8.20) with the following changes in the variables:

$$x \to X \quad u \to U \quad v \to \frac{1}{Re}$$
$$y \to Y \quad v \to V$$
$$t \to \tau \quad \omega \to \Omega$$

The central differencing can also be used to discretize all the space derivatives in the vorticity transport equation. For such a case, the first derivative terms $(\partial \omega/\partial x)_{i,j}^n$ and $(\partial \omega/\partial y)_{i,j}^n$ in equation (8.20) are discretized using the central difference formula. For such a case, the resulting finite difference equation has a second-order accurate truncation error in the space variables, that is, $0[(\Delta x)^2, (\Delta y)^2, \Delta t]$; but the stability constraint becomes more severe at a higher Reynolds number. Therefore, for a low Reynolds number, diffusion dominates and the central differencing can be used to discretize the convective terms because the scheme is more accurate. However, at a higher Reynolds number, convection dominates and the use of the upwind scheme is appropriate by stability considerations. The hybrid scheme, based on the grid Peclet (Pe) number, can be used so that when Pe tends toward zero, the central differencing scheme is recovered, whereas when Pe tends toward infinity, the upwind scheme is recovered. This will be discussed later in this chapter.

With the finite difference scheme considered here being explicit, the time step Δt must satisfy the stability constraint given by (Roache 1976)

$$\Delta t \leq \left\{ \frac{|u|}{\Delta x} + \frac{|v|}{\Delta y} + 2v \left[\frac{1}{(\Delta x)^2} + \frac{1}{(\Delta y)^2} \right] \right\}^{-1} \tag{8.23}$$

which is analogous to equation (6.18), with v replacing α.

8.1.2.2 Poisson's Equation for Stream Function

Poisson's equation (8.12) for the stream function ψ is readily discretized by using the central difference formula. We obtain

$$\frac{\psi_{i+1,j}^{n+1} - 2\psi_{i,j}^{n+1} + \psi_{i-1,j}^{n+1}}{(\Delta x)^2} + \frac{\psi_{i,j+1}^{n+1} - 2\psi_{i,j}^{n+1} + \psi_{i,j-1}^{n+1}}{(\Delta y)^2} = -\omega_{i,j}^{n+1} \tag{8.24a}$$

for the case of $\Delta x = \Delta y \equiv \delta$, this equation is solved for $\psi_{i,j}^{n+1}$ as

$$\psi_{i,j}^{n+1} = \frac{1}{4} [\psi_{i+1,j}^{n+1} + \psi_{i-1,j}^{n+1} + \psi_{i,j+1}^{n+1} + \psi_{i,j-1}^{n+1} + \delta^2 \omega_{i,j}^{n+1}] \tag{8.24b}$$

with a truncation error $0[(\Delta x)^2, (\Delta y)^2]$.

8.1.2.3 Poisson's Equation for Pressure

Poisson's equation (8.13) for pressure is discretized as

$$\frac{p_{i+1,j} - 2p_{i,j} + p_{i-1,j}}{(\Delta x)^2} + \frac{p_{i,j+1} - 2p_{i,j} + p_{i,j-1}}{(\Delta y)^2} = (s)_{i,j} \qquad (8.25a)$$

where

$$(s)_{i,j} = 2\rho_{i,j} \left[\left(\frac{\psi_{i+1,j} - 2\psi_{i,j} + \psi_{i-1,j}}{(\Delta x)^2} \right) \left(\frac{\psi_{i,j+1} - 2\psi_{i,j} + \psi_{i,j-1}}{(\Delta y)^2} \right) \right.$$
$$\left. - \left(\frac{\psi_{i+1,j+1} - \psi_{i+1,j-1} - \psi_{i-1,j+1} + \psi_{i-1,j-1}}{4\Delta x \Delta y} \right)^2 \right] \qquad (8.25b)$$

and the cross-derivative term appearing in s is discretized by using the second-order accurate formula given by equation (2.17b). Thus, the finite difference scheme has a truncation error $0[(\Delta x)^2, (\Delta y)^2]$.

For steady-state problems, the pressure equation is solved only once—after the steady-state values of ω and ψ have been computed; therefore, superscript n + 1 is omitted in equation (8.25a) for the pressure.

8.1.3 Method of Solution for ω and ψ

8.1.3.1 Solution for a Transient Problem

The basic steps for solving the finite difference equations (8.20) and (8.24) for a transient problem for the vorticity and the stream function are as follows.

Suppose the values of $\omega_{i,j}^n$ and $\psi_{i,j}^n$ are available at the time step n, then the velocity components $u_{i,j}^n$ and $v_{i,j}^n$ are available from equations (8.22a,b).

1. The vorticity $\omega_{i,j}^{n+1}$ at the time level (n + 1) is computed from equation (8.20).
2. Knowing $\omega_{i,j}^{n+1}$ at each grid point, Poisson's equation (8.24b) for the stream function, subject to appropriate boundary conditions, is solved iteratively and $\psi_{i,j}^{n+1}$ are determined at the grid points over the entire flow field.
3. The new values of the stream function $\psi_{i,j}^{n+1}$ are used to compute the corresponding velocity components $u_{i,j}$ and $v_{i,j}$ from equations (8.22a,b), and the latest values of $\omega_{i,j}$ and $\psi_{i,j}$ at the interior nodes are used to compute the new values of ω on the boundaries.

There are no definitive criterion for the convergence of the solution to the steady state. The usual criterion for convergence to the steady-state value of $\omega_{i,j}$ is to check the entire array for

$$\max_{i,j} \left| \omega_{i,j}^{n+1} - \omega_{i,j}^{n} \right| \leq \varepsilon \tag{8.26a}$$

or

$$\max_{i,j} \left| \frac{(\omega_{i,j}^{n+1} - \omega_{i,j}^{n})}{\omega_{i,j}^{n}} \right| \leq \varepsilon \tag{8.26b}$$

and similarly for $\psi_{i,j}$. Values of ε reported in the open literature have varied from $\varepsilon = 10^{-3}$ to 10^{-8}. Sometimes it is useful to examine the iterative behavior of the solution by plotting $\max_{i,j} |\omega_{i,j}^{n+1} - \omega_{i,j}^{n}|$ against n or time.

8.1.3.2 Solution for a Steady-State Problem

If only the steady-state solution is of interest, instead of marching the solution in time until it becomes invariant as described previously, the finite difference equations of the steady-state problem can be solved. We consider the dimensionless vorticity transport equation (8.14) and Poisson's equation (8.15) for the stream function. At steady state, the problem is obtained by dropping the time derivative term in the vorticity transport equation (8.14).

$$U \frac{\partial \Omega}{\partial X} + V \frac{\partial \Omega}{\partial Y} - \frac{1}{Re} \left(\frac{\partial^2 \Omega}{\partial X^2} + \frac{\partial^2 \Omega}{\partial Y^2} \right) = 0 \tag{8.27}$$

$$\frac{\partial^2 \Psi}{\partial X^2} + \frac{\partial^2 \Psi}{\partial Y^2} = -\Omega \tag{8.28}$$

These equations can be expressed in the finite difference form as discussed previously. The resulting system of algebraic equations can be solved with the methods presented in Chapter 3.

Here, we present the basic algorithm for solving such a system of algebraic equations. We adopt a finite difference notation

$$\phi^m(X, Y) = \phi^m(i\Delta X, j\Delta Y) \equiv \phi_{i,j}^m \tag{8.29a}$$

where superscript m denotes the m-th *iterate* and subscripts (i,j) denote the discrete grid points. In addition, we let

$$f_{i,j} = \text{the boundary values of } \Psi_{i,j} \tag{8.29b}$$

$$F_{i,j} = \text{the boundary values of } \Omega_{i,j} \tag{8.29c}$$

and assume that the quantities $\psi_{i,j}$ and $\Omega_{i,j}$ are known at the m-th iterate, that is

$$\Psi_{i,j}^m, \Omega_{i,j}^m = \text{known}$$

and $\Psi_{i,j}^{m+1}, \Omega_{i,j}^{m+1}$ are to be determined.

We also introduce the notation

$$\tilde{\Psi}_{i,j}^{m+1}, \tilde{\Omega}_{i,j}^{m+1}$$

where tilde denotes the approximate intermediate value. The basic steps for determining $\Psi_{i,j}^{m+1}$ and $\Omega_{i,j}^{m+1}$ are as follows.

1. The intermediate values $\tilde{\Psi}_{i,j}^{m+1}$ are computed from the solution of the finite difference form of Poisson's equation (8.28). Let the finite difference form of equation (8.28) be written formally as

$$\left(\frac{\partial^2 \tilde{\Psi}}{\partial X^2} + \frac{\partial^2 \tilde{\Psi}}{\partial Y^2}\right)_{i,j}^{m+1} = -\Omega_{i,j}^m \quad \text{in the region} \quad (8.30a)$$

$$\tilde{\Psi}_{i,j}^{m+1} = f_{i,j} \quad \text{on the boundaries} \quad (8.30b)$$

2. The final values $\Psi_{i,j}^m$ are determined by using the relaxation formula

$$\Psi_{i,j}^{m+1} = \alpha \tilde{\Psi}_{i,j}^{m+1} + (1-\alpha)\Psi_{i,j}^m \quad (8.31)$$

where the relaxation parameter α has values $0 < \alpha \leq 1$.

3. The intermediate values $\tilde{\Omega}_{i,j}^{m+1}$ are computed from the finite difference form of equation (8.27) given by

$$\left[U\frac{\partial \tilde{\Omega}}{\partial X} + V\frac{\partial \tilde{\Omega}}{\partial Y} - \frac{1}{Re}\left(\frac{\partial^2 \tilde{\Omega}}{\partial X^2} + \frac{\partial^2 \tilde{\Omega}}{\partial Y^2}\right)\right]_{i,j}^{m+1} = 0 \quad \text{in the region} \quad (8.32a)$$

$$\tilde{\Omega}_{i,j}^{m+1} = F_{i,j} \quad \text{on the boundaries} \quad (8.32b)$$

where the velocity components U and V are related to the stream function by equations (8.22a,b).

4. The final values $\Omega_{i,j}^{m+1}$ are determined by the following relaxation formulae:

$$\Omega_{i,j}^{m+1} = \beta \tilde{\Omega}_{i,j}^{m+1} + (1-\beta)\Omega_{i,j}^m \quad \text{in the region} \quad (8.33)$$

$$\Omega_{i,j}^{m+1} = \gamma \tilde{\Omega}_{i,j}^{m+1} + (1-\gamma)\Omega_{i,j}^m \quad \text{on the boundaries} \tag{8.34}$$

where the relaxation parameters have values $0 < \beta, \gamma \leq 1$.

The iterations are continued until desired convergence criteria are achieved. A simple convergence criteria can be taken as

$$\max_{i,j} |\Psi_{i,j}^{m+1} - \Psi_{i,j}^m| \leq \varepsilon_\psi \tag{8.35}$$

$$\max_{i,j} |\Omega_{i,j}^{m+1} - \Omega_{i,j}^m| \leq \varepsilon_\omega \tag{8.36}$$

8.1.4 Method of Solution for Pressure

Once the stream function ψ is available over the grid points, the source term s in the pressure equation (8.13) is considered known. Then, the finite difference equation (8.25) can be solved by any of the iterative methods applicable for the solution of Poisson's equation, and pressures $p_{i,j}$ are determined at the grid points over the entire flow field. The boundary conditions for the pressure equation are generally of the second kind (i.e., $\partial p / \partial \mathbf{n}$ is prescribed over the boundary). If the pressure is needed only for the steady-state condition, the pressure equation is solved only once by using the steady-state values of ω and ψ.

If the pressure is required only at the walls, there is no need to solve the pressure equation over the entire flow field. In such cases, a simple expression can be developed for determining the wall pressure by the application of the tangential momentum equation to the fluid adjacent to the wall surface. To illustrate the approach, we consider a wall aligned with the x axis as shown by Figure 8.1.

The x-momentum equation (8.2) is applied to the fluid adjacent to the wall. We find

$$\left.\frac{\partial p}{\partial x}\right|_{y=0} = \mu \left.\frac{\partial^2 u}{\partial y^2}\right|_{y=0} \tag{8.37}$$

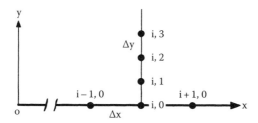

FIGURE 8.1
Notation for grid points in the fluid and on the wall aligned along the x axis.

We note that $u = v = 0$; hence $\partial u/\partial x = \partial v/\partial x = 0$ at the wall surface. When this result is applied to the definition of vorticity, $\omega = \partial v/\partial x - \partial u/\partial y$, we obtain $\omega = -\partial u/\partial y$. Then equation (8.37) is written in terms of vorticity as

$$\left.\frac{\partial p}{\partial x}\right|_{y=0} = -\mu \left.\frac{\partial \omega}{\partial y}\right|_{y=0} \tag{8.38}$$

This equation is discretized by using the central difference formula for the x derivative and the one-sided three-point formula given by equation (2.11a) for the y-derivative. We find

$$\frac{P_{i+1,0} - P_{i-1,0}}{2\Delta x} = -\mu \frac{-3\omega_{i,0} + 4\omega_{i,1} - \omega_{i,2}}{2\Delta y} \tag{8.39}$$

The vorticities on the right-hand side of this expression are known from the solution of the vorticity transport equation. Then the wall pressure can be calculated from equation (8.39). However, when all boundary conditions on pressure are of the second kind, as is the case of a container where velocities are imposed on the boundary, at least one pressure value must be specified or the pressure will be obtainable only to an arbitrary additive constant.

8.1.5 Treatment of Boundary Conditions

To solve the finite difference equations considered previously, boundary conditions are needed for the computational quantities ω and ψ, as well as for the pressure p. So far, we have presented the finite difference equations applicable in the flow field, but we did not examine the boundary conditions. A discussion of various approaches on the development of the computational boundary conditions is given by Roache (1976). Here, we present boundary conditions on velocity components and discuss the development of corresponding computational boundary conditions on ψ and ω.

8.1.5.1 Boundary Conditions on Velocity

The boundary conditions on velocity depend on the actual physical situations in the flow field.

For flow over a stationary impermeable wall, they are taken as

$$u = v = 0 \text{ at the wall surface} \tag{8.40}$$

If the wall is moving with a given velocity or porous with suction or injection of fluid through the pores, the velocity components are specified accordingly.

For an axis of symmetry or centerline of the flow aligned along the x axis, v is zero everywhere along the centerline; hence, $\partial v/\partial x = 0$. Also, with u being symmetric above and below the centerline, we have $\partial u/\partial y = 0$.

Different flow conditions can be specified for the upstream (i.e., inflow) and the outflow boundaries. Katnasis (1967) specified a uniform inflow velocity u

Multidimensional Incompressible Laminar Flow

and set v = 0. Thoman and Szewczyk (1966) used a less restrictive inflow condition by requiring v = 0 and allowing u to develop as a part of the solution.

8.1.5.2 Boundary Conditions on ψ

The stream function is a computational quantity, which satisfies the Poisson equation (8.12) for ψ, that is,

$$\frac{\partial^2 \psi}{\partial x^2} + \frac{\partial^2 \psi}{\partial y^2} = -\omega \qquad (8.41)$$

Therefore, the boundary conditions on stream function should be consistently simulated with the physical reality by utilizing the definition of ψ given by equations (8.6a,b). The specification of proper boundary conditions is extremely important because it is the boundary and initial conditions for the differential equation together with the flow parameters that distinguishes the flow field. There are numerous physical situations such as the *impermeable wall, no-slip at the wall, symmetry boundary*, and *upstream (inflow)* condition that require special attention. We will examine the simulation of such physical situations as boundary conditions on the stream function. The *downstream (outflow)* boundary condition will be presented together with the boundary conditions for the vorticity.

1. *Impermeable no-slip wall aligned with the x axis*: For such a situation, the velocity components at the wall vanish, that is

$$v = 0 \text{ and } u = 0 \text{ at the wall} \qquad (8.42)$$

   The first condition in terms of the stream function is given by

$$\frac{\partial \psi(x,0)}{\partial x} = -v(x,0) = 0 \qquad (8.43a)$$

   which implies that

$$\psi(x,0) = \text{constant} \qquad (8.43b)$$

   along the wall surface and represents an *impermeable* wall condition. The constant can be set equal to zero with no loss of generality.

   The second condition, u = 0, at the wall, in terms of ψ, becomes

$$\frac{\partial \psi(x,0)}{\partial y} = u(x,0) = 0 \qquad (8.43c)$$

   which represents *no-slip* condition at the wall.

   Either of the conditions given by equation (8.43b) or (8.43c) could be used as boundary condition on ψ in Poisson's equation (8.41); but both conditions, ψ = 0 and ∂ψ/∂y = 0, cannot be used along the same boundary because that would overspecify the problem. The condition

$(\partial \psi / \partial y)_{wall} = 0$ is required for boundary condition on vorticity ω; then the condition $\psi_{wall} = 0$ shall be used for Poisson's equation for ψ. This is the only correct distribution of these boundary conditions (Roache 1976).

2. *Symmetry boundary aligned with the x axis*: The normal velocity v is zero everywhere along the symmetry boundary. In terms of ψ we have

$$\frac{\partial \psi}{\partial x} = -v = 0 \tag{8.44a}$$

which implies that

$$\psi(x,0) = \text{constant} \tag{8.44b}$$

everywhere along the symmetry boundary.

3. *Inflow (upstream) boundary aligned with the y axis with a constant velocity u_0*: The inflow velocity u_0 in terms of the stream function is given by

$$\frac{\partial \psi(0,y)}{\partial y} = u_0 = \text{constant} \tag{8.45}$$

at $x = 0$ (i.e., upstream). Integrating equation (8.45) with respect to y and setting the integration constant equal to zero fixes the inflow condition on ψ as

$$\psi(0,y) = u_0 y \tag{8.46}$$

8.1.5.3 Boundary Condition on ω

The spatial and timewise variation of vorticity over the flow field is governed by the vorticity transport equation (8.11). The vorticity is produced at no-slip boundaries. It is the diffusion and subsequent convection of this wall-produced vorticity that actually drives the problem (Roache 1976).

For a wall aligned with the x axis, with u(x,y) and v(x,y) denoting the velocity components in the x and y directions, respectively, the vorticity $\omega(x,y)$ is defined by

$$\omega(x,y) = \frac{\partial v(x,y)}{\partial x} - \frac{\partial u(x,y)}{\partial y} \tag{8.47a}$$

where the velocity components are related to the stream function by

$$u(x,y) = \frac{\partial \psi(x,y)}{\partial y}, \quad -v(x,y) = \frac{\partial \psi(x,y)}{\partial x}. \tag{8.47b,c}$$

The symmetry, no-slip, impermeable, and moving wall conditions can be obtained readily by simulating the analytic boundary conditions; but for the upstream (inflow), downstream (outflow), and sharp corner boundaries, a variety of computational forms can be used. A review of such boundary conditions is given by Roache (1976). Here, we first discuss some of the possibilities for the upstream, downstream, and sharp corner boundary conditions and then present the development of boundary conditions for symmetry and a moving wall.

1. *Inflow (upstream) boundary*: Some of the approaches used to simulate the upstream boundary condition include:
   a. A complete specification of the inflow, for example, by assuming a fully developed Poiseuille flow that fixes both ψ and ω. Pao and Daugherty (1969) specified $\omega = 0$ and set $\partial \psi / \partial y = u_0$, thus fixing ψ.
   b. Greenspan (1969) fixed ψ, assumed $\partial v / \partial x = 0$, thus obtained $\omega = -\partial^2 \psi / \partial y^2$.

2. *Outflow (downstream) boundary*: If the test section is long and the flow is fully developed, the boundary conditions can be simply imposed setting $v = 0$ and $\partial u / \partial x = 0$ at the exit. However, for short distances between outflow and inflow, numerical computations show that instabilities may propagate from the outflow to the inflow. Some of the approaches used to simulate the outflow boundary condition include:
   a. Complete specification of the outflow conditions. This is safest from the stability point of view but is not suitable for separated flows. For example, Katnasis (1967) specified uniform outflow (and inflow) velocity with $u =$ constant and $v = 0$.
   b. Paris and Whitaker (1965) specified less restrictive boundary conditions by setting $v = -\dfrac{\partial \psi}{\partial x} = 0$ and $\dfrac{\partial \omega}{\partial x} = 0$.
   c. Thoman and Szewczyk (1966) used even less restrictive boundary conditions for the outflow by setting $\dfrac{\partial v}{\partial x} = 0$ and $\dfrac{\partial \omega}{\partial x} = 0$. Clearly, the first condition implies $\partial^2 \psi / \partial x^2 = 0$ because $v = \partial \psi / \partial x$.

3. *Sharp corners*: When examining the boundary conditions on the stream function and vorticity at a sharp corner, the cases of a sharp concave corner and a sharp convex corner as illustrated in Figure 8.2a,b should be distinguished.

   For a *sharp concave corner* C_1 shown in Figure 8.2a, both $\psi_c = 0$ and $\omega_c = 0$, regardless of whether both boundaries are no-slip walls or any one of them is a line of symmetry.

   In the case of a *sharp convex corner* C_2 shown in Figure 8.2b, the stream function is $\psi_c = 0$ or constant; but, several different ways of evaluating vorticity ω_c have been reported by Roache (1976):

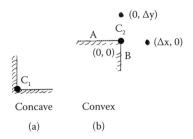

FIGURE 8.2
Sharp corners: (a) concave and (b) convex.

a. Vorticity at C_2 is assumed to be discontinuous, such that

$$\omega_c \equiv \omega_A = -2\psi(0,\Delta y)/(\Delta y)^2 \text{ as applied to wall A} \quad (8.48a)$$

$$\omega_c \equiv \omega_B = -2\psi(\Delta x,0)/(\Delta x)^2 \text{ as applied to wall B} \quad (8.48b)$$

b. Vorticity at C_2 is taken as arithmetic average of the two vorticities given by equations (8.48a,b), that is,

$$-\omega_c = \frac{\psi(0,\Delta y)}{(\Delta y)^2} + \frac{\psi(\Delta x,0)}{(\Delta x)^2} \quad (8.48c)$$

4. *Symmetry boundary aligned with the x axis*: The velocity component v being zero everywhere along the symmetry boundary, we have $\partial v/\partial x = 0$. The velocity component u being symmetric about the symmetry axis, we have $\partial u/\partial y = 0$. Then, by the definition of vorticity given by equation (8.47a), we have

$$\omega = 0 \text{ along the symmetry boundary} \quad (8.49)$$

Example 8.1

Develop a vorticity boundary condition for an impermeable wall moving with a constant velocity u_0 in the x direction.

Solution

For an impermeable wall, we have $v(x,0) = 0$; hence, $\partial v(x,0)/\partial x = 0$. Then, from the definition of vorticity given by equation (8.47a), we write

$$\omega(x,0) = -\frac{\partial u(x,0)}{\partial y} \quad (a)$$

or, in terms of the stream function,

$$\omega(x,0) = -\frac{\partial^2 \psi(x,0)}{\partial y^2} \quad (b)$$

To evaluate $\partial^2\psi(x,0)/\partial y^2$, we expand $\psi(x,\Delta y)$ about the point $(x,0)$ in a Taylor series

$$\psi(x, \Delta y) = \psi(x, 0) + \Delta y \frac{\partial \psi(x, 0)}{\partial y} + \frac{1}{2}(\Delta y)^2 \frac{\partial^2 \psi(x, 0)}{\partial y^2} + \cdots \qquad (c)$$

which is rearranged as

$$\frac{\partial^2 \psi(x, 0)}{\partial y^2} = \frac{2}{(\Delta y)^2}[\psi(x, \Delta y) - \psi(x, 0) - \Delta y\, u_0] \qquad (d)$$

because $\dfrac{\partial \psi(x, 0)}{\partial y} = u_0$.

Introducing equation (d) into equation (b), the wall vorticity $\omega(x,0)$ is determined as

$$\omega(x, 0) = -\frac{2}{(\Delta y)^2}[\psi(x, \Delta y) - \psi(x, 0) - \Delta y\, u_0] \qquad (e)$$

Using the finite difference notation shown in Figure 8.1, equation (e) becomes

$$\omega_{i,0} = -\frac{2}{(\Delta y)^2}(\psi_{i,1} - \psi_{i,0} - \Delta y\, u_0) \qquad (f)$$

For a stationary wall, this result simplifies to

$$\omega_{i,0} = -\frac{2}{(\Delta y)^2}(\psi_{i,1} - \psi_{i,0}) \qquad (g)$$

This example also shows that the value of ψ on the wall is not sufficient to determine wall vorticity; the information on $\partial\psi/\partial y$ at the wall is also needed.

8.1.5.4 Boundary Conditions on Pressure

The boundary conditions on pressure are determined by evaluating the steady-state momentum equations (8.2) and (8.3), at the boundary surfaces. The normal gradient of pressure $\partial p/\partial n$ is prescribed at the wall for all boundary surfaces. However, Poisson's equation for pressure with a second-kind boundary condition at all surfaces is unique only to within an additive constant; therefore, at least one value of pressure is needed in order to determine the pressure distribution over the solution domain.

8.1.5.5 Initial Condition

Because the vorticity transport equation (8.11) is time dependent, the initial condition needs to be specified for ω. However, for problems in which the steady-state solution is of interest, the initial condition becomes arbitrary. The ψ values at each time are utilized to determine the velocity components u and v, which are then introduced into the vorticity transport equation for use in the next time step.

8.1.6 Energy Equation

For incompressible flows of constant property fluids, once the velocity field is known from the solution of the equations of motion, the temperature field is determined from the solution of the energy equation. Here, we consider the unsteady energy equation for a two-dimensional, constant property, incompressible flow with no body forces given in the nonconservative form as (Schlichting 1979):

$$\rho C_p \left[\frac{\partial T}{\partial t} + u \frac{\partial T}{\partial x} + v \frac{\partial T}{\partial y} \right] = k \left(\frac{\partial^2 T}{\partial x^2} + \frac{\partial^2 T}{\partial y^2} \right) + \mu \Phi \quad (8.50a)$$

where Φ is the dissipation function defined by

$$\Phi = 2 \left[\left(\frac{\partial u}{\partial x} \right)^2 + \left(\frac{\partial v}{\partial y} \right)^2 \right] + \left(\frac{\partial v}{\partial x} + \frac{\partial u}{\partial y} \right)^2 \quad (8.50b)$$

These equations can be expressed in the dimensionless form as

$$\frac{\partial \Theta}{\partial \tau} + U \frac{\partial \Theta}{\partial X} + V \frac{\partial \Theta}{\partial Y} = \frac{1}{Pe} \left(\frac{\partial^2 \Theta}{\partial X^2} + \frac{\partial^2 \Theta}{\partial Y^2} \right) + \frac{Ec}{Re} \Phi^* \quad (8.51)$$

where

$$\Phi^* = 2 \left[\left(\frac{\partial U}{\partial X} \right)^2 + \left(\frac{\partial V}{\partial Y} \right)^2 \right] + \left(\frac{\partial V}{\partial X} + \frac{\partial U}{\partial Y} \right)^2 \quad (8.52)$$

and various dimensionless quantities are defined by

$$U = \frac{u}{u_0} \quad X = \frac{x}{L} \quad \Theta = \frac{T}{\Delta T_0} \quad \tau = \frac{u_0 t}{L}$$

$$V = \frac{v}{u_0} \quad Y = \frac{y}{L} \quad Re = \frac{\rho u_0 L}{\mu} \quad (8.53)$$

$$Pe = Re \, Pr = \left(\frac{\rho u_0 L}{\mu} \right) \left(\frac{C_p \mu}{k} \right) = \text{Peclet number}$$

$$Ec = \frac{u_0^2}{C_p \Delta T_0} = \text{Eckert number}$$

where
ΔT_0 = a reference, positive temperature difference
L = a characteristic length
u_0 = a characteristic velocity

The conservative form of equation (8.51) is given by

$$\frac{\partial \Theta}{\partial \tau} + \frac{\partial}{\partial X}(U\Theta) + \frac{\partial}{\partial Y}(V\Theta) = \frac{1}{Pe} \left(\frac{\partial^2 \Theta}{\partial X^2} + \frac{\partial^2 \Theta}{\partial Y^2} \right) + \frac{E}{Re} \Phi^* \quad (8.54)$$

where Φ^* is given by equation (8.52).

A comparison of the energy equation (8.51) with the vorticity transport equation (8.14) reveals that, except for the dissipation term, the two equations are of the same form, with Θ replacing Ω and Pe replacing Re. Therefore, all the finite difference expressions developed previously for the vorticity transport equation are applicable for the energy equation considered here. Furthermore, as the dissipation function Φ^* does not depend on temperature, the stability analysis developed for the vorticity transport equation also remains unaffected.

With the flow problem being decoupled from the temperature problem, the velocity components U and V determined from the flow problem become merely inputs into the energy equation. Then the energy equation can be solved for a variety of thermal boundary conditions and for different values of Pe and Ec.

We recall that the magnitude of the Reynolds number influenced the computational method to be used for the solution of the vorticity transport equation. In the case of the energy equation, the Peclet number plays the same role as the Reynolds number for the vorticity transport equation. In the case of gases, $Pr \cong 1$; hence $Pe = Re \cdot Pr \cong Re$. Then, the same method can be applicable for the solution of both the vorticity transport and the energy equations.

For oils, $Pr \gg 1$, whereas for liquid metals, $Pr \ll 1$; then, for such cases, different computational methods need to be considered for the solution of the vorticity transport and the energy equations for a given Reynolds number. Because the energy equation is linear and does not require the solution of Poisson's equation for the stream function at each time step, it requires much less computation time than the solution of the vorticity transport equation.

Finally, the energy equation given previously, without the dissipation term, can represent a mass diffusion problem if the temperature Θ is replaced by mass concentration C and the Peclet number Pe is replaced by the Schmidt number Sc. The finite difference solution of the energy equation with a known velocity field was examined in Section 6.1.2 and will not be repeated here.

8.2 Primitive Variables Formulation

When the problem is no longer two-dimensional, the application of the vorticity-stream function method loses the simplicity inherent in the two-dimensional case examined previously in this chapter. In this case, it is more appropriate to solve the original equations (8.1)–(8.3), formulated in terms of primitive variables. Consider the continuity, x- and y-momentum, and energy equations rewritten here in their conservative forms:

$$\text{Continuity: } \frac{\partial(\rho u)}{\partial x} + \frac{\partial(\rho v)}{\partial y} = 0 \qquad (8.55)$$

x-Momentum: $\dfrac{\partial(\rho u)}{\partial t} + \dfrac{\partial(\rho u u)}{\partial x} + \dfrac{\partial(\rho v u)}{\partial y} = -\dfrac{\partial p}{\partial x} + \mu\left(\dfrac{\partial^2 u}{\partial x^2} + \dfrac{\partial^2 u}{\partial y^2}\right)$ (8.56)

y-Momentum: $\dfrac{\partial(\rho v)}{\partial t} + \dfrac{\partial(\rho u v)}{\partial x} + \dfrac{\partial(\rho v v)}{\partial y} = -\dfrac{\partial p}{\partial y} + \mu\left(\dfrac{\partial^2 v}{\partial x^2} + \dfrac{\partial^2 v}{\partial y^2}\right)$ (8.57)

Energy: $\dfrac{\partial(\rho T)}{\partial t} + \dfrac{\partial(\rho u T)}{\partial x} + \dfrac{\partial(\rho v T)}{\partial y} = \dfrac{k}{C_p}\left(\dfrac{\partial^2 T}{\partial x^2} + \dfrac{\partial^2 T}{\partial y^2}\right) + g$ (8.58)

where the viscous dissipation term in the energy equation was neglected and μ is the absolute viscosity. Also, for writing the energy, x- and y-momentum conservation equations in conservative form, we used the continuity equation. These three equations can be written in the following general form

$$\frac{\partial}{\partial t}(\rho\phi) + \nabla \cdot (\rho\phi\mathbf{V}) = \nabla \cdot (\Gamma^\phi \nabla\phi) + S^\phi \qquad (8.59a)$$

or, for a two-dimensional case with constant Γ^ϕ,

$$\frac{\partial(\rho\phi)}{\partial t} + \frac{\partial(\rho u \phi)}{\partial x} + \frac{\partial(\rho v \phi)}{\partial y} = \Gamma^\phi\left(\frac{\partial^2 \phi}{\partial x^2} + \frac{\partial^2 \phi}{\partial y^2}\right) + S^\phi \qquad (8.59b)$$

In this equation, ϕ is a scalar that represents the quantity to be conserved, Γ^ϕ is the diffusion coefficient, and S^ϕ is the source term. Table 8.1 shows these variables for each equation considered. For discretizing these equations in this section, we will use the finite volume technique. The basic idea is to integrate these equations in the control volume around a point P represented in Figure 8.3. In this figure, the capital letters W, E, N, and S represent the center of the control volumes next to P in the west, east, north, and south directions, respectively. This figure also shows, in lowercase letters, the interfaces w, e, n, and s between P and the west, east, north, and south control volumes, respectively. In the finite volume method, discretized equations are derived for the

TABLE 8.1

Conserved Quantity, Diffusion Coefficient, and Source Term

Equation	ϕ	Γ^ϕ	S^ϕ
Continuity	1	0	0
x-momentum	u	μ	$-\dfrac{\partial p}{\partial x}$
y-momentum	v	μ	$-\dfrac{\partial p}{\partial y}$
Energy	T	k/C_p	0

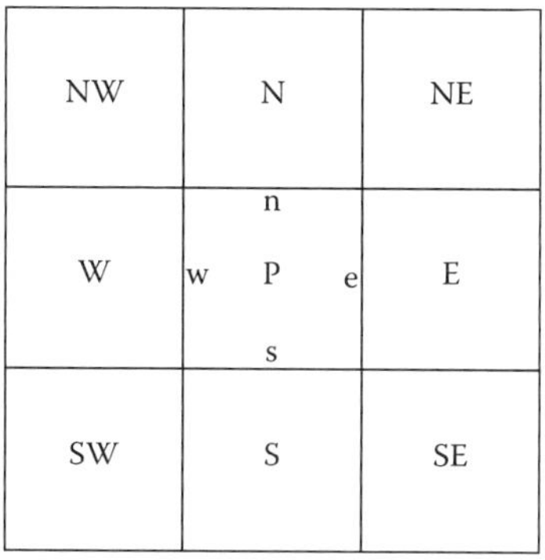

FIGURE 8.3
Control volume for P and its neighbors.

control volumes (capital letters) by integrating equation (8.59b) from time t to t + Δt in the control volume around P:

$$\int_{t}^{t+\Delta t} \int_{w}^{e} \int_{s}^{n} \left[\frac{\partial(\rho\phi)}{\partial t} + \frac{\partial(\rho u\phi)}{\partial x} + \frac{\partial(\rho v\phi)}{\partial y} \right] dy\,dx\,dt$$

$$= \int_{t}^{t+\Delta t} \int_{w}^{e} \int_{s}^{n} \left[\Gamma^{\phi} \left(\frac{\partial^2 \phi}{\partial x^2} + \frac{\partial^2 \phi}{\partial y^2} \right) + S^{\phi} \right] dy\,dx\,dt \quad (8.60)$$

We then obtain

$$[\rho\phi]_{t}^{t+\Delta t} \Delta y \Delta x + [\rho u\phi]_{w}^{e} \Delta y \Delta t + [\rho v\phi]_{s}^{n} \Delta x \Delta t$$

$$= \left[\Gamma^{\phi} \frac{\partial \phi}{\partial x} \right]_{w}^{e} \Delta y \Delta t + \left[\Gamma^{\phi} \frac{\partial \phi}{\partial y} \right]_{s}^{n} \Delta x \Delta t + S^{\phi} \Big|_{P} \Delta y \Delta x \Delta t \quad (8.61)$$

We can now define the following variables, which represent the mass within the control volume P

$$M_P = \rho \Delta x \Delta y \quad (8.62)$$

and the mass flow rate through the four boundaries of the control volume

$$\dot{M}_e = [\rho u]_e \Delta y \quad (8.63)$$

$$\dot{M}_w = [\rho u]_w \Delta y \quad (8.64)$$

$$\dot{M}_n = [\rho v]_n \Delta x \quad (8.65)$$

$$\dot{M}_s = [\rho v]_s \Delta x \quad (8.66)$$

We can also define the following coefficients, which will be part of the diffusive terms of the conservation equations

$$D_{11} = \Gamma^{\phi} \Delta y \quad (8.67)$$

$$D_{22} = \Gamma^\phi \Delta x \tag{8.68}$$

Replacing equations (8.62)–(8.68) into equation (8.61) and using a simple implicit time discretization we obtain

$$M_P \frac{(\phi_P^{m+1} - \phi_P^m)}{\Delta t} + \dot{M}_e \phi_e^{m+1} - \dot{M}_w \phi_w^{m+1} + \dot{M}_n \phi_n^{m+1} - \dot{M}_s \phi_s^{m+1}$$
$$= \left[D_{11} \frac{\partial \phi^{m+1}}{\partial x}\right]_e - \left[D_{11} \frac{\partial \phi^{m+1}}{\partial x}\right]_w + \left[D_{22} \frac{\partial \phi^{m+1}}{\partial y}\right]_n - \left[D_{22} \frac{\partial \phi^{m+1}}{\partial y}\right]_s + S^\phi|_P \Delta y \Delta x$$
$$\tag{8.69}$$

where the superscripts m and m + 1 refer to the quantities evaluated at times t and t + Δt, respectively.

From equation (8.69), it is possible to verify that the scalar ϕ, as well as its derivatives, must be calculated at the boundaries of the control volumes (locations indicated by lowercase letters w, e, n, and s). Thus, some interpolation functions are necessary because these quantities are only known at the centers of the control volumes.

In this chapter, we will use the same interpolation functions introduced in Chapter 4 for the weighted upstream differencing scheme (WUDS) method, where the value of ϕ and its derivative at the interface of the control volume are written, using the east face as an example, as

$$\phi_e = \left(\frac{1}{2} + \alpha_e\right)\phi_P + \left(\frac{1}{2} - \alpha_e\right)\phi_E \tag{8.70}$$

$$\left.\frac{\partial \phi}{\partial x}\right|_e = \beta_e \left(\frac{\phi_E - \phi_P}{\Delta x}\right) \tag{8.71}$$

The following coefficients, already presented in Chapters 4 and 6, are repeated here for the sake of convenience (for the east face):

$$\alpha_e = \frac{Pe_e^2}{10 + 2Pe_e^2} \operatorname{sign}(Pe_e) \tag{8.72}$$

$$\beta_e = \frac{1 + 0.005 Pe_e^2}{1 + 0.05 Pe_e^2} \tag{8.73}$$

$$Pe_e = \frac{\dot{M}_e}{D_{11}} \Delta x \tag{8.74}$$

Multidimensional Incompressible Laminar Flow

Applying equations (8.70) and (8.71) and the corresponding equations for the west, north, and south interfaces to equation (8.69), we obtain

$$M_P \frac{(\phi_P^{m+1} - \phi_P^m)}{\Delta t} + \dot{M}_e \left[\left(\frac{1}{2} + \alpha_e\right) \phi_P^{m+1} + \left(\frac{1}{2} - \alpha_e\right) \phi_E^{m+1} \right]$$
$$- \dot{M}_w \left[\left(\frac{1}{2} + \alpha_w\right) \phi_P^{m+1} + \left(\frac{1}{2} - \alpha_w\right) \phi_W^{m+1} \right]$$
$$+ \dot{M}_n \left[\left(\frac{1}{2} + \alpha_n\right) \phi_P^{m+1} + \left(\frac{1}{2} - \alpha_n\right) \phi_N^{m+1} \right]$$
$$- \dot{M}_s \left[\left(\frac{1}{2} + \alpha_s\right) \phi_P^{m+1} + \left(\frac{1}{2} - \alpha_s\right) \phi_S^{m+1} \right]$$
$$= D_{11e} \beta_e \left(\frac{\phi_E^{m+1} - \phi_P^{m+1}}{\Delta x} \right) - D_{11w} \beta_w \left(\frac{\phi_P^{m+1} - \phi_W^{m+1}}{\Delta x} \right)$$
$$+ D_{22n} \beta_n \left(\frac{\phi_N^{m+1} - \phi_P^{m+1}}{\Delta y} \right) - D_{22s} \beta_s \left(\frac{\phi_P^{m+1} - \phi_S^{m+1}}{\Delta y} \right) + S^\phi \bigg|_P \Delta y \Delta x$$
(8.75)

Defining the following coefficients

$$A_e = \dot{M}_e \left(\frac{1}{2} - \alpha_e\right) - \frac{D_{11e} \beta_e}{\Delta x} \tag{8.76}$$

$$A_w = -\dot{M}_w \left(\frac{1}{2} - \alpha_w\right) - \frac{D_{11w} \beta_w}{\Delta x} \tag{8.77}$$

$$A_n = \dot{M}_n \left(\frac{1}{2} - \alpha_n\right) - \frac{D_{22n} \beta_n}{\Delta y} \tag{8.78}$$

$$A_s = -\dot{M}_s \left(\frac{1}{2} - \alpha_s\right) - \frac{D_{22s} \beta_s}{\Delta y} \tag{8.79}$$

$$A_P = \frac{M_P}{\Delta t} + \dot{M}_e \left(\frac{1}{2} + \alpha_e\right) - \dot{M}_w \left(\frac{1}{2} + \alpha_w\right)$$
$$+ \dot{M}_n \left(\frac{1}{2} + \alpha_n\right) - \dot{M}_s \left(\frac{1}{2} + \alpha_s\right) \tag{8.80}$$
$$+ D_{11e} \frac{\beta_e}{\Delta x} + D_{11w} \frac{\beta_w}{\Delta x} + D_{22n} \frac{\beta_n}{\Delta y} + D_{22s} \frac{\beta_s}{\Delta y}$$

$$B_P = \frac{M_P}{\Delta t} \tag{8.81}$$

we can write equation (8.75) as

$$A_P \phi_P^{m+1} + A_e \phi_E^{m+1} + A_w \phi_W^{m+1} + A_n \phi_N^{m+1} + A_s \phi_S^{m+1} = B_P \phi_P^m + S^\phi \big|_P \Delta y \Delta x \tag{8.82}$$

which forms a linear system of equations composed of five diagonals that can be solved by well-established techniques. In the aforementioned equations, the pressure gradient, appearing in the $S^\phi|_P$ terms of the momentum equations, can be discretized using a central finite difference formula, that is,

$$\left.\frac{\partial p}{\partial x}\right|_P = \frac{p_E - p_W}{2\Delta x} \tag{8.83}$$

$$\left.\frac{\partial p}{\partial y}\right|_P = \frac{p_N - p_S}{2\Delta y} \tag{8.84}$$

8.2.1 Determination of the Velocity Field: The SIMPLEC Method

The momentum equations written in the form of equation (8.82) can be solved to find the velocity components $\phi = u$ and $\phi = v$ in the x and y directions, respectively. However, these equations also have another unknown—the pressure field. One of the possible ways to solve this problem is to use an equation of state to obtain the pressure field. Such a procedure is used for compressible flows, where the density ρ varies significantly with the pressure, as will be discussed in the next chapter. However, in incompressible flows, the density does not vary so much with the pressure. Therefore, small errors obtained in the computation of the density can cause large errors in the pressure field when using an equation of state. If this pressure is going to be introduced in the momentum conservation equations and the obtained velocities introduced in the mass conservation equation for the computation of ρ for the next time step, critical instabilities may occur. Such a problem is known in the literature as pressure–velocity coupling (Patankar 1980) because it is necessary to obtain a pressure field such that, when introduced in the momentum conservation equations, it will generate the velocity field that satisfies the mass conservation equation.

Several methods have been presented in the literature to solve this problem, such as the semi-implicit method for pressure-linked equations (SIMPLE) (Caretto et al. 1972; Patankar and Spalding 1972; Patankar 1975); SIMPLE revised (SIMPLER) (Patankar 1979); SIMPLE consistent (SIMPLEC) (Doormaal and Raithby 1984); and pressure implicit momentum explicit (PRIME) (Maliska 1981), among others. In this book, we will present the SIMPLEC method due to its wide use and excellent results for many problems of practical interest.

Let us suppose that some pressure field p^* is introduced into equation (8.82), written for the x-momentum equation (a similar procedure can be obtained

Multidimensional Incompressible Laminar Flow

for the y-momentum equation). Then, the velocity field component u* is obtained from

$$A_P u^{*m+1}_P + A_e u^{*m+1}_E + A_w u^{*m+1}_W + A_n u^{*m+1}_N + A_s u^{*m+1}_S = B_P u^m_P + S^{u^*}\Big|_P \Delta y \Delta x \tag{8.85}$$

where the term from the previous time step (denoted by the superscript m) was not modified. On the other hand, if the correct pressure field is introduced in this equation, we obtain

$$A_P u^{m+1}_P + A_e u^{m+1}_E + A_w u^{m+1}_W + A_n u^{m+1}_N + A_s u^{m+1}_S = B_P u^m_P + S^u\Big|_P \Delta y \Delta x \tag{8.86}$$

Subtracting equation (8.85) from equation (8.86), we obtain

$$A_P(u^{m+1}_P - u^{*m+1}_P) + A_e(u^{m+1}_E - u^{*m+1}_E) + A_w(u^{m+1}_W - u^{*m+1}_W) + \\ A_n(u^{m+1}_N - u^{*m+1}_N) + A_s(u^{m+1}_S - u^{*m+1}_S) = (S^u\Big|_P - S^{u^*}\Big|_P)\Delta y \Delta x \tag{8.87}$$

or yet

$$A_P u'^{m+1}_P + A_e u'^{m+1}_E + A_w u'^{m+1}_W + A_n u'^{m+1}_N + A_s u'^{m+1}_S = S^{u'}\Big|_P \Delta y \Delta x \tag{8.88}$$

where the superscript ' refers to the difference between the velocities obtained with the pressure fields p and p*.

In the SIMPLE method, the term $(A_e u'^{m+1}_E + A_w u'^{m+1}_W + A_n u'^{m+1}_N + A_s u'^{m+1}_S)$ is neglected. However, in the SIMPLEC method, in order to build a more consistent equation for the correction of the velocities, the term $(A_e u'^{m+1}_P + A_w u'^{m+1}_P + A_n u'^{m+1}_P + A_s u'^{m+1}_P)$ is subtracted from both sides of equation (8.88), resulting in

$$A_P u'^{m+1}_P + A_e(u'^{m+1}_E - u'^{m+1}_P) + A_w(u'^{m+1}_W - u'^{m+1}_P) \\ + A_n(u'^{m+1}_N - u'^{m+1}_P) + A_s(u'^{m+1}_S - u'^{m+1}_P) = S^{u'}\Big|_P \Delta y \Delta x \\ - (A_e u'^{m+1}_P + A_w u'^{m+1}_P + A_n u'^{m+1}_P + A_s u'^{m+1}_P) \tag{8.89}$$

Neglecting the differences $(u'^{m+1} - u'^{m+1}_P)$, we obtain

$$A_P u'^{m+1}_P + (A_e u'^{m+1}_P + A_w u'^{m+1}_P + A_n u'^{m+1}_P + A_s u'^{m+1}_P) = S^{u'}\Big|_P \Delta y \Delta x \tag{8.90}$$

or

$$(u^{m+1}_P - u^{*m+1}_P)(A_P + A_e + A_w + A_n + A_s) = S^{u'}\Big|_P \Delta y \Delta x \tag{8.91}$$

which is rewritten as

$$u_P^{m+1} = u_P^{*m+1} + S^{u'}\Big|_P \frac{\Delta y \Delta x}{[A_{nb}]_P} \tag{8.92a}$$

Performing the same steps for the y-momentum conservation equation, we obtain

$$v_P^{m+1} = v_P^{*m+1} + S^{v'}\Big|_P \frac{\Delta y \Delta x}{[A_{nb}]_P} \tag{8.92b}$$

Equations (8.92a,b) are called correction equations for the velocity field. In these equations,

$$[A_{nb}]_P = [A_P + A_e + A_w + A_n + A_s]_P \tag{8.93}$$

$$S^{u'}\Big|_P = -\frac{\partial p'}{\partial x}\Big|_P \tag{8.94}$$

$$S^{v'}\Big|_P = -\frac{\partial p'}{\partial y}\Big|_P \tag{8.95}$$

We still need to find an equation to p' in such a way that, when introduced into equations (8.92a,b), it originates velocities u and v that satisfy the continuity equation (8.55). We can write equations similar to equation (8.92) for each one of the four boundaries of the control volume P, that is,

$$u_e^{m+1} = u_e^{*m+1} - \frac{\partial p'}{\partial x}\Big|_e \frac{\Delta y \Delta x}{[A_{nb}]_e} \tag{8.96}$$

$$u_w^{m+1} = u_w^{*m+1} - \frac{\partial p'}{\partial x}\Big|_w \frac{\Delta y \Delta x}{[A_{nb}]_w} \tag{8.97}$$

$$v_n^{m+1} = v_n^{*m+1} - \frac{\partial p'}{\partial y}\Big|_n \frac{\Delta y \Delta x}{[A_{nb}]_n} \tag{8.98}$$

$$v_s^{m+1} = v_s^{*m+1} - \frac{\partial p'}{\partial y}\Big|_s \frac{\Delta y \Delta x}{[A_{nb}]_s} \tag{8.99}$$

where

$$\frac{\partial p'}{\partial x}\Big|_e = \frac{p'_E - p'_P}{\Delta x} \tag{8.100}$$

$$\frac{\partial p'}{\partial x}\Big|_w = \frac{p'_P - p'_W}{\Delta x} \tag{8.101}$$

$$\frac{\partial p'}{\partial y}\Big|_n = \frac{p'_N - p'_P}{\Delta y} \tag{8.102}$$

$$\left.\frac{\partial p'}{\partial y}\right|_s = \frac{p'_P - p'_S}{\Delta y} \tag{8.103}$$

and

$$[A_{nb}]_e = \frac{[A_{nb}]_E + [A_{nb}]_P}{2} \tag{8.104}$$

$$[A_{nb}]_w = \frac{[A_{nb}]_W + [A_{nb}]_P}{2} \tag{8.105}$$

$$[A_{nb}]_n = \frac{[A_{nb}]_N + [A_{nb}]_P}{2} \tag{8.106}$$

$$[A_{nb}]_s = \frac{[A_{nb}]_S + [A_{nb}]_P}{2} \tag{8.107}$$

From these equations, it is necessary to evaluate p' at the control volume P as well as in the four neighbor control volumes, E, W, N, and S. Such a value is called the *pressure correction* because p = p* + p'. We can integrate the continuity equation (8.55) in the control volume P shown on Figure 8.3 and obtain

$$\dot{M}_e - \dot{M}_w + \dot{M}_n - \dot{M}_s = 0 \tag{8.108}$$

Then, using equations (8.63)–(8.66), we obtain, considering velocities at time t + Δt

$$[u^{m+1}]_e \Delta y - [u^{m+1}]_w \Delta y + [v^{m+1}]_n \Delta x - [v^{m+1}]_s \Delta x = 0 \tag{8.109}$$

Then, using equations (8.96)–(8.103), equation (8.109) becomes

$$\left[u_e^{*m+1} - (p'_E - p'_P)\frac{\Delta y}{[A_{nb}]_e}\right]\Delta y - \left[u_w^{*m+1} - (p'_P - p'_W)\frac{\Delta y}{[A_{nb}]_w}\right]\Delta y$$
$$+ \left[v_n^{*m+1} - (p'_N - p'_P)\frac{\Delta x}{[A_{nb}]_n}\right]\Delta x - \left[v_s^{*m+1} - (p'_P - p'_S)\frac{\Delta x}{[A_{nb}]_s}\right]\Delta x = 0$$
$$\tag{8.110}$$

By defining the following coefficients

$$A_P^P = \frac{(\Delta y)^2}{[A_{nb}]_e} + \frac{(\Delta y)^2}{[A_{nb}]_w} + \frac{(\Delta x)^2}{[A_{nb}]_n} + \frac{(\Delta x)^2}{[A_{nb}]_s} \tag{8.111}$$

$$A_e^P = -\frac{(\Delta y)^2}{[A_{nb}]_e} \tag{8.112}$$

$$A_w^P = -\frac{(\Delta y)^2}{[A_{nb}]_w} \qquad (8.113)$$

$$A_n^P = -\frac{(\Delta x)^2}{[A_{nb}]_n} \qquad (8.114)$$

$$A_s^P = -\frac{(\Delta x)^2}{[A_{nb}]_s} \qquad (8.115)$$

$$B_P^P = -u_e^{*m+1}\Delta y + u_w^{*m+1}\Delta y - v_n^{*m+1}\Delta x + v_s^{*m+1}\Delta x \qquad (8.116)$$

we can write equation (8.110) as

$$A_P^P p'_P + A_e^P p'_E + A_w^P p'_W + A_n^P p'_N + A_s^P p'_S = B_P^P \qquad (8.117)$$

We can now establish a sequence of steps to calculate the velocity and pressure fields, using the SIMPLEC method.

1. Given some time t, estimate the velocity fields u and v and the pressure field p^* for the time $t + \Delta t$.
2. Calculate the coefficients of the momentum conservation equations, given by equations (8.76)–(8.81).
3. Solve the momentum conservation equations (8.82) for $\phi = u$ and $\phi = v$ using p^* in order to obtain u^* and v^*. Notice that the velocities obtained at this step are located at the centers of the control volumes. This system of equations can be solved using methods such as Gauss–Seidel, Thomas algorithm, biconjugate gradient, and so on (see Chapter 3).
4. Calculate the coefficients of the pressure equation, given by equations (8.111)–(8.116).
5. Solve equation (8.117) to obtain the pressure correction p'.
6. Correct the velocity field using equations (8.96)–(8.99) to obtain values that satisfy the mass conservation equation. Notice that the values with the superscript "*" must be evaluated at the interfaces of the control volumes and therefore some interpolation is needed.
7. Update the pressure field, making $p = p^* + p'$.
8. Make $p^* = p$ and return to Step 2 until convergence is achieved.
9. Solve the energy conservation equation (8.82) for $\phi = T$.
10. Return to Step 1 while the final time is not reached.

As pointed out in Step 6, some interpolation is needed to evaluate the velocity field at the interface of the control volumes. We follow the procedure suggested by Maliska (1981). Let us write equation (8.85) for the *center of the control volumes* P and E

$$(A_P)_P u^{*m+1}_P + (A_e u^{*m+1}_E + A_w u^{*m+1}_W + A_n u^{*m+1}_N + A_s u^{*m+1}_S)_P$$
$$= (B_P)_P u^m_P - \left(\frac{\partial p^*}{\partial x}\bigg|_P\right)_P \Delta y \Delta x \qquad (8.118)$$

$$(A_P)_E u^{*m+1}_E + (A_e u^{*m+1}_E + A_w u^{*m+1}_W + A_n u^{*m+1}_N + A_s u^{*m+1}_S)_E$$
$$= (B_P)_E u^m_E - \left(\frac{\partial p^*}{\partial x}\bigg|_P\right)_E \Delta y \Delta x \qquad (8.119)$$

where $(.)_P$ and $(.)_E$ means the quantity inside the parenthesis referenced to the locations P and E, respectively. We can also write a similar equation for the *interface* between control volumes P and E

$$(A_P)_e u^{*m+1}_e + (A_e u^{*m+1}_E + A_w u^{*m+1}_W + A_n u^{*m+1}_N + A_s u^{*m+1}_S)_e$$
$$= (B_P)_e u^m_e - \left(\frac{\partial p^*}{\partial x}\bigg|_P\right)_e \Delta y \Delta x \qquad (8.120)$$

Adopting an average process for each one of the terms in equation (8.120), based on equations (8.118) and (8.119), we can obtain

$$(A_P)_e = \frac{(A_P)_E + (A_P)_P}{2} \qquad (8.121)$$

$$(A_e u^{*m+1}_E + A_w u^{*m+1}_W + A_n u^{*m+1}_N + A_s u^{*m+1}_S)_e$$
$$= 0.5[(A_e u^{*m+1}_E + A_w u^{*m+1}_W + A_n u^{*m+1}_N + A_s u^{*m+1}_S)_E \qquad (8.122)$$
$$+ (A_e u^{*m+1}_E + A_w u^{*m+1}_W + A_n u^{*m+1}_N + A_s u^{*m+1}_S)_P]$$

$$(B_P)_e = \frac{(B_P)_E + (B_P)_P}{2} \qquad (8.123)$$

$$\left(\frac{\partial p^*}{\partial x}\bigg|_P\right)_e = \frac{1}{2}\left[\left(\frac{\partial p^*}{\partial x}\bigg|_P\right)_E + \left(\frac{\partial p^*}{\partial x}\bigg|_P\right)_P\right] \qquad (8.124)$$

Then, adopting a similar procedure for the west, north, and south interfaces, we can write for the four interfaces of the control volume P

$$u^{*m+1}_e = \frac{1}{(A_P)_E + (A_P)_P}$$
$$\left\{-0.5\left[(A_e u^{*m+1}_E + A_w u^{*m+1}_W + A_n u^{*m+1}_N + A_s u^{*m+1}_S)_E\right.\right.$$
$$\left.+ (A_e u^{*m+1}_E + A_w u^{*m+1}_W + A_n u^{*m+1}_N + A_s u^{*m+1}_S)_P\right] \qquad (8.125)$$
$$+ \left[\frac{(B_P)_E + (B_P)_P}{2}\right] u^m_e - \frac{1}{2}\left[\left(\frac{\partial p^*}{\partial x}\bigg|_P\right)_E + \left(\frac{\partial p^*}{\partial x}\bigg|_P\right)_P\right] \Delta y \Delta x\right\}$$

$$u^{*m+1}_w = \frac{1}{(A_P)_W + (A_P)_P}$$

$$\left\{ -0.5\left[(A_e u^{*m+1}_E + A_w u^{*m+1}_W + A_n u^{*m+1}_N + A_s u^{*m+1}_S)_W \right.\right.$$

$$\left.+ (A_e u^{*m+1}_E + A_w u^{*m+1}_W + A_n u^{*m+1}_N + A_s u^{*m+1}_S)_P \right]$$

$$\left. + \left[\frac{(B_P)_W + (B_P)_P}{2}\right] u^m_w - \frac{1}{2}\left[\left(\frac{\partial p^*}{\partial x}\bigg|_P\right)_W + \left(\frac{\partial p^*}{\partial x}\bigg|_P\right)_P\right] \Delta y \Delta x \right\}$$

(8.126)

$$v^{*m+1}_n = \frac{1}{(A_P)_N + (A_P)_P}$$

$$\left\{ -0.5\left[(A_e v^{*m+1}_E + A_w v^{*m+1}_W + A_n v^{*m+1}_N + A_s v^{*m+1}_S)_N \right.\right.$$

(8.127)

$$\left.+ (A_e v^{*m+1}_E + A_w v^{*m+1}_W + A_n v^{*m+1}_N + A_s v^{*m+1}_S)_P \right]$$

$$\left. + \left[\frac{(B_P)_N + (B_P)_P}{2}\right] u^m_n - \frac{1}{2}\left[\left(\frac{\partial p^*}{\partial y}\bigg|_P\right)_N + \left(\frac{\partial p^*}{\partial y}\bigg|_P\right)_P\right] \Delta y \Delta x \right\}$$

$$v^{*m+1}_s = \frac{1}{(A_P)_S + (A_P)_P}$$

$$\left\{ -0.5\left[(A_e v^{*m+1}_E + A_w v^{*m+1}_W + A_n v^{*m+1}_N + A_s v^{*m+1}_S)_S \right.\right.$$

(8.128)

$$\left.+ (A_e v^{*m+1}_E + A_w v^{*m+1}_W + A_n v^{*m+1}_N + A_s v^{*m+1}_S)_P \right]$$

$$\left. + \left[\frac{(B_P)_S + (B_P)_P}{2}\right] u^m_s - \frac{1}{2}\left[\left(\frac{\partial p^*}{\partial y}\bigg|_P\right)_S + \left(\frac{\partial p^*}{\partial y}\bigg|_P\right)_P\right] \Delta y \Delta x \right\}$$

This completes all the necessary steps to solve the velocity field using the SIMPLEC algorithm, except by the boundary conditions, which will be considered next.

8.2.2 Treatment of Boundary Conditions

Let us now analyze the boundary conditions for the solution of equation (8.69), which is the general conservation equation for the scalar ϕ. We also need to analyze the boundary condition of equation (8.117), which is the equation that will give the correction of the pressure field.

Here, we will consider, just for the illustration, a parallel plate channel where the fluid enters the channel through the left boundary at $x = 0$ and leaves through the right boundary at $x = L$. Figure 8.4 shows the problem under analysis. The upper plate is subjected to a heat flux q and the bottom

Multidimensional Incompressible Laminar Flow

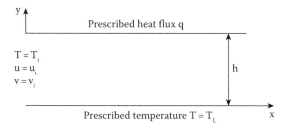

FIGURE 8.4
Geometry for the presentation of the boundary conditions.

plate is kept at a constant temperature T_L. The inlet temperature and velocity components are equal to T_i, u_i, and v_i, respectively. The initial conditions for the temperature and velocity components are equal to T_0, u_0, and v_0, respectively. Once the reader understands the basic concepts presented here, he/she can apply them to other problems.

A usual combination of boundary conditions for problems involving internal forced convection, where the duct is long enough, is to have the inlet velocity prescribed and a locally parabolic velocity profile specified at the outlet flow, without any inlet/outlet boundary conditions for the pressure field. Because the solution is obtained iteratively, the pressure field establishes itself automatically. This procedure, however, may result in some instabilities during the solution of the linear system due to the lack of dominance in the diagonal of the matrix for the system of equations for the pressure field. This problem can be remedied by specifying the pressure field at some point in the domain.

8.2.2.1 Pressure

The system of equations that need to be solved in order to determine the correction of the pressure field p' in Step 7 of the iterative procedure for the SIMPLEC method is a Poison type equation. This equation (8.117) was derived starting from the continuity equation.

8.2.2.1.1 Boundary $x = 0$

For this boundary, we consider a known inlet mass flow rate. The value of the velocity is known at the west interface of the control volume. Thus, $\dot{M}_w = [\rho u]_w \Delta y$ is known, and we can rewrite equation (8.110) as

$$\left[u_e^{*m+1} - (p'_E - p'_P) \frac{\Delta y}{[A_{nb}]_e} \right] \Delta y - u_w^{m+1} \Delta y$$
$$+ \left[v_n^{*m+1} - (p'_N - p'_P) \frac{\Delta x}{[A_{nb}]_n} \right] \Delta x \qquad (8.129)$$
$$- \left[v_s^{*m+1} - (p'_P - p'_S) \frac{\Delta x}{[A_{nb}]_s} \right] \Delta x = 0$$

Notice that in this equation the correction of the velocity at the west interface is not necessary anymore because we know the actual value of the velocity. For this interface, we can then define the following coefficients, similarly to equations (8.111)–(8.116)

$$A_P^P = \frac{(\Delta y)^2}{[A_{nb}]_e} + \frac{(\Delta x)^2}{[A_{nb}]_n} + \frac{(\Delta x)^2}{[A_{nb}]_s} \qquad (8.130)$$

$$A_e^P = -\frac{(\Delta y)^2}{[A_{nb}]_e} \qquad (8.131)$$

$$A_w^P = 0 \qquad (8.132)$$

$$A_n^P = -\frac{(\Delta x)^2}{[A_{nb}]_n} \qquad (8.133)$$

$$A_s^P = -\frac{(\Delta x)^2}{[A_{nb}]_s} \qquad (8.134)$$

$$B_P^P = -u_e^{*m+1}\Delta y + u_w^{m+1}\Delta y - v_n^{*m+1}\Delta x - v_s^{*m+1}\Delta x \qquad (8.135)$$

and, similarly to equation (8.117), we have

$$A_P^P p_P' + A_n^P p_N' + A_s^P p_s' = B_P^P \qquad (8.136)$$

Notice that because the velocity at the west interface is known, the A_w^P coefficient is zero. Also, the velocity u_w, entering in the source term given by equation (8.135), is the correct one (without the "*" superscript).

8.2.2.1.2 Boundary $x = L$

For this boundary, we have an outlet mass flow. If the duct is long enough, we can use the so-called locally parabolic boundary condition. In this case, for volumes next to the outlet boundary, the mass flow rate at the east boundary, $\dot{M}_e = [\rho u]_e \Delta y$, is known and it is the same as the mass flow rate at the west boundary, $\dot{M}_w = [\rho u]_w \Delta y$. Therefore, from equation (8.110), we have

$$u_e^{m+1}\Delta y - \left[u_w^{*m+1} - (p_P' - p_W')\frac{\Delta y}{[A_{nb}]_w}\right]\Delta y$$
$$+ \left[v_n^{*m+1} - (p_N' - p_P')\frac{\Delta x}{[A_{nb}]_n}\right]\Delta x - \left[v_s^{*m+1} - (p_P' - p_S')\frac{\Delta x}{[A_{nb}]_s}\right]\Delta x = 0$$

(8.137)

Multidimensional Incompressible Laminar Flow

For this boundary, the velocity correction at the east interface is not necessary anymore because we know the actual value of the velocity. Then, we can define the following coefficients, similarly to equations (8.111)–(8.116)

$$A_P^P = \frac{(\Delta y)^2}{[A_{nb}]_w} + \frac{(\Delta x)^2}{[A_{nb}]_n} + \frac{(\Delta x)^2}{[A_{nb}]_s} \quad (8.138)$$

$$A_e^P = 0 \quad (8.139)$$

$$A_w^P = -\frac{(\Delta y)^2}{[A_{nb}]_w} \quad (8.140)$$

$$A_n^P = -\frac{(\Delta x)^2}{[A_{nb}]_n} \quad (8.141)$$

$$A_s^P = -\frac{(\Delta x)^2}{[A_{nb}]_s} \quad (8.142)$$

$$B_P^P = -u_e^{m+1}\Delta y + u_w^{*m+1}\Delta y - v_n^{*m+1}\Delta x + v_s^{*m+1}\Delta x \quad (8.143)$$

and, similarly to equation (8.117), we have

$$A_P^P p'_P + A_n^P p'_N + A_s^P p'_S = B_P^P \quad (8.144)$$

Notice that because the velocity at the east interface is known, the A_e^P coefficient is zero. Also, the east velocity u_e, entering in the source term given by equation (8.143), is the correct one (without the "*" superscript).

8.2.2.1.3 Boundary $y = 0$

This boundary represents the bottom wall of the channel. Therefore, the mass flow rate at this boundary must be zero; that is, for volumes next to the bottom wall, $\dot{M}_s = [\rho v]_s \Delta x = 0$. We can then rewrite equation (8.110) as

$$\left[u_e^{*m+1} - (p'_E - p'_P)\frac{\Delta y}{[A_{nb}]_e} \right]\Delta y - \left[u_w^{*m+1} - (p'_P - p'_W)\frac{\Delta y}{[A_{nb}]_w} \right]\Delta y$$
$$+ \left[v_n^{*m+1} - (p'_N - p'_P)\frac{\Delta x}{[A_{nb}]_n} \right]\Delta x = 0 \quad (8.145)$$

or, defining the following coefficients

$$A_P^P = \frac{(\Delta y)^2}{[A_{nb}]_e} + \frac{(\Delta y)^2}{[A_{nb}]_w} + \frac{(\Delta x)^2}{[A_{nb}]_n} \quad (8.146)$$

$$A_e^P = -\frac{(\Delta y)^2}{[A_{nb}]_e} \quad (8.147)$$

$$A_w^P = -\frac{(\Delta y)^2}{[A_{nb}]_w} \tag{8.148}$$

$$A_n^P = -\frac{(\Delta x)^2}{[A_{nb}]_n} \tag{8.149}$$

$$A_s^P = 0 \tag{8.150}$$

$$B_P^P = -u_e^{*m+1}\Delta y + u_w^{*m+1}\Delta y - v_n^{*m+1}\Delta x \tag{8.151}$$

we can write

$$A_P^P p'_P + A_e^P p'_E + A_w^P p'_W + A_n^P p'_N = B_P^P \tag{8.152}$$

Notice that because the velocity at the south interface is zero, the A_s^P coefficient is also zero. Also, the south velocity v_s does not appear in the source term given by equation (8.151).

8.2.2.1.4 Boundary $y = H$

Again, this boundary represents a solid surface; and therefore, the mass flow rate must be zero. For volumes next to the top wall, $\dot{M}_n = [\rho v]_n \Delta x = 0$. We can then rewrite equation (8.110) as

$$\left[u_e^{*m+1} - (p'_E - p'_P)\frac{\Delta y}{[A_{nb}]_e}\right]\Delta y - \left[u_w^{*m+1} - (p'_P - p'_W)\frac{\Delta y}{[A_{nb}]_w}\right]\Delta y \\ - \left[v_s^{*m+1} - (p'_P - p'_S)\frac{\Delta x}{[A_{nb}]_s}\right]\Delta x = 0 \tag{8.153}$$

or, defining the following coefficients

$$A_P^P = \frac{(\Delta y)^2}{[A_{nb}]_e} + \frac{(\Delta y)^2}{[A_{nb}]_w} + \frac{(\Delta x)^2}{[A_{nb}]_s} \tag{8.154}$$

$$A_e^P = -\frac{(\Delta y)^2}{[A_{nb}]_e} \tag{8.155}$$

$$A_w^P = -\frac{(\Delta y)^2}{[A_{nb}]_w} \tag{8.156}$$

$$A_n^P = 0 \tag{8.157}$$

$$A_s^P = -\frac{(\Delta x)^2}{[A_{nb}]_s} \tag{8.158}$$

$$B_P^P = -u_e^{*m+1}\Delta y + u_w^{*m+1}\Delta y + v_s^{*m+1}\Delta x \qquad (8.159)$$

we can write

$$A_P^P p'_P + A_e^P p'_E + A_w^P p'_W + A_s^P p'_S = B_P^P \qquad (8.160)$$

Notice that because the velocity at the north interface is zero, the A_n^P coefficient is also zero. Also, the velocity v_n does not appear in the source term given by equation (8.159).

It is worth noticing that the linear system formed by equations (8.117), (8.136), (8.144), (8.152), and (8.160) is not diagonal dominant. In other words, the coefficient of the matrix multiplied by the pressure at the center of the control volume P is equal to the sum of the coefficients that are multiplied by the pressures at the center of the neighbor volumes around P. Thus, iterative methods of solution might have difficulty solving this system. This is the reason why it is advised to specify the value of the pressure in one or more points of the domain. In fact, according to Doormaal and Raithby (1984), the solution of the pressure field represents about 80% of the total computational cost of the problem. When one specifies the value of the pressure at some volume P, all neighbor coefficients vanish. Thus, for this volume, the system is diagonally dominant. Because the quantities that appear in the x- and y-momentum equations are the derivatives of the pressure in the x and y directions, respectively, any constant added to the pressure does not affect the final solution.

8.2.2.2 Momentum and Energy Equations

The physical boundary conditions for the momentum equations are known mass flow rate at x = 0 and non-slip and impermeable boundary conditions at y = 0 and y = H. The inlet mass flow rate corresponds to known u and v, while the non-slip and impermeable boundary conditions correspond to zero velocities on the walls. For an incompressible flow with specified inlet velocity, it is not mathematically necessary to specify the outlet velocities for the problem to be well-posed. However, for the application of the finite volume method, it is necessary to specify an outlet boundary condition, usually chosen as being locally parabolic. For the numerical locally parabolic boundary condition, it is considered that the derivative of the conserved property (u, v, or T) along the flow direction (the diffusive flux) is zero. Thus, the location of the outlet boundary must be carefully chosen at a downstream position where the flow is completely (hydrodynamically and thermally) developed and without recirculations.

For the energy equation, it is considered in this example that the inlet temperature is known. The boundary y = H is subjected to a heat flux q, while the lower boundary is kept at a constant temperature T_L. Such as with the hydrodynamic problem, the outflow boundary condition is considered to be locally parabolic.

For the application of the boundary conditions to equation (8.69), obtained from the integration of the conservation equation for a scalar ϕ in the control volume P, from time t to time t + Δt, let us consider that the convective and diffusive fluxes at each of the surfaces of the control volume P are given by

$$\text{east interface} \begin{cases} \text{convective flux} = \dot{M}_e \phi_e \\ \text{diffusive flux} = \left[D_{11} \dfrac{\partial \phi}{\partial x} \right]_e \end{cases} \quad (8.161)$$

$$\text{west interface} \begin{cases} \text{convective flux} = \dot{M}_w \phi_w \\ \text{diffusive flux} = \left[D_{11} \dfrac{\partial \phi}{\partial x} \right]_w \end{cases} \quad (8.162)$$

$$\text{north interface} \begin{cases} \text{convective flux} = \dot{M}_n \phi_n \\ \text{diffusive flux} = \left[D_{22} \dfrac{\partial \phi}{\partial y} \right]_n \end{cases} \quad (8.163)$$

$$\text{south interface} \begin{cases} \text{convective flux} = \dot{M}_s \phi_s \\ \text{diffusive flux} = \left[D_{22} \dfrac{\partial \phi}{\partial y} \right]_s \end{cases} \quad (8.164)$$

The diffusive and convective fluxes of equations (8.161)–(8.164) will be replaced by the information from the boundaries, while the other terms of the conservation equations will still be approximated by the WUDS interpolation function, as discussed earlier.

8.2.2.2.1 Boundary x = 0

As previously discussed, this boundary has inlet boundary conditions with known ρu and ρv. For this case, where the mass flow rate is specified on the inlet, the diffusive flux is zero at this boundary.

For this boundary, the physical boundary condition is

$$\dot{M}_w \phi_w = \text{known} \quad (8.165)$$

with

$$\left[D_{11} \dfrac{\partial \phi}{\partial x} \right]_w = 0 \quad (8.166)$$

Thus, replacing equations (8.165) and (8.166) into equation (8.69), we can obtain an equation for the volumes next to the boundary x = 0.

8.2.2.2.2 Boundary x = L

At this boundary, there exists an outflow condition, with unknown ρu and ρv. For this case, the usual boundary condition is locally parabolic, where a null diffusive flux is specified at the outlet (x = L). In this way, it is not necessary to specify the downstream boundary condition because the A_e coefficient of the conservation equation is set to zero once an upwind approximation is used. Thus, for this boundary, we have

$$\dot{M}_e \phi_e = \dot{M}_w \phi_w \qquad (8.167)$$

with

$$\left[D_{11} \frac{\partial \phi}{\partial x} \right]_w = 0 \qquad (8.168)$$

The value of ϕ at the outlet surface, which is taken as being equal to the value of ϕ at the center of the control volume, must be used in such a way that the derivative of the primitive variable in the flow direction is equal to zero. Therefore, replacing equations (8.167) and (8.168) into equation (8.69), we can obtain an equation for the boundary x = L.

8.2.2.2.3 Boundary y = 0

At this boundary, the following must be specified: impermeable condition for the momentum conservation equations (u = v = 0) and prescribed value of ϕ (temperature) for the energy conservation equation. For this case, the convective flux is equal to zero because the velocity is null at the boundary ($\dot{M} = 0$), and the diffusive flux must be approximated using the information from the boundary. Thus, using a forward second-order finite difference approximation for $\partial \phi / \partial y$ at y = 0, the physical boundary condition is

$$\dot{M}_s \phi_s = 0 \qquad (8.169)$$

with

$$-\left[D_{22} \frac{\partial \phi}{\partial y} \right]_s = -\left[D_{22s} \beta_s \frac{-3\phi_s + 4\phi_P - \phi_n}{2\frac{\Delta y}{2}} \right] = -\left[D_{22s} \beta_s \frac{-3\phi_s + 4\phi_P - \frac{\phi_N + \phi_P}{2}}{\Delta y} \right]$$

$$= -\left[D_{22s} \beta_s \frac{-6\phi_s + 7\phi_P - \phi_N}{2\Delta y} \right]$$

(8.170)

where $\phi_s = 0$ for the momentum conservation equation and $\phi_s = T_s = T_L$ for the energy conservation equation, where T_s is the prescribed value of $T = T_L$ at surface y = 0.

Therefore, replacing equations (8.169) and (8.170) into equation (8.69), we can obtain an equation for the volumes next to the boundary $y = 0$. Notice that because we know the values of ϕ_s at $y = 0$, this must be used in the approximation in order to transmit this information to the inner part of the domain.

8.2.2.2.4 Boundary $y = H$

At this boundary, the following conditions must be specified: impermeable boundary for the momentum conservation equation and prescribed flux of ϕ for the energy conservation equation.

For the momentum conservation equations, the convective flux is equal to zero because the velocity is null at the boundary, and the diffusive flux must be approximated using the information from the impermeable boundary. Thus, using a backward second-order finite difference approximation for $\partial \phi / \partial y$ at $y = H$, the physical boundary condition is

$$\dot{M}_n \phi_n = 0 \qquad (8.171)$$

with

$$\left[D_{22} \frac{\partial \phi}{\partial y} \right]_n = \left[D_{22s} \beta_s \frac{3\phi_n - 4\phi_P + \phi_S}{2\frac{\Delta y}{2}} \right] = -\left[D_{22s} \beta_s \frac{3\phi_n - 4\phi_P + \frac{\phi_P + \phi_S}{2}}{\Delta y} \right]$$

$$= \left[D_{22s} \beta_s \frac{6\phi_n - 7\phi_P + \phi_S}{2\Delta y} \right]$$

$$(8.172)$$

where $\phi_s = 0$ for the momentum conservation equation.

Therefore, replacing equations (8.171) and (8.172) into equation (8.69), we can obtain an equation for the volumes next to the boundary $y = H$ for the momentum conservation equations. Notice that $\phi_n = 0$ because $u = v = 0$ at this boundary.

For the energy conservation equation, the convective flux must also be specified as null because the velocity is equal to zero at this interface. Regarding the diffusive flux, we must use the heat flux specified at this boundary. Therefore, we have

$$\dot{M}_n \phi_n = 0 \qquad (8.173)$$

with

$$\left[D_{22} \frac{\partial \phi}{\partial y} \right]_n = \frac{q}{C_P} \Delta x \qquad (8.174)$$

where q is the prescribed heat flux at the boundary $y = H$. Replacing equations (8.173) and (8.174) into equation (8.69), we can obtain an equation for the volumes next to the boundary at $y = H$.

For the points located in the corners of the domain, a combination of the equations at adjacent boundaries must be utilized.

8.3 Two-Dimensional Steady Laminar Boundary Layer Flow

We now consider two-dimensional, steady laminar boundary layer flow of an incompressible, constant property fluid. The flow field is governed by the following continuity and momentum equations written in a rectangular coordinate system attached to the body surface, as shown by Figure 8.5:

$$\text{Continuity: } \frac{\partial u}{\partial x} + \frac{\partial v}{\partial y} = 0 \qquad (8.175a)$$

$$\text{Momentum: } u\frac{\partial u}{\partial x} + v\frac{\partial u}{\partial y} = -\frac{1}{\rho}\frac{dp(x)}{dx} + \upsilon\frac{\partial^2 u}{\partial y^2} \qquad (8.175b)$$

where u and v are the streamwise and normal velocity components, respectively, ρ is density, and υ is kinematic viscosity. In the boundary layer formulation of the flow problem as given previously, the momentum equation (8.175b) is converted from its original elliptic form into the parabolic form as a result of neglecting the diffusion term in the x direction. In addition, from the momentum equation in the y direction, transverse to the main flow, one can deduce that the pressure gradient is

$$-\frac{1}{\rho}\frac{dp(x)}{dx} = u_e(x)\frac{du_e(x)}{dx} \qquad (8.176)$$

Here, $u_e(x)$ is the velocity at the edge of the boundary layer and is assumed to be known from the solution of the inviscid flow field outside the boundary layer; hence, the pressure gradient term in the momentum equation (8.175b) is considered known. For a flat plate ($\theta = 0$ in Figure 8.5), where $u_e(x) = U_\infty = $ constant, the pressure gradient along the longitudinal x direction is null.

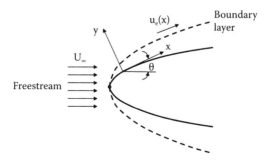

FIGURE 8.5
Velocity boundary layer over a body.

An explicit algorithm for the solution of boundary layer equations with finite differences is now illustrated, even though the size of the marching step Δx is limited by stability considerations.

We adopt the notation

$$u(x,y) = u(i\Delta x, j\Delta y) \equiv u_j^i \qquad (8.177a)$$

$$v(x,y) = v(i\Delta x, j\Delta y) \equiv v_j^i \qquad (8.177b)$$

where the x coordinate becomes analogous to the time coordinate used in the diffusion equation. Therefore, a discretization procedure similar to that used for the discretization of the diffusion equation can be used for the discretization of the momentum equation (8.175b).

The momentum equation (8.175b) is discretized with the simple explicit method as

$$u_j^i \frac{u_j^{i+1} - u_j^i}{\Delta x} + v_j^i \frac{u_{j+1}^i - u_{j-1}^i}{2\Delta y} = u_e^i \frac{u_e^{i+1} - u_e^i}{\Delta x} + \frac{\upsilon}{(\Delta y)^2}(u_{j+1}^i - 2u_j^i + u_{j-1}^i) \qquad (8.178)$$

with a truncation error $0[\Delta x, (\Delta y)^2]$. Here, central differencing is used to discretize the derivatives in the y variable in order to have second-order accuracy. Equation (8.178) provides an explicit expression for the computation of u_j^{i+1} at the level $i+1$ from the known values of u_{j+1}^i, u_j^i, u_{j-1}^i, and v_j^i.

The continuity equation (8.175a) is discretized as

$$\frac{(u_{j-1}^{i+1} - u_{j-1}^i) + (u_j^{i+1} - u_j^i)}{2\Delta x} + \frac{v_j^{i+1} - v_{j-1}^{i+1}}{\Delta y} = 0 \qquad (8.179)$$

where the approximation for $\partial u/\partial x$ is an average of two backward derivatives at $(j-1)\Delta y$ and $j\Delta y$ (Wu 1961).

Once the initial values for u_j^i are available, equation (8.178) is solved explicitly for u_j^{i+1}, starting from the wall and working outward until $(u_j^{i+1}/u_e^{i+1}) = 0.9995$, which is regarded as the edge of the boundary layer. Thus, the location of the edge of the boundary layer is established as the numerical computations proceed downstream. Equation (8.179) is used to compute v_j^{i+1}, starting from the node next to the plate and computing outward.

The method being explicit, the stability considerations lead to the following constraints on the step size Δx (Pletcher et al. 2012):

$$\frac{2\upsilon \Delta x}{u_j^i (\Delta y)^2} \leq 1 \quad \text{and} \quad \frac{(v_j^i)^2 \Delta x}{u_j^i \upsilon} \leq 2 \qquad (8.180a,b)$$

If in the momentum equation (8.175b), the term $v \partial u / \partial y$ is discretized with the upwind scheme in the following manner

$$v_j^i \frac{u_j^i - u_{j-1}^i}{\Delta y} \quad \text{for} \quad v_j^i > 0 \quad \text{(backward differencing)}$$

$$v_j^i \frac{u_{j+1}^i - u_j^i}{\Delta y} \quad \text{for} \quad v_j^i < 0 \quad \text{(forward differencing)}$$

instead of the central differencing used in equation (8.178), the following single stability criterion is applicable

$$\Delta x \leq \left[\frac{2\upsilon}{u_j^i (\Delta y)^2} + \frac{|v_j^i|}{u_j^i \Delta y} \right]^{-1} \tag{8.181}$$

but the order of the truncation error decreases to $0[\Delta x, \Delta y]$.

The use of a rectangular finite difference grid for the discretization of the boundary layer equations results in, at small x, too few grid points within the boundary layer, whereas there is excessively fine grid for the downstream region. Therefore, a rectangular grid is likely to be computationally inefficient. To alleviate such difficulties, Patankar and Spalding (1967a and 1967b) proposed the use of a coordinate transformation from the x,y coordinates into x,ω coordinates where ω is the normalized stream function defined by

$$\omega = \frac{\psi - \psi_I}{\psi_E - \psi_I} \tag{8.182}$$

where ψ is the local stream function, while ψ_I and ψ_E are the stream functions at the *inner* and *external* edges of the boundary layer, respectively, as illustrated in Figure 8.6. Thus, the value of ω always varies from zero to unity across the boundary layer, and the same number of grid points always are available everywhere within the boundary layer along the x axis. The transformed problem is then discretized in the x,ω grid and solved by finite differences. The details of the coordinate transformation, the discretization procedure, and the algorithm for computations are discussed by Patankar and Spalding (1967a and 1967b), Spalding (1977), and Patankar (1988). Other more general transformation of the governing equations is presented in Chapter 11.

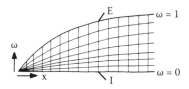

FIGURE 8.6
Normalized stream function coordinate system for boundary layers.

PROBLEMS

8.1. Consider the energy conservation equation in laminar forced convection inside a circular tube with fully developed velocity profile u(R)

$$\frac{\partial T}{\partial t} + u(R)\frac{\partial T}{\partial x} = \frac{\alpha}{R}\frac{\partial}{\partial R}\left(R\frac{\partial T}{\partial R}\right)$$

Write the finite difference approximation of this equation at the internal nodes, including the nodes along the axis of symmetry, by using the simple explicit scheme utilizing forward differencing in time, and apply the central differencing for discretizing both the convection and diffusion terms.

8.2. Repeat Problem 8.1 by applying upwind differencing for the convection term and central differencing for the diffusion term.

8.3. Consider the two-dimensional energy conservation equation in laminar forced convection inside a parallel plate channel with velocity components u and v available from the solution of the fluid mechanics problem. The energy equation is given in the nonconservative form as

$$\frac{\partial T}{\partial t} + u\frac{\partial T}{\partial x} + v\frac{\partial T}{\partial y} = \alpha\left(\frac{\partial^2 T}{\partial x^2} + \frac{\partial^2 T}{\partial y^2}\right)$$

Write the finite difference approximation of this equation using the simple explicit scheme in the time variable and central differencing for discretizing the convection and diffusion terms. What is the order of the truncation error?

8.4. Repeat Problem 8.3 using the upwind scheme for discretizing the convection term and central differencing for discretizing the diffusion terms. What is the order of the truncation error?

8.5. Consider two-dimensional transient laminar forced convection inside a circular tube. The velocity components u and v are available from the solution of the fluid mechanics problem. The governing energy equation in cylindrical symmetry is given by

$$\frac{\partial T}{\partial t} + u\frac{\partial T}{\partial x} + v\frac{\partial T}{\partial R} = \frac{\alpha}{R}\frac{\partial}{\partial R}\left(R\frac{\partial T}{\partial R}\right) + \alpha\frac{\partial^2 T}{\partial x^2}$$

Write the finite difference approximation of this equation at the internal nodes, including the nodes along the axis of symmetry, by using the simple explicit scheme utilizing forward differencing in time and the upwind scheme for the convection terms and central differencing for the diffusion terms.

8.6. Consider the energy equation given in the conservative form as

$$\frac{\partial T}{\partial t} + \frac{\partial}{\partial x}(uT) + \frac{\partial}{\partial y}(vT) = \alpha\left(\frac{\partial^2 T}{\partial x^2} + \frac{\partial^2 T}{\partial y^2}\right)$$

Write the finite difference form of this equation using explicit scheme and central differencing for all space variables.

8.7. Consider two-dimensional steady motion of an incompressible, constant property, viscous flow in a square cavity with its walls impermeable to flow. The fluid motion is caused by the uniform movement of the upper wall with a constant velocity u_0 as illustrated in the accompanying figure. Write the governing equations of motion in the vorticity-stream function formulation and the boundary conditions for the velocity components (u,v), stream function ψ, and vorticity ω.

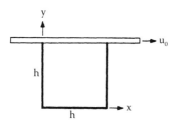

8.8. Consider two-dimensional incompressible steady flow of a constant property fluid through a channel with a varying cross section, as shown in the accompanying figure. Walls are impermeable to flow. We assume uniform velocity at the inlet and outlet. If this flow problem is to be solved with the vorticity-stream function formulation

   a. Write the appropriate vorticity and stream function equations for the determination of ω and ψ.
   b. Write the boundary conditions for the velocity components (u,v), the stream function ψ, and the vorticity ω.
   c. Write the finite difference approximation of these equations for the interior nodes using the upwind scheme for the convection terms.

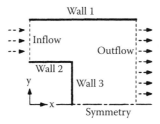

8.9. Consider two-dimensional, steady laminar flow inside a parallel plate channel of length L and spacing H. Fluid enters the channel at a constant temperature T_0, while the walls are kept at a constant temperature T_w. Viscous energy dissipation is important and hence should be included in the analysis. The velocity components u and v are available.

  a. Write the mathematical formulation of this energy problem in the nonconservative form.

  b. Write the finite difference representation of the energy equation by using central differences.

8.10. Consider a rectangular tank of height H, width W, and very long in the third dimension. There is an inflow from the top and outflow from the bottom of the tank as illustrated in the accompanying figure.

  a. Write the equations governing the velocity field in the tank by using the vorticity-stream function formulation.

  b. Write the appropriate boundary conditions.

  c. Formulate a numerical scheme for solving these equations.

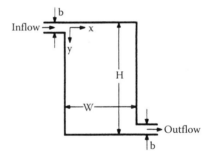

8.11. Consider the cavity Problem 8.7 written in the dimensionless vorticity-stream function formulation together with the appropriate boundary conditions. Solve this problem for a Reynolds number 30 using the explicit scheme and upwind differencing for the vorticity transport equation and the successive-over-relaxation for the stream function equation. Use a 7 × 7 grid for the first computations.

8.12. Consider flow of water in a parallel plate channel, where the distance between the plates is h = 0.05 m and the total length of the channel is L = 7 m. The upper and bottom walls are subjected to a prescribed heat flux q_w = 100 W/m² each and the inlet temperature is equal to 25°C. Solve this problem using the finite volume method with the WUDS interpolation scheme on a mesh with 100 × 100 volume. Consider hydrodynamically developed flow, where the velocity is given by

where

$$u(y) = -\frac{6\bar{u}}{h^2}(y^2 - hy), \quad v = 0$$

$$\bar{u} = \frac{\text{Re}\mu}{2\rho h}$$

Let Re = 100, ρ = 1000.52 kg/m³, μ = 0.001 kg/(m s), k = 0.597 W/(m K), and C_p = 4.1818 × 10³ J/(kg°C). Plot your results for the Nusselt number, where

$$\text{Nu}(Z) = \frac{2hq_w}{k[T_w(x) - T_{av}(x)]}$$

$$Z = \frac{\alpha x}{4\bar{u}h^2}$$

T_w is the wall temperature along the x axis, $\alpha = k/(\rho C_p)$ and

$$T_{av}(x) = \frac{\int_{y=0}^{h} u(y)T(x,y)dy}{\bar{u}h}$$

8.13. Consider a square cavity where the top and bottom walls are thermally insulated; the left wall is subjected to a temperature T_h = 12°C and the right wall is subjected to a temperature T_c = 2°C. The cavity is filled with air, with the following properties: ρ = 1.19 kg/m³, μ = 1.8 × 10⁻⁵ kg/(m s), β = 0.00341 1/K, Pr = 0.71, k = 0.02624 W/(m K), and C_p = 1035.0222 J/(kg°C). Consider g = 9.81 m/s² and solve the steady natural convection problem for three different Rayleigh (Ra) numbers: 10⁴, 10⁵, and 10⁶, where

$$\text{Ra} = \frac{\rho g \beta (T_h - T_c)L^3}{\mu \alpha}$$

and L is the length of the cavity. Solve this problem with the finite volume method, using the SIMPLEC algorithm and the WUDS interpolation scheme for a grid with 80 × 80 volumes. Plot the streamlines and isotherms for each Rayleigh number. Also consider the following dimensionless variables

$$\{x,y\} = \frac{1}{L}\{x',y'\}$$

$$\{u,v\} = \frac{L}{\alpha}\{u',v'\}$$

$$T = \frac{T' - T_c}{T_h - T_c}$$

where the variables without the superscript " ′ " mean dimensionless quantities. Calculate the maximum and minimum values of the u component of the velocity field along $x = 1/2$. Calculate the maximum and minimum value of the v component of the velocity field along $y = 1/2$. Also calculate the following local and mean Nusselt numbers:

$$Nu(x) = \int_{y=0}^{1} \left\{ uT - \frac{\partial T}{\partial x} \right\} dy$$

$$Nu_{av} = \int_{x=0}^{1} Nu(x) dx$$

$$Nu_{max} = \left\{ uT - \frac{\partial T}{\partial x} \right\} \bigg|_{max.\ at\ x=0}$$

$$Nu_{min} = \left\{ uT - \frac{\partial T}{\partial x} \right\} \bigg|_{min.at\ x=0}$$

$$Nu_{1/2} = Nu(x = 1/2)$$

$$Nu_0 = Nu(x = 0)$$

8.14. Consider a square cavity where the top and bottom walls are kept thermally insulated, the left wall is subjected to a temperature $T_h = 12°C$, and the left wall is subjected to a temperature $T_c = 2°C$. The cavity is filled with air, with the following properties: $\rho = 1.19$ kg/m^3, $\mu = 1.8 \times 10^{-5}$ kg/(m s), $\beta = 0.00341$ 1/K, Pr $= 0.71$, $k = 0.02624$ W/(m K), and $C_p = 1035.0222$ J/(kg°C). Let $g = 9.81$ m/s^2 and solve the transient natural convection problem for two different Rayleigh (Ra) numbers: 10^4, 10^5, where

$$Ra = \frac{\rho g \beta (T_h - T_c) L^3}{\mu \alpha}$$

and L is the length of the cavity. Solve this problem with the finite volume method, using the SIMPLEC algorithm and the WUDS interpolation scheme for a 50×50 grid. Consider the initial condition for temperature $T_0 = 2°C$. Consider also the following dimensionless variables

$$t = \frac{\alpha}{L^2} t'$$

$$x = \frac{x'}{L}$$

$$y = \frac{y'}{L}$$

$$u = \frac{u'L}{\alpha}$$

$$v = \frac{v'L}{\alpha}$$

$$Nu = \int_{y=0}^{1} \left[uT - k\frac{\partial T}{\partial x} \right] dy$$

$$T = \frac{T' - T_c}{T_h - T_c}$$

where the variables without the superscript " ' " are dimensionless.

a. Plot T versus x for y = 1/2 and Ra = 10^4, considering t = 0.01, t = 0.05, and t = 0.20.
b. Plot v versus x for y = 1/2 and Ra = 10^4, considering t = 0.01, t = 0.05, and t = 0.20.
c. Plot Nu versus t for x = 0 and x = 1/2, considering Ra = 10^4.
d. Plot T versus x for y = 1/2 and Ra = 10^6, considering t = 0.01, t = 0.05, and t = 0.20.
e. Plot v versus x for y = 1/2 and Ra = 10^6, considering t = 0.01, t = 0.05, and t = 0.20.
f. Plot Nu versus t for x = 0 and x = 1/2, considering Ra = 10^6.

9

Compressible Flow

The unsteady compressible Navier–Stokes equations are a set of hyperbolic–parabolic equations in time. The unsteady compressible Euler equations constitute a hyperbolic system. In the case of steady state, they become a mixed set of hyperbolic–elliptic equations in the streamwise direction. Problems modeled by hyperbolic equations have applications in a variety of fields. For example, high-speed flow with shock waves and flow over the boundaries of wings are important in aeronautical applications.

In this chapter, we focus attention on the finite difference approximations of the compressible flow problems that are governed by a system of partial differential equations that are hyperbolic in nature. We start with a simple situation involving the so-called quasi one-dimensional flow and then move to a case involving multidimensional compressible flow. Discretization schemes appropriate for these types of problems are presented.

9.1 Quasi-One-Dimensional Compressible Flow

The designation "quasi-one-dimensional flow" refers to a simplified model that applies to compressible flows in pipes and nozzles in which the fluid velocity and properties are assumed to be uniform at the cross section. Hence, such quantities vary spatially only along the longitudinal direction, despite the fact that the cross section may gradually vary as illustrated by Figure 9.1 (Anderson 1990, 1995).

The mass, momentum, and energy conservation equations for the quasi-one-dimensional compressible flow in a duct with a circular cross section can be written in vector form as (Anderson 1990, 1995):

$$\frac{\partial \mathbf{U}}{\partial t} + \frac{\partial \mathbf{F}(\mathbf{U})}{\partial x} = \mathbf{S}(\mathbf{U}) \qquad (9.1)$$

where t is time, x is the longitudinal space variable, and

$$\mathbf{U} \equiv \begin{bmatrix} \rho \\ \rho u \\ \rho\left(e + \dfrac{u^2}{2}\right) \end{bmatrix} \qquad (9.2a)$$

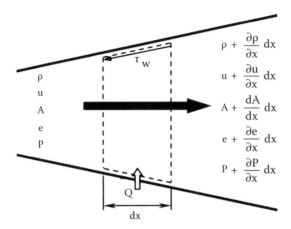

FIGURE 9.1
Properties are assumed to be uniform across the cross section, but the cross section may vary longitudinally in the quasi-one-dimensional compressible flow.

$$\mathbf{F(U)} \equiv \begin{bmatrix} \rho u \\ \rho u^2 + p \\ \rho u \left(e + \dfrac{u^2}{2} + \dfrac{p}{\rho} \right) \end{bmatrix} \quad (9.2b)$$

$$\mathbf{S(U)} \equiv -\dfrac{1}{A} \begin{bmatrix} \rho u \dfrac{dA}{dx} \\ \rho u^2 \left(\dfrac{C_f \pi D}{2} + \dfrac{dA}{dx} \right) \\ \rho u \dfrac{dA}{dx} \left(e + \dfrac{u^2}{2} + \dfrac{p}{\rho} \right) - Q \pi D \end{bmatrix} \quad (9.2c)$$

In equations (9.1) and (9.2), **U**, **F(U)**, and **S(U)** are the vectors of conserved variables, flux, and source term, respectively, while the scalar quantities are as follows: ρ is density; u is the longitudinal velocity; e is the specific internal energy; p is pressure; D and A are the local diameter and cross-section area, respectively; Q is local heat flux imposed at the surface of the duct; and C_f is the friction factor.

We note that the system of equations (9.1) and (9.2) contains four unknowns, given by ρ, u, P, and e, but three equations. Therefore, constitutive equations are needed to make the number of unknowns and number of equations equal.

Compressible Flow

An ideal gas model with constant specific heat is used here, so that we can write (Van Wylen et al. 1994)

$$e = C_v T \qquad (9.3a)$$

and

$$p = \rho R_g T \qquad (9.3b)$$

where C_v is the specific heat at constant volume, T is the temperature, and R_g is the gas constant; that is,

$$R_g = \frac{R_u}{M} \qquad (9.3c)$$

where $R_u = 8.314$ J/(mol. K) is the universal gas constant and M is the gas molecular mass.

Equations (9.3a,b) can be rewritten as

$$p = (\gamma - 1)\rho e \qquad (9.4a)$$

and

$$T = \frac{p}{\rho R_g} \qquad (9.4b)$$

where $\gamma = C_p/C_v$ and C_p is the specific heat at constant pressure. With equations (9.4a,b), the system now contains five equations and five unknowns, namely: ρ, u, p, e, and T; thus, it can be solved.

The well-posedness of the quasi-one-dimensional flow problem depends on the appropriate specification of initial and boundary conditions. The initial conditions can be specified as the values of u, T, and p in the domain, while the initial conditions for density and internal energy are then calculated by satisfying the constitutive equations (9.4a,b). On the other hand, the specification of the boundary conditions for a well-posed problem depends on whether the local flow is subsonic or supersonic. By following Yee (1981), the analytical boundary conditions for a well-posed quasi-one-dimensional flow problem are specified as follows:

- *Subsonic Inlet*: Density, ρ, needs to be specified. Thus, one can specify the sets [ρ, u] or [ρ, p] but not [p, u]. Note that one can also specify [T, p] because temperature and pressure are more easily measured, which in turn signifies specifying [ρ, p] because ρ relates to T and p by the equation of state (9.4b).
- *Subsonic Outlet*: One of the three variables ρ, u, or p needs to be specified.
- *Supersonic Inlet*: The three variables ρ, u, and p need to be specified.
- *Supersonic Outlet*: No variable needs to be specified.

Variables that are not specified at the boundary are numerically treated as follows.

9.1.1 Solution with MacCormack's Method

The finite difference approximation of the vector equation (9.1), using MacCormack's explicit predictor–corrector scheme, is given by:
Predictor:

$$\overline{U_i^{n+1}} = U_i^n - \frac{\Delta t}{\Delta x}(F_{i+1}^n - F_i^n) + \Delta t \ S_i^n \quad (9.5a)$$

Corrector:

$$U_i^{n+1} = \frac{1}{2}\left[U_i^n + \overline{U_i^{n+1}} - \frac{\Delta t}{\Delta x}(\overline{F_i^{n+1}} - \overline{F_{i-1}^{n+1}}) + \Delta t \ \overline{S_i^{n+1}}\right] \quad (9.5b)$$

where the following notation is used.

$$G(x,t) = G(i\Delta x, n\Delta t) = G_i^n \quad G \equiv U, F \text{ or } S \quad (9.6)$$

Here, the predictor step constructs approximate values for U_i^{n+1}, denoted by $\overline{U_i^{n+1}}$, for each point by using forward differences for the spatial derivative. These approximate values are then used to calculate $\overline{F_i^{n+1}}$ and $\overline{S_i^{n+1}}$. The solution for U_i^{n+1} is computed in the corrector step, which uses backward differences in the spatial derivative. The system is accurate to $0[(\Delta x)^2, (\Delta t)^2]$.

With the scheme being explicit, the maximum time step Δt should satisfy the stability criterion. In the present case, it is not possible to obtain an explicit expression for the stability criterion. Instead, the following empirical formula has been proposed (Tannehill et al. 1975; Pletcher et al. 2012):

$$\Delta t \leq \frac{\sigma(\Delta t)_{CFL}}{1 + \frac{2}{Re_\Delta}} \quad (9.7)$$

where σ is a safety factor ($\cong$ 0.7 to 0.9) and $(\Delta t)_{CFL}$ is the inviscid Courant, Friedrichs, and Lewy (CFL) condition given by (MacCormack 1971):

$$(\Delta t)_{CFL} \leq \left(\frac{|u|}{\Delta x} + \frac{a}{\Delta x}\right)^{-1} \quad (9.8)$$

The grid Reynolds number Re_Δ is given by

$$Re_\Delta = \frac{\rho|u|\Delta x}{\mu} \quad (9.9)$$

where μ is the fluid dynamic viscosity. In equation (9.8), a is the local speed of sound, that is,

$$a = \sqrt{\frac{\gamma p}{\rho}} \qquad (9.10)$$

The specification of the boundary conditions for the well-posed problem, as discussed previously, requires that some of the variables be extrapolated to the boundaries for the implementation of the MacCormack method. The variables are extrapolated to the boundary nodes from their values in the interior nodes. We consider the variables to be specified and extrapolated as follows (the subscripts designate the node index):

- *Subsonic Inlet*: Specify T and p, then extrapolate u, that is, for a node i = 0 at the inlet boundary make $u_0 = u_1$.
- *Subsonic Outlet*: Specify p and extrapolate T and u, that is, $T_I = T_{I-1}$ and $u_I = u_{I-1}$, where node I is at the outlet boundary.
- *Supersonic Inlet*: The three variables ρ, u, and p are specified and there is no need for extrapolation.
- *Supersonic Outlet*: No variables are specified and p, T and u are extrapolated, that is, $p_I = p_{I-1}$, $T_I = T_{I-1}$ and $u_I = u_{I-1}$, where node I is at the outlet boundary.

The examples presented in Section 5.2.1, for the linear wave equation with constant velocity, demonstrate that the MacCormack method, being second order, exhibits oscillations near solution discontinuities. We also noted in that section that such oscillations result from the fact that the modified equation of the MacCormack method [see equation (5.126)] has the highest order term in the truncation error with an odd derivative. Therefore, the method lacks an artificial viscosity that would induce dissipation instead of dispersion in the numerical solution. For more general models, such as the nonlinear quasi-one-dimensional compressible flow that is now examined, an artificial viscosity is usually added to the MacCormack method in order to reduce dispersion in the neighborhood of solution discontinuities. Several artificial viscosity schemes have been proposed in the literature (Hirsch 1990; Anderson 1995; Laney 1998; Pletcher et al. 2012). Here, we present the artificial viscosity term proposed by MacCormack and Baldwin (1975), which is very effective in reducing oscillations and at the same time does not add too much dissipation to the numerical solution. Such a term is given by

$$\mathbf{M} = -\varepsilon \Delta x^3 \frac{|u|+a}{4p} \left| \frac{\partial^2 p}{\partial x^2} \right| \frac{\partial \mathbf{U}}{\partial x} \qquad (9.11)$$

where $0 \leq \varepsilon \leq 0.5$ (Pletcher et al. 2012).

The artificial viscosity term is large in regions of large second derivatives of pressure that represent oscillations in the solution, but small when the solution is smooth. Equation (9.11) is added to the flux terms in equations (9.5a,b), which are now rewritten as

Predictor:

$$\overline{U_i^{n+1}} = U_i^n - \frac{\Delta t}{\Delta x}(F_{i+1}^n + M_{i+1}^n - F_i^n - M_i^n) + \Delta t\ S_i^n \qquad (9.12a)$$

Corrector:

$$U_i^{n+1} = \frac{1}{2}\left[U_i^n + \overline{U_i^{n+1}} - \frac{\Delta t}{\Delta x}(\overline{F_i^{n+1}} + \overline{M_i^{n+1}} - \overline{F_{i-1}^{n+1}} - \overline{M_{i-1}^{n+1}}) + \Delta t\ \overline{S_i^{n+1}}\right]$$

(9.12b)

The pressure derivative in the artificial viscosity is discretized with central differences, while the finite difference approximation for the first derivative of U should follow the opposite direction of the derivative of the flux **F** in each step of MacCormack's method (Pletcher et al. 2012). We note, however, that the use of central differences has also been suggested for the first derivative of U (Anderson 1995). By using backward differences in the predictor step and forward differences in the corrector step for the first derivative of U, the discretized artificial viscosity terms are given by

$$M_i^n = -\varepsilon(|u|+a)_i^n \frac{|p_{i+1}^n - 2p_i^n + p_{i-1}^n|}{p_{i+1}^n + 2p_i^n + p_{i-1}^n}(U_i^n - U_{i-1}^n) \qquad (9.13a)$$

$$\overline{M_i^{n+1}} = -\varepsilon\ \overline{(|u|+a)_i^{n+1}} \frac{|\overline{p_{i+1}^{n+1}} - 2\overline{p_i^{n+1}} + \overline{p_{i-1}^{n+1}}|}{\overline{p_{i+1}^{n+1}} + 2\overline{p_i^{n+1}} + \overline{p_{i-1}^{n+1}}}(\overline{U_{i+1}^{n+1}} - \overline{U_i^{n+1}}) \qquad (9.13b)$$

Example 9.1

A shock tube is a tube with the two ends closed, filled with a gas initially at rest, with regions of high pressure (left) and low pressure (right) separated by a diaphragm (see Figure 9.2a). The system is disturbed at time t = 0 by rupturing the diaphragm, and flow is established without wall friction and wall heat transfer. Then, the pressure discontinuity propagates to the right as a normal shock wave, while an expansion wave propagates to the left (see Figure 9.2b). Although the flow is isentropic and the pressure variation is continuous in the expansion wave, entropy and pressure varies discontinuously across the shock wave. Pressure remains at the initial values ahead of the shock wave (Region D) and to the left of the expansion wave (Region E), where the fluid is still at rest. The pressure and the velocities are equal; that is, $p_2 = p_3$ and $u_2 = u_3$, in the region between these two waves. On the other hand, since Region 2 results from a shock and Region 3 from an expansion, entropy, density, and temperature vary discontinuously from Region 2 to Region 3.

This point where such variables are discontinuous, but pressure and velocity are continuous, is called a contact surface. The shock-tube problem can be solved analytically (Anderson 1990; Hirsch 1990; Anderson 1995; Laney 1998). If the velocities at the initial time, in the regions of high and low pressure, are not zero, the shock tube is referred to as the Riemann problem (Laney 1998). Apply MacCormack's method to the shock-tube problem by using the quasi-one-dimensional flow model and examine the effect of the artificial viscosity on the numerical solution. Use the input data given in Table 9.1 (Hirsch 1990).

Solution

The number of mesh points used in the domain discretization was 100 and the time step was computed by using equation (9.7) with $\sigma = 0.9$. MacCormack's method was initially applied without the artificial viscosity, but the numerical solution was unstable due to the large dispersion in the neighborhood of discontinuities. Figures 9.2c,d present the numerical and the analytical solutions for pressure and temperature, respectively; the numerical solution was obtained with a coefficient of artificial viscosity $\varepsilon = 0.01$. Dispersion effects are clear in these figures, due to such small value of artificial viscosity. Dispersion is more apparent in Figure 9.2d, in Region 2 behind the contact surface. The numerical results present a much better agreement with the exact solution for $\varepsilon = 0.5$, as demonstrated by Figure 9.2e and 9.2f. Almost no dispersion is observed for pressure (Figure 9.2e), although this large value of artificial viscosity induces dissipation that tends to smear out the discontinuities. On the other hand, dispersion is still noticed for temperature (Figure 9.2f), especially behind the contact surface, but dissipation is also apparent ahead of the contact surface and at the shock wave.

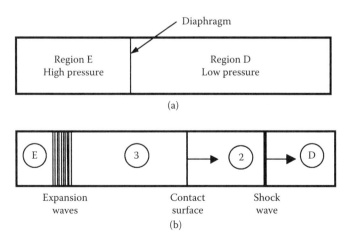

FIGURE 9.2
Shock tube: (a) Initial state; (b) Behavior after the rupture of the diaphragm. *(Continued)*

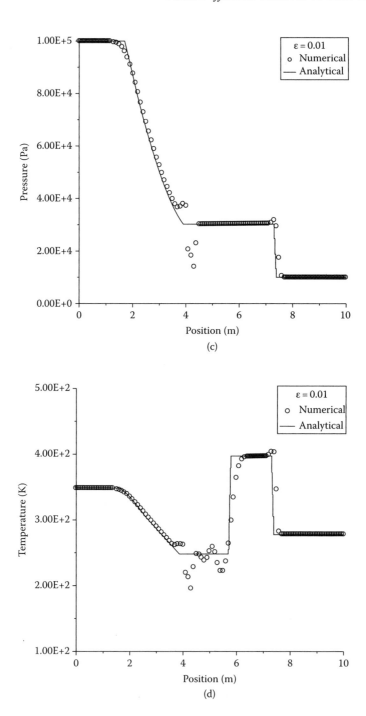

FIGURE 9.2 (CONTINUED)
MacCormack's solution for the shock-tube problem with artificial viscosity $\varepsilon = 0.01$: (c) pressure and (d) temperature. *(Continued)*

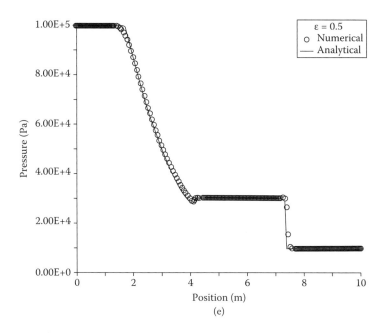

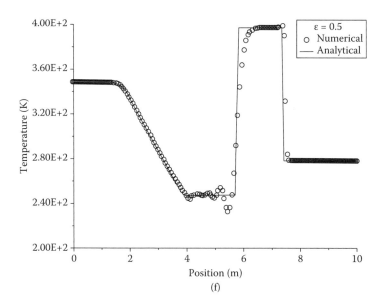

FIGURE 9.2 (CONTINUED)
MacCormack's solution for the shock-tube problem with artificial viscosity $\varepsilon = 0.5$: (e) pressure and (f) temperature.

TABLE 9.1

Input Data for the Shock-Tube Problem in Example 9.1

Pressure in Region E (Pa)	10^5
Pressure in Region D (Pa)	10^4
Density in Region E (kg/m^3)	1.00
Density in Region D (kg/m^3)	0.125
Length of the tube (m)	10
Diaphragm position (m)	4
Tube diameter	1

9.1.2 Solution with WAF-TVD Method

As discussed earlier, the MacCormack method can result in accurate solutions in smooth regions but has poor stability near shocks, which become apparent in the form of spurious oscillations and overshoots. Although such instabilities can be controlled by artificial viscosity, the choice of the parameter that gives its magnitude is empirical, and its successful use is highly dependent on the user's experience. Methods developed more recently are solution-sensitive and combine different methods based on solution features (e.g., gradients). Such methods are commonly classified as flux-averaged methods or solution-averaged methods. Flux-averaged methods can still be further classified as flux-limited, flux-corrected, and self-adjusting hybrid methods. For linear systems of equations, there is no fundamental distinction between flux averaging and solution averaging, but such is not the case for nonlinear systems (Laney 1998). In this section, a flux-averaged scheme will be applied to the quasi-one-dimensional flow equations. It is the weighted average flux–total variation diminishing (WAF-TVD) scheme that was presented by Toro (1999).

For linear systems, stability can be established through Lax's equivalence theorem (see Chapter 2), but nonlinear systems rely on the following conditions for stability: monotonicity preservation, TVD, range diminishing, positivity, upwind range, total variation boundedness, essentially nonoscillatory (ENO), and contraction, as discussed in detail by Laney (1998). Some of these conditions are stronger than others, in the sense that if they are satisfied, other weaker conditions are satisfied as well. For example, the positivity condition implies that the TVD condition is satisfied. Furthermore, a method that satisfies the TVD condition also satisfies the ENO condition, which in turn implies that the total variation is bounded. The upwind range condition is the strongest of those listed here, while total variation boundedness is the weakest. Stronger stability conditions imply that overshoots and oscillations decrease, but the solution accuracy decreases at extrema or even globally.

The TVD property implies that for a solution u to a one-dimensional problem,

$$TV[u(x, t_2)] \leq TV[u(x, t_1)] \qquad (9.14a)$$

where x is the position, t is time, and $t_2 > t_1$. Alternatively, we can write

$$TV(u^{n+1}) \leq TV(u^n) \tag{9.14b}$$

for all n. The total variation is defined as

$$TV(u^n) = \sum_{\text{all possible } i} |u_{i+1}^n - u_i^n| \tag{9.15}$$

where the subscript i indicates the discrete grid point that corresponds to the position x_i, that is, $u_i^n = u(x_i, t_n)$. The so-called TVD methods aim at satisfying the TVD condition.

The quasi-one-dimensional flow equations

$$\frac{\partial U}{\partial t} + \frac{\partial F(U)}{\partial x} = S(U) \tag{9.16}$$

are non-homogeneous due to the source terms $S(U)$ for the general case examined here that involves variation of the cross section, wall friction, and wall heat transfer [see equation (9.2c)]. For the application of the WAF-TVD scheme to these flow equations, the non-homogeneous system of nonlinear partial differential equations must be split into a non-homogeneous system of ordinary differential equations and a homogenous system of partial differential equations (Toro 1999).

A first-order splitting is given by (Toro 1999)

$$C^{(\Delta t)}: \left.\begin{array}{l} \dfrac{\partial U}{\partial t} + \dfrac{\partial F(U)}{\partial x} = 0 \\ U(x,t^n) = U^n \end{array}\right\} \Rightarrow \overline{U}^{n+1} \tag{9.17a}$$

$$S^{(\Delta t)}: \left.\begin{array}{l} \dfrac{dU}{dt} = S(U) \\ U(x,t^n) = \overline{U}^{n+1} \end{array}\right\} \Rightarrow U^{n+1} \tag{9.17b}$$

Note that the initial condition of the system of ordinary differential equations, given by equation (9.17b), is the solution of the homogeneous system (9.17a) at each time step. The splitting given by equations (9.17a,b) can be written in compact form as

$$U^{n+1} = S^{(\Delta t)} C^{(\Delta t)}(U^n) \tag{9.18}$$

where $C^{(\Delta t)}$ and $S^{(\Delta t)}$ are the operator forms of the system of partial differential equation (9.17a) and of the system of ordinary differential equation (9.17b), respectively.

A second-order splitting procedure can also be used, as given by (Toro 1999)

$$\mathbf{U}^{n+1} = S^{(\frac{1}{2}\Delta t)} C^{(\Delta t)} S^{(\frac{1}{2}\Delta t)} (\mathbf{U}^n) \qquad (9.19)$$

The solution of the system of ordinary differential equations involved in the operator $S^{(\Delta t)}$ can be obtained with well-tested subroutines available in commercial packages. Therefore, the remainder of this section is concerned with the solution of the homogeneous system of transient partial differential equations, given by equation (9.17a). This system appears both in the first- and second-order splitting procedures, given by equations (9.18) and (9.19), respectively.

The integration of equation (9.17a) in a control volume around node i gives

$$\mathbf{U}_i^{n+1} = \mathbf{U}_i^n + \frac{\Delta t}{\Delta x}(\mathbf{F}_{i+1/2}^n - \mathbf{F}_{i-1/2}^n) \qquad (9.20)$$

where the time integration was performed explicitly and we assumed a constant control volume size Δx. In equation (9.20), $\mathbf{F}_{i+1/2}$ and $\mathbf{F}_{i-1/2}$ denote the flux vector given by equation (9.2b) at the right and left boundaries of the control volume around node i, respectively. These fluxes are approximated by the WAF-TVD scheme given by Toro (1999). For the flux $\mathbf{F}_{i+1/2}$ we have

$$\mathbf{F}_{i+1/2} = \frac{1}{2}(\mathbf{F}_i + \mathbf{F}_{i+1}) - \frac{1}{2}\sum_{k=1}^{N}\text{sign}(c_k)\phi_{i+1/2}^{(k)}\Delta\mathbf{F}_{i+1/2}^{(k)} \qquad (9.21)$$

where the flux $\mathbf{F}_i$ is calculated with $\mathbf{U}_i$, while the flux $\mathbf{F}_{i+1}$ is calculated with $\mathbf{U}_{i+1}$. N = 3 is the number of waves in the solution of Riemann's problem and

$$\phi_{i+1/2}^{(k)} \equiv \phi_{i+1/2}(r^{(k)}, |c_k|) \qquad (9.22)$$

is the limiter function calculated with the flow parameter $r^{(k)}$ and the absolute value of c_k, where c_k denotes the Courant number with speed S_k for the wave k (=1,2,3), that is,

$$c_k = \frac{\Delta t\, S_k}{\Delta x} \qquad (9.23)$$

with $c_0 = -1$ and $c_{N+1} = 1$. The sign function for c_k is given by

$$\text{sign}(c_k) = \frac{c_k}{|c_k|} \qquad (9.24)$$

The flux jump across wave k is given by

$$\Delta\mathbf{F}_{i+1/2}^{(k)} = \mathbf{F}_{i+1/2}^{(k+1)} - \mathbf{F}_{i+1/2}^{(k)} \qquad (9.25)$$

where $\mathbf{F}_{i+1/2}^{(k)} = \mathbf{F}_{i+1/2}(\mathbf{U}^{(k)})$,

$$\mathbf{U}^{(1)} = \mathbf{U}_i^n \tag{9.26a}$$

$$\mathbf{U}^{(2)} = \mathbf{U}_{*E} \tag{9.26b}$$

$$\mathbf{U}^{(3)} = \mathbf{U}_{*D} \tag{9.26c}$$

$$\mathbf{U}^{(4)} = \mathbf{U}_{i+1}^n \tag{9.26d}$$

and the vectors $\mathbf{U}_{*E}$ e $\mathbf{U}_{*D}$ are given by an approximate Riemann solver, described as follows.

The parameter $r^{(k)}$ used in the limiter equation (9.22) is obtained with wave k in Riemann's problem. It is given by

$$r^{(k)} = \begin{cases} \dfrac{\Delta q_{i-1/2}^{(k)}}{\Delta q_{i+1/2}^{(k)}} & \text{if } c_k > 0 \\[2ex] \dfrac{\Delta q_{i+3/2}^{(k)}}{\Delta q_{i+1/2}^{(k)}} & \text{if } c_k < 0 \end{cases} \tag{9.27}$$

where the quantity q is selected as a variable that changes across each wave family in Riemann's problem, usually taken as the density, ρ, or the specific internal energy, e; $\Delta q_{i-1/2}^{(k)}$ is calculated with the solution $\mathbf{U}_{i-1/2}(x,t)$ of Riemann's problem with the vectors $\mathbf{U}_{i-1}$ and $\mathbf{U}_i$, while $\Delta q_{i+3/2}^{(k)}$ is calculated with the solution $\mathbf{U}_{i+3/2}(x,t)$ of Riemann's problem with the vectors $\mathbf{U}_{i+1}$ and $\mathbf{U}_{i+2}$. Analogously, $\Delta q_{i+1/2}^{(k)}$ is obtained from the solution $\mathbf{U}_{i+1/2}(x,t)$ with $\mathbf{U}_i$ and $\mathbf{U}_{i+1}$.

Different limiter functions can be found in the literature. One of the functions recommended by Toro (1999) is

$$\phi(r^{(k)}, |c_k|) = \begin{cases} 1 & \text{if } r^{(k)} \leq 0 \\[1ex] 1 - \dfrac{(1-|c_k|)2r^{(k)}}{1+r^{(k)}} & \text{if } r^{(k)} > 0 \end{cases} \tag{9.28}$$

For the approximate Riemann solver, we follow Toro (1999) and use the HLLC scheme, which is a modification of the HLL scheme proposed by Harten, Lax, and van Leer. Details are avoided here because they can be found in Toro (1999). In the HLLC scheme, the approximations for $\mathbf{U}_{*E}$ and $\mathbf{U}_{*D}$ are given by

$$\mathbf{U}_{*K} = \rho_K \left(\frac{S_K - u_K}{S_K - S_*}\right) \begin{bmatrix} 1 \\ S_* \\ \frac{E_K}{\rho_K} + (S_* - u_K)\left[S_* + \frac{p_K}{\rho_K(S_K - u_K)}\right] \end{bmatrix} \quad (9.29)$$

for the subscripts K = E and K = D, with

$$S_E = u_E - a_E f_E \quad (9.30a)$$

$$S_* = u_* \quad (9.30b)$$

$$S_D = u_D + a_D f_D \quad (9.30c)$$

where

$$f_K = \begin{cases} 1 & \text{if } p_* \leq p_K \\ \left[1 + \frac{\gamma + 1}{2\gamma}\left(\frac{p_*}{p_K} - 1\right)\right]^{1/2} & \text{if } p_* > p_K \end{cases} \quad (9.31)$$

and

$$p_* = \frac{1}{2}(p_E + p_D) - \frac{1}{2}(u_D - u_E)\bar{\rho}\,\bar{a} \quad (9.32a)$$

$$u_* = \frac{1}{2}(u_E + u_D) - \frac{1}{2}\frac{(p_D - p_E)}{\bar{\rho}\,\bar{a}} \quad (9.32b)$$

$$\bar{\rho} = \frac{1}{2}(\rho_E + \rho_D) \quad (9.32c)$$

$$\bar{a} = \frac{1}{2}(a_E + a_D) \quad (9.32d)$$

In these equations, p, ρ, and a refer to pressure, density, and sound velocity, respectively.

Toro (1999) recommended that the time step for the WAF-TVD scheme be selected with the following empirical criterion

$$\Delta t = \frac{\sigma \Delta x}{S_{max}^n} \quad (9.33a)$$

The safety factor can be taken as $\sigma = 0.9$, although it might need to be reduced to values around 0.2 in the first few time steps for problems with initial conditions involving discontinuities, such as the shock tube. The maximum wave speed of all control volumes in the domain is given by

$$S^n_{max} = \max_i \left[|u_i^n| + a_i^n\right] \quad (9.33b)$$

Example 9.2

Solve the shock-tube problem by using the WAF-TVD scheme described earlier for the conditions specified in Table 9.2.

TABLE 9.2

Input Data for the Shock-Tube Problem in Example 9.2

Pressure in Region E (Pa)	10^5
Pressure in Region D (Pa)	10^4
Density in Region E (kg/m^3)	1.00
Density in Region D (kg/m^3)	0.125
Length of the tube (m)	10
Diaphragm position (m)	5
Tube diameter	1

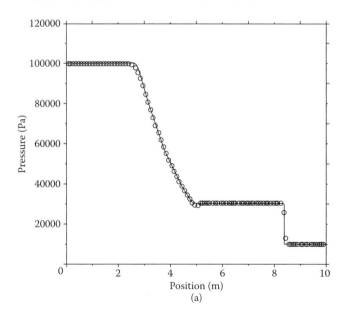

FIGURE 9.3
WAF-TVD solution for the shock-tube problem: (a) pressure. *(Continued)*

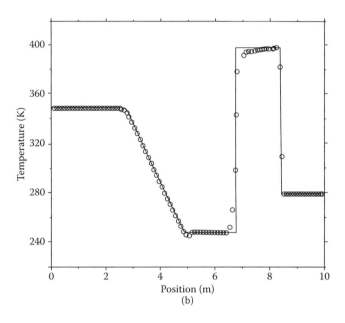

FIGURE 9.3 (CONTINUED)
WAF-TVD solution for the shock-tube problem: (b) temperature.

Solution

The WAF-TVD results for pressure and temperature for this example are presented by Figure 9.3a and b, respectively, at time t = 6.1 ms. This solution was obtained with 100 control volumes used for the discretization of the spatial domain. An analysis of Figure 9.3a and b reveals that the numerical solution is in excellent agreement with the exact one. In fact, the WAF-TVD solution does not exhibit oscillations near the discontinuities, although some dispersion is noticed in the temperature variation presented by Figure 9.3b. The dispersion effects are much smaller than those observed with MacCormack's method (see the results for Example 9.1). Furthermore, the WAF-TVD does not require an empirical artificial viscosity as in MacCormack's method.

9.2 Two-Dimensional Compressible Flow

This section illustrates the application of the WAF-TVD scheme to the compressible axisymmetric flow of a gas, modeled by the time-dependent Euler equations (Anderson 1990, 1995). Euler's equations form a system of nonlinear hyperbolic conservation laws that govern the dynamics of a compressible material, such as gas or liquid at high pressure, for which the effects of body forces, viscous stresses, and heat conduction are neglected.

Compressible Flow

The two-dimensional transient Euler equations in cylindrical axisymmetric coordinates are given by

$$\frac{\partial \mathbf{U}}{\partial t} + \frac{\partial \mathbf{F}(\mathbf{U})}{\partial R} + \frac{\partial \mathbf{G}(\mathbf{U})}{\partial x} = \mathbf{S}(\mathbf{U}) \tag{9.34a}$$

$$\mathbf{U} = \begin{bmatrix} \rho \\ \rho u \\ \rho v \\ E \end{bmatrix} \quad \mathbf{F} = \begin{bmatrix} \rho u \\ \rho u^2 + p \\ \rho u v \\ u(E+p) \end{bmatrix} \tag{9.34b,c}$$

$$\mathbf{G} = \begin{bmatrix} \rho v \\ \rho u v \\ \rho v^2 + p \\ v(E+p) \end{bmatrix} \quad \mathbf{S} = -\frac{1}{R}\begin{bmatrix} \rho u \\ \rho u^2 \\ \rho u v \\ u(E+p) \end{bmatrix} \tag{9.34d,e}$$

where $\rho(x,R,t)$ = density, $p(x,R,t)$ = pressure, $u(x,R,t)$ = radial velocity, and $v(x,R,t)$ = axial velocity, and the total energy is given by

$$E = \rho\left[\frac{1}{2}(u^2 + v^2) + e\right] \tag{9.35}$$

where e is the specific internal energy.

Such as for the one-dimensional case of the previous section, the system given by equations (9.34) and (9.35) for the compressible flow is not sufficient to completely describe the physical processes involved. This is because there are more unknowns than equations. The unknowns are ρ, u, v, e, and p. Thus, closure conditions are required. Physically, such conditions are statements related to the nature of the medium (Van Wylen et al. 1994). For ideal gases, one of the closure conditions is given by the equation of state

$$p = \rho R_g T \tag{9.36}$$

where R_g is the gas constant, that is,

$$R_g = \frac{R_u}{M} \tag{9.37}$$

R_u = 8.314 J/(mol. K) is the universal gas constant, and M is the gas molecular mass. The other closure condition takes into account the fact that the internal energy of an ideal gas is a function of temperature only. For a gas with constant specific heat, the second closure condition is given by

$$e = \frac{p}{(\gamma - 1)\rho} \tag{9.38}$$

where γ is the ratio of specific heats at constant pressure and constant volume, that is,

$$\gamma = \frac{C_p}{C_v} \tag{9.39}$$

Different equations of state can be found in the literature for real gases. One example is Van der Waal's equation of state given in the form

$$(p + \rho^2 c)\left(\frac{1}{\rho} - b\right) = R_g T \tag{9.40}$$

where b and c are constants for the gas. In real gases, the specific energy is obtained by integrating the following equation

$$de = C_v\, dT - \frac{1}{\rho^2}\left[T\left(\frac{\partial p}{\partial T}\right)_\rho - p\right] d\rho \tag{9.41}$$

The system given by equations (9.34) and (9.35), together with the closure conditions, which can be obtained by assuming an ideal gas model (see equations (9.36) and (9.38) or one of the models for real gases, [see, for example, equations (9.40) and (9.41)], is numerically integrated by using appropriate initial and boundary conditions for the problem. The one-dimensional WAF-TVD scheme presented in the previous section is now extended for the two-dimensional axisymmetric flow problem.

A first-order splitting for equation (9.34a) is used for the time integration, given by (Toro 1999)

$$\left. \begin{array}{c} \dfrac{\partial U}{\partial t} + \dfrac{\partial F(U)}{\partial R} + \dfrac{\partial G(U)}{\partial x} = 0 \\ U(x, R, t^n) = U^n \end{array} \right\} \Rightarrow \overline{U}^{n+1} \tag{9.42a}$$

$$\left. \begin{array}{c} \dfrac{dU}{dt} = S(U) \\ U(x, R, t^n) = \overline{U}^{n+1} \end{array} \right\} \Rightarrow U^{n+1} \tag{9.42b}$$

which can be written in compact form as:

$$U^{n+1} = S^{(\Delta t)} C^{(\Delta t)}(U^n) \tag{9.43}$$

where $C^{(\Delta t)}$ and $S^{(\Delta t)}$ are the operator forms of the system of partial differential equation (9.42a) and of the system of ordinary differential equation (9.42b), respectively. Similarly to the one-dimensional case presented in the previous section, a second-order splitting can be used in the form (Toro 1999):

$$U^{n+1} = S^{(\frac{1}{2}\Delta t)} C^{(\Delta t)} S^{(\frac{1}{2}\Delta t)}(U^n) \tag{9.44}$$

Compressible Flow

The solution of the homogeneous system given by equation (9.42a) can be obtained by dimensional splitting, which permits the extension of methods developed for one-dimensional systems to be used for the solution of two-dimensional systems. In the dimensional splitting, the system given by equation (9.42a) is replaced by a pair of one-dimensional systems. In order to evolve the solution of system (9.42a) from time t^n to time t^{n+1}, the simplest version of this approach is given by the following sequence:

$$\left. \begin{array}{c} \dfrac{\partial U}{\partial t} + \dfrac{\partial F(U)}{\partial R} = 0 \\ U(x, R, t^n) = U^n \end{array} \right\} \overset{\Delta t}{\Rightarrow} U^{n+\frac{1}{2}} \qquad (9.45a)$$

$$\left. \begin{array}{c} \dfrac{\partial U}{\partial t} + \dfrac{\partial G(U)}{\partial x} = 0 \\ U(x, R, t^n) = U^{n+\frac{1}{2}} \end{array} \right\} \overset{\Delta t}{\Rightarrow} \overline{U}^{n+1} \qquad (9.45b)$$

In the first step of the sequence, a one-dimensional system in the R direction is solved with a time step Δt. This is called the R sweep, and its solution is denoted by $U^{n+1/2}$. In the next step, a one-dimensional system is solved for the x direction and with a time step Δt, by using as the initial condition the solution of the R sweep, $U^{n+1/2}$. This second step is called the x sweep, and its solution is the solution of the system (9.42a) at time t^{n+1}, that is, $\overline{U}^{n+1}$.

The splitting procedure given by equations (9.45a,b) can be written in compact form as

$$\overline{U}^{n+1} = X^{(\Delta t)} \Psi^{(\Delta t)} (U^n) \qquad (9.46)$$

where $\Psi^{(\Delta t)}$ and $X^{(\Delta t)}$ are the operator forms of the one-dimensional systems (9.45a) and (9.45b), respectively. The scheme given by equation (9.46) is *first-order accurate* in time if the operators $\Psi^{(\Delta t)}$ and $X^{(\Delta t)}$ are at least first-order accurate in time (Toro 1999). A *second-order accurate* splitting is given by (Toro 1999)

$$\overline{U}^{n+1} = \Psi^{(\frac{1}{2}\Delta t)} X^{(\Delta t)} \Psi^{(\frac{1}{2}\Delta t)} (U^n) \qquad (9.47)$$

if the one-dimensional operators are at least second order in time.

With the splitting given by equation (9.46) or (9.47), the solution of the two-dimensional homogeneous system for Euler equations is obtained with an extension of the one-dimensional technique described in the previous section, with the one-dimensional vectors added with the elements due to the two-dimensional effects. For the implementation of the splitting procedure, it is more convenient to rewrite the flux vectors **F** and **G**, so that a single one-dimensional subroutine needs to be used. The R and x sweeps are then rewritten, respectively, as

$$\frac{\partial}{\partial t}\begin{bmatrix} \rho \\ \rho u \\ E \\ \rho v \end{bmatrix} + \frac{\partial}{\partial R}\begin{bmatrix} \rho u \\ \rho u^2 + p \\ u(E+p) \\ \rho u v \end{bmatrix} = 0 \qquad (9.48a)$$

$$\frac{\partial}{\partial t}\begin{bmatrix} \rho \\ \rho v \\ E \\ \rho u \end{bmatrix} + \frac{\partial}{\partial x}\begin{bmatrix} \rho v \\ \rho v^2 + p \\ v(E+p) \\ \rho u v \end{bmatrix} = 0 \qquad (9.48b)$$

In the R sweep, equation (9.48a) differs from the pure one-dimensional problem in two aspects, namely: (i) There is one additional equation for momentum in the x direction and (ii) the total energy E contains contributions from the velocity component in the x direction, v, for the kinetic energy. For the R sweep, the axial velocity component v is said to be passively advected with radial velocity u. Analogously, for the x sweep, the radial velocity component u is said to be passively advected with axial velocity v.

PROBLEMS

9.1. Fanno's line gives the solutions for compressible one-dimensional adiabatic flows in a duct with friction. Use MacCormack's method to compute the solution for the following conditions in a circular duct:

Inlet pressure (Pa)	101,300
Outlet pressure (Pa)	63,210
Inlet temperature (K)	273
Wall friction coefficient	0.005
Safety factor for Δt	0.9
Coefficient of artificial viscosity	0
Number of spatial grid points	100
Wall heat flux (W/m^2)	0
Duct length (m)	30
Duct diameter (m)	0.015

9.2. Rayleigh's line gives the solutions for compressible one-dimensional flows in a duct without friction but with wall heat transfer. Use MacCormack's method to compute the solution for the following conditions in a circular duct:

Inlet pressure (Pa)	137895.2
Outlet pressure (Pa)	68947.6
Inlet temperature (K)	333
Wall friction coefficient	0
Safety factor for Δt	0.9
Coefficient of artificial viscosity	0
Number of spatial grid points	100
Wall heat flux (W/m^2)	436379.35
Duct length (m)	10
Duct diameter (m)	0.1719

9.3. Apply MacCormack's method to solve the compressible quasi-one-dimensional flow in Shubin's divergent nozzle, with the flow area given by $A(x) = 1.398 + 0.347 \tanh(0.8x-4)$, where x denotes the distance from the nozzle inlet. Use the following parameters for the problem, with supersonic inlet conditions:

Inlet pressure (Pa)	38,048
Outlet pressure (Pa)	73418.6
Inlet temperature (K)	228.57
Inlet velocity (m/s)	379
Wall friction coefficient	0
Safety factor for Δt	0.95
Coefficient of artificial viscosity	0.5
Number of spatial grid points	81
Wall heat flux (W/m^2)	0
Nozzle length (m)	10

9.4. Apply MacCormack's method to solve the compressible quasi-one-dimensional flow in a convergent divergent nozzle, with the flow area given by

$$A(x) = \begin{cases} 1 + 1.5[1-(x/5)]^2, & x < 5 \\ 1 + 0.5[1-(x/5)]^2, & x \geq 5 \end{cases}$$

Use the following parameters for the problem and compute the solutions with outlet pressures of 95,000 Pa and 90,000 Pa, with subsonic inlet conditions.

Inlet pressure (Pa)	101,300
Inlet temperature (K)	300
Wall friction coefficient	0
Safety factor for Δt	0.9
Coefficient of artificial viscosity	0 and 1.5
Number of spatial grid points	160
Wall heat flux (W/m²)	0
Nozzle length (m)	10

9.5. Repeat problem 9.1 by solving the one-dimensional flow with the WAF-TVD scheme.

9.6. Repeat problem 9.2 by solving the one-dimensional flow with the WAF-TVD scheme.

9.7. Repeat problem 9.3 for the Shubin nozzle by using the WAF-TVD scheme.

9.8. Use the WAF-TVD scheme to solve the shock-tube problem with the following conditions:

Pressure in Region E (Pa)	10^5
Pressure in Region D (Pa)	10^3
Density in Region E (kg/m³)	1.00
Density in Region D (kg/m³)	0.01
Length of the tube (m)	10
Diaphragm position (m)	5
Tube diameter	1

10
Phase Change Problems

The transient heat transfer problems involving melting or solidification are generally referred to as "phase change" or "moving boundary" problems. Sometimes, they are referred to as "Stefan problems," with reference to the pioneering work of J. Stefan (1891) in connection with the melting of the polar ice cap (Vuik 1993). Phase change problems have numerous applications in such areas as the making of ice, the freezing of food, the solidification of metals in castings, the cooling of large masses of igneous rock, and in many others.

The mathematical formulation of phase change problems is governed by the partial differential equation of parabolic type; but because the location of the moving solid–liquid interface is not known a priori, it has to be determined as a part of the solution. As a result, the phase change boundary problems are nonlinear, and their analytic solution is very difficult. A limited number of exact analytic solutions for idealized situations can be found in textbooks, such as by Carslaw and Jaeger (1959) and Özişik (1993). Purely numerical methods of solving phase change problems have been reported by a number of investigators, including Crank (1957), Murray and Landis (1959), Douglas and Gallie (1955), Goodling and Khader (1974a, 1974b), Crank and Gupta (1972), Gupta (1974), Gupta and Kumar (1980, 1981), Pham (1985), Tacke (1985), and Poirier and Salcudean (1988).

The numerical methods of solving phase change problems may be categorized as:

1. *Fixed grid methods* are those in which the space and time domains are subdivided into a finite number of equal grids Δx, Δt for all times. Then the moving solid–liquid interface will, in general, lie somewhere between two grid points at any given time. The methods of Crank (1957) and Ekrlick (1958) may be the examples for estimating the location of the interface by a suitable interpolation formula as a part of the solution.

2. *Variable grid methods* are those in which the space and time domains are subdivided into equal intervals in one direction only, and the corresponding grid side in the other direction is determined so that the moving boundary always remains at a grid point. For example, Murray and Landis (1959) choose equal steps Δt in the time variable and kept the number of space intervals fixed, which in turn allowed the size of the space interval Δx to be changed (decreased or increased) as the interface moved. In an

alternative approach, the space domain can be subdivided into fixed equal intervals Δx and the time step Δt allowed to vary in such a manner that the moving interface always remains at a grid point at the end of each time interval Δt. Several variations of such a variable time step approach have been reported by Douglas and Gallie (1955), Goodling and Khader (1974a and 1974b), Crank and Gupta (1972), Gupta (1974), and Gupta and Kumar (1980, 1981).

3. *Front fixing methods,* used in one-dimensional problems, are essentially coordinate transformation schemes that immobilize the moving front; hence, they alleviate the need for tracking the moving front at the expense of solving a more complicated problem by the numerical scheme. The works of Crank (1957) and Furzeland (1980) are examples of such an approach.

4. *Adaptive grid generation methods,* advanced by Brackbill and Saltzman (1982), Saltzman and Brackbill (1982), and Brackbill (1982), may also be used for solving multidimensional moving boundary problems in bodies having irregular shapes. In this approach, the numerical grid generation is applied to map the irregular region into a regular-shaped region in the computational domain where the problem is solved with finite differences and the results are transformed back into the physical domain. Because the grids are generated at each successive time step, the method requires considerable amounts of computer time.

5. *Enthalpy methods* have been used by several investigators, including Price and Slack (1954), Rose (1960), Oleinik (1960), Meyer (1973), Shamsundar and Sparrow (1975), Crowley (1978), Voller and Cross (1981, 1983), Tacke (1985), Colaço et al. (2003, 2004, 2005), Colaço and Dulikravich (2007), and Dulikravich et al. (2004) to solve phase change problems, even in situations in which the material does not have a distinct solid–liquid interface, that is, the melting or solidification takes place over an extended range of temperatures. The solid and liquid phases are separated by a two-phase moving region. In this approach, an enthalpy function, H(T), which is the total heat content of the substance, is used as a dependent variable along with the temperature.

In this chapter, we review the mathematical formulation of phase change problems and then discuss the numerical methods of solution of one-dimensional cases, with both the variable time step and the enthalpy methods. Sections 10.1 to 10.4 describe some methods for a purely diffusive problem, and at the end of the chapter we will present a formulation for a diffusive–convective problem, using the finite volume methodology presented in Chapter 8.

10.1 Mathematical Formulation of Phase Change Problems

To illustrate the mathematical formulation of phase change problems involving solidification or melting, we consider the following one-dimensional solidification problem.

A liquid confined to a semi-infinite region $x > 0$ is initially at a uniform temperature, T_i^*, higher than the melting temperature, T_m^*, of the liquid. At time $t = 0$, the temperature at the boundary surface $x = 0$ is lowered to T_0, which is less than the melting temperature, T_m^*. As a result, the freezing starts at $x = 0$ and the location of the solid–liquid interface, $S(t)$, moves in the positive x direction as illustrated in Figure 10.1a. Assuming no fluid motion, the heat transfer through the liquid and solid phases takes place by conduction. The mathematical formulation of this phase change problem consists of parabolic transient heat conduction equations for the liquid and solid phases given by

$$\frac{\partial^2 T_s}{\partial x^2} = \frac{1}{\alpha_s}\frac{\partial T_s}{\partial t} \quad \text{for the solid phase} \tag{10.1a}$$

$$\frac{\partial^2 T_l}{\partial x^2} = \frac{1}{\alpha_l}\frac{\partial T_l}{\partial t} \quad \text{for the liquid phase} \tag{10.1b}$$

where $T_s(x,t)$ and $T_l(x,t)$ are the temperatures in the solid and liquid regions, respectively. Equations (10.1a,b) are coupled at the solid–liquid interface as described next.

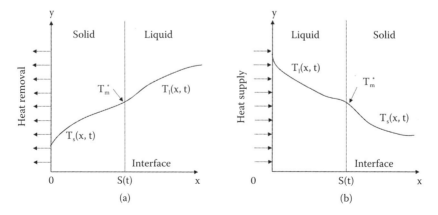

FIGURE 10.1
One-dimensional (a) solidification and (b) melting problems.

10.1.1 Interface Condition

Figure 10.1a and b shows the location, S(t), of the solid–liquid interface for the cases of solidification and melting, respectively. The energy balance equation for the solid–liquid interface can be stated as

$$\begin{pmatrix} \text{Rate of heat} \\ \text{removed from the} \\ \text{solid phase in the} \\ \text{negative x direction} \end{pmatrix} = \begin{pmatrix} \text{Rate of heat supplied} \\ \text{to the interface from} \\ \text{the liquid phase in the} \\ \text{negative x direction} \end{pmatrix} + \begin{pmatrix} \text{Rate of heat} \\ \text{liberated at the} \\ \text{interface during} \\ \text{solidification} \end{pmatrix} \quad (10.2a)$$

Introducing the equivalent mathematical expressions for each of these three terms, we obtain

$$k_s \frac{\partial T_s}{\partial x} = k_l \frac{\partial T_l}{\partial x} + \rho_s L \frac{dS(t)}{dt} \quad (10.2b)$$

where L is the latent heat of solidification (or melting) per unit mass, J/kg; k_s and k_l are the thermal conductivities of the solid and liquid phases, respectively; ρ_s is the density of the solid phase; and S(t) is the location of the solid–liquid interface.

In general, the solid and liquid densities are not the same. Therefore, motion of liquid resulting from the density change is expected in actual situations. Usually the solid density ρ_s is greater than the liquid density ρ_l (except for water, bismuth, and antimony) at the melting point and, as a result, liquid motion is expected toward the interface during solidification. However, it can be shown that convection effects due to density differences cancel out if the solid phase density ρ_s is used in the interface energy balance equation (10.2b). An additional boundary condition for the solid–liquid interface is obtained by requiring that the solid and liquid phase temperatures should be equal to the melting (or solidification) temperature.

Summarizing, for the problem of one-dimensional planar solidification illustrated in Figure 10.1, the boundary conditions for the solid–liquid interface are taken as

$$T_s(x,t) = T_l(x,t) = T_m^* \quad \text{at } x = S(t) \quad (10.3a)$$

$$k_s \frac{\partial T_s}{\partial x} - k_l \frac{\partial T_l}{\partial x} = \rho_s L \frac{dS(t)}{dt} \quad \text{at } x = S(t) \quad (10.3b)$$

The boundary condition (10.3b) is derived for the solidification problem illustrated in Figure 10.1a; it can be shown that it is also applicable for the melting problem in Figure 10.1b. We note that, in both the solidification and melting problems shown in Figure 10.1, the solid–liquid interface is moving in the positive x direction.

Phase Change Problems

The nonlinearity of the interface energy balance boundary condition (10.3b) can be shown by relating $dS(t)/dt$ to the derivative of temperatures by taking the total derivative of the interface condition (10.3a) as discussed by Özişik (1993).

10.1.2 Generalization to Multidimensions

The interface boundary conditions given by equations (10.3a,b) can be generalized to the multidimensional situation illustrated in Figure 10.2. Let the solid and liquid phases be separated by an interface defined by the equation

$$F(x,y,z,t) = 0 \qquad (10.4)$$

Assuming that the densities of the solid and liquid phases are the same, the three-dimensional forms of the interface boundary conditions become

$$T_s(x,y,z,t) = T_l(x,y,z,t) = T_m^* \quad \text{at} \quad F(x,y,z,t) = 0 \qquad (10.5a)$$

$$k_s \frac{\partial T_s}{\partial n} - k_l \frac{\partial T_l}{\partial n} = \rho L v_n \quad \text{at} \quad F(x,y,z,t) = 0 \qquad (10.5b)$$

where $\frac{\partial}{\partial n}$ denotes derivatives along the direction of the normal vector $\mathbf{n}$ to the surface, and v_n is the velocity of the interface at the location P in the direction $\mathbf{n}$.

The interface boundary conditions (10.5) are not in a suitable form for analytic or numerical solutions. Alternative forms of these equations are given by Özişik (1993) as

$$T_s(x,y,z,t) = T_l(x,y,z,t) = T_m^* \quad \text{at} \quad z = S(x,y,t) \qquad (10.6a)$$

$$\left[1 + \left(\frac{\partial S}{\partial x}\right)^2 + \left(\frac{\partial S}{\partial y}\right)^2\right]\left[k_s \frac{\partial T_s}{\partial z} - k_l \frac{\partial T_l}{\partial z}\right] = \rho L \frac{\partial S(x,y,t)}{\partial t} \quad \text{at} \quad z = S(x,y,t) \qquad (10.6b)$$

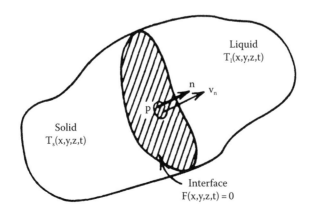

FIGURE. 10.2
Solid–liquid interface for a multidimensional situation.

where
$$F(x,y,z,t) \equiv z - S(x,y,t) = 0 \tag{10.7}$$

Clearly, for the two-dimensional case, (x,z), equation (10.6b) simplifies to

$$\left[1+\left(\frac{\partial S}{\partial x}\right)^2\right]\left[k_s\frac{\partial T_s}{\partial z}-k_l\frac{\partial T_l}{\partial z}\right]=\rho L\frac{\partial S(x,t)}{\partial t} \quad \text{at } z=S(x,t) \tag{10.8a}$$

where
$$F(x,z,t) \equiv z - S(x,t) = 0 \tag{10.8b}$$

And, for the one-dimensional case, (z), equations (10.8a,b) become

$$k_s\frac{\partial T_s}{\partial z}-k_l\frac{\partial T_l}{\partial z}=\rho L\frac{dS(t)}{dt} \quad \text{at } z=S(t) \tag{10.9a}$$

where
$$F(z,t) \equiv z - S(t) = 0 \tag{10.9b}$$

Equation (10.9a) is similar to equation (10.3b) with z replacing x.

10.1.3 Dimensionless Variables

The dimensionless variables associated with the phase change problem can be illustrated by considering the interface energy balance equation (10.3b)

$$k_s\frac{\partial T_s}{\partial x}-k_l\frac{\partial T_l}{\partial x}=\rho_s L\frac{dS(t)}{dt} \quad \text{at } x=S(t) \tag{10.10}$$

This equation can be written in the dimensionless form as

$$\frac{\partial \theta_s}{\partial \eta}-\frac{k_l}{k_s}\frac{\partial \theta_l}{\partial \eta}=\frac{1}{\text{Ste}}\frac{d\delta(t)}{d\tau} \quad \text{at } \eta=\delta(t) \tag{10.11}$$

where the dimensionless variables are defined as

$$\theta_j(\tau,\eta)=\frac{T_j(x,t)-T_m^*}{T_m^*-T_0}, \quad j=s \text{ or } l \tag{10.12a}$$

$$\eta=\frac{x}{B}, \quad \delta(\tau)=\frac{S(t)}{B} \tag{10.12b}$$

$$\tau=\frac{\alpha t}{B^2}, \quad \text{Ste}=\frac{C_{ps}(T_m^*-T_0)}{L} \tag{10.12c}$$

Here, B is a reference length, T_m^* is the melting temperature, T_0 is a reference temperature, S(t) is the location of the solid–liquid interface, and Ste is the Stefan number (named after J. Stefan). The Stefan number signifies the importance of sensible heat relative to latent heat. If the Stefan number is small, say, less than about 0.1, the heat released or absorbed by the interface during phase change is affected very little as a result of the variation of the sensible heat content of the substance during the phase change process. For materials such as aluminum, copper, iron, lead, nickel, tin, and so on, the Stefan number based on the temperature difference between the melting temperature and the room temperature varies from about 1 to 3. For freezing and melting of water in lakes, rivers, and so on for a temperature difference of 10 °C, the value of the Stefan number is 0.06 for freezing and 0.12 for thawing problems. For thermal storage problems, the range of the Stefan number is $0 \leq \text{Ste} \leq 0.1$.

10.1.4 Mathematical Formulation

We now present the mathematical formulation of the one-dimensional solidification problem discussed earlier.

We assume heat transfer through the solid and liquid phases takes place by conduction only and thermal properties are constant. The mathematical formulation of this solidification problem is given by:

The solid phase:

$$\frac{\partial^2 T_s}{\partial x^2} = \frac{1}{\alpha_s}\frac{\partial T_s}{\partial t} \quad \text{in } 0 < x < S(t), \quad t > 0 \tag{10.13a}$$

$$T_s(x,t) = T_0 \quad \text{at } x = 0, \quad t > 0 \tag{10.13b}$$

The liquid phase:

$$\frac{\partial^2 T_l}{\partial x^2} = \frac{1}{\alpha_l}\frac{\partial T_l}{\partial t} \quad \text{in } S(t) < x < \infty, \quad t > 0 \tag{10.14a}$$

$$T_l \to T_i^* \quad x \to \infty, \quad t > 0 \tag{10.14b}$$

$$T_l(x,t) = T_i^* \quad \text{for } t = 0, \quad \text{in } 0 \leq x < \infty \tag{10.14c}$$

where T_i^* is the initial temperature of the liquid phase. The interface conditions are

$$T_s(x,t) = T_l(x,t) = T_m^* \quad \text{at} \quad x = S(t), \quad t > 0 \tag{10.15a}$$

$$k_s\frac{\partial T_s}{\partial x} - k_l\frac{\partial T_l}{\partial x} = \rho L\frac{dS(t)}{dt} \quad \text{at} \quad x = S(t), \quad t > 0 \tag{10.15b}$$

Note that no initial condition is specified for the solid phase because no solid phase exists for $t = 0$. The aforementioned set of equations provides the complete mathematical formulation of this solidification problem.

10.2 Variable Time Step Approach for Single-Phase Solidification

Several different versions of the variable time step approach have been proposed by various investigators for solving one-dimensional phase change problems with finite differences. Here, we present the modified variable time step (MVTS) method proposed by Gupta and Kumar (1981), which is based on the modification of the discretization procedure of Goodling and Khader (1974a and 1974b).

We consider the solidification of a liquid initially at the melting temperature T_m^*, confined to the region $0 \leq x \leq B$. For times $t > 0$, the boundary surface at $x = 0$ is subjected to convective cooling into an ambient at a constant temperature T_∞ with a heat transfer coefficient h, while the boundary surface at $x = B$ is kept insulated or satisfies the symmetry condition. The solidification starts at the boundary surface $x = 0$, and the solid–liquid interface moves in the x direction as illustrated in Figure 10.3.

Temperature T(x,t) varies only in the solid phase because the liquid region is at the melting temperature T_m^*. We are concerned with the determination of the temperature distribution T(x,t) in the solid phase and the location of the interface as a function of time. The mathematical formulation of this solidification problem is given by

Solid region:

$$\frac{\partial^2 T}{\partial x^2} = \frac{1}{\alpha}\frac{\partial T}{\partial t} \quad \text{in} \quad 0 < x < S(t), \quad t > 0 \tag{10.16}$$

$$-k\frac{\partial T}{\partial x} + hT = hT_\infty \quad \text{at} \quad x = 0, \quad t > 0 \tag{10.17}$$

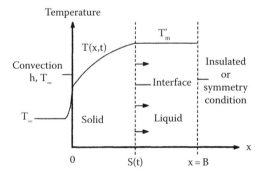

FIGURE 10.3
Geometry and coordinate for single-phase solidification.

Phase Change Problems

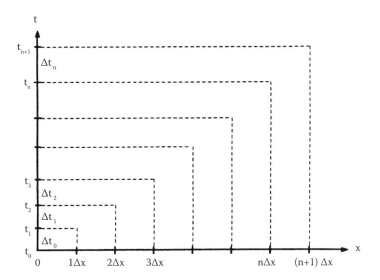

FIGURE 10.4
Subdivision of x and t domains using constant Δx and variable Δt.

Interface:

$$T(x,t) = T_m^* \quad \text{at} \quad x = S(t), \quad t > 0 \tag{10.18}$$

$$k\frac{\partial T}{\partial x} = \rho L \frac{dS(t)}{dt} \quad \text{at} \quad x = S(t), \quad t > 0 \tag{10.19}$$

where h is the heat transfer coefficient, S(t) is the location of solid–liquid interface, ρ is the density, L is the latent heat of solidification (or melting), k is the thermal conductivity, and α is the thermal diffusivity.

To solve this problem with finite differences, the x and t domains are subdivided into small intervals of constant Δx in space and variable Δt in time, as illustrated in Figure 10.4. The variable time step approach requires that, at each time level t_n, the time step Δt_n is so chosen that the interface moves exactly a distance Δx; hence, it always stays on the node. Therefore, we are concerned with the determination of the time step $\Delta t_n = t_{n+1} - t_n$ such that, in the time interval from t_n to t_{n+1}, the interface moves from the position $n\Delta x$ to the next position $(n+1)\Delta x$. Next, we describe the finite difference approximation of this solidification problem and then the determination of the time step Δt_n.

10.2.1 Finite Difference Approximation

The finite difference approximation of equations (10.16)–(10.19) is described as follows.

10.2.1.1 Differential Equation

This differential equation can be approximated with finite differences using either the simple implicit scheme or the combined method. Here, we prefer the simple implicit method and write equation (10.16) in finite difference form as

$$\frac{T_{i-1}^{n+1} - 2T_i^{n+1} + T_{i+1}^{n+1}}{(\Delta x)^2} = \frac{1}{\alpha}\frac{T_i^{n+1} - T_i^n}{\Delta t_n} \quad (10.20a)$$

where the following notation is adopted

$$T(x_i, t_n) = T(i\Delta x, t_n) \equiv T_i^n \quad (10.20b)$$

Equation (10.20a) is rearranged as

$$\left[-r_n T_{i-1}^{n+1} + (1 + 2r_n)T_i^{n+1} - r_n T_{i+1}^{n+1}\right]^{(p)} = T_i^n \quad (10.21a)$$

where the superscript p over the bracket refers to the p-th iteration, and the parameter r_n is defined as

$$r_n = \frac{\alpha \Delta t_n}{(\Delta x)^2}, \quad n = 1, 2, 3, \ldots \quad \text{and} \quad \Delta t_n = t_{n+1} - t_n \quad (10.21b,c)$$

10.2.1.2 Boundary Condition at x = 0

The convection boundary condition (10.17) is rearranged

$$\frac{\partial T}{\partial x} = HT - HT_\infty \quad \text{where} \quad H = h/k \quad (10.22a)$$

and then discretized

$$\frac{T_1^{n+1} - T_0^{n+1}}{\Delta x} = HT_0^{n+1} - HT_\infty \quad (10.22b)$$

This result is rearranged in the form

$$[T_1^{n+1} - (1 + H\Delta x)T_0^{n+1}]^{(p)} = -H\Delta x\, T_\infty \quad (10.23)$$

where superscript p over the bracket denotes the p-th iteration. The finite difference equation (10.23) is first-order accurate. A second-order expression can be developed either using a fictitious node concept as discussed in Chapter 2 or by applying the control volume approach in order to include the heat capacity of the control volume.

10.2.1.3 Interface Conditions

The condition of continuity of temperature at the interface, equation (10.18), is written as

$$T_{n+1}^{n+1} = T_m^* = \text{melting temperature,} \qquad (10.24)$$

which is valid for all times. The interface energy balance equation (10.19) is discretized as

$$\frac{T_{n+1}^{n+1} - T_n^{n+1}}{\Delta x} = \frac{\rho L}{k} \frac{\Delta x}{\Delta t_n}, \qquad (10.25a)$$

which is rearranged in the form

$$[\Delta t_n]^{(p+1)} = \frac{\rho L}{k} \left[\frac{(\Delta x)^2}{T_m^* - T_n^{n+1}} \right]^{(p)} \qquad (10.25b)$$

because $T_{n+1}^{n+1} = T_m^* = $ melting temperature.

10.2.2 Determination of Time Steps

We now describe the algorithm for the determination of time step Δt_n such that, during this time step, the interface moves exactly a distance Δx.

10.2.2.1 Starting Time Step Δt_0

An explicit expression can be developed for the calculation of the first time step Δt_0 as follows: Set $n = 0$ in equations (10.23) and (10.25b) and eliminate T_0^1 between the resulting two equations; note that $T_1^1 \equiv T_m^*$. The following explicit expression is obtained for Δt_0.

$$\Delta t_0 = \frac{\rho L}{k} \frac{\Delta x (1 + H \Delta x)}{H (T_m^* - T_\infty)} \qquad (10.26)$$

where $\Delta t_0 \equiv t_1 - t_0$.

10.2.2.2 Time Step Δt_1

We set $i = 1$, $n = 1$ in equation (10.21a) and note that $T_1^1 = T_2^2 = T_m^*$. Then equation (10.21a) becomes

$$\left[-r_1 T_0^2 + (1 + 2r_1) T_1^2 \right]^{(p)} = (1 + r_1^{(p)}) T_m^* \qquad (10.27a)$$

and from the boundary condition (10.23), for $n = 1$, we obtain

$$\left[-(1 + H \Delta x) T_0^2 + T_1^2 \right]^{(p)} = -H \Delta x \, T_\infty \qquad (10.27b)$$

To solve equations (10.27a) and (10.27b) for T_0^2 and T_1^2, the value of $r_1^{(p)}$ is needed; but $r_1^{(p)}$ as defined by equation (10.21b) depends on $\Delta t_1^{(p)}$. Therefore, iteration is needed for their solution. To start iterations, we set

$$\Delta t_1^{(0)} = \Delta t_0$$

Then, $r_1^{(0)}$ is determined from equation (10.21b); using this value of $r_1^{(0)}$, equations (10.27a,b) are solved for T_0^2 and T_1^2. Knowing T_1^2, $\Delta t_1^{(1)}$ is computed from equation (10.25b), calculations are continued until the difference between two consecutive iterations

$$|\Delta t_1^{p+1} - \Delta t_1^p|$$

satisfies a specified convergence criterion.

10.2.2.3 Time Step Δt_n

The previous results are now used in the following algorithm to calculate the time steps Δt_n at each time level t_n, $n = 2,3,\ldots$.

1. The time steps Δt_n at the time levels t_n, $n = 2,3,\ldots$ are calculated by iteration. A guess value $\Delta t_n^{(0)}$ is chosen as

   $$\Delta t_n^{(0)} = \Delta t_{n-1}, \qquad n = 2, 3, \ldots \tag{10.28a}$$

   The system of finite difference equations (10.21a) and (10.23), together with the condition [equation (10.24)] are solved for $i = 1,2,3,\ldots,n$ by setting $p = 0$, and a first estimate is obtained for the nodal temperatures

   $$[T_i^{n+1}]^{(0)} \quad \text{for} \quad i = 1, 2, \ldots, n \tag{10.28b}$$

   We note that the system of equations is tridiagonal; hence, it is readily solvable.

2. The values of $[T_i^{n+1}]^{(0)}$ obtained from equation (10.28b) are introduced into equation (10.25b) for $p = 0$ and a first estimate for the time step $\Delta t_n^{(1)}$ is determined.

3. $\Delta t_n^{(1)}$ is used as a guess value and Steps (i) and (ii) are repeated to calculate a second estimate for the time step $\Delta t_n^{(2)}$.

4. The Steps (i), (ii), and (iii) are repeated until the difference between two consecutive iterations

   $$|\Delta t_n^{(p+1)} - \Delta t_n^{(p)}|$$

   satisfies a specified convergence criterion.

Example 10.1

Consider a single-phase solidification problem for a liquid initially at the melting temperature T_m^*, confined to the region $0 \leq x \leq 1$. Solidification takes place as a result of convective cooling at the boundary surface $x = 0$, while the boundary surface at $x = 1$ is kept insulated. The mathematical formulation of this problem is given in the dimensionless form as follows.

Solid region:

$$\frac{\partial^2 T}{\partial x^2} = \frac{\partial T}{\partial t} \quad \text{in } 0 < x < S(t), \quad t > 0$$

$$-\frac{\partial T}{\partial x} + 10T = 0 \quad \text{at } x = 0, \quad t > 0$$

$$\frac{\partial T}{\partial x} = 0 \quad \text{at } x = 1, \quad t > 0$$

Interface:

$$T(x, t) = 1 \quad \text{at } x = S(t), \quad t > 0$$

$$\frac{\partial T}{\partial x} = \frac{dS}{dt} \quad \text{at } x = S(t), \quad t > 0$$

Calculate the time step Δt required for the solid–liquid interface $S(t)$ to move one space interval $\Delta x = 0.1$ and the temperature of the boundary surface at $x = 0$ for the interface positions $S(t) = 0.1, 0.2, 0.3, \ldots, 1.0$.

Solution

This problem has been solved by Gupta and Kumar (1981) using the variable time step approach described earlier, and their results are listed in Table 10.1. For example, the first time step Δt_0, needed for the

TABLE 10.1

Time Step Δt Required for the Interface Position to Move By One Space Interval Δx and Temperature of the Boundary Surface at $x = 0$

Interface Position S(t)	Time Step Δt	T(0,t)	Number of Iterations
0.1	0.0200	0.5000	0
0.2	0.0356	0.3596	4
0.3	0.0494	0.2770	4
0.4	0.0627	0.2242	4
0.5	0.0759	0.1879	4
0.6	0.0890	0.1616	4
0.7	0.1021	0.1416	4
0.8	0.1152	0.1260	4
0.9	0.1282	0.1135	4
1.0	0.1413	0.1032	4

interface to move from S(t) = 0 to S(t) = 0.1, is determined directly from equation (10.26). The numerical values of various parameters appearing in this equation are determined by comparing the mathematical formulation of this example with that given by equations (10.16)–(10.19). We find

$$\alpha = 1, \quad T_m^* = 1, \quad H = \frac{h}{k} = 10, \quad T_\infty = 0, \quad \text{and} \quad \frac{\rho L}{k} = 1$$

and the grid size is chosen as $\Delta x = 0.1$. Introducing these numerical values into equation (10.26), the starting time step Δt_0 is determined as

$$\Delta t_0 = \frac{\rho L}{k} \frac{\Delta x (1 + H \Delta x)}{H(T_m - T_\infty)} = \frac{0.1(1 + 10 \times 0.1)}{10 \times (1 - 0)} = 0.020$$

The next time step Δt_1 needed for the interface to move from the position S(t) = 0.1 to the position S(t) = 0.2 is determined by an iterative procedure described previously. According to Table 10.1, a value of Δt_1 = 0.0356 is obtained with a maximum tolerance of 0.05%. The remaining time steps are determined iteratively and listed in Table 10.1. Also included in this Table is T(0,t), the temperature of the boundary surface at x = 0.

A computer program written in Fortran is presented in Appendix E at the end of the book for solving the one-dimensional solidification problem considered in Example 10.1 for the cases of materials with a single phase-change temperature. The dimensionless INPUT data for this example include: the number of space intervals, N = 10; relative tolerance TOL = 10^{-4}; maximum number of iterations MAXIT = 20; heat transfer coefficient, H = 10; surrounding temperature, TINF = 0; thermal diffusivity, ALPA = 1; melting temperature, TMELT = 1; and thickness of the region, BB = 1. Solutions for other values of the parameters are readily obtained by setting these parameters accordingly.

10.3 Variable Time Step Approach for Two-Phase Solidification

Consider the solidification of a liquid initially at a uniform temperature θ_i^*, which is higher than the melting temperature T_m^* of the fluid, and confined to the region $0 \leq x \leq B$. For times t > 0, the boundary surface at x = B is kept insulated. The solidification starts at x = 0 as a result of cooling, and the solid–liquid interface moves in the positive x direction. The temperature T(x,t) of the solid phase, the temperature $\theta(x,t)$ of the liquid phase, and the location of the solid–liquid interface S(t) are the unknowns to be determined.

Figure 10.5 illustrates the temperature profiles in the solid and liquid phases. The mathematical formulation of this solidification problem is given as follows.

Phase Change Problems

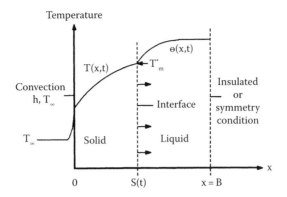

FIGURE 10.5
Geometry and coordinates for two-phase solidification.

Solid phase:

$$\frac{\partial^2 T}{\partial x^2} = \frac{1}{\alpha_s}\frac{\partial T}{\partial t} \quad \text{in } 0 < x < S(t), \quad t > 0 \tag{10.29a}$$

$$-k_s\frac{\partial T}{\partial x} + hT = hT_\infty \quad \text{at } x = 0, \quad t > 0 \tag{10.29b}$$

Liquid phase:

$$\frac{\partial^2 \theta}{\partial x^2} = \frac{1}{\alpha_l}\frac{\partial \theta}{\partial t} \quad \text{in } S(t) < x < B, \quad t > 0 \tag{10.30a}$$

$$\frac{\partial \theta}{\partial x} = 0 \quad \text{at } x = B, \quad t > 0 \tag{10.30b}$$

$$\theta = \theta_i^* = \text{constant for } t = 0, \quad \text{in } 0 < x < B \tag{10.30c}$$

Interface:

$$T(x,t) = \theta(x,t) = T_m^* \quad \text{at } x = S(t), \quad t > 0 \tag{10.31a}$$

$$k_s\frac{\partial T}{\partial x} - k_l\frac{\partial \theta}{\partial x} = \rho_s L \frac{dS(t)}{dt} \quad \text{at } x = S(t), \quad t > 0 \tag{10.31b}$$

where θ_i^* is the initial temperature for the liquid region, and the subscripts s and l refer to the solid and liquid phases, respectively.

To solve this problem with finite differences, the x and t domains are subdivided into small intervals of constant Δx in space; but the time step Δt is allowed to vary. At each time level t_n, the time step $\Delta t_n = t_{n+1} - t_n$ is to be selected so that, in the time interval from t_n to t_{n+1}, the interface moves from the position $n\Delta x$ to the position $(n + 1)\Delta x$. We now present the finite difference form of the governing differential equations and an algorithm for the calculation of the variable time step Δt_n at each time level t_n.

10.3.1 Finite Difference Approximation

The differential equations (10.29)–(10.31) are represented in finite difference form as follows.

10.3.1.1 Equation for the Solid Phase

The differential equation (10.29a) can be discretized using either the implicit or the combined scheme. For simplicity, we prefer the implicit method and write

$$\frac{T_{i-1}^{n+1} - 2T_i^{n+1} + T_{i+1}^{n+1}}{(\Delta x)^2} = \frac{1}{\alpha_s} \frac{T_i^{n+1} - T_i^n}{\Delta t_n} \qquad (10.32a)$$

where the following notation is adopted

$$T(x, t_n) = T(i\Delta x, t_n) \equiv T_i^n \qquad (10.32b)$$

Equation (10.32a) is rearranged in the form

$$\left[-r_{n,s} T_{i-1}^{n+1} + (1 + 2r_{n,s}) T_i^{n+1} - r_{n,s} T_{i+1}^{n+1} \right]^{(p)} = T_i^n \qquad (10.33a)$$

where

$$r_{n,s} = \frac{\alpha_s \Delta t_n}{(\Delta x)^2}, \quad i = 1, 2, 3, \ldots \qquad (10.33b)$$

$$\Delta t_n = t_{n+1} - t_n$$

The subscript "s" refers to the solid phase, and the superscript "p" over the bracket denotes the p-th iteration.

10.3.1.2 Boundary Condition at x = 0

The boundary condition (10.29b) for the solid phase is discretized as

$$\left[T_1^{n+1} - (1 + H\Delta x) T_0^{n+1} \right]^{(p)} = -H\Delta x\, T_\infty \qquad (10.34)$$

which is similar to that given by equation (10.23).

10.3.1.3 Equation for the Liquid Phase

The differential equation (10.30a) is discretized using the simple implicit scheme to give

$$\left[-r_{n,l}\theta_{i+1}^{n+1} + (1+2r_{n,l})\theta_i^{n+1} - r_{n,l}\theta_{i+1}^{n+1}\right]^{(p)} = \theta_i^n \tag{10.35a}$$

which is similar to that given by equation (10.33a). The parameter $r_{n,l}$ is defined by

$$r_{n,l} = \frac{\alpha_l \Delta t_n}{(\Delta x)^2} \tag{10.35b}$$

where subscript "l" refers to the liquid phase.

10.3.1.4 Interface Conditions

The condition of continuity of temperature at the interface is written as

$$T_{n+1}^{n+1} = \theta_{n+1}^{n+1} = T_m^* \tag{10.36}$$

which is valid for all times.

The interface energy balance, equation (10.31b), is discretized as

$$k_s \frac{T_{n+1}^{n+1} - T_n^{n+1}}{\Delta x} - k_l \frac{\theta_{n+2}^{n+1} - \theta_{n+1}^{n+1}}{\Delta x} = \rho_s L \frac{\Delta x}{\Delta t_n} \tag{10.37a}$$

which is written in the form

$$k_s \frac{T_m^* - T_n^{n+1}}{\Delta x} - k_l \frac{\theta_{n+2}^{n+1} - T_m^*}{\Delta x} = \rho_s L \frac{\Delta x}{\Delta t_n} \tag{10.37b}$$

because $T_{n+1}^{n+1} = \theta_{n+1}^{n+1} = T_m^*$. equation (10.37b) is rearranged in the form

$$[\Delta t_n]^{(p+1)} = \frac{\rho L}{k_s} \left[\frac{(\Delta x)^2}{(T_m^* - T_n^{n+1}) - \frac{k_l}{k_s}(\theta_{n+2}^{n+1} - T_m^*)} \right]^{(p)} \tag{10.38}$$

When the liquid phase is at the melting temperature, T_m^*, we always have $\theta = T_m^*$; then equation (10.38) reduces to the single-phase solidification given by equation (10.25b).

10.3.2 Determination of Time Steps

We now describe the determination of time steps Δt_n such that the interface moves exactly a distance Δx from time t_n to time t_{n+1}.

10.3.2.1 Starting Time Step Δt_0

Figure 10.6 shows that the solid–liquid interface moved from the position $S(t) = 0$ to the position $S(t) = \Delta x$ during the time period Δt_0. The solid region has only one unknown temperature, T_0^1. Hence, there is no need to solve finite difference equations; it can be determined directly from equation (10.34) by setting n = 0. We find

$$T_0^1 = (T_m^* + H\Delta x T_\infty)/(1 + H\Delta x) \tag{10.39}$$

since $T_1^1 \equiv T_m^*$. This value of T_0^1, together with the approximation $\theta_2^1 \simeq \theta_i^*$, where θ_i^* is the initial temperature, is used in equation (10.38) with n = 0 to obtain an initial guess for the starting time step $\Delta t_0^{(0)}$. The actual value of Δt_0 is obtained by iteration as follows.

i. Using this value of $\Delta t_0^{(0)}$, solve the finite difference equations (10.35) and (10.36), for the liquid phase subject to appropriate boundary and initial conditions and obtain an estimate for the liquid phase nodal temperatures $[\theta_i^1]^{(0)}$.

ii. Using T_0^1 and $[\theta_2^1]^{(0)}$ in equation (10.38) with n = 0, obtain a new estimate for $\Delta t_0^{(1)}$.

iii. Using $\Delta t_0^{(1)}$, repeat Steps (i) and (ii) and obtain a second estimate $\Delta t_0^{(2)}$.

iv. Carry out the calculations until the difference between two consecutive iterations

$$|\Delta t_0^{(p+1)} - \Delta t_0^{(p)}|$$

satisfies a specified convergence criterion.

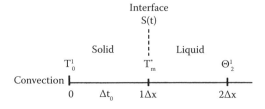

FIGURE 10.6
Interface at $S(t) = \Delta x$.

10.3.2.2 Time Step Δt_1

Figure 10.7 shows the solid–liquid interface moved from the position $S(t) = \Delta x$ to the position $S(t) = 2\Delta x$ during the time step Δt_1. An initial guess for this time step is taken as

$$\Delta t_1^{(0)} = \Delta t_0 \tag{10.40}$$

The solid phase has only two unknown node temperatures, T_0^2 and T_1^2; hence, there is no need for finite difference calculations. Two relations needed for their determination are obtained from equations (10.33) and (10.34) by setting

$$n = 1, \quad i = 1, \quad T_1^1 = T_2^2 \equiv T_m^* \quad \text{and} \quad \Delta t_1^{(0)} = \Delta t_0 \tag{10.41}$$

The resulting equations are

$$-r_{1,s} T_0^2 + (1 + 2r_{1,s}) T_1^2 = (1 + r_{1,s}) T_m^* \tag{10.42a}$$

$$T_1^2 - (1 + H\Delta x) T_0^2 = -H\Delta x T_\infty \tag{10.42b}$$

where

$$r_{1,s} = \frac{\alpha_s \Delta t_1^{(0)}}{(\Delta x)^2} \tag{10.42c}$$

A simultaneous solution of equations (10.42a,b) gives the temperatures T_0^2 and T_1^2. The actual value of the time step Δt_1 is obtained by iteration as follows.

i. Using $\Delta t_1^{(0)} = \Delta t_0$, solve the finite difference equations (10.35) and (10.36), for the liquid phase subject to appropriate boundary and initial conditions and determine an estimate for the liquid phase nodal temperatures $[\theta_i^2]^{(0)}$.

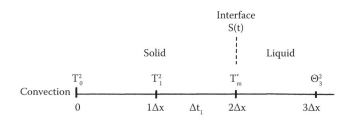

FIGURE 10.7
Interface at $S(t) = 2\Delta x$.

ii. Using T_1^2 and $[\theta_3^2]^{(0)}$ in equation (10.38) with n = 1, obtain a first estimate for the time step $\Delta t_1^{(1)}$.

iii. Using $\Delta t_1^{(1)}$, repeat Steps (i) and (ii) to obtain $\Delta t_1^{(2)}$.

iv. Carry out the calculations until the difference between two consecutive iterations satisfies a specified convergence criterion.

10.3.2.3 Time Steps Δt_n, $(2 \leq n \leq N - 4)$

Figure 10.8 shows that the interface has moved from the position $S(t) = n\Delta x$ to the position $S(t) = (n + 1)\Delta x$ during the time step Δt_n. An initial guess for this time step is taken as

$$\Delta t_n^{(0)} = \Delta t_{n-1} \tag{10.43}$$

The actual value of the time step Δt_n is determined by iteration as follows.

i. Using $\Delta t_n^{(0)} = \Delta t_{n-1}$, solve the finite difference equations for the solid and liquid phases and determine a first estimate for the nodal temperatures

$$[T_i^{n+1}]^{(0)} \quad \text{and} \quad [\theta_i^{n+1}]^{(0)} \tag{10.44}$$

ii. Using the estimates $[T_i^{n+1}]^{(0)}$ and $[\theta_{n+2}^{n+1}]^{(0)}$ in equation (10.38) with p = 0, obtain a first estimate for the time step $\Delta t_n^{(1)}$.

iii. Using $\Delta t_n^{(1)}$, repeat Steps (i) and (ii) to obtain $\Delta t_n^{(2)}$.

iv. Carry out the iteration until the difference between two consecutive iterations satisfies a specified convergence criterion.

10.3.2.4 Time Step Δt_{N-3}

Figure 10.9 shows that the solid–liquid interface has moved from the position $S(t) = (N-3)\Delta x$ to the position $S(t) = (N-2)\Delta x$ during the time step Δt_{N-3}. An initial guess for this time step is made as

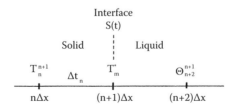

FIGURE 10.8
Interface at $S(t) = (n + 1)\Delta x$.

Phase Change Problems

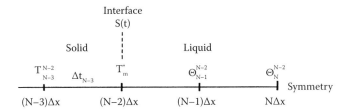

FIGURE 10.9
Interface at $S(t) = (N-2)\Delta x$.

$$\Delta t_{N-3}^{(0)} = \Delta t_{N-4} \qquad (10.45a)$$

For this case, the liquid phase has only two unknown node temperatures θ_{N-1}^{N-2} and θ_{N}^{N-2}. Two relations are needed for their determination. One relation is obtained from equation (10.35a) by setting $n+1 = N-2$, $i = N-1$ and noting that $\theta_{N-2}^{N-2} = T_m^*$ as

$$-r_{N-3,1}T_m^* + (1+2r_{N-3,1})\theta_{N-1}^{N-2} - r_{N-3,1}\theta_N^{N-2} = \theta_{N-1}^{N-3} \qquad (10.45b)$$

The second relation is obtained by setting in equation (10.35a), $n+1 = N-2$, $i = N$ and noting that $\theta_{N-1}^{N-2} = \theta_{N+1}^{N-2}$ by the symmetry boundary condition. We find

$$-2r_{N-3,1}\,\theta_{N-1}^{N-2} + (1+2r_{N-3,1})\theta_N^{N-2} = \theta_N^{N-3} \qquad (10.45c)$$

where $r_{N-3,1}$ is calculated using the above value of $\Delta t_{N-3}^{(0)}$. A simultaneous solution of equations (10.45b) and (10.45c) gives a first estimate for

$$[\theta_{N-1}^{N-2}]^{(0)}$$

The solid-phase nodal temperatures are determined by solving the finite difference equations for the solid phase subject to appropriate boundary conditions, and a first estimate is obtained for

$$[T_{N-3}^{N-2}]^{(0)}$$

These results are used in equation (10.38), and a first estimate is obtained for the time step $\Delta t_{N-3}^{(1)}$. This procedure is repeated until $\Delta t_{N-3}^{(P)}$ satisfies a specified convergence criterion.

10.3.2.5 Time Step Δt_{N-2}

Figure 10.10 shows that the solid–liquid interface has moved from the position $S(t) = (N-2)\Delta x$ to the position $(N-1)\Delta x$ during the time step Δt_{N-2}. An initial guess for this time step is taken as

$$\Delta t_{N-2}^{(0)} = \Delta t_{N-3} \qquad (10.46)$$

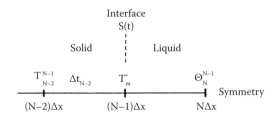

FIGURE 10.10
Interface at $S(t) = (N-1)\Delta x$.

The liquid phase contains only one unknown node temperature θ_N^{N-1}. An expression can be developed for its determination by setting $i = N$, $n + 1 = N - 1$ in the finite difference equation (10.35a) and noting that $\theta_{N-1}^{N-1} = T_m^*$. We find

$$-r_{N-2,1}T_m^* + (1 + 2r_{N-2,1})\theta_N^{N-1} - r_{N-2,1}\theta_{N+1}^{N-1} = \theta_N^{N-2} \qquad (10.47a)$$

The thermal insulation boundary condition at $N\Delta x$ gives

$$\theta_{N+1}^{N-1} = \theta_{N-1}^{N-1} \equiv T_m^* \qquad (10.47b)$$

Eliminating the fictitious node temperature θ_{N+1}^{N-1} between equations (10.47a) and (10.47b), and solving the resulting expression for θ_N^{N-1}, we obtain

$$\theta_N^{N-1} = \frac{\theta_N^{N-2} + 2r_{N-2,1}T_m^*}{1 + 2r_{N-2,1}} \qquad (10.48)$$

For the solid phase, the finite difference equations are solved and an estimate is obtained for $[T_{N-2}^{N-1}]^{(0)}$. These results are then used in equation (10.38), and a first estimate is obtained for $\Delta t_{N-2}^{(1)}$. Iterations are continued until $\Delta t_{N-2}^{(p)}$ satisfies a specified convergence criterion.

10.3.2.6 Time Step Δt_{N-1}

This is the last time step during which the solid–liquid interface moves from the position $S(t) = (N-1)\Delta x$ to the position $S(t) = N\Delta x$; hence, there is no liquid region. Taking the initial guess for the time step Δt_{N-1} as

$$\Delta t_{N-1}^{(0)} = \Delta t_{N-2} \qquad (10.49)$$

the finite difference equations for the solid region are solved subject to appropriate boundary and initial conditions, and a first estimate is obtained for the nodal temperature T_{N-1}^N. Then, a first estimate for the time step $\Delta t_{N-1}^{(1)}$ is obtained from equation (10.38) by setting n = N − 1 in this equation and replacing θ_{N+1}^N by θ_{N-1}^N because of symmetry at the boundary.

The final time step Δt_{N-1} is determined by repeating the previous procedure and carrying out the iteration until the difference between the two consecutive values of time step satisfies a specified convergence criterion.

10.4 Enthalpy Method

The mathematical formulation of phase change problems considered previously consisted of transient heat conduction equations for the solid and liquid phases, an interface energy balance equation, and appropriate boundary and initial conditions. In such formulations, temperature is the sole dependent variable, and the position of the solid–liquid interface is to be determined as a part of the solution. Therefore, an accurate tracking of the interface position is essential for accurate solution of the problem.

In many industrial problems, phase change occurs over a temperature range rather than at a specified temperature. For such situations, the schemes used for solving the problems involving a single discrete phase change temperature are not applicable. The so-called *enthalpy method* has been used to solve phase change problems for such situations as well as for those dealing with a single phase-change temperature. In the enthalpy formulation, the enthalpy function H(T), which is the total heat content of the substance, enters the problem as a dependent variable along with the temperature.

The enthalpy formulation of the phase change problem is given by

$$\rho \frac{\partial H(T)}{\partial t} = \nabla \cdot (k \nabla T) + g(\mathbf{r}, t) \tag{10.50}$$

which is considered valid over the entire solution domain, including both the solid and the liquid phases as well as the solid–liquid interface. The energy generation term $g(\mathbf{r},t)$ is omitted if the problem involves no internal energy generation.

Therefore, the method is attractive in that the solution of the phase change problem is reduced to the solution of a single equation in terms of enthalpy. There are no boundary conditions to be satisfied at the solid–liquid interface; there is no need to accurately track the phase change boundary; there is no need to consider liquid and solid regions separately; and any numerical scheme can be used for its solution.

Figure 10.11 shows enthalpy-temperature relations for (a) pure crystalline substances and eutectics and (b) glassy substances and alloys. For pure substances, the phase change takes place at a discrete temperature and hence is

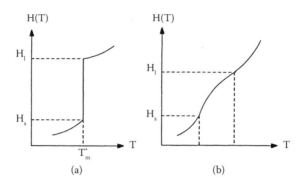

FIGURE 10.11
Enthalpy–temperature relationship for (a) pure crystalline substances and eutectics and (b) glassy substances and alloys.

associated with the latent heat L. Therefore, in Figure 10.11a, a jump discontinuity occurs at the melt temperature T_m^*; hence, $\partial H/\partial T$ becomes infinite and the energy equation apparently is not meaningful at this point. However, Shamsundar and Sparrow (1975) used the divergence theorem to confirm that the integral form of the energy equation (10.50) with no generation is equivalent to the classical formulation of the phase change problem.

Therefore, the enthalpy method is applicable for the solution of phase change problems involving both a distinct phase change at a discrete temperature as well as phase change taking place over an extended range of temperatures. To illustrate the physical significance of the enthalpy function H(T) shown in Figure 10.11a, we consider a pure substance having a melting point temperature T_m^*, with equal specific heats for the solid and liquid phases. In the solid state at temperature T, the substance contains a sensible heat per unit mass $C_p(T-T_m^*)$, where the melting point temperature T_m^* is taken as the reference temperature. In the liquid state, it contains latent heat L per unit mass in addition to the sensible heat; that is, $C_p(T-T_m^*) + L$. For the specific case considered here, the enthalpy is related to temperature by

$$H = \begin{cases} C_p(T-T_m^*) & \text{for } T \leq T_m^* \quad (10.51a) \\ C_p(T-T_m^*) + L & \text{for } T > T_m^* \quad (10.51b) \end{cases}$$

Conversely, given the enthalpy of the substance, the corresponding temperature is determined from

$$T = \begin{cases} T_m^* + \dfrac{H}{C_p} & \text{for } H < 0 \quad (10.52a) \\ T_m^* & \text{for } 0 \leq H \leq L \quad (10.52b) \\ T_m^* + \dfrac{H-L}{C_p} & \text{for } H > L \quad (10.52c) \end{cases}$$

Phase Change Problems

A more general expression for the enthalpy function H(T), for a pure substance having a latent heat L, can be written as

$$H(T) = \int_{T_0}^{T} C(T')dT', \quad T < T_m^* \tag{10.53a}$$

$$H(T) = \int_{T_0}^{T} C(T')dT' + L, \quad T > T_m^* \tag{10.53b}$$

$$\int_{T_0}^{T} C(T')dT' \leq H(T) \leq \int_{T_0}^{T} C(T')dT' + L, \quad T = T_m^* \tag{10.53c}$$

where the lower limit of the integration is the reference temperature.

In the case of glassy substances and alloys, there is no discrete melting point temperature because the phase change takes place over an extended range of temperatures, as illustrated in Figure 10.11b. Such relationship between H(T) and T is obtained either from experimental data or standard physical tables. In general, enthalpy is a nonlinear function of temperature.

We now illustrate the application of the enthalpy method for the solution of one-dimensional solidification or melting problems using both explicit and implicit finite difference schemes. The problems involving phase change over a temperature range are much easier to handle with this method than those with a single melting point.

10.4.1 Explicit Enthalpy Method: Phase Change with Single Melting Temperature

We consider one-dimensional solidification of a liquid having a single melting point temperature T_m^* and confined to the region $0 \leq x \leq B$. Initially, the liquid is at a uniform temperature T_0 that is higher than the melting temperature T_m^* of the liquid. For times $t > 0$, the boundary surface at $x = 0$ is kept at a temperature f_0, which is lower than the melting temperature T_m^* of the substance. The boundary condition at $x = B$ is insulated. For simplicity, the properties are assumed to be constant.

The enthalpy formulation of this phase change problem is given by

$$\rho \frac{\partial H}{\partial t} = k \frac{\partial^2 T}{\partial x^2} \quad \text{in} \quad 0 < x < B, \quad t > 0 \tag{10.54}$$

$$T = f_0 \quad \text{at } x = 0, \quad t > 0 \tag{10.55a}$$

$$\frac{\partial T}{\partial x} = 0 \quad \text{at} \quad x = B, \quad t > 0 \tag{10.55b}$$

$$T = T_0 \text{ (or } H = H_0) \quad \text{for} \quad t = 0, 0 \leq x \leq B \tag{10.55c}$$

To approximate this problem with finite differences, the region $0 \leq x \leq B$ is subdivided into M equal parts, each of width $\Delta x = B/M$, and the simple explicit scheme is used. The differential equation (10.54) becomes

$$\rho \frac{H_i^{n+1} - H_i^n}{\Delta t} = k \frac{T_{i-1}^n - 2T_i^n + T_{i+1}^n}{(\Delta x)^2} \tag{10.56}$$

which is solved for H_i^{n+1} as

$$H_i^{n+1} = H_i^n + \frac{k}{\rho} \eta (T_{i-1}^n - 2T_i^n + T_{i+1}^n) \tag{10.57a}$$

where

$$\eta = \frac{\Delta t}{(\Delta x)^2} \tag{10.57b}$$

$i = 1,2,\ldots,M-1$ and $n = 0,1,2,\ldots$.

The following notation is adopted:

$$T(x,t) = T(i\Delta x, n\Delta t) \equiv T_i^n \tag{10.58}$$

The boundary and initial conditions are discretized as

$$T_0^n = f_0 \quad \text{(Boundary at } x = 0\text{)} \tag{10.59a}$$

$$T_{M-1}^n = T_M^n \quad \text{(Thermal insulation condition)} \tag{10.59b}$$

$$T_i^0 = T_0 \quad \text{and} \quad H_i^0 = H_0 \quad \text{(Initial condition)} \tag{10.59c}$$

Finally, with equation (10.57a) being explicit, the following condition should be satisfied for stability

$$\eta = \frac{\Delta t}{(\Delta x)^2} < \frac{\rho C_p}{2k} \tag{10.60}$$

which is equivalent to the condition $r = \frac{\alpha \Delta t}{(\Delta x)^2} < \frac{1}{2}$. Equations (10.57) and (10.59), together with the temperature–enthalpy relation given by equations (10.52a–c), provide the complete finite difference approximation of the enthalpy formulation of the phase change problem described previously.

10.4.1.1 Algorithm for Explicit Method

The following algorithm can be used to solve the explicit finite difference equations for the enthalpy method, where linearization is applied by lagging the evaluation of properties by one time step.

Phase Change Problems

We assume that the numerical solution has progressed as far as the time level $t = n\Delta t$ and that the numerical values of T_i^n and H_i^n ($0 \le i \le M$) are known. The enthalpies and temperatures at the next time level $t = (n + 1)\Delta t$ are calculated as follows.

i. Compute H_i^{n+1} for $i = 1,2,...,M-1$ from equation (10.57a).
ii. Compute the corresponding T_i^{n+1} from one of the three relationships given by equations (10.52a–c) appropriate to the value of H_i^{n+1}.
iii. Compute the surface temperature from equation (10.59b).

The calculations are started with $n = 0$, which corresponds to the initial condition, and the procedure is repeated for each successive time step. Clearly, the enthalpy approach removes the need to trace the position of the interface; hence, it simplifies the numerical procedure. The position of the interface has to be estimated in retrospect by inspection of the computed values of H_i and T_i, such that $T = T_m^*$. However, the application of the explicit enthalpy method for the solution of phase change problems involving a single melting point temperature often leads to oscillations in the prediction of both temperature histories and position of the interface. The oscillations have their source at the volume containing the solid–liquid interface. An improved interpretation of enthalpy is needed in order to develop an improved algorithm for accurate prediction of temperature and the interface position.

10.4.1.2 Interpretation of Enthalpy Results

Voller and Cross (1981) advanced explanations of the oscillatory behavior of the basic enthalpy solutions for both discontinuous and smoothed enthalpy functions, and they proposed an interpretation of the numerical solution that leads to a more accurate evaluation of the boundary movement and temperature history at any point.

Consider an element e_i of width Δx associated with the node i as illustrated in Figure 10.12, containing the solid–liquid interface at the position $x = S(t)$ at any time t. Let f_s be the fraction of this element that is solid and $1-f_s$ the fraction of the element that is liquid. The phase front is moving in the positive x direction. The total heat content of the element e_i at any time t can be approximated as

$$H_i \Delta x \tag{10.61a}$$

where H_i is the nodal enthalpy. This total heat of the element may also be approximated as the sum of the heat in the solid and liquid parts of the element, that is

$$[C_p T_i f_s + (C_p T_i + L)(1 - f_s)]\Delta x \tag{10.61b}$$

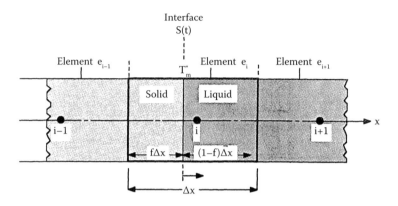

FIGURE 10.12
An element e_i of width Δx associated with the node i.

Equating equations (10.61a) and (10.61b), we obtain

$$H_i = C_p T_i f_s + (C_p T_i + L)(1 - f_s) \qquad (10.62)$$

when the front reaches the node i, we have $f_{s,i} = \frac{1}{2}$ and $T_i = T_m^*$, then equation (10.62) reduces to

$$H_i = C_p T_m^* + \frac{1}{2} L \qquad (10.63)$$

This result implies that, when any nodal enthalpy H_i satisfies the relation given by equation (10.63), the interface can be regarded located approximately on the node i.

10.4.1.3 Improved Algorithm for Explicit Method

Based on the interpretation of enthalpy defined by equations (10.62) and (10.63), Voller and Cross (1981) proposed the following improved algorithm.

i. The enthalpy and a "working" temperature are calculated at each time step from the solution of explicit finite difference equations (10.57a,b) and equation (10.52).

ii. Whenever the enthalpy at the node point i is such that at the time level n + 1

$$H_i^n > \left(C_p T_m^* + \frac{L}{2} \right) \qquad (10.64a)$$

and

$$H_i^{n+1} < \left(C_p T_m^* + \frac{L}{2} \right) \qquad (10.64b)$$

then, for a *freezing* problem, the phase change boundary has passed through the nodal point i in the preceding time interval Δt.

iii. Assuming that enthalpy changes linearly in any time interval, the time t_i at which the interface is on the node i (i.e., when $H_i = C_p T_m^* + \frac{L}{2}$) is given by

$$t_i = (n + \beta)\Delta t \tag{10.65a}$$

where $\beta < 1$ is estimated via linear interpolation as

$$\beta = \frac{(C_p T_m^* + \frac{L}{2}) - H_i^n}{H_i^{n+1} - H_i^n} \tag{10.65b}$$

Then at time t_i the temperature at the node i is T_m^*.

iv. The temperatures at the other node points k are easily estimated using a linear interpolation given by

$$T_k^{n+\beta} = \beta(T_k^{n+1} - T_k^n) + T_k^n, \quad (k \neq i) \tag{10.66}$$

If the thermal conductivities differ between the solid and liquid phases, the previous algorithm is modified by replacing the explicit finite difference equation (10.57a) by

$$H_i^{n+1} = H_i^n + \frac{\Delta t}{\rho(\Delta x)^2}[k_{i-1/2}(T_{i-1}^n - T_i^n) - k_{i+1/2}(T_i^n - T_{i+1}^n)] \tag{10.67}$$

To verify the accuracy of the improved algorithm, the classical one-dimensional freezing problem, for which analytic solution is available, is solved numerically. Comparison is based on the frost penetration problem studied by Goodrich (1978) because it offers a more sensitive test of the quality of the method than the usual comparison of temperature profiles. A comparison of the numerical results with the exact analytic solution revealed that the relative error between the predicted and analytic results for the location of the solid–liquid interface was of the order of 0.1%. Such an error is compatible with the error found in a typical finite difference solution of heat conduction problems involving no phase change. The predicted temperature history was also very accurate.

10.4.2 Implicit Enthalpy Method: Phase Change with Single Melting Temperature

The drawback of the explicit enthalpy method is that the time steps should satisfy the stability criterion. The implicit enthalpy method algorithm that will now be described here removes such a shortcoming. The essential feature of the algorithm is that, at each time level, a time step Δt is chosen so as to ensure the interface moves from one nodal point to the next. Thus, the position of the interface is known at each time step, and it is always on a grid point. Such a requirement is satisfied by ensuring that at each time step one and only one nodal enthalpy has the value

$$C_p T_m^* + \frac{1}{2} L$$

We now present the algorithm for the solution of one-dimensional phase change problem with the implicit enthalpy method.

10.4.2.1 Algorithm for Implicit Method

Consider a temperature–enthalpy relation for a substance having a single melting point temperature given in the form

$$T = \begin{cases} \dfrac{H}{C_p} & H < C_p T_m^* & (10.68a) \\ T_m^* & C_p T_m^* \leq H \leq (C_p T_m^* + L) & (10.68b) \\ \dfrac{H - L}{C_p} & H > (C_p T_m^* + L) & (10.68c) \end{cases}$$

The difference between these equations and equation (10.52) is that in the latter the melting temperature T_m^* is used as the reference temperature. Then the relation between temperature and enthalpy can be written formally as

$$T = F(H) \tag{10.69}$$

Assuming constant properties, the finite difference approximation of the one-dimensional enthalpy equation (10.54) with the simple implicit scheme is obtained as

$$\rho \frac{H_i^{n+1} - H_i^n}{\Delta t} = k \frac{T_{i-1}^{n+1} - 2T_i^{n+1} + T_{i+1}^{n+1}}{(\Delta x)^2} \tag{10.70}$$

Solving for H_i^{n+1} and utilizing the notation given by equation (10.69) for the temperature–enthalpy relation, equation (10.70) is rewritten as

$$H_i^{n+1} = H_i^n + \frac{k \Delta t}{\rho (\Delta x)^2} [F(H_{i-1}^{n+1}) - 2F(H_i^{n+1}) + F(H_{i+1}^{n+1})] \tag{10.71}$$
$$i = 1, 2, 3, \ldots, M - 1$$

These equations are written more compactly in the vector form as

$$\mathbf{H}^{n+1} = \mathbf{H}^n + \Delta t \mathbf{F}^*(\mathbf{H}^{n+1}) \tag{10.72a}$$

where $\mathbf{H}$ is a vector whose components are the nodal enthalpies H_i, and $\mathbf{F}^*$ is a function with i-th component given by

$$F_i^*(\mathbf{H}) = \frac{k}{\rho(\Delta x)^2}[F(H_{i-1}) - 2F(H_i) + F(H_{i+1})] \qquad (10.72b)$$

The system of finite difference equations (10.71), together with the finite difference representation of the boundary and initial conditions for the problem, constitute the complete set of equations for the determination of the nodal enthalpies H_i^{n+1} at the time level n + 1 from the knowledge of the enthalpies in the previous time level n.

Suppose the enthalpies H_i^n are known for all nodal points and the interface is located at the node k at the time level n. The algorithm for the determination of Δt_k, the time step during which the interface moves from the node k to k + 1, is as follows.

i. Take the initial guess for the time step Δt_k as

$$\Delta t_k^0 = \Delta t_{k-1} \qquad (10.73)$$

ii. Using this initial guess, solve the nonlinear implicit finite difference equations (10.71) subject to appropriate boundary and initial conditions and determine a first estimate for the nodal enthalpies

$$[H_i^{n+1}]^1$$

Knowing the nodal enthalpies, determine a first estimate for the nodal temperatures

$$[T_i^{n+1}]^1$$

according to equation (10.69).

iii. Calculate the successive estimates, Δt_k^p, from the following iterative formula

$$\Delta t_k^{p+1} = \Delta t_k^p + \omega \Delta t_k^p \left(\frac{H_{k+1}^{t+\Delta t_k^p}}{C_p T_m^* + \tfrac{1}{2}L} - 1 \right) \qquad (10.74)$$

where superscript p refers to the p-th iteration and ω is a relaxation factor.

Clearly, when $H_{k+1}^{t+\Delta t_k^p}$ has converged to $C_p T_m^* + \tfrac{1}{2}L$, the corresponding enthalpy values at all nodes are adopted as the solution for the time $t + \Delta t_k^p$. The corresponding values of nodal temperatures are determined from the temperature–enthalpy relation (10.69).

Voller and Cross (1981) repeated the frost penetration problem considered by Goodrich (1978) using both the explicit and implicit methods and concluded that the implicit algorithm was stable, accurate, and fast. The algorithm is faster because there are no restrictions on the time step. Furthermore, the algorithm can also be applied with a fixed Δt that ensures stable and accurate solutions for the problem. In this case, convergence is verified over the enthalpies calculated at subsequent iterations.

10.4.3 Explicit Enthalpy Method: Phase Change over a Temperature Range

We now examine the application of the enthalpy method to one-dimensional solidification (or melting) with phase change taking place over a temperature range as illustrated in Figure 10.11b. The mathematical formulation of this problem is similar to that described by equations (10.54) and (10.55). The finite difference representation of these equations is the same as that given by equations (10.57)–(10.60), but the computational procedure is simpler because the interpretation of the enthalpy results is easier.

The following algorithm can be used. We assume that at the time level $n\Delta t$ the numerical values of T_i^n and $H_i^n (0 \leq i \leq M)$ are known.

i. Compute H_i^{n+1} for $i = 1,2,\ldots,M-1$ from equation (10.57a).
ii. Compute the corresponding T_i^{n+1} from the enthalpy–temperature relation, that is, Figure 10.11b.
iii. For the thermal insulation condition considered here, determine the temperature T_M^{n+1} at $x = B$ according equation (10.59b).

The calculations are started with $n = 0$, which corresponds to the initial condition and the previous procedure is repeated for each successive time step.

10.5 Phase Change Model for Convective–Diffusive Problems

In this section, we will address a phase change problem with convective–diffusive effects. Let us consider a binary alloy whose phase diagram is given in Figure 10.13. The ordinate of this diagram is the temperature of the alloy, while the abscissa is the concentration of the solute in the alloy. The origin point in the x axis corresponds to a pure solvent.

There are three important regions in this diagram. Region 1 is the liquid region; Region 2, named the mushy region, contains both liquid and solid phases; and Region 3 is the solid region. Several models exist to represent the mushy region. Some of them assume that this region is a porous media with variable porosity, while others consider the solid fully dispersed within the liquid, with both having the same velocity, which is the viscosity variable within this phase.

Phase Change Problems

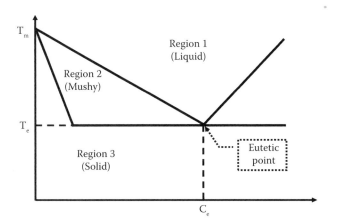

FIGURE 10.13
Binary phase diagram. (From Colaço, M.J., Dulikravich, G.S., *Mater. Manuf. Process.*, 22, 594–606, 2007.)

The liquid and mushy regions are separated by a curve, named the liquidus curve, which is simplified in Figure 10.13 by a straight line. In the same way, the solidus line is the line that separates the solid and mushy regions. Two temperatures are important in this diagram: The first one is the melting temperature of a pure substance, represented by T_m in Figure 10.13. The other important point is the eutectic temperature T_e, corresponding to the eutectic concentration C_e. The liquidus line connects the point of melting temperature and zero concentration to the point of eutectic temperature and eutectic concentration.

To show how to model a phase change problem with liquid movement, we will consider an incompressible laminar natural convection problem of a Newtonian fluid undergoing a solidification process. The fluid physical properties are assumed constant within each phase. The energy source term resulting from viscous dissipation is neglected, and buoyancy effects are represented by the Oberbeck–Boussinesq approximation. Radiative heat transfer, Soret, and Dufour effects are neglected (Bennacer and Gobin 1996; Gobin and Bennacer 1996). For this analysis, we will consider a hypoeutectic alloy, where the initial concentration is below the eutectic value.

The general conservation equation for liquid and solid phases can be written as (Voller et al. 1989)

$$\frac{\partial}{\partial t}(g_s \rho_s \phi_s) + \nabla \cdot (g_s \rho_s \mathbf{V}_s \phi_s) = \nabla \cdot (g_s \Gamma_s^\phi \nabla \phi_s) + F_s + P \qquad (10.75)$$

$$\frac{\partial}{\partial t}(g_l \rho_l \phi_l) + \nabla \cdot (g_l \rho_l \mathbf{V}_l \phi_l) = \nabla \cdot (g_l \Gamma_l^\phi \nabla \phi_l) + F_l - P \qquad (10.76)$$

where ϕ is the quantity being conserved, g is the volume fraction in a representative region, ρ is the density, Γ is the diffusion coefficient, **V** is the velocity vector, F is the body force, P represents the interphase source terms, and the subscripts s and l refer to the solid and liquid phases, respectively. Notice that the interphase source terms have opposite signs and will cancel out in an additive combination of the phases. The volume fractions are related to the mass fractions, f_s and f_l, by (Bennacer and Gobin 1996; Gobin and Bennacer 1996)

$$\rho f_s = \rho_s g_s \tag{10.77}$$

$$\rho f_l = \rho_l g_l \tag{10.78}$$

where

$$\rho = \rho_s g_s + \rho_l g_l \tag{10.79}$$

The overall conservation equation can be obtained by summing equations (10.75) and (10.76)

$$\frac{\partial}{\partial t}(g_s \rho_s \phi_s + g_l \rho_l \phi_l) + \nabla \cdot (g_s \rho_s \mathbf{V}_s \phi_s + g_l \rho_l \mathbf{V}_l \phi_l) \\ = \nabla \cdot (g_s \Gamma_s^\phi \nabla \phi_s + g_l \Gamma_l^\phi \nabla \phi_l) + F_s + F_l \tag{10.80}$$

For the columnar dendritic zone, a porous media model (Swaminathan and Voller 1997; Voller et al. 1989; Voller 2004; Zabaras and Samanta 2004) must be utilized, such that the velocity of the solid phase is imposed as zero. Also, the dissipative interfacial stress usually is modeled in an analogy with Darcy's law, where the permeability is commonly approximated using the Kozeny–Carman equation (Voller et al. 1989; Zabaras and Samanta 2004). This porous media model will not be discussed in this section. For the mushy zone model (Ghosh 2001; Voller et al. 1989), which is applicable to amorphous materials (waxes and glasses), and the equiaxed zone of metal casting, the solid is assumed to be fully dispersed within the liquid such that

$$\mathbf{V} = \mathbf{V}_s = \mathbf{V}_l \tag{10.81}$$

Therefore, the velocity within the solid phase is reduced by imposing a large difference of viscosity between the solid and liquid phases, such that

$$\frac{\mu_s}{\mu_l} \geq 10^5 \tag{10.82}$$

Then, equation (10.80) reduces to

$$\frac{\partial}{\partial t}(g_s \rho_s \phi_s + g_l \rho_l \phi_l) + \nabla \cdot [(g_s \rho_s \phi_s + g_l \rho_l \phi_l)\mathbf{V}] \\ = \nabla \cdot (g_s \Gamma_s^\phi \nabla \phi_s + g_l \Gamma_l^\phi \nabla \phi_l) + F_s + F_l \tag{10.83}$$

If there is equilibrium among the solid and liquid phases such that $\phi_l = \phi_s$ in a representative elementary volume (e.g., in the energy equation written

Phase Change Problems

for the temperature) (Zabaras and Samanta 2004), we obtain, after using equation (10.79)

$$\frac{\partial}{\partial t}(\rho\phi) + \nabla \cdot (\rho\phi \mathbf{V}) = \nabla \cdot (\Gamma^\phi \nabla\phi) + F \tag{10.84}$$

where

$$\Gamma^\phi = g_s \Gamma_s^\phi + g_l \Gamma_l^\phi \tag{10.85}$$

$$F = F_s + F_l \tag{10.86}$$

$$\rho = g_s \rho_s + g_l \rho_l \tag{10.87}$$

Notice that equation (10.84) is similar to equation (8.59a). Thus, when defining proper mixture quantities, given by equations (10.85)–(10.87), the same methodology derived earlier in Section 8.2 can be used to solve equation (10.84). We still need to include the effects of the latent heat in the energy equation and add the possibility to analyze binary alloys. This will be done in the next subsections.

10.5.1 Model for the Passive Scalar Transport Equation

If we are dealing with problems where no phase change takes place or with problems where a binary phase diagram does not apply, such as with the solidification of a pure substance or a mixture, equation (10.84) can be directly applied by simply replacing ϕ with the concentration of the solute C, Γ with the mass diffusivity coefficient, and F with zero.

However, if we are dealing with a solidification of an alloy, the solute concentrations of the liquid and solid phases, C_l and C_s, must follow the binary diagram, and equation (10.84) is no longer valid because ϕ_l is different from ϕ_s. Thus, for the concentration equation, we must use equation (10.83) in its full form. For the sake of simplicity, let us consider $\rho_s = \rho_l$ (Voller et al. 1989; Zabaras and Samanta 2004). In the context of solidification, such an assumption implies no shrinkage induced by solidification (Zabaras and Samanta 2004). Then, from equations (10.77), (10.78), and (10.83), it follows that

$$\frac{\partial}{\partial t}[\rho(f_s\phi_s + f_l\phi_l)] + \nabla \cdot [\rho(f_s\phi_s + f_l\phi_l)\mathbf{V}] \\ = \nabla \cdot (f_s \Gamma_s^\phi \nabla\phi_s + f_l \Gamma_l^\phi \nabla\phi_l) + F \tag{10.88}$$

For the mushy region (the region between the solidus and liquidus lines of the binary phase diagram), the concentrations of the liquid and solid phases are related through the partition coefficient K_p (Rappaz 1989), which is defined as

$$C_s = K_p C_l \tag{10.89}$$

where $0 < K_p < 1$.

Thus, substituting equation (10.89) into equation (10.88) written for concentration, and knowing that $f_s = (1-f_l)$, it follows, after some manipulations, that (Colaço and Dulikravich 2007)

$$\frac{\partial}{\partial t}[\rho C_s] + \nabla \cdot [\rho C_s \mathbf{V}]$$
$$= \nabla \cdot (D^+ \nabla C_s) - \frac{\partial}{\partial t}\left[f_l\left(\frac{1}{K_p} - 1\right)\rho C_s\right] - \nabla \cdot \left[f_l\left(\frac{1}{K_p} - 1\right)\rho C_s \mathbf{V}\right] \quad (10.90)$$

where

$$D^+ = f_s \rho_s D_s + \frac{f_l \rho_l D_l}{K_p} \quad (10.91)$$

Equation (10.90) is the same equation proposed by Voller et al. (1989). Notice, however, that it is valid only for the mushy region and not for the solid and liquid regions. Thus, in order to obtain a more general model, let us define

$$C = f_s C_s + f_l C_l \quad (10.92)$$

$$D = f_s \rho_s D_s + f_l \rho_l D_l \quad (10.93)$$

Equation (10.92) assumes equilibrium (i.e., reversible) solidification. Equilibrium solidification assumes complete mixing in both liquid and solid phases at every stage of the cooling. It also assumes equilibrium at the interface of solid and liquid. Because the interfacial process during solidification is very simple from an atomistic point of view, assumption of interfacial equilibrium is justified (Ghosh 2001).

Using equations (10.92) and (10.93), it is possible to rewrite equation (10.88) for the concentration as

$$\frac{\partial}{\partial t}(\rho C) + \nabla \cdot (\rho C \mathbf{V})$$
$$= \nabla \cdot (D \nabla C) + \nabla \cdot [f_s \rho_s D_s \nabla (C_s - C)] + \nabla \cdot [f_l \rho_l D_l \nabla (C_l - C)] \quad (10.94)$$

where the last two terms can be written as a source term

$$S^C = \nabla \cdot [f_s \rho_s D_s \nabla (C_s - C)] + \nabla \cdot [f_l \rho_l D_l \nabla (C_l - C)] \quad (10.95)$$

The values of the coefficients in the source term depend on the solid and liquid fractions f_s and f_l. In this equation, one must also determine the concentration of liquid and solid phases at a given temperature. Considering the liquidus line as a straight line, one can obtain the following equations for the mushy zone by inspecting the binary diagram shown in Figure 10.13

$$C_l = \frac{T_m - T}{T_m - T_e} C_e \quad (10.96)$$

TABLE 10.2

Coefficients for the Source Term of the Concentration Equation

Region	f_s	f_l	C_s	C_l
Liquid	0	1	0	C
Solid	1	0	C	0
Mushy	$0 < f_s < 1$	$(1-f_s)$	Equation 10.97	Equation 10.96

$$C_s = \frac{T_m - T}{T_m - T_e} K_p C_e \qquad (10.97)$$

Therefore, we have three different possible values for the coefficients in the source term given by equation (10.95), depending on which region (liquid, mushy zone, or solid) exists locally. These values are summarized in Table 10.2.

One can check that equation (10.94) reduces to equation (10.90) when $0 < f_s < 1$. The solid fraction can be modeled by the Lever rule (Rappaz 1989)

$$f_s = \frac{1}{1 - K_p} \left(\frac{T_l - T}{T_m - T} \right) \qquad (10.98)$$

which assumes that the diffusion of the solid is infinite. Here, the partition coefficient, K_p, is given by equation (10.89). Mixing in the liquid is done primarily by convection; hence, it may take place quite fast. Therefore, an assumption of complete mixing in the liquid is often reasonable. Mixing in the solid is by diffusion, which is a very slow process (Ghosh 2001). Therefore, an assumption of complete mixing in the solid is not always valid. With the assumption of complete mixing in the liquid and no diffusion in the solid, the solid fraction can be described by the Scheil's model (Rappaz 1989)

$$f_s = 1 - \left(\frac{T_m - T}{T_m - T_l} \right)^{\frac{1}{K_p - 1}} \qquad (10.99)$$

In either the Lever rule or Sheil's model, one must determine the value of the temperature at the liquidus curve for a given concentration, which is represented as T_l in these equations. Considering the liquidus line as a straight line, one can obtain the following equation by inspection of the binary diagram shown in Figure 10.13

$$T_l = T_m - (T_m - T_e) \frac{C}{C_e} \qquad (10.100)$$

$$T_s = \text{MAX} \left[T_e, T_m - (T_m - T_e) \frac{C}{K_p C_e} \right] \qquad (10.101)$$

It is worth noting that the definition of T_s is more elaborate than the definition of T_l because the solid–mushy interface in Figure 10.13 is limited by two straight-line segments, while the mushy–liquid interface is limited by only one straight line.

Notice that, as the solidification begins, the solid phase rejects solute and the concentration in the remaining liquid and mushy regions increases. Thus, the solid and liquid temperatures at each point of the domain must be determined as the solidification front advances. This indeed induces a buoyancy effect represented by the Oberbeck–Boussinesq equation. The rejection of the solute by the solid often leads to secondary reactions, such as formation of oxides, sulfides, oxysulfides, and nitrides during solidification, which can significantly alter microsegregation patterns (Ghosh 1990). Such reactions will not be treated in this book.

With the inclusion of the aforementioned derivation for the concentration equation, Table 8.1 can now be modified in order to represent the conserved quantity, diffusion coefficient, and source term to be used in the general equation (8.59a) for the conservation of mass, concentration, and momentum in the x and y directions. The new coefficients are given in Table 10.3, where the y axis is aligned with the negative gravity acceleration direction. In this table, C is given by equation (10.92), D is given by equation (10.93), C_l is given by equation (10.96), C_s is given by equation (10.97), f_s is given by equation (10.98) or (10.99), the viscosity ratio is given by equation (10.82), and ρ is given by equation (10.87). The energy equation will be discussed in the next subsection.

10.5.2 Model for the Energy Equation

In this subsection, we will use the enthalpy method, previously discussed, where we can represent the time variation of enthalpy as

$$\frac{\partial H}{\partial t} = C_p \frac{\partial T}{\partial t} \qquad (10.102)$$

Therefore, for the one-dimensional version of the energy equation for a purely diffusive problem addressed in the previous section, we have

TABLE 10.3

Modified Conserved Quantity, Diffusion Coefficient, and Source Term

Equation	ϕ	Γ^ϕ	S^ϕ
Continuity (Mass)	1	0	0
Concentration	C	D	$\nabla \cdot [f_s \rho_s D_s \nabla (C_s - C)] + \nabla \cdot [f_l \rho_l D_l \nabla (C_l - C)]$
x-momentum	u	μ	$-\dfrac{\partial p}{\partial x}$
y-momentum	v	μ	$-\dfrac{\partial p}{\partial y} - \rho g[1 - \beta(T - T_{ref}) - \beta_s(C - C_{ref})]$

Phase Change Problems

$$\rho \frac{\partial H}{\partial t} = \frac{\partial}{\partial x}\left(k \frac{\partial T}{\partial x}\right) \tag{10.103}$$

In this method, the energy conservation equation appears as a mixed enthalpy-temperature equation. Writing the general equation (10.80) for the energy balance results in

$$\frac{\partial}{\partial t}(g_s\rho_s H_s + g_l\rho_l H_l) + \nabla \cdot (g_s\rho_s \mathbf{V}_s H_s + g_l\rho_l \mathbf{V}_l H_l)$$
$$= \nabla \cdot (g_s k_s \nabla T_s + g_l k_l \nabla T_l) \tag{10.104}$$

which is valid in the whole region.

Notice that, for the energy conservation coefficients, given in Table 8.1, the diffusion coefficient Γ^ϕ was given as k/C_p. Thus, for the derivation of the previous equation, the entire equation was multiplied by C_p, in such a way that in the transient and convective terms the temperature appears multiplied by the specific heat at constant pressure, which was then replaced by the enthalpy. The diffusive term was kept as a function of the temperature only. Thus, we have an equation for two variables: temperature and enthalpy. However, these quantities are a function of each other.

Invoking the hypothesis of thermodynamic equilibrium for the temperature, constant density, and a fully dispersed solid within the liquid in the mushy zone (as used for the concentration equation), and defining

$$H = g_s H_s + g_l H_l \tag{10.105}$$

$$k = g_s k_s + g_l k_l \tag{10.106}$$

it is possible to rewrite the energy equation as

$$\frac{\partial}{\partial t}(\rho H) + \nabla \cdot (\rho \mathbf{V} H) = \nabla \cdot (k \nabla T) \tag{10.107}$$

For the enthalpy method, we will use the same methodology described in the previous subsections. However, the transient and convective terms in the energy equation, which contain the product $C_p T$, will be rewritten for the variable H, while the diffusive terms will be written for the temperature. Therefore, the energy equation will have a mixture of temperature and enthalpy as variables. In terms of the general equation (8.59a), slightly modified here, we can define the conserved quantity, the diffusion coefficient, and the source term as given by Table 10.4. Notice that a new variable ϕ^* appears in the equation due to the presence of two different dependent variables: temperature and enthalpy. Also notice that the mass, concentration, and momentum equations are the same as those derived previously. Only the energy equation needs to be modified. The specific heat at constant pressure now appears inside the transient and convective terms. Therefore, the transport coefficient Γ^ϕ now includes only the thermal conductivity.

TABLE 10.4

Conserved Quantity, Diffusion Coefficient, and Source Term for the Energy Equation

Equation	ϕ	ϕ^*	Γ^ϕ	S^ϕ
Energy	H	T	k	0

$$\frac{\partial}{\partial t}(\rho\phi) + \nabla \cdot (\rho\phi \mathbf{V}) = \nabla \cdot (\Gamma^\phi \nabla \phi^*) + S^\phi \qquad (10.108)$$

The general computational procedure to solve the energy equation by advancing from time t_n to time t_{n+1}, is summarized as follows:

1. Use the temperatures at time t_n as an initial guess for the temperatures at time t_{n+1}.
2. Replace T into the diffusive terms of the energy equation.
3. Solve the energy equation for the enthalpy.
4. Using the enthalpy H, calculate the new temperature T.
5. Calculate the void fraction, given by equation (10.98) or (10.99).
6. Check the convergence. If it is not satisfied, return to Step 2. Otherwise, advance in time and go to Step 1.

Step 4 of the previous procedure does not calculate the temperature explicitly but as a function of the enthalpy. For the case of a pure substance, we have

- If $H < H_s \Rightarrow H = C_{ps}T$
- If $H > H_l \Rightarrow H = C_{ps}T_s + C_{pl}(T-T_s) + L$

where the subscripts s and l refer to the liquid and solid phases, respectively, and L is the latent heat of solidification/melting. For pure substances, the phase change occurs at a single temperature. Therefore, there is no mushy zone, and $T_l = T_s = T_m$ at the phase change front.

Thus, we have, if $H < H_s$

$$T = \frac{H}{C_{ps}} \qquad (10.109)$$

or, if $H > H_l$

$$T = \frac{H + T_s(C_{pl} - C_{ps}) - L}{C_{pl}} \qquad (10.110)$$

or yet, if $H_s < H < H_l$

$$T = T_m = T_s = T_l \qquad (10.111)$$

Phase Change Problems

Equations (10.109)–(10.111) are the expressions used in Step 4 of the previously presented computational procedure.

Following the procedure presented in Chapter 8, for the control volume P and its neighbors, presented in Figure 8.3, the following equation, similar to equation (8.82), can be obtained

$$A_P^D \phi_P^{*m+1} + A_e^D \phi_E^{*m+1} + A_w^D \phi_W^{*m+1} + A_n^D \phi_N^{*m+1} + A_s^D \phi_S^{*m+1}$$
$$+ A_P^C \phi_P^{m+1} + A_e^C \phi_E^{m+1} + A_w^C \phi_W^{m+1} + A_n^C \phi_N^{m+1} + A_s^C \phi_S^{m+1} = B_P \phi_P^m + S^\phi|_P \Delta y \Delta x$$
(10.112)

Equation (10.112) can also be written in the following form, after using Table 10.4,

$$A_P^D T_P^{m+1} + A_e^D T_E^{m+1} + A_w^D T_W^{m+1} + A_n^D T_N^{m+1} + A_s^D T_S^{m+1}$$
$$+ A_P^C H_P^{m+1} + A_e^C H_E^{m+1} + A_w^C H_W^{m+1} + A_n^C H_N^{m+1} + A_s^C H_S^{m+1} = B_P H_P^m + S^H|_P \Delta y \Delta x$$
(10.113)

where

$$A_e^D = -\frac{D_{11e} \beta_e}{\Delta x} \quad (10.114)$$

$$A_w^D = -\frac{D_{11w} \beta_w}{\Delta x} \quad (10.115)$$

$$A_n^D = -\frac{D_{22n} \beta_n}{\Delta y} \quad (10.116)$$

$$A_s^D = -\frac{D_{22s} \beta_s}{\Delta y} \quad (10.117)$$

$$A_P^D = D_{11e} \frac{\beta_e}{\Delta x} + D_{11w} \frac{\beta_w}{\Delta x} + D_{22n} \frac{\beta_n}{\Delta y} + D_{22s} \frac{\beta_s}{\Delta y} \quad (10.118)$$

$$A_e^C = \dot{M}_e \left(\frac{1}{2} - \alpha_e \right) \quad (10.119)$$

$$A_w^C = -\dot{M}_w \left(\frac{1}{2} - \alpha_w \right) \quad (10.120)$$

$$A_n^C = \dot{M}_n \left(\frac{1}{2} - \alpha_n \right) \quad (10.121)$$

$$A_s^C = -\dot{M}_s \left(\frac{1}{2} - \alpha_s \right) \quad (10.122)$$

$$A_P^C = \frac{M_P}{\Delta t} + \dot{M}_e\left(\frac{1}{2}+\alpha_e\right) - \dot{M}_w\left(\frac{1}{2}+\alpha_w\right) + \\ \dot{M}_n\left(\frac{1}{2}+\alpha_n\right) - \dot{M}_s\left(\frac{1}{2}+\alpha_s\right) \tag{10.123}$$

$$B_P = \frac{M_P}{\Delta t} \tag{10.124}$$

and the other terms were defined in equations (8.70)–(8.74). Notice that equation (10.113) now includes both temperature and enthalpy. In order to solve it for enthalpy, equations (10.109)–(10.111) must be used to transform temperatures into enthalpies for each of the control volumes. However, since equations (10.109)–(10.111) depend upon the value of the enthalpy for each control volume, equation (10.113) must be solved iteratively.

Let us suppose, for example, the value of enthalpy H, in some iterative step at the east volume is less than H_s. In this case, the east temperature appearing multiplied by the diffusive terms in equation (10.113) is replaced by equation (10.109); that is, $T_E = H_E/C_{ps}$. If the enthalpy is greater than H_l, however, equation (10.110) shall be used instead of equation (10.109). In this case, we can write $T_E = H_E/C_{ps}$ and add the term $A_e^D[T_s(C_{pl}-C_{ps})-L]/C_{ps}$ to the source term S^H. For the third possibility, where $H_s < H < H_l$, equation (10.111) must be used, where the diffusive term at the east volume $A_e^D T_m$ is then moved to the source term S^H. After performing all substitutions for all control volumes, the system is then solved for the enthalpy, and the iterative procedure is repeated until the enthalpy is converged for all control volumes. Only after this procedure is done, time is advanced.

For Step 4 of the computational procedure presented previously, we simply calculate T_P directly from equations (10.109)–(10.111), after H is obtained from equation (10.113) in Step 3.

The physical properties are evaluated using the temperature as shown in Table 10.5. For a binary alloy, instead of a material with a single phase-change temperature, we have a range of temperatures where the solidification

TABLE 10.5

Physical Properties as Function of Temperature for a Pure Substance

Property	$T < T_m$	$T > T_m$
ρ	ρ_s	ρ_l
D	D_s	D_l
K	K_s	K_l
C_p	C_{ps}	C_{pl}
μ	μ_s	μ_l

occurs. Everything is the same as in the previous case (pure substance), except that, if $H_s < H < H_l$, we have

$$H = f_s C_{ps} T + (1 - f_s)[C_{ps} T_s + C_{pl}(T - T_s) + L] \qquad (10.125)$$

and

$$T = \frac{H + [T_s(C_{pl} - C_{ps}) - L](1 - f_s)}{C_{pl} + f_s(C_{ps} - C_{pl})} \qquad (10.126)$$

where the solid fraction f_s is given by equation (10.98) or (10.99) for the Lever rule or Scheil's model, respectively. Note that equation (10.125) is equal to equation (10.109) multiplied by f_s plus equation (10.110) multiplied by $(1-f_s)$.

Note that if $T < T_s$, then f_s must be set to unity. If $T > T_l$, then f_s must be set to zero. The other thermal properties can be approximated, for example, as linear functions within the mushy region ($T_s < T < T_l$) and kept constant within each phase. Thus, in the mushy region

$$\psi = f_s \psi_s + (1 - f_s) \psi_l \qquad (10.127)$$

where ψ can represent, for example, the density, thermal conductivity, or viscosity. For the specific heat at constant pressure within the mushy region, the following approximation can be used (Rappaz 1989)

$$C_p = \frac{\partial H}{\partial T} \approx \frac{\sqrt{\left(\frac{\partial H}{\partial x}\right)^2 + \left(\frac{\partial H}{\partial y}\right)^2}}{\sqrt{\left(\frac{\partial T}{\partial x}\right)^2 + \left(\frac{\partial T}{\partial y}\right)^2}} \qquad (10.128)$$

Note that enthalpy is a function of the temperature, which is a function of the solid fraction that is itself a function of the temperature. Thus, if $H_s < H < H_l$, we must solve a nonlinear equation for T. From equations (10.98), (10.99), and (10.126), it follows that

$$T - \frac{H + [T_s(C_{pl} - C_{ps}) - L]\left[1 - \frac{1}{1-K_p}\left(\frac{T_l - T}{T_m - T}\right)\right]}{C_{pl} + \frac{1}{1-K_p}\left(\frac{T_l - T}{T_m - T}\right)(C_{ps} - C_{pl})} = 0 \qquad (10.129)$$

for the Lever rule, equation (10.98), or

$$T - \frac{H - [T_s(C_{pl} - C_{ps}) - L]\left(\frac{T_m - T}{T_m - T_l}\right)^{\frac{1}{K_p - 1}}}{C_{pl} + \left[1 - \left(\frac{T_m - T}{T_m - T_l}\right)^{\frac{1}{K_p - 1}}\right](C_{ps} - C_{pl})} = 0 \qquad (10.130)$$

for Scheil's method, equation (10.99), respectively. These equations can be solved for T by the secant method. It should be pointed out that in this section we have not considered viscous dissipation in the mushy region. We also did not utilize adaptive grids in order to resolve the details of the mushy region.

Once T is obtained, and knowing the value of C, the values of the liquid and solid concentrations can be obtained by inspecting the binary diagram given by Figure 10.13 [see equations (10.96) and (10.97)]. After calculating T and f_s, if the fraction of the solid is equal to zero, then the concentration of the liquid C_l is set to C and the concentration of the solid C_s is set to zero. Otherwise, if the solid fraction is equal to one, then the concentration of the liquid C_l is set to zero and the concentration of the solid C_s is set to C [see equation (10.92) and the source term appearing in Table 10.3]. Note that, in order to avoid oscillations, f_s must be set to zero if $H < H_s$ and to one if $H > H_l$.

Example 10.2 Melting in a Heat Convection Problem (Colaço et al. 2005)

In this example, we will present the results for a melting problem in a closed cavity undergoing natural convection. Results of this problem using several different techniques were presented by Bertrand et al. (1999). The problem is composed of a square cavity filled with a solid material, as shown in Figure 10.14. The initial temperature was set just below the melting temperature $T_m = 232°C$. The right wall was subjected to the melting temperature, $T_0 = T_m$, and the left wall was subjected to a temperature equal to 235 °C, which was above the melting temperature. The natural convection initiates when the left wall is exposed to a temperature greater than the melting temperature. Then, the material starts to melt and the natural convection initiates at the left boundary. The top and bottom walls were kept insulated. The physical properties for molten steel used by Bertrand et al. (1999) are:

- $\rho = 7500 \text{ kg/m}^3$
- $\upsilon = 8 \times 10^{-7} \text{ m}^2/\text{s}$
- $\beta = 8.3 \times 10^{-4} \text{ 1/K}$

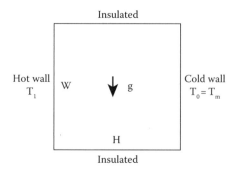

FIGURE 10.14
Geometry for the heat convection problem with phase change (Colaço et al. 2005).

- $k = 60$ W/(m K)
- $C_p = 200$ J/(kg K)
- $T_m = 232°C$
- $g = 9.81$ m/s^2
- $\alpha = 4 \times 10^{-5}$ m^2/s
- $L = 6 \times 10^4$ J/kg
- $\Delta T = 3°C$
- $H = W = 0.1$ m

Solution
The solution procedure outlined in this section converged to the results presented in Figure 10.15a–d. These figures show the comparison between the present results and those presented by Bertrand et al. (1999) for a grid with 90 × 90 volume at t = 100 s, 250 s, 1000 s, and 2500 s, respectively. In each figure, the position of the solidification front is plotted inside the cavity, whose dimensions are normalized by the length, H. The results obtained by the finite volume method are plotted as a thicker dashed line. In Bertrand's paper, several solutions from different methods were presented, corresponding to each of the other curves shown in these figures. One can notice a good agreement between the results obtained by the current technique and the solutions presented by Bertrand et al. (1999). The time increment used for the finite volume method was $\Delta t = 10^{-5}$ s.

Example 10.3 Transient solidification of a binary alloy (Colaço and Dulikravich 2007)

This example will present a thermosolutal problem with solidification in a square cavity of size 0.025 m, where all surfaces were kept insulated, except the left wall, which was suddenly cooled to a temperature below the melting temperature. Results are presented for the methodology described previously. The SIMPLEC method was used for dealing with the pressure velocity coupling, in a grid with 30 × 30 volumes.

The initial concentration and temperature inside the cavity were 0.1 kg/m^3 and 600 K, respectively (Voller et al. 1989). The temperature of the left wall was kept at 400 K. All walls were impermeable to mass. The fluid was a mixture of ammonium chloride and water (NH$_4$Cl-H$_2$O), with the following physical properties:

- $\beta = 4 \times 10^{-5}$ 1/K
- $\beta_s = 0.025$ m^3/kg
- $k = 0.4$ W/(m K)
- $C_p = 3000$ J/(kg K)
- $T_m = 630$ K
- $T_e = 250$ K

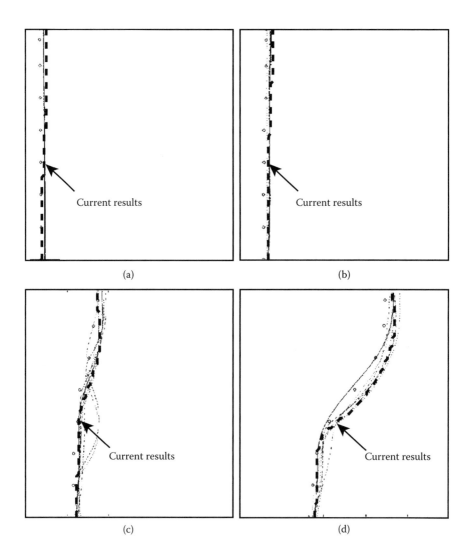

FIGURE 10.15
Comparison between the current (thicker line) and several other methodologies presented by Bertrand et al. (1999) for the heat convection problem with melting: (a) t = 100 s, (b) t = 250 s, (c) t = 500 s, and (d) t = 1000 s (Colaço and Dulikravich 2005).

- $C_e = 0.8$ kg/m^3
- $K_p = 0.3$
- $\mu_l = 0.001$ kg/(m s)
- $\mu_s = 10000$ kg/(m s)
- $D = 4.8 \times 10^{-9}$ m^2/s
- $g = 9.81$ m/s^2

Phase Change Problems

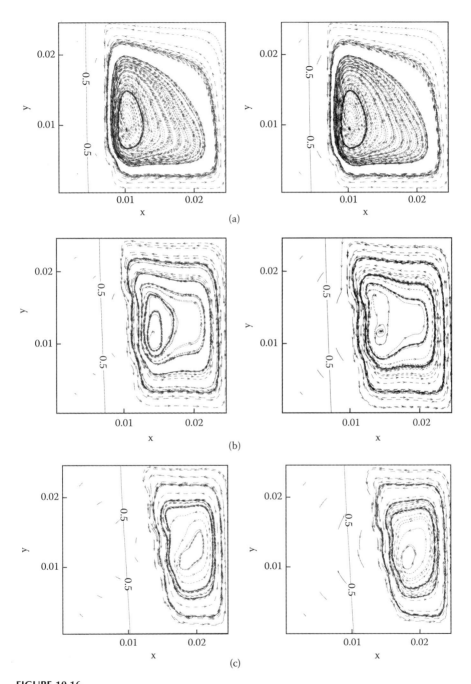

FIGURE 10.16
Results for $f_s = 0.5$ at different times, considering equation (10.90) (left) and equation (10.94) (right): (a) 100 s, (b) 250 s, and (c) 500 s. (From Colaço, M.J., Dulikravich, G.S., *Mater. Manuf. Process.*, 22, 594–606, 2007.)

Solution
In the present results, the time step for the numerical method was taken as 0.001 s. The stopping criterion at each iteration was taken as the one used by Voller et al. (1989): The mass source within each control volume must be less than 8×10^{-6}, and the error in the overall energy and solute balance must drop below 10^{-2}% and 10^{-4}%, respectively.

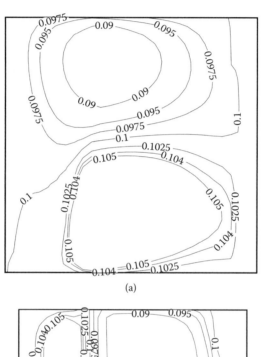

(a)

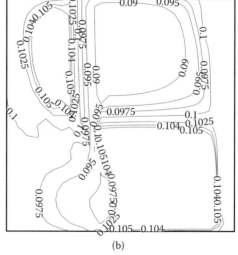

(b)

FIGURE 10.17
Results for the macrosegregation profiles at t = 3000 s considering (a) equation (10.90) and (b) equation (10.94). (From Colaço, M.J., Dulikravich, G.S., *Mater. Manuf. Process.*, 22, 594–606, 2007.)

Figure 10.16 presents the results for the fraction of solid equal to 0.5 at different times: (a) 100 s, (b) 250 s, and (c) 500 s. In this figure, we also present the results obtained with equation (10.90), used by Voller et al. (1989), and equation (10.94) derived here, which, as discussed previously, is supposed to be more robust. Figure 10.17 present the results for the macrosegregation profiles at t = 3000 s considering (a) equation (10.90) and (b) equation (10.94).

PROBLEMS

10.1. Consider the single-phase solidification problem defined by equations (10.16)–(10.19). Applying the variable time step approach described in Section 10.2, but using the Crank–Nicolson method, develop the finite difference representation of this phase change problem.

10.2. Solve the phase change problem described in Example 10.1 and reproduce the results given in Table 10.1.

10.3. Develop the finite difference formulation of the two-phase solidification problem considered in Section 10.3 using the combined method to discretize the governing differential equations.

10.4. Solve the phase change problem described in Example 10.1 using the explicit enthalpy method as applied to materials with a single melting temperature.

10.5. Solve the phase change problem described in Example 10.1 using the implicit enthalpy method as applied to materials with a single melting temperature.

10.6. Develop the finite difference formulation of the variable time step approach described in Section 10.2 for the case of single phase melting.

10.7. Pure water (melting temperature: $T_m = 0°C$) occupies the region $x > 0$. The initial temperature of water is uniform and equal to 2°C when the temperature of the boundary at $x = 0$ is lowered to $-10°C$. The latent heat of solidification is 100 MJ/kg, while the volumetric heat capacities of the solid and liquid phases are 4 MJ/m^3°C. The thermal conductivities of the solid and liquid phases are 0.6 W/m°C, respectively. Solve this phase change problem using the explicit and the implicit enthalpy methods.

11
Numerical Grid Generation

The traditional finite difference methods have computational simplicity when they are applied for the solution of problems involving regular geometries with uniformly distributed grids over the region. However, their major drawbacks include their inability to effectively handle the solution of problems over arbitrarily shaped complex geometries. When the geometry is irregular, difficulties arise from the boundary conditions, because interpolation is needed between the boundaries and the interior points in order to develop finite difference expressions for nodes at the boundaries. Such interpolations produce large errors in the vicinity of strong curvatures and sharp discontinuities. Therefore, it is difficult and inaccurate to solve problems with regular grids over regions with irregular geometries. The use of coordinate transformation and mapping the irregular region into a regular one over the computational domain is not new. Many transformations are available in which the physical and computational coordinates are related with algebraic expressions. However, such transformations are very difficult to construct, except for some relatively simple situations; for most multidimensional cases, it is impossible to find a transformation. The coordinate transformation technique advanced by Thompson (1977) alleviates such difficulties because the transformation is obtained automatically from the solution of partial differential equations in a regular computational domain. Therefore, the scheme includes the geometrical flexibility of the finite element method while maintaining the simplicity of the conventional finite difference technique. In this approach, a curvilinear mesh is generated over the physical domain such that one member of each family of curvilinear coordinate lines is coincident with the boundary contour of the physical domain. Therefore, the scheme is also called the *boundary-fitted coordinate method*.

To illustrate the basic concepts in the implementation of this technique, we consider a two-dimensional situation with x,y being the coordinates in the physical plane and ξ,η in the computational plane. The basic steps in Thompson's approach can be summarized as follows:

1. Transformation relations for mapping from the x,y plane to the ξ,η plane (or vice versa) are determined from the numerical solution of two elliptic partial differential equations of the Laplace or the Poisson type. Parabolic- and hyperbolic-type differential equations have also been used for numerical grid generation; however,

elliptic equations are preferred because of their smoothing effect in spreading out the boundary slope irregularities.

2. The irregular physical region is mapped from the x,y physical plane to the computational ξ,η plane as a regular region. We note that the transformed regular region in the computational domain can be in the Cartesian, cylindrical, or polar coordinate systems, depending on the original geometry.
3. The partial differential equations governing the physical phenomena are transformed from the x,y independent variables of the physical domain to the ξ,η independent variables of the computational domain. Hence, the finite difference techniques described in this book are used to solve the governing partial differential equations in a regular geometry in the computational plane.
4. Once the transformed field equations are solved in the computational domain, the solution is transformed from the ξ,η computational plane to the x,y physical plane by the transformation relations developed in step 2.

Such a procedure devised by Thomson (1977) results in a structured grid, in which all cells are quadrilaterals (parallelepipeds in three-dimensional problems). Unstructured grids, in which the cells are triangles (tetrahedrals in three-dimensional problems), with an irregular pattern can also be numerically generated. As expected, unstructured grids offer more geometrical flexibility than structured grids but require a lot of computational bookkeeping for appropriately taking into account the connection between the cells for discretization of the governing equations. Only structured grids in two dimensions are considered in this introductory test, because they readily allow the extension of the classical finite difference techniques to many practical problems with irregular geometries (see, e.g., Goldman and Kao 1981; Johnson 1982; Masiulaniec et al. 1984; Thompson 1982, 1983, 1984, 1985; Ushikawa and Takeda 1985; Eiseman 1982, 1985; Guceri 1988; Coulter and Guceri 1988; Subbiah et al. 1989; Knupp and Steinberg 1993; Colaço and Orlande 2001a, 2001b, 2004; Dulikravich et al. 2004; Colaço and Dulikravich 2006, 2007; Dulikravich and Colaço 2006; Colaço et al. 2010). More extensive reviews of numerical grid generation that include unstructured grids as well as time-dependent adaptive grids can be found in Thompson (1982), Carey (1997), Thomson et al. (1999), and Liseikin (2010).

Since the transformation of the partial differential equations from the x,y coordinates to the ξ,η coordinates is an integral part of the numerical grid-generation technique, we first present an overview of the coordinate transformation relations appropriate for the transformation of partial differential equations encountered in heat transfer applications.

11.1 Coordinate Transformation Relations

Consider a partial differential equation given in the x,y independent variables in the physical domain. We seek the transformation of this partial differential equation from the x,y to the ξ,η independent variables. The transformation from the x,y to the ξ,η variables can be expressed as

$$\xi \equiv \xi(x,y) \tag{11.1a}$$

$$\eta \equiv \eta(x,y) \tag{11.1b}$$

and the inverse transformation is given by

$$x \equiv x(\xi,\eta) \tag{11.2a}$$

$$y \equiv y(\xi,\eta) \tag{11.2b}$$

Basic to the transformation of the partial differential equations is the availability of the relations for the transformation of various differential operators, such as the first derivative, gradient, and Laplacian. Therefore, in this section, we present such relations for use as a ready reference in later sections.

The Jacobian of the transformation J is given by Courant (1956) as

$$J = J\left(\frac{x,y}{\xi,\eta}\right) = \begin{vmatrix} x_\xi & y_\xi \\ x_\eta & y_\eta \end{vmatrix} = x_\xi y_\eta - x_\eta y_\xi \neq 0 \tag{11.3}$$

where the subscripts denote differentiation with respect to the variable considered, that is,

$$x_\xi = \frac{\partial x}{\partial \xi}, \quad y_\eta = \frac{\partial y}{\partial \eta}, \text{etc} \tag{11.4}$$

The transformation relations can be developed by application of the chain rule of differentiation. Consider, for example, the first derivatives $\partial T/\partial x$ and $\partial T/\partial y$. By the chain rule of differentiation, we write

$$\frac{\partial T}{\partial x} = \xi_x \frac{\partial T}{\partial \xi} + \eta_x \frac{\partial T}{\partial \eta} \tag{11.5a}$$

$$\frac{\partial T}{\partial y} = \xi_y \frac{\partial T}{\partial \xi} + \eta_y \frac{\partial T}{\partial \eta} \tag{11.5b}$$

Interchanging x and ξ, as well as y and η, we obtain

$$\frac{\partial T}{\partial \xi} = x_\xi \frac{\partial T}{\partial x} + y_\xi \frac{\partial T}{\partial y} \tag{11.6a}$$

$$\frac{\partial T}{\partial \eta} = x_\eta \frac{\partial T}{\partial x} + y_\eta \frac{\partial T}{\partial y} \qquad (11.6b)$$

The solution of equations (11.6a,b) for $\partial T/\partial x$ and $\partial T/\partial y$ with Cramer's rule gives the transformation relations for the first derivatives as

$$\frac{\partial T}{\partial x} = \frac{1}{J}\left(y_\eta \frac{\partial T}{\partial \xi} - y_\xi \frac{\partial T}{\partial \eta}\right) \qquad (11.7a)$$

$$\frac{\partial T}{\partial y} = \frac{1}{J}\left(-x_\eta \frac{\partial T}{\partial \xi} + x_\xi \frac{\partial T}{\partial \eta}\right) \qquad (11.7b)$$

where the Jacobian of the transformation is defined by equation (11.3).
A comparison of equations (11.5a,b) and (11.7a,b) gives

$$\xi_x = \frac{1}{J}y_\eta, \quad \xi_y = -\frac{1}{J}x_\eta \qquad (11.8a,b)$$

$$\eta_x = -\frac{1}{J}y_\xi, \quad \eta_y = \frac{1}{J}x_\xi \qquad (11.8c,d)$$

In the above relations, the derivatives ξ_x, ξ_y, η_x, and η_y are called *metrics*, and the derivatives y_ξ, y_η, x_ξ, and x_η are called *computational derivatives*, which are related by equations (11.8a–d).

Example 11.1

Transform the continuity equation

$$\frac{\partial u}{\partial x} + \frac{\partial v}{\partial y} = 0$$

from the x,y coordinates of the physical plane to the ξ,η coordinates of the computational plane.

Solution
The transformation of the first derivatives is given by equations (11.7a,b). Then, the continuity equation transformed from the x,y coordinates to the ξ,η coordinates becomes

$$\frac{1}{J}\left(y_\eta \frac{\partial u}{\partial \xi} - y_\xi \frac{\partial u}{\partial \eta}\right) + \frac{1}{J}\left(-x_\eta \frac{\partial v}{\partial \xi} + x_\xi \frac{\partial v}{\partial \eta}\right) = 0$$

The transformation of second derivatives can be obtained by utilizing the transformation relations for the first derivatives and the chain rule of differentiation. Thompson et al. (1985) presented extensive relations for the transformation of the divergence, gradient, Laplacian, and so on, for both conservative and nonconservative cases, from the Cartesian coordinates to general curvilinear coordinates. Here we present, for ready reference, some of these transformation relations from the x,y coordinates to the coordinates in both the conservative and nonconservative forms. It is to be noted that the nonconservative forms are obtained by expanding all derivatives and cancelling the identical terms.

11.1.1 Gradient

The gradient components in the x and y directions are given, respectively, by

Conservative form:

$$T_x = \frac{1}{J}[(y_\eta\, T)_\xi - (y_\xi T)_\eta] \qquad (11.9a)$$

$$T_y = \frac{1}{J}[-(x_\eta\, T)_\xi + (x_\xi T)_\eta\,] \qquad (11.9b)$$

Nonconservative form:

$$T_x = \frac{1}{J}\,(y_\eta\, T_\xi - y_\xi T_\eta) \qquad (11.10a)$$

$$T_y = \frac{1}{J}\,(-x_\eta\, T_\xi + x_\xi\, T_\eta) \qquad (11.10b)$$

where the Jacobian J is defined by equation (11.3). Note that when the product derivative terms in the conservative form are expanded, the identity terms cancel out, and equations (11.9a,b) reduce to the nonconservative form given by equations (11.10a,b).

11.1.2 Divergence

We consider the vector quantity **T**

$$\mathbf{T} = \mathbf{i}T_1 + \mathbf{j}T_2 \qquad (11.11)$$

where **i** and **j** are unit direction vectors. Then, the transformation of the divergence of **T**, in conservative and nonconservative forms, is given as

Conservative form:

$$\nabla \cdot \mathbf{T} = \frac{1}{J}[(y_\eta\, T_1 - x_\eta\, T_2)_\xi + (-y_\xi\, T_1 + x_\xi\, T_2)_\eta] \qquad (11.12)$$

Nonconservative form:

$$\nabla \cdot \mathbf{T} = \frac{1}{J}[y_\eta\,(T_1)_\xi - x_\eta(T_2)_\xi - y_\xi\,(T_1)_\eta + x_\xi\,(T_2)_\eta] \qquad (11.13)$$

11.1.3 Laplacian

We consider the Laplacian operator

$$\nabla^2 \equiv \frac{\partial^2}{\partial x^2} + \frac{\partial^2}{\partial y^2} \tag{11.14}$$

The transformation relations for this operator in conservative and nonconservative forms are given by

Conservative form:

$$J\nabla^2 T = \left\{ \frac{1}{J} y_\eta [(y_\eta T)_\xi - (y_\xi T)_\eta] - \frac{1}{J} x_\eta [-(x_\eta T)_\xi + (x_\xi T)_\eta] \right\}_\xi$$

$$+ \left\{ -\frac{1}{J} y_\xi [(y_\eta T)_\xi + (y_\xi T)_\eta] + \frac{1}{J} x_\xi [-(x_\eta T)_\xi + (x_\xi T)_\eta] \right\}_\eta \tag{11.15}$$

Nonconservative form:

$$\nabla^2 T = \frac{1}{J^2} [\alpha T_{\xi\xi} - 2\beta T_{\xi\eta} + \gamma T_{\eta\eta}] + [(\nabla^2 \xi) T_\xi + (\nabla^2 \eta) T_\eta] \tag{11.16}$$

where

$$\alpha = x_\eta^2 + y_\eta^2 \quad \beta = x_\xi x_\eta + y_\xi y_\eta \quad \gamma = x_\xi^2 + y_\xi^2 \quad J = x_\xi y_\eta - x_\eta y_\xi \tag{11.17}$$

11.1.4 Normal Derivatives

Conservative form:

The normal derivative of T to the ξ-constant line along the normal n_3 shown in Figure 11.1 is given by

$$\frac{\partial T}{\partial n^{(\xi)}} = \frac{1}{J\alpha^{1/2}} \left\{ y_\eta [(y_\eta T)_\xi - (y_\xi T)_\eta] - x_\eta [-(x_\eta T)_\xi + (x_\xi T)_\eta] \right\} \tag{11.18a}$$

and to the η-constant line along the normal n_4 shown in Figure 11.1 is given by

$$\frac{\partial T}{\partial n^{(\eta)}} = \frac{1}{J\gamma^{1/2}} \left\{ -y_\xi [(y_\eta T)_\xi - (y_\xi T)_\eta] + x_\xi [-(x_\eta T)_\xi + (x_\xi T)_\eta] \right\} \tag{11.18b}$$

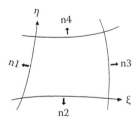

FIGURE 11.1
Outward drawn unit normal vectors to ξ = constant and η = constant lines.

Numerical Grid Generation

Nonconservative form:
The normal derivative of T to the ξ-constant line along the normal $\mathbf{n}_3$ is given by

$$\frac{\partial T}{\partial \mathbf{n}^{(\xi)}} = \frac{1}{J\alpha^{1/2}}(\alpha T_\xi - \beta T_\eta) \tag{11.19a}$$

and to the η-constant line along the normal $\mathbf{n}_4$ is given by

$$\frac{\partial T}{\partial \mathbf{n}^{(\eta)}} = \frac{1}{J\gamma^{1/2}}(-\beta T_\xi + \gamma T_\eta) \tag{11.19b}$$

where α, β, γ, and J are defined by equation (11.17).

11.1.5 Tangential Derivatives

Conservative form:
The tangential derivative of T to the ξ-constant line is given by

$$\frac{\partial T}{\partial \tau^{(\xi)}} = \frac{1}{J\alpha^{1/2}}\left\{x_\eta[(y_\eta T)_\xi - (y_\xi T)_\eta] - y_\eta[(x_\eta T)_\xi - (x_\xi T)_\eta]\right\} \tag{11.20a}$$

and to the η-constant line given by

$$\frac{\partial T}{\partial \tau^{(\eta)}} = \frac{1}{J\gamma^{1/2}}\left\{x_\xi[(y_\eta T)_\xi - (y_\xi T)_\eta] - y_\xi[(x_\eta T)_\xi - (x_\xi T)_\eta]\right\} \tag{11.20b}$$

Nonconservative form:
The tangential derivative of T to the ξ-constant line is given by

$$\frac{\partial T}{\partial \tau^{(\xi)}} = \frac{1}{\alpha^{1/2}} T_\eta \tag{11.21a}$$

and to the η-constant line is given by

$$\frac{\partial T}{\partial \tau^{(\eta)}} = \frac{1}{\gamma^{1/2}} T_\xi \tag{11.21b}$$

The reader should refer to Thompson et al. (1985) for the transformation relations for other partial derivatives, such as $\partial^2/\partial x \partial y$ and $\partial^2/\partial x^2$.

Although the above transform relations are presented for Cartesian coordinates, similar expressions could be derived for transformations in cylindrical or polar coordinates. The notes at the end of this chapter summarize the Laplacians in computational domains transformed from physical domains in cylindrical and polar coordinates.

Example 11.2

Consider the two-dimensional steady-state energy equation in the physical plane given by

$$u\frac{\partial T}{\partial x} + v\frac{\partial T}{\partial y} = \alpha_t \left(\frac{\partial^2 T}{\partial x^2} + \frac{\partial^2 T}{\partial y^2}\right)$$

where α_t is the thermal diffusivity. Transform this equation from the x,y independent variables of the physical domain to the ξ,η independent variables of the computational domain.

Solution
The first derivatives $\partial T/\partial x$ and $\partial T/\partial y$ are transformed by utilizing Equations (11.7a,b), and the Laplacian term is transformed by utilizing equation (11.16). We obtain

$$\frac{u}{J}(y_\eta T_\xi - y_\xi T_\eta) + \frac{v}{J}(-x_\eta T_\xi + x_\xi T_\eta) = \frac{\alpha_t}{J^2}(\alpha T_{\xi\xi} - 2\beta T_{\xi\eta} + \gamma T_{\eta\eta})$$
$$+ \alpha_t[(\nabla^2\xi)T_\xi + (\nabla^2\eta)T_\eta]$$

where α_t is the thermal diffusivity, and α, β, γ, and J are defined by

$$\alpha = x_\eta^2 + y_\eta^2, \quad \beta = x_\xi x_\eta + y_\xi y_\eta, \quad \gamma = x_\xi^2 + y_\xi^2, \quad \text{and} \quad J = x_\xi y_\eta - x_\eta y_\xi.$$

Example 11.3
The two-dimensional advection–diffusion conservation equation can be written in the following general form (see Section 8.2):

$$\frac{\partial}{\partial t}(\rho\phi) + \nabla \cdot (\rho\phi \mathbf{V}) = \nabla \cdot (\Gamma^\phi \nabla\phi) + S^\phi$$

where $\mathbf{V}$ is the velocity vector, while the conserved quantities, the diffusion coefficients, and the source terms are given in Table 8.1 for the continuity, x and y momenta, and energy conservation equations. By using the conservative transform relations presented above, transform this equation from the x,y physical domain to the ξ,η computational domain.

Solution
Equations (11.9) and (11.12) are used for the transformation, which, after some straightforward manipulations, yields

$$\frac{\partial(J\rho\phi)}{\partial t} + \frac{\partial(\tilde{U}\rho\phi)}{\partial \xi} + \frac{\partial(\tilde{V}\rho\phi)}{\partial \eta} = \frac{\partial}{\partial \xi}\left\{J\Gamma^\phi\left[a\frac{\partial\phi}{\partial\xi} + d\frac{\partial\phi}{\partial\eta}\right]\right\}$$
$$+ \frac{\partial}{\partial \eta}\left\{J\Gamma^\phi\left[d\frac{\partial\phi}{\partial\xi} + b\frac{\partial\phi}{\partial\eta}\right]\right\} + JS$$

where

$$a = \xi_x^2 + \xi_y^2; \quad b = \eta_x^2 + \eta_y^2; \quad d = \xi_x\eta_x + \xi_y\eta_y; \quad J = x_\xi y_\eta - x_\eta y_\xi;$$
$$\tilde{U} = J(u\xi_x + v\xi_y); \quad \tilde{V} = J(u\eta_x + v\eta_y)$$

$\tilde{U}$ and $\tilde{V}$ denote the contravariant velocities in the ξ and η directions, respectively, which are the velocity components in these directions, orthogonal to the surfaces of a finite control volume. The metrics required

Numerical Grid Generation

for the computation can be conveniently calculated with equations (11.8a–d). Issues related to the calculation of the metrics are discussed further in this chapter.

11.2 Basic Ideas in Simple Transformations

A variety of approaches have been reported in the literature for the transformation of irregularly shaped regions into simple regular regions such as a square and a rectangle. The basic theory behind such transformations is quite old. For example, conformal transformations have been widely used in classical analysis; the Schwarz–Christoffel transformation is well known for conformal mapping of regions with polynomial boundaries onto the upper-half plane. A comprehensive study on constructing conformal mapping using the Schwarz–Christoffel formula is given by Trefethen (1980), and a dictionary of conformal transformations is compiled by Kober (1957). Details of application of conformal transformation with complex variable technique can be found in the standard texts by Milne-Thompson (1950), Churchill (1948), and Davies (1979).

Before presenting the numerical grid-generation technique, we illustrate in this section the basic concepts in grid generation and mapping, by considering a one-dimensional simple transformation utilizing algebraic relations.

Consider a two-dimensional, steady, boundary-layer flow over a flat plate, mathematically modeled in the physical plane using x,y Cartesian coordinates. To solve such flow problems with finite differences, customarily a rectangular grid is constructed over the solution domain, and the nodes are concentrated near the wall where the gradients are large, as illustrated in Figure 11.2a; that is, a uniform grid is constructed in the x-direction, but a nonuniform grid is used in the y-direction. To alleviate the difficulties associated with the use

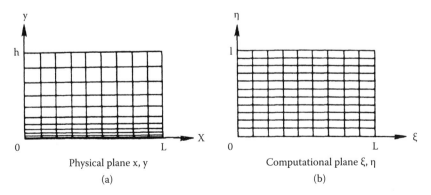

FIGURE 11.2
One-dimensional stretching transformation: (a) Nonuniform grid in the physical plane; (b) Uniform grid in the computational plane.

of nonuniform grids, the problem is transformed from the physical x-y plane to the computational ξ-η plane with a coordinate transformation that allows the use of uniform grids in both the ξ and η directions, as illustrated in Figure 11.2b.

The coordinate transformation that transforms a nonuniform grid spacing in the y-direction into a uniform grid spacing in the η-direction, but allows the grid spacing in the x-direction remain unchanged, is given by Roberts (1971) in the form

$$\xi = x \qquad (11.22a)$$

$$\eta = 1 - \frac{\ln[A(y)]}{\ln B} \qquad (11.22b)$$

where

$$A(y) = \frac{\beta + \left(1 - \frac{y}{h}\right)}{\beta - \left(1 - \frac{y}{h}\right)}, \qquad B = \frac{\beta + 1}{\beta - 1} \qquad (11.23a,b)$$

Here, β is the *stretching parameter* that assumes values $1 < \beta < \infty$. As β approaches unity, more grid points are clustered near the wall in the physical domain. The inverse transform is given by

$$x = \xi \qquad (11.24a)$$

$$y = \frac{(\beta+1) - (\beta-1)B^{1-\eta}}{1 + B^{1-\eta}} h \qquad (11.24b)$$

To illustrate the grid concentration as $\beta \to 1$, we set, for example, $\eta = 0.4$ and calculate y for different values of β.

β	1.5	1.1	1.01
y	0.327	0.205	0.0705

Once the relations for the coordinate transformation are established, the differential equations governing the physical phenomena must be transformed from the x,y independent variables of the physical domain to the ξ,η independent variables of the computational domain under the same transformation, since all numerical computations will be performed in the ξ,η computational domain. To illustrate the transformation of the governing partial differential equations, we consider the continuity equation given by

$$\frac{\partial u}{\partial x} + \frac{\partial v}{\partial y} = 0 \qquad (11.25)$$

The transformation of this equation from the x,y to the ξ,η variables, under the general transformation defined by equations (11.1a,b), was given in Example 11.1 by

$$\frac{1}{J}\left(y_\eta \frac{\partial u}{\partial \xi} - y_\xi \frac{\partial u}{\partial \eta}\right) + \frac{1}{J}\left(-x_\eta \frac{\partial v}{\partial \xi} + x_\xi \frac{\partial v}{\partial \eta}\right) = 0 \tag{11.26}$$

The computational derivatives y_η, y_ξ, x_η, and x_ξ are expressed in terms of the metrics ξ_x, ξ_y, η_x, and η_y, according to equations (11.8a–d). Then, the transformed equation (11.26) takes the form

$$\left(\xi_x \frac{\partial u}{\partial \xi} + \eta_x \frac{\partial u}{\partial \eta}\right) + \left(\xi_y \frac{\partial v}{\partial \xi} + \eta_y \frac{\partial v}{\partial \eta}\right) = 0 \tag{11.27}$$

For the specific problem considered here, the transformation relations are given by equations (11.22a,b). Then, metrics ξ_x, η_x, ξ_y, and η_y become

$$\xi_x = 1, \quad \xi_y = 0 \tag{11.28a,b}$$

$$\eta_x = 0, \quad \eta_y = \frac{2\beta}{h\,\ln(B)} \frac{1}{\beta^2 - \left(1 - \frac{y}{h}\right)^2} \tag{11.28c,d}$$

Introducing equations (11.28a–d) into equation (11.27), the transformed continuity equation takes the form

$$\frac{\partial u}{\partial \xi} + \eta_y \frac{\partial v}{\partial \eta} = 0 \tag{11.29}$$

where η_y is defined by equation (11.28d).

We note that the transformed continuity equation (11.29) retains its original general form, except for the coefficient η_y accompanying the $\partial v/\partial \eta$ term. Therefore, the transformed equation (11.29) is slightly more complicated than its original form given by equation (11.25); however, it will be solved with uniform grid, both in the ξ and η directions, in the computational domain using the ξ,η rectangular coordinates. Clearly, the finite difference solution in the ξ,η computational domain with a uniform grid is much easier than solving the problem in the original physical domain with nonuniform grid. If the problem involves other partial differential equations, they are also transformed into the ξ,η computational domain in a similar manner.

Once the problem is solved in the computational domain, the results are transformed back into the physical domain by using the inverse transformation given by equations (11.24a,b), from each ξ,η location to the corresponding x,y location.

Roberts (1971) and other investigators have proposed numerous other simple stretching transformations. However, it is difficult to develop analytic transformations capable of clustering grids around arbitrary locations, whereas the numerical grid-generation technique provides a unified approach for developing transformations capable of dealing with more general situations.

11.3 Basic Ideas in Numerical Grid Generation and Mapping

In finite difference solution of partial differential equations over regions having regular shapes, such as a rectangle, cylinder, or sphere, the discretization can be made to conform to the boundaries of the region. As a result, the boundary interpolation is avoided. For regions having an arbitrary and irregular shape, this is not possible. One way to overcome such difficulties is to map the region, with a suitable transformation, into the computational domain where the geometry becomes regular, say, rectangular, and the problem is solved over the rectangular region with a square mesh by using the conventional finite difference techniques. The solution developed in the computational domain is then transformed back into the physical domain.

To illustrate the basic concepts in the mapping and development of curvilinear coordinates, we consider a two-dimensional physical domain in the x,y Cartesian coordinates and a computational domain in the ξ,η Cartesian coordinates. The transformation between x,y and ξ,η coordinates should be such that the boundaries of the physical domain must be coincident with the curvilinear coordinates ξ,η; thus, there is no need for boundary node interpolation.

The physical regions to be transformed into the computational domain can be identified in the following two categories: The *simply connected region* and the *multiply connected region*.

Simply Connected Region. Consider an irregular region ABCDA in the physical plane in x,y Cartesian coordinates as illustrated in Figure 11.3a. The region is called simply connected because it contains no obstacles within the region. This region is to be mapped into the computational domain in the ξ,η Cartesian coordinates in such a manner that the mapped region will have a rectangular shape and allow the construction of a square mesh over it. In addition, the boundaries of the physical domain will be coincident with the ξ,η coordinate lines of the boundaries of the transformed region in the computational domain. One way to accomplish such a mapping is to set the values of ξ,η along the boundaries of the physical region in the following manner:

Set η = constant and ξ monotonically varying along the boundary segments AB and DC of the physical region;

Set ξ = constant and η monotonically varying along the boundary segments AD and BC of the physical region.

Clearly, with such requirements on the values of ξ and η along the boundaries of the physical region, the segments AB and DC of the physical region are mapped into the computational domain as horizontal lines, while the segments AD and BC are mapped into the computational

Numerical Grid Generation 423

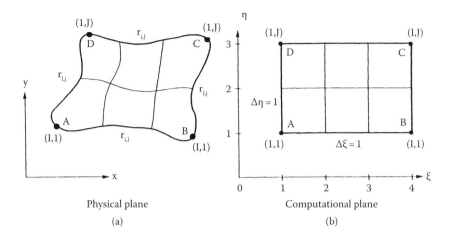

FIGURE 11.3
Mapping an irregular simply connected region (a) into the computational domain as a rectangle (b).

domain as vertical lines, as illustrated in Figure 11.3b. Notice that each boundary segment of the irregular region in the physical domain is mapped into the sides of the rectangular region in the computational domain. Furthermore, the mapping should satisfy the following requirements.

1. The mapping of the physical to the computational domain must be one-to-one.
2. Coordinate lines of the same family (i.e., ξ or η) must not cross.
3. The lines of different families must not cross more than once.

To accomplish these requirements, a proper organization of the grid points along the boundaries of the physical region is needed. If I points are placed along the bottom boundary segment AB of the physical domain, I grid points must also be placed along the opposing top boundary segment DC of the physical domain. Similarly, if J grid points are placed along the right boundary segment BC of the physical domain, J grid points must also be placed along the left segment AD. The actual values of ξ and η in the computational domain are immaterial because they do not appear in the final expressions. Therefore, without a loss of generality, we can select the coordinates of the node A in the computational domain as $\xi = \eta = 1$ and the mesh size as $\Delta\xi = \Delta\eta = 1$. Thus, in the computational domain, we can construct a square mesh over the rectangular transformed region. Hence, the identification of the I grid points placed along the bottom boundary segment AB and the I grid points placed along the top boundary segment DC should satisfy the following organizational matter:

The values of the position vector $\mathbf{r}_{i,j}$ along the boundary segment AB are selected as

$$\mathbf{r}_{i,1} \ (i = 1, 2, \ldots, I),$$

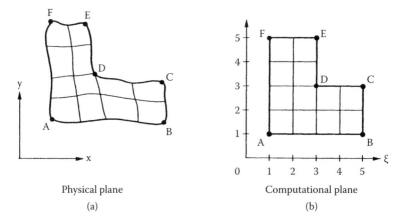

FIGURE 11.4
Mapping the L-shaped irregular region (a) into a L-shaped regular region (b).

and those along the boundary segment DC as

$$\mathbf{r}_{i,J} \quad (i = 1, 2, \ldots, I)$$

The identification of the J grid points along the boundary segments BC and AD is also done in a similar manner. These points may be located along the boundary segments in any arbitrary distribution, but it is required that they progress on the boundary without reversals as the index increases. In fact, if the regions of high gradients are known *a priori* in the physical domain, the grids can be concentrated in such areas.

In the previous illustration of mapping, an irregular region in the physical domain is mapped as a rectangular region into the computational domain. Depending on the choice of the values of ξ, η along the boundary segments of the physical region, a variety of other acceptable configurations can be generated in the computational domain. To illustrate this, we consider an L-shaped irregular region ABCDEFA in the physical domain as shown in Figure 11.4a. One possibility for the mapping is to map the region into an L-shaped regular region as illustrated in Figure 11.4b. This is accomplished by choosing the values of ξ, η along various boundary segments of the physical domain in the following manner:

Set η = constant and ξ monotonically varying along the boundary segments AB, DC, and FE of the physical region;

Set ξ = constant and η monotonically varying along the boundary segments AF, DE, and BC of the physical region.

Clearly, under such conditions, the L-shaped irregular region is mapped into the computational domain as an L-shaped regular region. Without loss of generality, the coordinates of the node A in the computational domain can be chosen as $\xi = \eta = 1$ and the mesh size taken as $\Delta\xi = \Delta\eta = 1$.

The second possibility for the mapping of the L-shaped irregular region, as illustrated in Figure 11.5a, is to map it as a rectangular region as shown in

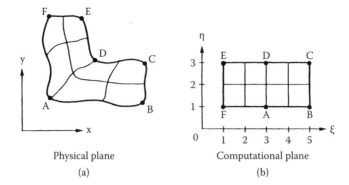

FIGURE 11.5
Mapping the L-shaped irregular region (a) into a rectangular region (b).

Figure 11.5b. This is accomplished by choosing the ξ and η values along various boundary segments of the physical region in the following manner:

Set η = constant and ξ monotonically varying along the boundary segments FAB and EDC of the physical region;

Set ξ = constant and η monotonically varying along the boundary segments FE and BC of the physical region.

We note that the nodes D and A have the same ξ value.

Figure 11.6 shows two different arrangements for the mapping of an irregular region from the physical x,y plane to the computational ξ,η plane, as a square region with a square mesh (i.e., $\Delta\xi = \Delta\eta = 1$) drawn over it. In both cases, 16 × 16 subdivisions are considered.

In the first arrangement, shown in (1a) and (1b) of Figure 11.6, the transformation of various boundary segments from the physical to the computational plane is organized as follows:

Side AB (16 units)..........as η = constant boundary
Side BC (16 units)..........as ξ = constant boundary
Side DC (16 units)..........as η = constant boundary
Sides DE, EF, FG, and GA (each 4 units)..........as ξ = constant boundary

In the second arrangement shown in (2a) and (2b) of Figure 11.6, the transformation of various boundary segments from the physical to the computational domain is arranged as follows:

Sides AB, BC (each 8 units)..........as ξ = constant boundary
Sides ED, DC (each 8 units)..........as η = constant boundary
Sides EF, FG (each 8 units)...........as ξ = constant boundary
Side GA (16 units).....................as η = constant boundary

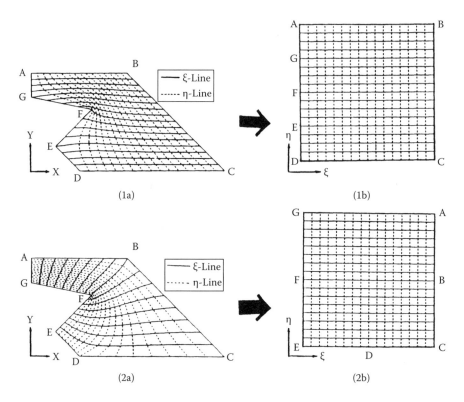

FIGURE 11.6
Different arrangements for mapping from the physical domain into the computational domain: region (1a) is mapped into region (1b) and region (2a) is mapped into region (2b).

An examination of the resulting ξ,η curvilinear coordinate lines constructed over the physical region reveals that, despite the fact that 16×16 subdivisions are used over the computational domain for both cases, the distribution of the ξ, η curvilinear grid lines constructed over the physical region is significantly different. Therefore, in the mapping of an irregular region from the physical to the computational domain, it is desirable to examine different mapping arrangements in order to find the one that will yield the most suitable grid distribution in the physical domain, where the solution of the physical problem is of interest.

Multiply Connected Region. We now consider an irregular region in the physical domain with one obstacle in the interior part, as illustrated in Figure 11.7a. The region shown in Figure 11.7a is *a doubly connected region* because there is only one obstacle within the region. When there is more than one obstacle within the region, the region is called the multiply connected region. We consider the following two possibilities for the mapping of the doubly connected region.

In the first approach, the doubly connected irregular physical region is mapped into the computational domain as a rectangular doubly connected

Numerical Grid Generation

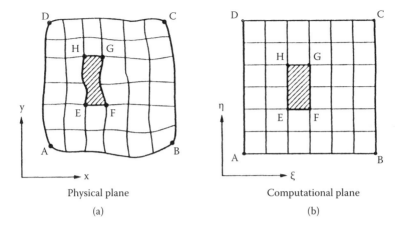

FIGURE 11.7
Mapping of a doubly connected irregular region (a) into a rectangular region with a rectangular window (b).

region with a rectangular window as illustrated in Figure 11.7b. This is accomplished by setting the values of ξ,η along the boundaries of the physical region in the following manner:

Set η = constant and ξ monotonically varying along the boundaries AB, DC, EF, and HG of the physical region;

Set ξ = constant, η monotonically varying along the boundaries AD, BC, EH, and FG of the physical region.

In the second approach, the doubly connected irregular region is mapped into the computational domain as a simply connected rectangular region by a branch cut, as illustrated in Figure 11.8. The top sketch in Figure 11.8 shows how two pseudo boundaries BC and AD are generated by the branch cut. The middle sketch illustrates the stretching procedure, and the bottom one shows the final mapping in the form of a simply connected rectangular region. The figure illustrates how I grid points are selected on the inner and outer boundaries, while J grid points are selected on the pseudo boundaries BC and AD. The position vectors along the inner and outer boundaries are identified as $r_{i,1}$ and $r_{i,J}$, respectively, for i = 1, 2,..., I. Similarly, the position vectors on the pseudo boundaries AD and BC are identified as $r_{1,j}$ and $r_{I,j}$, respectively, for j = 1, 2,..., J. We note that there are J−2 circumferential lines between the inner and outer boundaries. The pseudo boundaries on the left and right of the rectangular region in the computational domain correspond to the same curve in the physical domain; hence, we write $r_{1,j} = r_{I,j}$ for j = 1, 2,..., J. Across the branch cut the indexes of ξ are continuous, and along the branch cut the indexes of η are continuous.

Finally, Figure 11.9 illustrates the mapping of a multiply connected region with circular obstacles in its interior. Only one-third of the physical region is

428   *Finite Difference Methods in Heat Transfer*

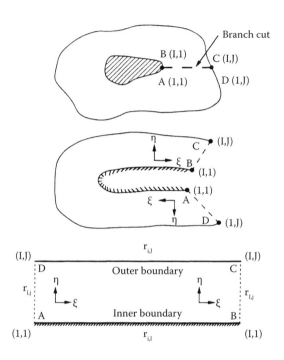

FIGURE 11.8
Mapping of a doubly connected region into a simply connected region by using a branch cut.

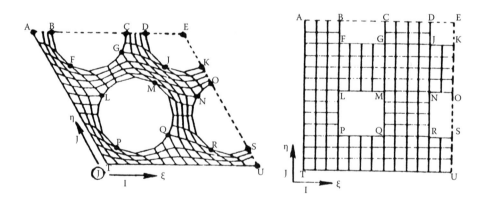

FIGURE 11.9
Mapping of a multiply connected region. (From Thompson, J.F., *Elliptic Grid Generation*, Elsevier Publishing Co. Inc., New York, pp. 79–105, 1982.)

considered because of symmetry considerations. We note that the outer contour of the physical domain is mapped into a square in the computational plane in ξ,η Cartesian coordinates. The circular obstacle is mapped into a square window and the semicircular obstacles into rectangular windows.

Also, shown over the physical domain are the constant ξ and η lines corresponding to those of the computational domain.

11.4 Boundary Value Problem of Numerical Grid Generation

The numerical grid generation and mapping technique advanced by Thompson et al. (1977) differs from the conventional analytic mapping in that the coordinate transformation relations are determined from the finite difference solution of a set of partial differential equations. Generally, elliptic partial differential equations of the Poisson or Laplace type are used for such transformations because of the smoothing properties of the elliptic equations; that is, the solutions obtained from the elliptic equations do not propagate the boundary shape discontinuities into the coordinate field.

Here, we consider elliptic equations for two-dimensional grid generation associated with the mapping from the physical domain in the x,y Cartesian coordinates to the computational domain in the ξ,η Cartesian coordinates. Such transformations are also possible for cylindrical and polar coordinates. The elliptic equations for numerical grid generation in such coordinate systems are given in Notes at the end of this chapter.

The mathematical problem of coordinate transformation becomes one of determining the correspondence between the coordinates x,y and ξ,η of the physical and computational regions. To establish the basis for such a coordinate transformation, we let the coordinates ξ,η satisfy the following two Poisson's equations in the physical domain:

$$\frac{\partial^2 \xi}{\partial x^2} + \frac{\partial^2 \xi}{\partial y^2} = P(\xi, \eta) \qquad (11.30a)$$

$$\frac{\partial^2 \eta}{\partial x^2} + \frac{\partial^2 \eta}{\partial y^2} = Q(\xi, \eta) \qquad (11.30b)$$

where $P(\xi,\eta)$ and $Q(\xi,\eta)$ are nonhomogeneous source terms. As such, they act by concentrating lines of ξ and η, respectively, and are called *grid control functions*. By proper selection of the $P(\xi,\eta)$ and $Q(\xi,\eta)$ functions, the coordinate lines ξ and η can be concentrated toward a specified coordinate line or about a specific grid point. In the absence of these functions, that is, $P = Q = 0$, the coordinate lines will tend to be equally spaced in the regions away from the boundaries, regardless of the concentration of the grid points along the boundaries.

While equations (11.30a,b) describe the basic coordinate transformation between x,y and ξ,η coordinate systems, all numerical computations of the governing differential equations for the physical problem are to be performed in the computational ξ,η plane since the transformed region has simple regular geometry. Then, the problem becomes one of seeking the (x,y) values of the physical plane corresponding to the known (ξ,η) grid locations of the computational plane. For this reason, equations (11.30a,b) should be transformed into the computational domain by interchanging the roles of the dependent and the independent variables. This yields the following two elliptic equations to be solved in the computational domain for the determination of the unknown x,y:

$$\alpha \frac{\partial^2 x}{\partial \xi^2} - 2\beta \frac{\partial^2 x}{\partial \xi \partial \eta} + \gamma \frac{\partial^2 x}{\partial \eta^2} + J^2 \left(P(\xi,\eta) \frac{\partial x}{\partial \xi} + Q(\xi,\eta) \frac{\partial x}{\partial \eta} \right) = 0 \qquad (11.31)$$

$$\alpha \frac{\partial^2 y}{\partial \xi^2} - 2\beta \frac{\partial^2 y}{\partial \xi \partial \eta} + \gamma \frac{\partial^2 y}{\partial \eta^2} + J^2 \left(P(\xi,\eta) \frac{\partial y}{\partial \xi} + Q(\xi,\eta) \frac{\partial y}{\partial \eta} \right) = 0 \qquad (11.32)$$

where the geometric coefficients α, β, γ, and the Jacobian J are given by

$$\alpha = \left(\frac{\partial x}{\partial \eta}\right)^2 + \left(\frac{\partial y}{\partial \eta}\right)^2 \qquad (11.33a)$$

$$\beta = \frac{\partial x}{\partial \xi}\frac{\partial x}{\partial \eta} + \frac{\partial y}{\partial \xi}\frac{\partial y}{\partial \eta} \qquad (11.33b)$$

$$\gamma = \left(\frac{\partial x}{\partial \xi}\right)^2 + \left(\frac{\partial y}{\partial \xi}\right)^2 \qquad (11.33c)$$

$$J = \frac{\partial x}{\partial \xi}\frac{\partial y}{\partial \eta} - \frac{\partial x}{\partial \eta}\frac{\partial y}{\partial \xi} \qquad (11.33d)$$

The mathematical problem defined by the partial differential equations (11.31) and (11.32), subject to appropriate boundary conditions, constitute the boundary value problem of elliptic numerical grid generation. The solution of this problem establishes the values of the x,y coordinates at each ξ,η grid point in the computational domain. Such calculations are generally performed with finite differences using central difference formulas to yield coupled algebraic expressions, which are solved for $x_{i,j}$ and $y_{i,j}$ at each ξ,η grid point; successive over-relaxation can be used to solve for the $x_{i,j}$ and $y_{i,j}$ values. Once the correspondence between x,y and ξ,η coordinate values are known at each grid point in the computational domain, the results can be transformed to the physical domain. Hence, ξ,η curvilinear coordinate lines can be constructed over the physical domain.

Numerical Grid Generation

The boundary conditions needed to solve equations (11.31) and (11.32) can be prescribed by specifying the values of ξ and η over each boundary segment or, alternatively, by making the grid lines orthogonal to the boundaries, as described later. Besides that, cases involving branch cuts need special consideration.

i. *The Boundary Condition of the First Kind:* As the values of the x,y coordinates of the boundaries of the physical domain are known from its geometry, they can be assigned to specific values of the computational coordinates ξ,η. Then, the grid-generation problem becomes one of solving the grid-generation equations (11.31) and (11.32) over the regular computational domain, subject to the prescribed values of x,y at the boundaries.

ii. *Orthogonality of Grid Lines:* If the values of the x,y coordinates are fixed over the boundary with the first-kind boundary condition, there might be grid lines that intersect the boundaries at a small angle, thus increasing the discretization error. An alternative boundary condition to such cases is to make the grid lines intersect the boundaries at a specified angle, ϕ (see Figure 11.10), for example, $\phi = \dfrac{\pi}{2}$. To establish the mathematical expression for implementing such a requirement, we consider the gradient of ξ and η, defined by

$$\nabla \xi = \xi_x \mathbf{i} + \xi_y \mathbf{j} \tag{11.34}$$

$$\nabla \eta = \eta_x \mathbf{i} + \eta_y \mathbf{j} \tag{11.35}$$

where $\mathbf{i}$ and $\mathbf{j}$ are the unit direction vectors. The dot product of $\nabla \xi$ and $\nabla \eta$, that is,

$$\nabla \xi \cdot \nabla \eta = \xi_x \eta_x + \xi_y \eta_y \tag{11.36}$$

represents the cosine of the angle ϕ. Introducing ξ_x, η_x, ξ_y, and η_y from equations (11.8) into equation (11.36), we obtain

$$\nabla \xi \cdot \nabla \eta = -\frac{1}{J^2}(x_\xi x_\eta + y_\xi y_\eta) \tag{11.37}$$

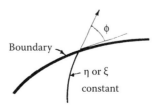

FIGURE 11.10
The angle of intersection ϕ.

In the case of orthogonality, we have $\phi = \dfrac{\pi}{2}$ or $\cos\phi = 0$; then, equation (11.37) reduces to

$$x_\xi x_\eta + y_\xi y_\eta = 0 \tag{11.38}$$

This is the criterion to be implemented in the computational domain whenever the ξ (or η) constant grid lines are required to intersect the physical boundary orthogonally.

iii. *Branch Cut:* In the case of a branch cut as illustrated in Figure 11.8, a doubly connected region is mapped into a simply connected rectangular region in the computational domain. The branch cut shown in the physical domain corresponds to two different $\xi =$ constant boundaries, AD and BC in the computational domain; conversely, the boundaries AD and BC of the computational domain represent the same coordinate line along the branch cut in the physical domain.

Therefore, when the partial derivatives are discretized along the boundaries corresponding to a branch cut, the continuity of the function and its derivative across the branch cut should be imposed. That is, the finite difference expression at one side of the branch cut, say boundary AD, should be identical to that at the other side, that is, BC. This can be accomplished by applying the discretization only to one side of the branch cut and setting the values of the variables at the other side equal to those at the first side.

Grid Control Functions. The user specified grid control functions $P(\xi,\eta)$ and $Q(\xi,\eta)$ are useful to concentrate the interior grid lines in regions where large gradients occur. For example, in problems of natural convection, large gradients occur near the walls; hence, grid points need to be concentrated in such locations. Thompson (1982) specified the $P(\xi,\eta)$ and $Q(\xi,\eta)$ functions in the form:

$$P(\xi,\eta) = -\sum_{i=1}^{n} a_i\, \mathrm{sign}(\xi-\xi_i)\exp(-C_i|\xi-\xi_i|) - \sum_{i=1}^{m} b_i\, \mathrm{sign}(\xi-\xi_i)$$
$$\exp\!\left(-d_i\sqrt{(\xi-\xi_i)^2+(\eta-\eta_i)^2}\right) \tag{11.39}$$

and

$$Q(\xi,\eta) = -\sum_{i=1}^{n*} a_i^*\, \mathrm{sign}(\eta-\eta_i)\exp(-C_i^*|\eta-\eta_i|) - \sum_{i=1}^{m*} b_i^*\, \mathrm{sign}(\eta-\eta_i)$$
$$\exp\!\left(-d_i^*\sqrt{(\xi-\xi_i)^2+(\eta-\eta_i)^2}\right) \tag{11.40}$$

We note that $P(\xi,\eta)$ and $Q(\xi,\eta)$ functions have similar forms, except that ξ and η are interchanged. The physical significance of various terms in these

Numerical Grid Generation

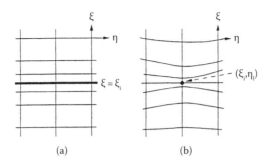

FIGURE 11.11
The attraction of ξ = constant lines toward (a) the coordinate line $\xi = \xi_i$ and (b) the point (ξ_i, η_i).

equations is as follows. In equation (11.39), in the first summation, the amplitude a_i is to attract ξ = constant lines toward the $\xi = \xi_i$ line; and in the second summation, the amplitude b_i is to attract ξ = constant lines toward the point (ξ_i, η_i). Figure 11.11 illustrates the control of ξ = constant lines toward the coordinate line $\xi = \xi_i$ and a point (ξ_i, η_i).

The summation indices n and m (or n* and m*) denote the number of lines and points of grid attraction, respectively. The sign function "sign $(\xi-\xi_i)$" ensures that attraction of ξ lines occurs on both sides of the $\xi = \xi_i$ line or at the ξ_i, η_i point. Without the sign function, the attraction occurs only on the side toward increasing ξ, with repulsion occurring on the other side. The C_i, C_i^* and d_i, d_i^* are the coefficients that control the decay of attraction with the distance, while a_i, a_i^* and b_i, b_i^* are the amplitude coefficients.

Figure 11.12 illustrates the effects of these coefficients in controlling the concentration of ξ, η curvilinear grid lines toward the boundaries over the physical domain between two confocal ellipses. Because of symmetry, only half of the region is shown in these figures. Figure 11.12a and b show how the semi-elliptical physical domain is mapped into a rectangular region in the computational domain. Figure 11.12c,d is intended to show the effects of various coefficients in the $P(\xi, \eta)$ and $Q(\xi, \eta)$ functions in controlling the ξ, η grid line distribution over the physical domain. In Figure 11.12c and d, the cases with *no internal grid control* show that the elliptic grid-generating system used here tends to make the η = constant coordinate lines to become more closely spaced over the convex boundary and spread out over the concave boundary. The effect of the coefficients a_1, a_2, b_1, b_2, and so on in controlling the grid line distribution in the interior of the physical domain is illustrated in these figures for the cases with 11 × 21 grid lines and 31 × 51 grid lines, respectively.

Considerable amount of experimentation is recommended with different values of the grid control coefficients before selecting any particular combination.

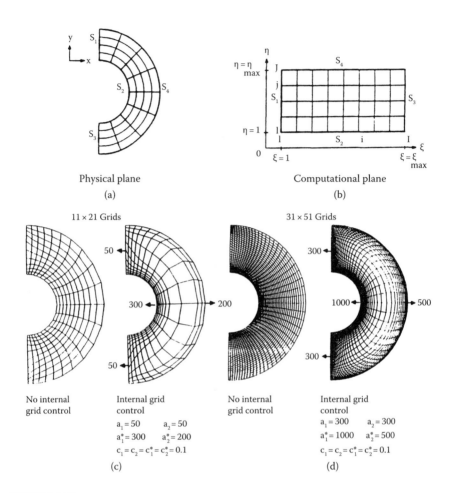

FIGURE 11.12
(a) Region in the physical domain; (b) Rectangular region in the computational domain; (c) and (d) Effects of control coefficients on the grid concentration in the interior of the region. (From Elshamy, M.M., et al., *Numer. Heat Transf.*, 18, 95–112, 1990.)

Boundary Grid Point Control. There are situations that require the control of the concentration of grid points at specific locations along the contour of the physical region. For example, Figure 11.13 shows the concentration of grid points on the boundary of the physical domain about the points C and D. This concentration is accomplished by selecting the location of the grid points along the boundary according to the following relations (Coulter and Güceri 1988).

$$S = S_{max} \frac{\tan^{-1}\left\{B_w \left(\frac{i-1}{M-1}\right)\right\}}{\tan^{-1}(B_w)}, \quad i = 1, 2, \ldots, M \qquad (11.41)$$

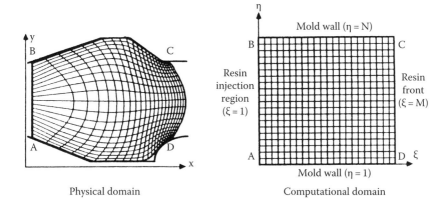

FIGURE 11.13
Concentration of grid points on the boundary about the points C and D. (From Coulter, J.P. & Guceri, S.I., *J. Reinforc. Plast. Compos.*, 7, 200–219, 1988.)

along the boundary contours BC and AD, and

$$S = S_{max} \left\{ 0.5 + \left[\frac{\tan^{-1}\left\{ B_f \left[\left(\frac{2j-2}{N-1}\right) - 1 \right] \right\}}{2 \tan^{-1}(B_f)} \right] \right\}, \quad j = 1, 2, \ldots, N \quad (11.42)$$

along the boundary contour CD. Here, S refers to the curvilinear distance along the boundary segment and S_{max} is the maximum of S in the physical domain. The constants B_w and B_f are the concentration control coefficients. The indices $i = 1, 2, \ldots, M$ and $j = 1, 2, \ldots, N$ are node indices in the ξ and η directions, respectively. The boundary concentration shown in Figure 11.13 is achieved by setting the values of the constants $B_w = 2.5$ and $B_f = 2.0$.

Effects of Nonorthogonality. In the curvilinear coordinates constructed over the physical domain as illustrated in Figures 11.12 and 11.13, the ξ and η coordinate lines do not necessarily intersect each other orthogonally. Severe departure from orthogonality introduces truncation error in finite difference expressions. Mastin (1982) examined the sources of truncation error in the numerical solution of partial differential equations on curvilinear coordinate system and concluded that the truncation error is not only dependent on the higher order derivatives and the local grid spacing but also on the rate of change of the grid spacing and the departure of the grid spacing from orthogonality. A slight degree of nonorthogonality has a negligible effect on the truncation error, but the truncation error does increase with the degree of nonorthogonality.

11.5 Finite Difference Representation of Boundary Value Problem of Numerical Grid Generation

We consider the transformation equations (11.31) and (11.32) for the determination of the x,y coordinate values over the computational ξ, η plane, that is,

$$\alpha \frac{\partial^2 x}{\partial \xi^2} - 2\beta \frac{\partial^2 x}{\partial \xi \partial \eta} + \gamma \frac{\partial^2 x}{\partial \eta^2} + J^2 \left(P \frac{\partial x}{\partial \xi} + Q \frac{\partial x}{\partial \eta} \right) = 0 \qquad (11.43)$$

$$\alpha \frac{\partial^2 y}{\partial \xi^2} - 2\beta \frac{\partial^2 y}{\partial \xi \partial \eta} + \gamma \frac{\partial^2 y}{\partial \eta^2} + J^2 \left(P \frac{\partial y}{\partial \xi} + Q \frac{\partial y}{\partial \eta} \right) = 0 \qquad (11.44)$$

where the geometric coefficients α, β, γ, and the Jacobian J are obtained from equation (11.33), rewritten here for convenience, that is,

$$\alpha = \left(\frac{\partial x}{\partial \eta} \right)^2 + \left(\frac{\partial y}{\partial \eta} \right)^2 \qquad (11.45a)$$

$$\beta = \frac{\partial x}{\partial \xi} \frac{\partial x}{\partial \eta} + \frac{\partial y}{\partial \xi} \frac{\partial y}{\partial \eta} \qquad (11.45b)$$

$$\gamma = \left(\frac{\partial x}{\partial \xi} \right)^2 + \left(\frac{\partial y}{\partial \xi} \right)^2 \qquad (11.45c)$$

$$J = \frac{\partial x}{\partial \xi} \frac{\partial y}{\partial \eta} - \frac{\partial x}{\partial \eta} \frac{\partial y}{\partial \xi} \qquad (11.45d)$$

First, we examine the discretization of the differential equation (11.43) or (11.44) at the internal grid points of the rectangular computational domain shown in Figure 11.3 and then discuss the boundary conditions associated with the problem.

Discretization of the Grid-Generation Differential Equation. We consider the computational domain shown in Figure 11.3 and assume a square network with unit grid spacing, that is, $\Delta \xi = \Delta \eta = 1$. Various derivatives appearing in equation (11.43) are discretized as

$$(f_\xi)_{i,j} = \frac{1}{2}(f_{i+1,j} - f_{i-1,j}) \qquad (11.46a)$$

$$(f_\eta)_{i,j} = \frac{1}{2}(f_{i,j+1} - f_{i,j-1}) \qquad (11.46b)$$

$$(f_{\xi\xi})_{i,j} = (f_{i+1,j} - 2f_{i,j} + f_{i-1,j}) \tag{11.46c}$$

$$(f_{\eta\eta})_{i,j} = (f_{i,j+1} - 2f_{i,j} + f_{i,j-1}) \tag{11.46d}$$

$$(f_{\xi\eta})_{i,j} = \frac{1}{4}(f_{i+1,j+1} - f_{i-1,j+1} - f_{i+1,j-1} + f_{i-1,j-1}) \tag{11.46e}$$

where $f \equiv x$ or y, and the indices i and j are related to ξ and η, respectively.

The finite difference expressions given by equations (11.46a–e) are introduced into equation (11.43) or (11.44), and the following successive-substitution formula is obtained for the determination of x and y.

$$f_{i,j} = \frac{1/2}{\alpha_{i,j} + \gamma_{i,j}} [\alpha_{i,j}(f_{i+1,j} + f_{i-1,j}) - 1/2\, \beta_{i,j}(f_{i+1,j+1} - f_{i-1,j+1} - f_{i+1,j-1} + f_{i-1,j-1})$$
$$+ \gamma_{i,j}(f_{i,j+1} + f_{i,j-1}) + 1/2\, J_{i,j}^2\, P_{i,j}\, (f_{i+1,j} - f_{i-1,j}) + 1/2\, J_{i,j}^2\, Q_{i,j}\, (f_{i,j+1} - f_{i,j-1})] \tag{11.47}$$

where $f \equiv x$ or y. The quantities α, β, γ, and J are treated as coefficients and calculated from their finite difference representations by lagging one iteration step. For example,

$$\alpha_{i,j} = (x_\eta\, x_\eta + y_\eta\, y_\eta)_{i,j} \tag{11.48a}$$

where $(x_\eta)_{i,j}$ and $(y_\eta)_{i,j}$ are computed from

$$(x_\eta)_{i,j} = \frac{1}{2}(x_{i,j+1} - x_{i,j-1}) \tag{11.48b}$$

$$(y_\eta)_{i,j} = \frac{1}{2}(y_{i,j+1} - y_{i,j-1}) \tag{11.48c}$$

Similar expressions are written for β, γ, and J.

Boundary Conditions. We now examine the boundary conditions for the numerical grid-generation equation considered above.

i. **Boundary Condition of the First Kind:** In boundary conditions for equations (11.43) and (11.44) of the first kind, the x,y coordinate values are specified at every grid point along the boundary contour of the computational domain shown in Figure 11.3. For such a case, equation (11.47) is sufficient for solving the problem, say, by performing the calculations over the ranges

   i: from 2 to I – 1

   j: from 2 to J – 1.

If the successive over-relaxation is used, the iterations are performed as:

$$f_{i,j}^{(k+1)} = \omega\{\text{RHS of equation } (11.47)\} + (1-\omega)f_{i,j}^{(k)} \qquad (11.49)$$

where $f_{i,j} \equiv x_{i,j}$, or $y_{i,j}$, the superscript k denotes the k-th iteration, and ω is the relaxation parameter. Clearly, $\omega = 1$ corresponds to the Gauss–Seidel iteration (see Chapter 3).

ii. **Orthogonality of Grid Lines:** As discussed previously, the requirement that the ξ (or η) grid lines should intersect some portion of the boundary in the physical domain *normally* is another possibility as a boundary condition for the solution of the numerical grid-generation problem. Such a requirement is implemented if the condition given by equation (11.38), that is,

$$x_\xi x_\eta + y_\xi y_\eta = 0 \qquad (11.50)$$

is satisfied at the boundary of the computational domain that corresponds to the physical boundary where ξ (or η) grid lines are orthogonal to the boundary.

Therefore, the finite difference form of this equation is needed at such a boundary in the computational domain. To illustrate the discretization procedure, we consider the rectangular computational domain shown in Figure 11.3 and require that the orthogonality condition given by equation (11.50) be satisfied at the grid points along the boundary $\eta = 1$. We need the finite difference form of the derivatives x_ξ, x_η, y_ξ, and y_η at the grid points ($\xi = i, \eta = 1$) for $i = 2$ to $I - 1$. The derivatives x_ξ and y_ξ are discretized by using central differences, while the derivatives x_η and y_η are discretized by the second-order accurate one-sided differencing [see equation (2.11a,b)]. We obtain

$$x_\xi = \frac{x_{i+1,1} - x_{i-1,1}}{2} \qquad \text{central} \qquad (11.51a)$$

$$x_\eta = \frac{-3x_{i,1} + 4x_{i,2} - x_{i,3}}{2} \qquad \text{one-sided forward} \qquad (11.51b)$$

$$y_\xi = \frac{y_{i+1,1} - y_{i-1,1}}{2} \qquad \text{central} \qquad (11.51c)$$

$$y_\eta = \frac{-3y_{i,1} + 4y_{i,2} - y_{i,3}}{2} \qquad \text{one-sided forward} \qquad (11.51d)$$

where we have taken $\Delta\xi = \Delta\eta = 1$.

Substituting equations (11.51a–d) into equation (11.50) and solving the resulting expression for $x_{i,1}$, we obtain

$$x_{i,1} = \frac{(x_{i+1,1} - x_{i-1,1})(4x_{i,2} - x_{i,3}) + (y_{i+1,1} - y_{i-1,1})(-3y_{i,1} + 4y_{i,2} - y_{i,3})}{3(x_{i+1,1} - x_{i-1,1})} \quad (11.52)$$

The corresponding $y_{i,1}$ value is immediately obtained from the relationship between the x,y coordinates of the boundary contour in the physical domain, that is, $y \equiv y(x)$. Depending on the boundary contour, it can be more convenient to solve equation (11.50) for $y_{i,1}$ and then compute $x_{i,1}$ from the function that defines the boundary surface, that is, $x \equiv x(y)$.

Expressions similar to that given by equation (11.52) can readily be developed for other boundary segments where the orthogonality condition should be satisfied.

11.6 Steady-State Heat Conduction in Irregular Geometry

To illustrate the basic steps in the application of numerical grid-generation technique in the solution of field problems over irregular geometry, we consider a simple steady-state heat conduction problem over a semi-elliptical physical domain shown in Figure 11.14a. The mathematical formulation of this heat conduction problem is given by

$$\frac{\partial^2 T}{\partial x^2} + \frac{\partial^2 T}{\partial y^2} + \frac{1}{k} g(x,y) = 0 \quad \text{in the region} \quad (11.53a)$$

subject to the boundary conditions

$$\frac{\partial T}{\partial x} = 0 \quad \text{on } S_1 \quad (11.53b)$$

$$T = f(x,y) \quad \text{on } S_2 \quad (11.53c)$$

$$\frac{\partial T}{\partial x} = 0 \quad \text{on } S_3 \quad (11.53d)$$

$$T = 0 \quad \text{on } S_4 \quad (11.53e)$$

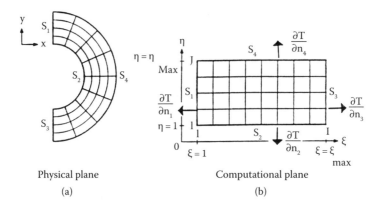

FIGURE 11.14
The physical (a) and computational (b) domains.

Basic steps in the solution of this problem are as follows:

i. **The Mapping of the Irregular Region:** The irregular semi-elliptical region in the x,y physical plane shown in Figure 11.14a is mapped into the ξ,η computational plane as a rectangular region as shown in Figure 11.14b by applying the mapping procedure described previously. In this mapping, the S_2 and S_4 curved boundaries of the physical domain correspond to the bottom and top boundaries, respectively, of the rectangular region in the computational domain. The insulated boundaries S_1 and S_3 correspond, respectively, to the sides S_1 and S_3 of the rectangular region. A square net of unit spacing, that is, $\Delta\xi = \Delta\eta = 1$ is considered over the computational domain. There are i = 1 to I grid points along the ξ direction and j = 1 to J grid points along the η direction.

ii. **The Grid-Generation Problem:** To establish the transformation relation between the ξ,η and x,y coordinate values in the interior regions of the computational and the physical domains, the numerical grid-generation problem defined by the grid-generation equations (11.43) and (11.44) should be solved with finite differences over the computational domain shown in Figure 11.14b. These equations are discretized as described in the previous section, and the successive-substitution formula given by equation (11.47) is obtained. If the boundary conditions for the grid-generation problem are all of the first kind, equation (11.47) is solved iteratively by sweeping

   i: from 2 to I − 1
   j: from 2 to J − 1

If the orthogonality of the ξ (or η) grid lines to any one of the physical boundaries is invoked, then the orthogonality condition given by equation (11.50) should be implemented at the corresponding boundary in the computational domain. The finite difference equation analogous to that given by equation (11.52) is developed and solved iteratively together with the successive-substitution formula given by equation (11.47).

iii. **The Transformation of the Field Equation:** The heat conduction equations (11.53a–e) should be transformed from the x,y coordinates to the ξ,η coordinates of the computational domain. The transformation of the two-dimensional Laplacian $\nabla^2 T$ is given by equation (11.16). By utilizing this relation, the heat conduction equations (11.53a–e) are transformed to the ξ,η independent variables as

$$\frac{1}{J^2}(\alpha T_{\xi\xi} - 2\beta T_{\xi\eta} + \gamma T_{\eta\eta}) + (P\ T_\xi + Q\ T_\eta) + \frac{g(\xi,\eta)}{k} = 0 \qquad (11.54)$$

where

$$\alpha = x_\eta^2 + y_\eta^2 \qquad (11.55a)$$

$$\beta = x_\xi x_\eta + y_\xi y_\eta \qquad (11.55b)$$

$$\gamma = x_\xi^2 + y_\xi^2 \qquad (11.55c)$$

$$J = x_\xi y_\eta - x_\xi y_\eta \qquad (11.55d)$$

and P and Q are the grid control functions.

iv. **Transformation of Boundary Conditions:** The boundary conditions for the field equation should also be transformed from the x,y coordinates to the ξ,η coordinates. For generality, we consider here the boundary conditions of the first, second, and third kinds. The transformation of the boundary conditions for the considered problem are then obtained as special cases.

First-Kind Boundary Condition: If the boundary condition is of the first kind, that is, the value of temperature is prescribed at the boundary in the physical domain, no transformation is needed. It should be clear that, in this case, the values of T specified along the boundaries in the physical domain remain the same at the corresponding grid locations along the boundaries in the computational domain.

Second-Kind Boundary Condition: If the boundary condition is of the second kind, that is, the value of the heat flux is prescribed at the boundary, the boundary condition involves the derivative of the field variable (i.e., temperature) in the direction normal to the boundary. For such a case, the normal derivative $\partial T/\partial n$ should be expressed in terms of the ξ, η independent variables of the computational domain. It is to be noted that $\partial T/\partial \mathbf{n}$ at a boundary in the physical plane need to be related to $\partial T/\partial \eta$ and/or $\partial T/\partial \xi$ in the computational domain, because ξ and η coordinate lines are not necessarily normal to the physical boundaries.

In the computational plane shown in Figure 11.14a, the normal derivatives $\partial T/\partial \mathbf{n}_i$ (i = 1, 2, 3, 4), along the outward drawn unit vectors can be obtained from the coordinate transformation relations given in Section 11.1. For example, the normal derivatives $\partial T/\partial \mathbf{n}_4$ and $\partial T/\partial \mathbf{n}_3$ at the surfaces S_4 and S_3 are obtained from equations (11.19a) and (11.19b), respectively, as

$$\text{on } S_4: \quad \frac{\partial T}{\partial \mathbf{n}_4} = \frac{1}{J\sqrt{\gamma}}(\gamma T_\eta - \beta T_\xi) \tag{11.56a}$$

$$\text{on } S_3: \quad \frac{\partial T}{\partial \mathbf{n}_3} = \frac{1}{J\sqrt{\alpha}}(\alpha T_\xi - \beta T_\eta) \tag{11.56b}$$

We note that these derivatives are along the positive η and ξ directions, respectively. The normal derivatives at the surfaces S_2 and S_1 are in the negative η and ξ directions, respectively. Therefore, they are obtained by changing the sign of $\partial T/\partial \mathbf{n}_4$ and $\partial T/\partial \mathbf{n}_3$, that is,

$$\text{on } S_2: \quad \frac{\partial T}{\partial \mathbf{n}_2} = \frac{-1}{J\sqrt{\gamma}}(\gamma T_\eta - \beta T_\xi) \tag{11.56c}$$

$$\text{on } S_1: \quad \frac{\partial T}{\partial \mathbf{n}_1} = \frac{-1}{J\sqrt{\alpha}}(\alpha T_\xi - \beta T_\eta) \tag{11.56d}$$

where the coefficients α, β, γ, and J are as defined by equations (11.55a–d).

Third-Kind Boundary Condition: The boundary condition of the third kind in heat transfer problems represents a convection boundary condition, which is given in the general form as

$$k\frac{\partial T}{\partial \mathbf{n}} + hT = f \tag{11.57}$$

where k is the thermal conductivity, h is the heat transfer coefficient, f is a prescribed quantity, which is related to the ambient temperature (see Chapter 2), and $\partial T/\partial \mathbf{n}$ is the normal derivative of temperature at the boundary in the outward direction. Clearly, the case h = 0 corresponds to the boundary condition of the second kind discussed above. To transform equation (11.57) to the ξ,η variables in the computational domain, all we need is to transform the normal derivative $\partial T/\partial \mathbf{n}$ according to the relations given by equations (11.56a–d).

We are now in a position to transform the boundary conditions [equation (11.53)] to the ξ,η independent variables of the computational domain. The boundary conditions (11.53c) and (11.53e) do not require any transformation because the values of temperature are specified along the boundaries S_2 and S_4. The derivative boundary conditions on S_1 and S_3 are transformed by utilizing the relation given by equations (11.56d and 11.56b), respectively. We obtain

$$\alpha T_\xi - \beta T_\eta = 0 \quad \text{on } S_1 \qquad (11.58a)$$

$$T = f \quad \text{on } S_2 \qquad (11.58b)$$

$$\alpha T_\xi - \beta T_\eta = 0 \quad \text{on } S_3 \qquad (11.58c)$$

$$T = 0 \quad \text{on } S_4 \qquad (11.58d)$$

v. **Finite Difference Representation of the Transformed Problem:** The transformed field equation (11.54) together with the transformed boundary conditions [equation (11.58)] constitutes the complete mathematical formulation of the original physical problem in the ξ,η independent variables in the computational domain. We summarize the transformed problem:

$$\alpha T_{\xi\xi} - 2\beta T_{\xi\eta} + \gamma T_{\eta\eta} + J^2(P\ T_\xi + Q\ T_\eta) + \frac{J^2}{k}g = 0 \qquad (11.59)$$

to be solved over the rectangular region in the computational domain as shown in Figure 11.14b, subject to the following boundary conditions:

$$\alpha T_\xi - \beta T_\eta = 0 \quad \text{on } S_1 \qquad (11.60a)$$

$$T = f \quad \text{on } S_2 \qquad (11.60b)$$

$$\alpha T_\xi - \beta T_\eta = 0 \quad \text{on } S_3 \qquad (11.60c)$$

$$T = 0 \quad \text{on } S_4 \qquad (11.60d)$$

The differential equation (11.59) is similar to the grid-generation equation (11.43), except for the energy generation term. Therefore, the finite difference form of equation (11.59) is readily determined and the successive-substitution formula is immediately obtained from equation (11.47) as

$$T_{i,j} = \frac{1}{2(\alpha_{i,j} + \gamma_{i,j})} \left[\alpha_{i,j}(T_{i+1,j} + T_{i-1,j}) - \frac{\beta_{i,j}}{2}(T_{i+1,j+1} - T_{i-1,j+1} - T_{i+1,j-1} \right.$$

$$+ T_{i-1,j-1}) + \gamma_{i,j}(T_{i,j+1} + T_{i,j-1}) + \frac{J_{i,j}^2 P_{i,j}}{2}(T_{i+1,j} - T_{i-1,j})$$

$$\left. + \frac{J_{i,j}^2 Q_{i,j}}{2}(T_{i,j+1} - T_{i,j-1}) + \frac{J_{i,j}^2}{k} g_{i,j} \right] \quad (11.61)$$

for i = 2 to I − 1 and j = 2 to J − 1

If the boundary conditions for equation (11.59) were all of the first kind, the above successive-substitution formula would be sufficient to solve this equation. However, two of the boundary conditions, equations (11.60a) and (11.60c), being of the derivative type, they should also be discretized as described below.

First, we develop the finite difference form of the first derivatives T_ξ and T_η along the boundaries S_1 and S_3 shown in Figure 11.14b by using one-sided and central differencing.

Boundary S_1:

$$T_\xi = \frac{1}{2}(-3T_{1,j} + 4T_{2,j} - T_{3,j}) \quad (11.62a)$$

$$T_\eta = \frac{1}{2}(T_{1,j+1} - T_{1,j-1}) \quad (11.62b)$$

for j = 2 to J − 1.

Boundary S_3:

$$T_\xi = \frac{1}{2}(3T_{I,j} - 4T_{I-1,j} + T_{I-2,j}) \quad (11.62c)$$

$$T_\eta = \frac{1}{2}(T_{I,j+1} - T_{I,j-1}) \quad (11.62d)$$

for j = 2 to J − 1.

The finite difference form of the derivative-type boundary conditions is obtained by substituting the above finite difference expressions into equations (11.60a) and (11.60c). We find

$$\alpha_{i,j}(-3T_{1,j} + 4T_{2,j} - T_{3,j}) - \beta_{i,j}(T_{1,\,j+1} - T_{1,j-1}) = 0 \text{ on } S_1 \quad (11.63a)$$

$$\alpha_{i,j}(3T_{I,j} - 4T_{I-1,j} + T_{I-2,j}) - \beta_{i,j}(T_{I,\,j+1} - T_{I,j-1}) = 0 \text{ on } S_3 \quad (11.63b)$$

for $j = 2$ to $J - 1$

Equation (11.61), together with equations (11.63a) and (11.63b) and the first-kind boundary conditions given by equations (11.60b,d), constitutes the complete finite difference representation of the problem to be solved in the computational domain. Although one-sided differences have been used in this example for the discretization of the boundary conditions involving normal derivatives, the concept of the fictitious node, presented in Chapter 2 and applied to several examples in this book, could also be applied here to the equations in the computational domain.

In these equations, the coefficients α, β, γ, J, P, and Q are considered known, since they have been already calculated as a part of the numerical grid-generation problem.

vi. **Transformation of the Solution to the Physical Domain:** The nodal temperatures $T_{i,j}$, calculated at each grid point ξ_i, η_j from the solution of the above finite difference problem are transformed to the corresponding x_i, y_j locations in the physical plane, since the transformation between ξ_i, η_j and x_i, y_j are known from the solution of the grid-generation problem.

11.7 Steady-State Laminar Free Convection in Irregular Enclosures—Vorticity-Stream Function Formulation

Natural convection within enclosures is of interest in many engineering applications. In many situations, the geometry of the enclosure is irregular. The objective of this section is to illustrate the application of the numerical grid-generation technique to transform the irregular physical region into a regular one in the computational domain and solve the free convection problem with finite differences over the regular region. The specific problem considered here consists of an air-filled cylindrical enclosure with its longitudinal axis oriented horizontally. A flat, rectangular plate with a rounded end and width less than the diameter of the cylinder is placed inside the cylinder horizontally as illustrated in Figure 11.15. It is assumed that the plate is hot and maintained at a uniform temperature T_h, while the cylindrical enclosure

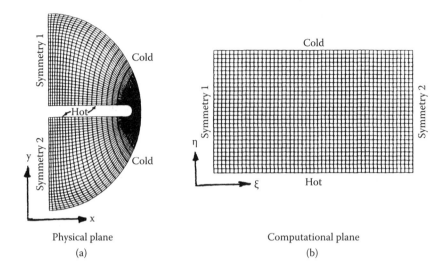

FIGURE 11.15
Free convection inside an irregular enclosure: (a) Physical domain; (b) Computational domain.

is cold and maintained at a uniform temperature T_c. The physical problem is concerned with the determination of the free convection heat transfer coefficient from the hot plate to the air filling the enclosure.

Figure 11.15a shows the physical domain, whereas Figure 11.15b shows the mapping of this irregular geometry into a regular rectangular one in the computational domain.

Assuming incompressible, constant-property fluid with Boussinesq approximation, the two-dimensional vorticity–stream function formulation of this free convection problem is given in dimensionless form by (Elshamy and Özişik 1990):

$$\frac{\partial^2 \psi}{\partial y^2} + \frac{\partial^2 \psi}{\partial y^2} = -\omega \tag{11.64}$$

$$\left[\frac{\partial}{\partial x}\left(\omega \frac{\partial \psi}{\partial y}\right) - \frac{\partial}{\partial y}\left(\omega \frac{\partial \psi}{\partial x}\right)\right] - \frac{\partial}{\partial x}\left(\frac{\partial \omega}{\partial x}\right) - \frac{\partial}{\partial y}\left(\frac{\partial \omega}{\partial x}\right) - \mathrm{Gr}\frac{\partial T}{\partial x} = 0 \tag{11.65}$$

$$\left[\frac{\partial}{\partial x}\left(T\frac{\partial \psi}{\partial y}\right) - \frac{\partial}{\partial y}\left(T\frac{\partial \psi}{\partial x}\right)\right] - \frac{\partial}{\partial x}\left(\frac{1}{\mathrm{Pr}}\frac{\partial T}{\partial x}\right) - \frac{\partial}{\partial y}\left(\frac{1}{\mathrm{Pr}}\frac{\partial T}{\partial y}\right) = 0 \tag{11.66}$$

subject to the boundary conditions.

$$\psi = 0 \text{ on all the boundaries} \tag{11.67a}$$

$$\omega = 0 \text{ on the symmetry line} \tag{11.67b}$$

Numerical Grid Generation 447

$\omega = \omega_{wall}$ on the flat plate and the circular surface (11.67c)

$T = 1$ on the flat plate surface (hot) (11.67d)

$T = 0$ on the cylinder surface (cold) (11.67e)

$\dfrac{\partial T}{\partial \mathbf{n}} = 0$ on the symmetry line (11.67f)

where

$$\omega_{wall} = -2 \dfrac{\psi_{c1} - \psi_{wall}}{(L_{c1})^2} \qquad (11.67g)$$

Various quantities are defined as

$\partial/\partial \mathbf{n}$	normal derivatives
Gr	Grashof number $(g\beta_t d^3(T_h - T_c)/\upsilon^2)$
Pr	Prandtl number (υ/α_t)
T	dimensionless temperature $[(T^* - T_c)/(T_h - T_c)]$
x, y	dimensionless coordinates $(x = x^*/d, y = y^*/d)$
d	diameter of the cylinder
υ	kinematic viscosity
ψ	dimensionless stream function (ψ^*/υ)
α_t	thermal diffusivity
β_t	coefficient of thermal expansion
ω	dimensionless vorticity $(d^2\omega^*/\upsilon)$
L_{c1}	physical distance between the closest interior node and the wall node

We now present the basic steps to be followed in the solution of this problem, with the finite control volume approach in the computational domain.

   i. **The Mapping of the Irregular Region:** The irregular region shown in Figure 11.15a is mapped into the ξ,η computational domain as a rectangular region, as shown in Figure 11.15b, by applying the mapping technique described previously. We note that, for this particular case, the hot and cold surfaces of the physical domain are mapped as the bottom and top surfaces of the rectangular computational domain, whereas the surfaces of symmetry are mapped as the two side surfaces of the rectangle.
  ii. **The Numerical Grid Generation:** A square ξ,η mesh with unit grid spacing, that is, $\Delta\xi = \Delta\eta = 1$, is constructed over the rectangular computational domain in order to solve the grid-generation equations (11.43) and (11.44) with finite differences. The discretization of these equations leads to a successive-substitution formula given by equation (11.47). If all the boundary conditions are of the first

kind (i.e., x,y values are specified at every ξ,η grid point at the boundaries of the computational domain), equation (11.47) is sufficient to solve the numerical grid-generation problem iteratively. For the irregular physical domain shown in Figure 11.15a, it is preferable to impose the orthogonality of grid lines at the symmetry boundaries and at the straight portions of the hot plate, and use the boundary condition of the first kind at the cold cylindrical surface and at the curved portion of the hot plate. For such a case, the finite difference expressions should be developed for the grid points at the boundaries where the orthogonality condition is imposed. These finite difference equations should be solved together with the successive-substitution formula (11.47) in order to calculate the x,y coordinate values at each ξ,η grid point.

iii. **The Transformation of the Field Equation:** The field equations (11.64)–(11.66) for the field unknowns ω, ψ, and T need to be transformed from the x,y to the ξ,η independent variables. By using the transformation relations given in Section 11.1, the resulting transformed equations in conservation form become

$$\nabla^2 \psi = \omega \qquad (11.68)$$

$$\frac{1}{J}[(\psi_\eta \omega)_\xi - (\psi_\xi \omega)_\eta] - \nabla^2 \omega - \frac{Gr}{J}((y_\eta T)_\xi - (y_\xi T)_\eta) = 0 \qquad (11.69)$$

$$\frac{1}{J}[(\psi_\eta T)_\xi - (\psi_\xi T)_\eta] - \frac{1}{Pr}\nabla^2 T = 0 \qquad (11.70)$$

where

$$\nabla^2 = \left(\frac{\alpha}{J^2}\frac{\partial^2}{\partial \xi^2} - \frac{2\beta}{J^2}\frac{\partial^2}{\partial \xi \partial \eta} + \frac{\gamma}{J^2}\frac{\partial^2}{\partial \eta^2} + P\frac{\partial}{\partial \xi} + Q\frac{\partial}{\partial \eta}\right) \qquad (11.71)$$

and the coefficients α, β, γ, and J, defined by equation (11.17) are considered known quantities. Equations (11.68)–(11.71) can be combined into the following single general elliptic equation (Coulter and Guceri 1985; Pletcher et al. 2012):

$$\underbrace{\frac{a_\phi}{J}[(\psi_\eta \phi)_\xi - (\psi_\xi \phi)_\eta]}_{\text{Convection}} + \underbrace{b_\phi[\nabla^2(c_\phi \phi)]}_{\text{Diffusion}} + \underbrace{d_\phi}_{\text{Source}} = 0 \qquad (11.72)$$

where the Laplacian is defined by equation (11.71). The generic variable φ represents any one of the three dependent variables ω, ψ, and T; the coefficients a_ϕ, b_ϕ, c_ϕ, and d_ϕ associated with these variables are given in Table 11.1.

TABLE 11.1
Definitions of the Generic Variables a_ϕ, b_ϕ, c_ϕ, and d_ϕ

ϕ	a_ϕ	b_ϕ	c_ϕ	d_ϕ
ψ	0	1	1	ω
ω	1	−1	1	$-\dfrac{Gr}{J}\left((y_\eta T)_\xi - (y_\xi T)_\eta\right)$
T	1	−1/Pr	1	0

iv. **The Transformation of Boundary Conditions for the Field Equations:** The boundary conditions (11.67) for the field equations are transformed from the x,y physical plane to the ξ,η computational plane as

$$\xi = 1: \quad \psi = 0, \quad \omega = 0, \quad \alpha T_\xi - \beta T_\eta = 0 \qquad (11.73\text{a–c})$$

$$\xi = I: \quad \psi = 0, \quad \omega = 0, \quad \alpha T_\xi - \beta T_\eta = 0 \qquad (11.74\text{a–c})$$

$$\eta = 1: \quad \psi = 0, \quad \omega = \omega_{\text{wall}}, \quad T = 1 \qquad (11.75\text{a–c})$$

$$\eta = J: \quad \psi = 0, \quad \omega = \omega_{\text{wall}}, \quad T = 0 \qquad (11.76\text{a–c})$$

We note that the boundary conditions (11.73c) and (11.74c) correspond to the transformation of the symmetry condition (11.67f), which implies that the normal derivative in the ξ direction should vanish. The transformation relation is obtained from equations (11.56a–d).

v. **Finite Difference Representation of the Transformed Field Equations and Boundary Conditions:** The control volume approach is used to discretize the differential equation (11.72). That is, this equation is integrated over an integration cell shown by dotted lines in Figure 11.16. In this figure, the capital letters E, W, S, and N refer to the centers of the four neighboring volumes surrounding the center node P. The lower case characters e, w, s, and n refer to the limits of the integration, that is, the surfaces of the control volume with center P.

The differential equation (11.72) contains three distinct groups, namely *convection*, *diffusion*, and the *source*. To illustrate the discretization procedure, we consider in detail the integration of only one representative term from each of these three different groups.

Equation (11.72) is written as

$$\frac{a_\phi}{J}\left[\frac{\partial}{\partial \xi}\left(\phi \frac{\partial \psi}{\partial \eta}\right) - \frac{\partial}{\partial \eta}\left(\phi \frac{\partial \psi}{\partial \xi}\right)\right] + b_\phi[\nabla^2(c_\phi \phi)] + d_\phi = 0 \qquad (11.77)$$

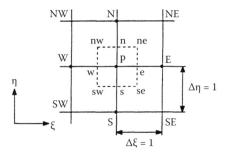

FIGURE 11.16
The integration cell in the computational domain.

Equation (11.77) is integrated over the dotted integration cell of Figure 11.16, and only one term is considered from each group, that is,

$$[I_{conv,1} - .....] + [I_{diff,1} - .....] + I_{sour} = 0 \qquad (11.78)$$

where

$$I_{conv,1} \equiv \int_s^n \int_w^e \frac{a_\phi}{J} \frac{\partial}{\partial \xi}\left(\phi \frac{\partial \psi}{\partial \eta}\right) d\xi \, d\eta \qquad (11.79a)$$

$$I_{diff,1} \equiv \int_s^n \int_w^e \frac{\alpha b_\phi}{J^2} \frac{\partial^2}{\partial \xi^2}(c_\phi \phi) d\xi \, d\eta \qquad (11.79b)$$

$$I_{sour} \equiv \int_s^n \int_w^e d_\phi \, d\xi \, d\eta \qquad (11.79c)$$

Convection Terms

We consider the term $I_{conv,1}$ given by equation (11.79a)

$$I_{conv,1} \equiv \int_s^n \int_w^e \frac{a_\phi}{J} \frac{\partial}{\partial \xi}\left(\phi \frac{\partial \psi}{\partial \eta}\right) d\xi \, d\eta \qquad (11.80)$$

Assuming that the average value of (a_ϕ/J) over the integration cell can be taken as its value at the central node P, the inner integral is performed. Equation (11.80) becomes

Numerical Grid Generation

$$I_{conv,1} \equiv \left(\frac{a_\phi}{J}\right)_P \underbrace{\int_s^n \phi_e \left(\frac{\partial \psi}{\partial \eta}\right)_e d\eta}_{I^{(1)}_{conv,1}} - \left(\frac{a_\phi}{J}\right)_P \underbrace{\int_s^n \phi_w \left(\frac{\partial \psi}{\partial \eta}\right)_w d\eta}_{I^{(2)}_{conv,1}} \quad (11.81)$$

where $(a_\phi/J)_P$ refers to the values of a_ϕ and J evaluated at the center node P. Assuming that ϕ and ψ are well-behaved functions, the integrals $I^{(1)}_{conv,1}$ and $I^{(2)}_{conv,1}$ are evaluated as

$$I^{(1)}_{conv,1} \cong \left(\frac{a_\phi}{J}\right)_P \overline{\phi}_e (\psi_{ne} - \psi_{se}) \quad (11.82)$$

$$I^{(2)}_{conv,1} \cong \left(\frac{a_\phi}{J}\right)_P \overline{\phi}_w (\psi_{nw} - \psi_{sw}) \quad (11.83)$$

where $\overline{\phi}_e$ is the average value of ϕ_e defined by

$$\overline{\phi}_e = \frac{\int_s^n \phi_e \left(\frac{\partial \psi}{\partial \eta}\right)_e d\eta}{\int_s^n \left(\frac{\partial \psi}{\partial \eta}\right)_e d\eta} \quad (11.84)$$

In equation (11.82), the quantities $\overline{\phi}_e, \psi_{ne}$, and ψ_{se} need to be expressed in terms of the values of these parameters at the grid points of the control volume cell. To achieve this, the following assumptions are made:

1. ϕ is uniform within each cell and its value can be taken as the value at the central node P.
2. If $(\psi_{ne} - \psi_{se}) > 0$, the flow is from P to E and $\overline{\phi}_e = \phi_P$.

   If $(\psi_{ne} - \psi_{se}) < 0$, the flow is from E to P and $\overline{\phi}_e = \phi_E$.
3. The value of the stream function ψ_{se} at the corner of the cell is equal to the average values of ψ at the four neighboring nodes, that is

$$\psi_{se} = \frac{1}{4}(\psi_{SE} + \psi_E + \psi_P + \psi_S) \quad (11.85)$$

We note that the assumptions (2) imply the *upstream differencing*. Consider, for example, $I^{(1)}_{conv,1}$, given by equation (11.83). If this equation should satisfy the upstream differencing requirement, it is convenient to write it in the form

$$I^{(1)}_{conv,1} \cong \left(\frac{a_\phi}{J}\right)_P \left[\phi_E \frac{(\psi_{ne} - \psi_{se}) - |\psi_{ne} - \psi_{se}|}{2} + \phi_P \frac{(\psi_{ne} - \psi_{se}) + |\psi_{ne} - \psi_{se}|}{2}\right]$$
$$(11.86)$$

Clearly, for $\psi_{ne} - \psi_{se} > 0$, equation (11.86) reduces to

$$I_{conv,1}^{(1)} \cong \left(\frac{a_\phi}{J}\right)_P \phi_P(\psi_{ne} - \psi_{se}) \qquad (11.87a)$$

and for $\psi_{ne} - \psi_{se} < 0$, it becomes

$$I_{conv,1}^{(1)} \cong \left(\frac{a_\phi}{J}\right)_P \phi_E(\psi_{ne} - \psi_{se}) \qquad (11.87b)$$

In the above discretization procedure, we considered only one of the convection integral terms. If all the convection terms are integrated in a similar manner and the results are combined, the finite difference form of the convection term in equation (11.72) can be expressed in the form (Coulter and Guceri 1985):

$$\sum_i I_{conv,i} \equiv A_E(\phi_P - \phi_E) + A_W(\phi_P - \phi_W) + A_N(\phi_P - \phi_N) + A_S(\phi_P - \phi_S) \qquad (11.88a)$$

where

$$A_E = \frac{1}{8}\left(\frac{a_\phi}{J}\right)_P \{(\psi_{SE} + \psi_S - \psi_{NE} - \psi_N) + |\psi_{SE} + \psi_S - \psi_{NE} - \psi_N|\} \qquad (11.88b)$$

$$A_W = \frac{1}{8}\left(\frac{a_\phi}{J}\right)_P \{(\psi_{NW} + \psi_N - \psi_{SW} - \psi_S) + |\psi_{NW} + \psi_N - \psi_{SW} - \psi_S|\} \qquad (11.88c)$$

$$A_N = \frac{1}{8}\left(\frac{a_\phi}{J}\right)_P \{(\psi_{NE} + \psi_E - \psi_{NW} - \psi_W) + |\psi_{NE} + \psi_E - \psi_{NW} - \psi_W|\} \qquad (11.88d)$$

$$A_S = \frac{1}{8}\left(\frac{a_\phi}{J}\right)_P \{(\psi_{SW} + \psi_W - \psi_{SE} - \psi_E) + |\psi_{SW} + \psi_W - \psi_{SE} - \psi_E|\} \qquad (11.88e)$$

Diffusion Terms

To illustrate the discretization of the diffusion term by integration over the control volume, we consider only the first diffusion term given by equation (11.79b) as

$$I_{diff,1} \equiv \int_s^n \int_w^e \frac{\alpha b_\phi}{J^2} \frac{\partial^2(c_\phi \phi)}{\partial \xi^2} \, d\xi \, d\eta \qquad (11.89)$$

The terms under the integral are rearranged as

$$\frac{b_\phi \alpha}{J^2}\frac{\partial^2(c_\phi \phi)}{\partial \xi^2} \equiv \frac{\partial}{\partial \xi}\left[\frac{b_\phi \alpha}{J^2}\frac{\partial(c_\phi \phi)}{\partial \xi}\right] - \frac{\partial}{\partial \xi}\left(\frac{b_\phi \alpha}{J^2}\right)\frac{\partial(c_\phi \phi)}{\partial \xi} \tag{11.90}$$

Equation (11.90) is introduced into equation (11.89).

$$I_{\text{diff},1} \equiv \underbrace{\int_s^n \int_w^e \left\{\frac{\partial}{\partial \xi}\left[\left(\frac{b_\phi \alpha}{J^2}\right)\frac{\partial(c_\phi \phi)}{\partial \xi}\right]\right\} d\xi d\eta}_{I^{(1)}_{\text{diff},1}} - \underbrace{\int_s^n \int_w^e \left\{\frac{\partial}{\partial \xi}\left(\frac{b_\phi \alpha}{J^2}\right)\frac{\partial(c_\phi \phi)}{\partial \xi}\right\} d\xi d\eta}_{I^{(2)}_{\text{diff},1}}$$

$$\tag{11.91}$$

The inner integration is formally performed in the first of these two integrals. Thus, we obtain

$$I^{(1)}_{\text{diff},1} \equiv \int_s^n \left\{\left(\frac{b_\phi \alpha}{J^2}\right)_e \left[\frac{\partial(c_\phi \phi)}{\partial \xi}\right]_e - \left(\frac{b_\phi \alpha}{J^2}\right)_w \left[\frac{\partial(c_\phi \phi)}{\partial \xi}\right]_w\right\} d\eta \tag{11.92}$$

Assuming $(b_\phi \alpha/J^2)$ and $(c_\phi \phi)$ vary linearly with ξ over the cell, equation (11.92) is formally integrated

$$I^{(1)}_{\text{diff},1} \equiv \left[\frac{1}{2}\left\{\left(\frac{b_\phi \alpha}{J^2}\right)_E + \left(\frac{b_\phi \alpha}{J^2}\right)_P\right\}\right]\left[\frac{(c_\phi \phi)_E - (c_\phi \phi)_P}{\xi_E - \xi_P}\right]\left(\frac{\eta_N - \eta_S}{2}\right)$$
$$- \left[\frac{1}{2}\left\{\left(\frac{b_\phi \alpha}{J^2}\right)_W + \left(\frac{b_\phi \alpha}{J^2}\right)_P\right\}\right]\left[\frac{(c_\phi \phi)_P - (c_\phi \phi)_W}{\xi_P - \xi_W}\right]\left(\frac{\eta_N - \eta_S}{2}\right)$$

$$\tag{11.93}$$

We note that

$$\Delta\xi = \Delta\eta = 1 \quad \text{and} \quad \xi_E - \xi_P = \xi_P - \xi_W = \frac{1}{2}(\eta_N - \eta_S) = 1 \tag{11.94}$$

and rearrange the terms. Then, equation (11.93) is written as

$$I^{(1)}_{\text{diff},1} \equiv B_{1E}[(c_\phi \phi)_E - (c_\phi \phi)_P] + B_{1W}[(c_\phi \phi)_W - (c_\phi \phi)_P] \tag{11.95a}$$

where

$$B_{1E} = \frac{1}{2}\left\{\left(\frac{b_\phi \alpha}{J^2}\right)_E + \left(\frac{b_\phi \alpha}{J^2}\right)_P\right\} \tag{11.95b}$$

$$B_{1W} = \frac{1}{2}\left\{\left(\frac{b_\phi \alpha}{J^2}\right)_W + \left(\frac{b_\phi \alpha}{J^2}\right)_P\right\} \tag{11.95c}$$

Other diffusion terms are discretized in a similar manner.

Source Term

The integration of the source term over the integration cell gives

$$I_{sour} = \int_s^n \int_w^e d_\phi \, d\xi \, d\eta \tag{11.96}$$

If the source term d_ϕ is smooth over the cell, it is evaluated at the center node P, and the integration over the cell is performed.

$$I_{sour} = d_{\phi,P} \int_s^n \int_w^e d\xi \, d\eta = d_{\phi,P} \tag{11.97}$$

since $\Delta\xi = \Delta\eta = 1$. Here, $d_{\phi,P}$ is the value of d_ϕ evaluated at the center node, P.

The same techniques presented in Chapter 2 are applied for the control volumes with surfaces that coincide with the boundaries of the body. The finite difference expressions developed as described above for the convection, diffusion, and the source terms are assembled by substituting them into the original differential equation (11.72), and the resulting expression is solved for ϕ_P. Here, the function ϕ_P can be ψ_P, ω_P, or T_P, as given by Table 11.1.

11.7.1 The Nusselt Number

Once the temperature distribution in the fluid is available, the local Nusselt number along the hot surface is determined from its definition

$$Nu \equiv \frac{hd}{k} = \frac{\partial T}{\partial \mathbf{n}} \tag{11.98}$$

where $\partial/\partial \mathbf{n}$ is the dimensionless derivative along the direction of the outward drawn normal to the hot surface and d is the diameter of the cylinder. Since all the calculations have been performed on the computational domain, this derivative $\partial/\partial \mathbf{n}$ should be expressed in terms of the ξ, η independent variables. Figure 11.15b shows that the hot surface is mapped as the bottom surface of the rectangular computational domain, and $\partial/\partial \mathbf{n}$ at this surface is obtained from equation (11.56c) as

$$\frac{\partial T}{\partial \mathbf{n}} = \frac{-1}{J\sqrt{\gamma}} (\gamma T_\eta - \beta T_\xi) \tag{11.99}$$

Introducing equation (11.99) into equation (11.98), the local Nusselt number at the hot surface becomes

$$Nu = \frac{1}{J\sqrt{\gamma}} (\beta T_\xi - \gamma T_\eta) \tag{11.100}$$

Numerical Grid Generation

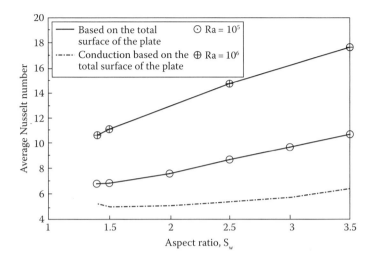

FIGURE 11.17
Effect of the aspect ratio S_w on the average Nusselt number for $S_t = 0.06$.

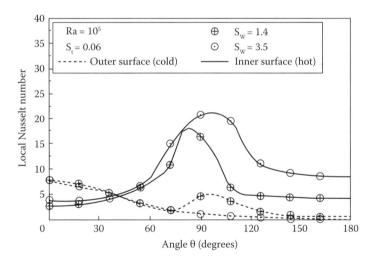

FIGURE 11.18
Variation of local Nusselt number with angle θ.

11.7.2 Results

The computed average Nusselt number obtained with a 51 × 31 mesh was compared with the experimental results of Singh and Liburdy (1986), and the agreement was good. Figure 11.17 shows the effects of aspect ratio S_w on the average Nusselt number for two different Rayleigh numbers, 10^5 and 10^6. As expected, the average Nusselt number increases as Rayleigh

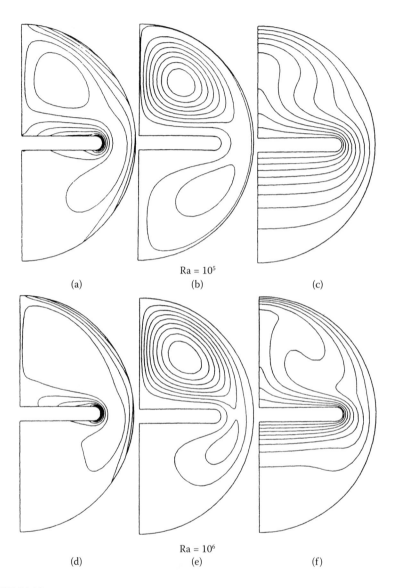

FIGURE 11.19
Effect of Rayleigh number on the flow patterns $S_w = 1.4$.

number increases. Figure 11.18 shows the variation of the Nusselt number with the angle θ around the cylindrical surface, with $\theta = 0$ corresponding to the top of the cylindrical surface. The effects of Rayleigh number on vorticity, streamlines, and isotherms for $Ra = 10^5$ and 10^6 are shown in Figure 11.19.

11.8 Transient Laminar Free Convection in Irregular Enclosures—Primitive Variables Formulation

As the previous section, this one deals with the solution of laminar natural convection problems in irregular enclosures. However, we now consider transient problems written in terms of the primitive variables given by the x and y velocity components, u and v, respectively, and by the temperature T. The surface for the cavity is assumed to be defined by four segments as illustrated by Figure 11.20, which are transformed into the computational domain as the surfaces $\xi = 1$, $\xi = I$, $\eta = 1$, and $\eta = J$ (see Figure 11.3). The fluid is initially at rest and at the temperature T_c, which is also assumed to be the temperature at the surface $\eta = 1$. At time zero, the surface at $\eta = J$ has its temperature changed to T_h. The other two surfaces at $\xi = 1$ and $\xi = I$ are either kept insulated or they correspond to conditions of symmetry, depending on the application of interest; in both cases, the boundary condition is given by zero normal derivative of temperature. The fluid properties are assumed constant, except for the density in the buoyancy term, where we consider Boussinesq's approximation as valid (Colaço and Orlande 2002). The mathematical formulation for this physical problem can be written in terms of the following conservation equation in generalized Cartesian coordinates (see Example 11.3):

$$\frac{\partial (J\rho\varphi)}{\partial t} + \frac{\partial (\tilde{U}\rho\varphi)}{\partial \xi} + \frac{\partial (\tilde{V}\rho\varphi)}{\partial \eta} = \frac{\partial}{\partial \xi}\left\{J\Gamma^\varphi\left[a\frac{\partial \varphi}{\partial \xi} + d\frac{\partial \varphi}{\partial \eta}\right]\right\}$$
$$+ \frac{\partial}{\partial \eta}\left\{J\Gamma^\varphi\left[d\frac{\partial \varphi}{\partial \xi} + b\frac{\partial \varphi}{\partial \eta}\right]\right\} + JS \tag{11.101}$$

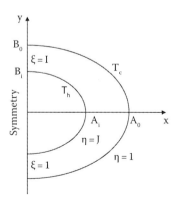

FIGURE 11.20
Annular elliptical cavity.

TABLE 11.2
Conservation Variable, Diffusion Coefficient, and Source Term

Conservation of	ϕ	Γ^ϕ	S^ϕ
Mass	1	0	0
x-Momentum	u	μ	$-\partial P/\partial x$
y-Momentum	v	μ	$-\partial P/\partial y - \rho g [1 - \beta (T - T_{ref})]$
Energy	T	K/C_p	0

where

$$a = \xi_x^2 + \xi_y^2; \quad b = \eta_x^2 + \eta_y^2; \quad d = \xi_x \eta_x + \xi_y \eta_y; \quad J = x_\xi y_\eta - x_\eta y_\xi;$$
$$\tilde{U} = J(u\xi_x + v\xi_y); \quad \tilde{V} = J(u\eta_x + v\eta_y) \quad (11.102\text{a–f})$$

The general conservation variable, as well as the diffusion coefficient and the source term, is found in Table 11.2 for the mass, momentum, and energy conservation equations. These equations are solved, subject to the following boundary and initial conditions:

$$u = v = 0 \quad \text{at } \eta = 1 \text{ and } \eta = N, \text{ for } t > 0 \quad (11.103\text{a})$$

$$u = v = 0 \quad \text{at } \xi = 1 \text{ and } \xi = M, \text{ for } t > 0; \text{ for cases with insulated boundaries} \quad (11.103\text{b})$$

$$D_{11}\frac{\partial u}{\partial \xi} + D_{12}\frac{\partial u}{\partial \eta} = D_{11}\frac{\partial v}{\partial \xi} + D_{12}\frac{\partial v}{\partial \eta} = 0 \quad \text{for cases with symmetry at } \xi = 1 \text{ and } \xi = M \quad (11.103\text{c})$$

$$T = T_h \quad \text{at } \eta = N, \text{ for } t > 0 \quad (11.103\text{d})$$

$$T = T_c \quad \text{at } \eta = 1, \text{ for } t > 0 \quad (11.103\text{e})$$

$$D_{11}\frac{\partial T}{\partial \xi} + D_{12}\frac{\partial T}{\partial \eta} = 0 \quad \text{at } \xi = 1 \text{ and } \xi = M, \text{ for } t > 0 \quad (11.103\text{f})$$

$$u = v = 0 \quad \text{for } t = 0 \text{ in the region} \quad (11.103\text{g})$$

$$T = T_c \quad \text{for } t = 0 \text{ in the region} \quad (11.103\text{h})$$

where

$$D_{11} = a J \Gamma^\phi; \quad D_{12} = b J \Gamma^\phi \quad (11.104\text{a,b})$$

We note in Table 11.2 that the positive y-axis in the physical domain is supposed to be aligned with the opposite direction of the gravitational acceleration vector.

The transient natural convection problem was solved *via* finite volumes, by utilizing the WUDS interpolation scheme and the SIMPLEC method for the treatment of the pressure–velocity coupling, as described in detail in Chapter 8. Two cases are considered here, involving annular elliptical and circular cavities.

For the case of the elliptical cavity, the geometrical parameters were taken as $A_0 = 0.0230$ m, $A_i = 0.0108$ m, $B_0 = 0.0214$ m, and $B_i = 0.0055$ m. The temperatures at the walls were taken as $T_h = 34°C$ and $T_c = 10°C$. The Rayleigh number for this case is 10^4, where the characteristic length was taken as $B_0 - B_i$. Figure 11.21 presents a comparison of the results obtained

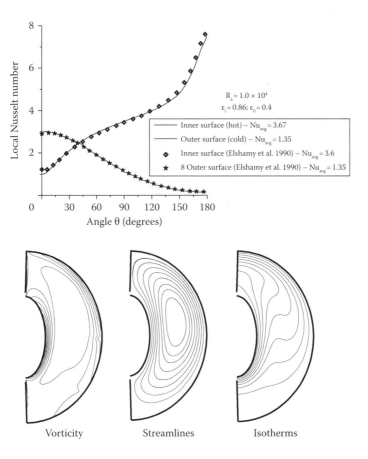

FIGURE 11.21
Local Nusselt number for the annular elliptical cavity.

here with those of Elshamy and Özişik (1990) for the steady-state local Nusselt number at the inner and outer surfaces of the cavity. For the numerical results computed here, we have used 67 × 67 volumes and a time increment of 5×10^{-3} s. The computations were run until steady state was reached at 100 s. An examination of Figure 11.21 reveals a very good agreement between the present results and those of Elshamy and Özişik (1990), which were obtained by using the same technique described in the previous section. Indeed, the discrepancies between the present results and those of Elshamy and Özişik (1990) were smaller than 0.3% for the average Nusselt number.

We now turn our attention to the case of an annular circular cavity. The results obtained here are for the natural convection of air with properties $\rho = 1.19$ kg/m^3, $\mu = 1.8 \times 10^{-5}$ kg/ms, $\beta = 0.00341$ K^{-1}, Pr = 0.71, k = 0.02624 W/mK, and $C_p = 1035.02$ J/kg°C. The geometrical dimensions were taken as $A_i = B_i = 0.0228$ m and $A_0 = B_0 = 0.0594$ m, while the temperatures at the walls were taken as $T_h = 30$°C and $T_c = 20$°C. For this case, the Rayleigh number is 5×10^4. The numerical results obtained with finite volumes are compared with those of Pereira et al. (2000) obtained with the generalized integral transform technique (see Chapter 12). Figure 11.22 presents a comparison of the results obtained here with those of Pereira et al. (2000) for the steady-state local Nusselt number at the hot and cold surfaces. The agreement between the two solutions is excellent. The numerical results were obtained using a mesh with 80 × 80 volumes and a time increment of 5×10^{-3} s. The computations were run until steady state was reached at 100 s.

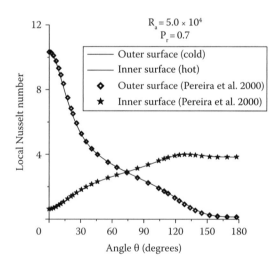

FIGURE 11.22
Local Nusselt number for annular circular cavity.

11.9 Computational Aspects for the Evaluation of Metrics

The numerical solution of partial differential equations within irregular domains, using the finite volume discretization of equations in conservative form and grid-generation techniques, requires the evaluation of the transformation metrics at the volume center, as well as at each volume surface center points. For regularly spaced structured grids, the determination of the positions of the points where metrics need to be calculated becomes straightforward. However, that is not the case when the grid is distorted because of irregular domains or concentration of grid points at regions of large gradients.

Transformation metrics and positions of volume and surface center points are usually calculated during the grid-generation procedure. Indeed, in order to avoid artificial numerical source terms, the metrics at the control volume surface center shall not be calculated by averaging the metrics at the center of the neighboring volumes. Instead, the positions of the control volume surface centers shall be calculated by averaging the positions of the center points of neighboring volumes, and the metrics are then calculated by central finite differences (Thompson et al. 1985). Despite satisfying metric identities and avoiding spurious artificial source terms, this traditional approach still introduces numerical errors by using grid point positions for the averaging procedure. These errors are related to the grid spacing and grid line angles. We note that metrics evaluation and discretization errors for the solution of partial differential equations are independent and can be addressed individually (Thompson et al. 1985).

In this section, a technique is introduced to reduce errors due to the computation of the transformation metrics when the finite control volume approach is used for the discretization of equations in conservative form. The technique advanced by Moreira Filho et al. (2002) makes use of an auxiliary grid, with twice the number of points used for the solution of the governing partial differential equation, for the evaluation of grid geometrical parameters and transformation metrics within the solution domain. Therefore, positions of the control volume center and the control volume surface center are evaluated in the transformed domain during the grid-generation procedure, thus avoiding the use of coordinate averaging within the physical domain. The reduction of numerical errors by using this technique is illustrated in the following sections with two examples (Moreira Filho et al. 2002).

11.9.1 One-Dimensional Advection–Diffusion Equation

In this example, an analytical transformation is applied to the steady-state advection–diffusion one-dimensional equation (see Chapter 4). Control volume centers can be readily obtained from the analytical transformation. Despite being analytically available, transformation metrics are numerically evaluated to allow the comparison of the traditional approach based on the averaging of positions of the centers of the neighboring volumes and the approach presented here that is based on a grid with twice the number of points.

The one-dimensional convection–diffusion model problem is written in the conservative form as

$$\frac{\partial}{\partial y}(\rho v \varphi) = \frac{\partial}{\partial y}(\Gamma \partial_y \varphi), \quad 0 < y < h \tag{11.105a}$$

with boundary conditions

$$\varphi = 1, \quad y = 0 \tag{11.105b}$$

$$\varphi = 0, \quad y = h \tag{11.105c}$$

where the velocity v is constant.

The convection–diffusion equation and boundary conditions are transformed to the computational domain using the logarithmic transformation defined by equations (11.22a,b) and (11.23a,b). The analytic metric is given by

$$\eta_y = \frac{2\beta}{h\{\beta^2 - [1-(y/h)]^2\}\ln\left[\frac{\beta+1}{\beta-1}\right]} \tag{11.106}$$

while equation (11.105a) is written in the transformed domain as

$$\frac{\partial}{\partial \eta}(\eta_n \rho v \varphi) = \frac{\partial}{\partial \eta}\left[\eta_y \Gamma \frac{\partial}{\partial \eta}(\eta_y \varphi)\right], \quad 0 < \eta < 1 \tag{11.107a}$$

with boundary conditions

$$\varphi = 1, \quad \eta = 0 \tag{11.107b}$$

$$\varphi = 0, \quad \eta = h \tag{11.107c}$$

Applying the finite control volume method with the WUDS scheme for the discretization of equation (11.107a), the resulting algebraic equation for the internal volumes can be written as

$$a_P \varphi_P - a_E \varphi_E - a_W \varphi_W = 0 \tag{11.108}$$

where

$$a_E = -(\rho v)_e (1/2 - \alpha_e) + \frac{\beta_e (\Gamma \eta_y)_e}{\Delta \eta} \tag{11.109a}$$

$$a_W = (\rho v)_w (1/2 - \alpha_w) + \frac{\beta_w (\Gamma \eta_y)_w}{\Delta \eta} \tag{11.109b}$$

$$a_P = a_E + a_W \qquad (11.109c)$$

It is clear from equations (11.109a) and (11.109b) that the finite volume discretization of an equation in conservative form transformed to a regular computational domain requires the computation of the metrics (η_y in this case) at the surfaces of the control volumes (represented by the subscripts e and w).

Tables 11.3 through 11.5 present results for pure diffusion ($v = 0$), obtained for different values of the clustering parameter β. These tables show the numerical solutions obtained using the double-grid and the coordinate-average approaches. The analytic solution and the relative errors obtained with each numerical approach are also included in these tables. Results shown in Tables 11.3 through 11.5 were obtained using five control volumes. Numerical and analytical solutions were evaluated at the center of each control volume. Positions where the solutions are computed are displaced toward $x = 0$ as β approaches unity.

Tables 11.3 through 11.5 show that the two techniques used for computation of the metrics are equivalent for regular meshes ($\beta \to \infty$), since the center point of each volume coincides with the center point obtained with the double-grid approach. On the contrary, Tables 11.3 through 11.5 reveal a general

TABLE 11.3

Steady-State One-Dimensional Pure Diffusion ($v = 0$) with $\beta \to \infty$ (No Grid Distortion)

	ϕ			Relative Error	
Position	DG Approach	CA Approach	Analytic	DG Approach	CA Approach
0.1000	0.9000	0.9000	0.9000	0.0000	0.0000
0.3000	0.7000	0.7000	0.7000	0.0000	0.0000
0.5000	0.5000	0.5000	0.5000	0.0000	0.0000
0.7000	0.3000	0.3000	0.3000	0.0000	0.0000
0.9000	0.1000	0.1000	0.1000	0.0000	0.0000

Note: CA, coordinate average; DG, double grid.

TABLE 11.4

Steady-State One-Dimensional Pure Diffusion ($v = 0$) with $\beta = 2.0$

	ϕ			Relative Error	
Position	DG Approach	CA Approach	Analytic	DG Approach	CA Approach
0.0846	0.9175	0.9395	0.9154	0.0023	0.0257
0.2668	0.7350	0.7811	0.7332	0.0024	0.0612
0.4641	0.5372	0.5861	0.5359	0.0024	0.0856
0.6734	0.3274	0.3616	0.3266	0.0025	0.0967
0.8902	0.1100	0.1205	0.1098	0.0025	0.0895

Note: CA, coordinate average; DG, double grid.

TABLE 11.5

Steady-State One-Dimensional Pure Diffusion (v = 0) with β = 1.01

Position	ϕ			Relative Error	
	DG Approach	CA Approach	Analytic	DG Approach	CA Approach
0.0070	0.9947	0.9999	0.9931	0.0017	0.0069
0.0382	0.9648	0.9989	0.9619	0.0030	0.0372
0.1231	0.8827	0.9828	0.8769	0.0066	0.1077
0.3319	0.6777	0.6298	0.6681	0.0142	0.1948
0.7383	0.2681	0.2766	0.2617	0.0236	0.0537

Note: CA, coordinate average; DG, double grid.

TABLE 11.6

Steady-State One-Dimensional Advection–Diffusion Problem with v = 0.1 and β →∞ (No Grid Distortion)

Position	ϕ			Relative Error	
	DG Approach	CA Approach	Analytical	DG Approach	CA Approach
0.0250	0.9858	0.9858	0.9853	0.0006	0.0006
0.2250	0.8539	0.8539	0.8532	0.0008	0.0008
0.4250	0.6927	0.6927	0.6918	0.0013	0.0013
0.6250	0.4957	0.4957	0.4947	0.0021	0.0021
0.8250	0.2552	0.2552	0.2539	0.0049	0.0049
0.9750	0.0405	0.0405	0.0391	0.0366	0.0366

Note: CA, coordinate average; DG, double grid.

increase in the relative errors of the two numerical solutions when the mesh distortion is increased, that is, β→1. However, the relative errors of the double-grid approach are, for most of the examined points, one order of magnitude smaller than those obtained from the averaging technique of metrics evaluation, because the metrics are more accurately calculated.

In order to verify the influence of convective effects, cases with v = 0.1 and different clustering parameters were also analyzed. Results obtained with 20 control volumes used for the spatial discretization are presented in Tables 11.6 through 11.8. Similar to the pure diffusion problem, the results obtained with both numerical approaches are identical for v = 0.1, when a regular mesh (β→∞) is used (see Table 11.6). As the grid is distorted by reducing the values of β, errors associated with the coordinate averaging approach substantially increase, as can be observed from the results in Tables 11.7 and 11.8. On the contrary, the relative error between the analytical solution and the numerical results obtained from the double-grid approach is practically unaffected by mesh distortion. In fact, it is observed in Tables 11.7 and 11.8 that the errors obtained with the double-grid approach are at least one

TABLE 11.7
Steady-State One-Dimensional Advection–Diffusion Problem with $v = 0.1$ and $\beta = 2.0$

	ϕ			Relative Error	
Position	DG Approach	CA Approach	Analytical	DG Approach	CA Approach
0.0207	0.9883	0.9924	0.9878	0.0005	0.0046
0.1965	0.8743	0.9077	0.8736	0.0008	0.0390
0.4387	0.6804	0.7339	0.6795	0.0014	0.0801
0.5938	0.5293	0.5814	0.5282	0.0022	0.1007
0.8083	0.2774	0.3093	0.2759	0.0053	0.1209
0.9725	0.0446	0.0498	0.0429	0.0399	0.1617

Note: CA, coordinate average; DG, double grid.

TABLE 11.8
Steady-State One-Dimensional Advection–Diffusion Problem with $v = 0.1$ and $\beta = 1.01$

	ϕ			Relative Error	
Position	DG Approach	CA Approach	Analytical	DG Approach	CA Approach
0.0014	0.9992	0.9999	0.9992	0.0000	0.0008
0.2332	0.8492	0.9797	0.8472	0.0025	0.1564
0.4600	0.6643	0.8646	0.6601	0.0064	0.3099
0.5623	0.5661	0.7663	0.5608	0.0094	0.3665
0.8017	0.2923	0.4134	0.2845	0.0273	0.4531
0.9331	0.1115	0.1555	0.1023	0.0902	0.5202

Note: CA, coordinate average; DG, double grid.

order of magnitude smaller than the errors obtained with the traditional averaging technique. For many positions in the solution domain, the relative errors associated with the double-grid technique are two orders of magnitude smaller than those for coordinate averaging.

11.9.2 Two-Dimensional Heat Conduction in a Hollow Sphere

We now consider axisymmetric transient heat conduction in a spherical shell. Although this consists of a problem in a regular domain in spherical coordinates, in order to explore the double-grid approach when applied to irregular domains, the heat conduction problem is written here in terms of cylindrical coordinates, where the spherical shell is an irregular geometry. Elliptic grid generation in axisymmetric cylindrical coordinates is used, and different distortion levels are analyzed by making use of grid control functions for attraction to coordinate lines. The elliptic scheme of numerical grid generation was presented above in Cartesian coordinates. Equations for elliptic grid generation in axisymmetric cylindrical coordinates and in polar coordinates can be found in the Notes section at the end of this chapter.

The heat conduction problem for the spherical shell with internal radius R_i, external radius R_e, and center at $R = 0$ and $z = R_e$, which is represented by the region

$$\Omega = \begin{cases} 0 < z < (R_e - R_i) & \text{and} \quad 0 < R < \sqrt{R_e^2 - (z - R_e)^2} \\ (R_e - R_i) < z < (R_e + R_i) & \text{and} \quad \sqrt{R_i^2 - (z - R_e)^2} < R < \sqrt{R_e^2 - (z - R_e)^2} \\ (R_e + R_i) < z < 2R_e & \text{and} \quad 0 < R < \sqrt{R_e^2 - (z - R_e)^2} \end{cases}$$

(11.110)

is written in cylindrical coordinates as

$$\rho C_p \frac{\partial T}{\partial t} = \frac{1}{R} \frac{\partial}{\partial R} \left(k R \frac{\partial T}{\partial R} \right) + \frac{\partial}{\partial z} \left(k \frac{\partial T}{\partial z} \right) \quad \text{in } \Omega \text{ for } t > 0 \qquad (11.111)$$

The problem examined here involves boundary conditions given by prescribed temperature $T = 0$ on the internal and externals surfaces of the shell and symmetry at the axis $R = 0$. The initial condition is given by uniform temperature T_0 in the region. The problem is solved in dimensionless form for $R_i = 0.1$, $R_e = 1$, and $T_0 = 10$. In order to evaluate the double-grid approach, numerical results are compared with the analytic solution of the problem, which was obtained by classical integral transform technique (Özişik 1993). The number of eigenvalues used in the analytic solution guarantees convergence of four significant digits on the final solution.

Figure 11.23 illustrates a typical mesh used for this test case, involving 30 × 30 volumes in the R and z directions, respectively. It should be noticed in this figure that the attraction of grid lines toward the domain boundaries was used during grid generation in order to generate a distorted mesh. Although such a distorted mesh was prepared with the purpose of illustrating the application of the double-grid approach for computing the metrics and geometrical characteristics of the mesh, it is typical of practical situations that involve large gradients near the boundaries, and grid points need to be concentrated in these regions. Figure 11.24 shows the transient variation of the maximum relative errors for the numerical solutions obtained with the double-grid and coordinate-averaging techniques, for meshes with different number of control volumes. The results presented in Figure 11.24 show that the double-grid approach leads to a reduction of the maximum relative errors for all the grids examined. It should also be noted in Figure 11.24, smaller relative errors for the double-grid approach with 50 × 50 control volumes than the ones for 80 × 80 grids with the coordinate average approach. Therefore, for a specified tolerance, computational costs can be reduced with the double-grid approach because it requires fewer volumes.

In order to further test the double-grid approach, a mesh with a large distortion is now examined. Figure 11.25 shows a highly distorted grid used in the

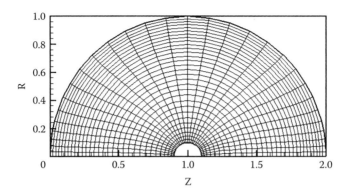

FIGURE 11.23
Typical low distortion mesh with 30 × 30 control volumes.

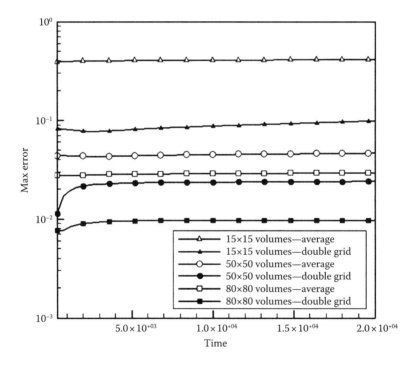

FIGURE 11.24
Maximum relative error with meshes of low distortion.

calculations. Results for the maximum relative error are shown in Figure 11.26 for different grids. This figure indicates an improvement in the maximum relative error when the double-grid approach was used. It is also interesting to note in Figure 11.26 that using 15 × 15 volume grids and the double-grid approach lead to similar maximum errors as grids with 50 × 50 volumes when

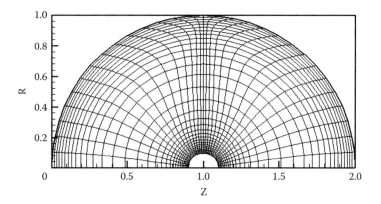

FIGURE 11.25
Typical high distortion mesh with 30 × 30 control volumes.

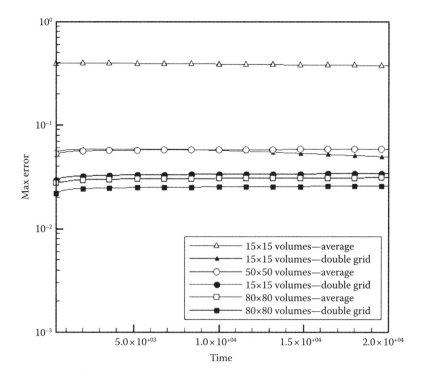

FIGURE 11.26
Maximum relative error with meshes of high distortion.

metrics are evaluated using the coordinate average approach. The same equivalence of maximum errors is observed for 50 × 50 double-grid and 80 × 80 coordinate-average results. Therefore, the use of the double-grid approach for distorted grids can result in substantial reduction of the number of control volumes required to reach an *a priori* established accuracy level.

PROBLEMS

11.1. Transform the following partial differential equation from the physical plane in the x,y coordinates to the computational plane in the ξ,η coordinates using the transformation ξ = ξ(x,y) and η = η(x,y).

$$\frac{\partial u}{\partial x} + c\frac{\partial u}{\partial y} + y = 0$$

11.2. The velocity components u and v are related to the stream function ψ by

$$u = \frac{\partial \psi}{\partial y}, \qquad v = -\frac{\partial \psi}{\partial x}$$

Transform these derivatives from the x,y physical plane to the computational ξ,η plane.

11.3. By utilizing the transformation equations in Section 11.1, derive the advection–diffusion conservation equation of Example 11.3 in the computational domain.

11.4. Using central differences, write the finite difference representation of the first, second, and mixed derivatives appearing in the transformed equations (11.31) and (11.32).

11.5. Utilizing the results obtained in Problem 11.4, write equation (11.31) in the finite difference form by noting that in the computational domain Δξ=Δη=1. The mixed derivative $x_{\xi\eta}$ need not be expanded since it does not contain the term of interest $x_{i,j}$.

11.6. Utilizing the results obtained in Problem 11.4, write equation (11.31) in finite difference form by setting the control functions P and Q equal to zero and Δξ = Δη = 1. The mixed derivative $x_{\xi\eta}$ need not be expanded since it does not contain the term of interest $x_{i,j}$. It is simpler to compute $x_{\xi\eta}$ prior to solving for the actual unknown grid coordinate. Also α, β, and γ can be computed separately.

11.7. Transform the following differential equation

$$\frac{\partial^2 T}{\partial x^2} + \frac{\partial^2 T}{\partial y^2} + AT = f$$

from the physical x,y plane to the computational ξ,η plane.

11.8. Consider a three-dimensional transformation from the x,y,z Cartesian coordinates of the physical plane to the ξ,η,υ coordinates of the computational domain. Starting from the following expressions obtained by the chain rule of transformation,

$$T_\xi = x_\xi \left(\frac{\partial T}{\partial x}\right) + y_\xi \left(\frac{\partial T}{\partial y}\right) + z_\xi \left(\frac{\partial T}{\partial z}\right)$$

$$T_\eta = x_\eta \left(\frac{\partial T}{\partial x}\right) + y_\eta \left(\frac{\partial T}{\partial y}\right) + z_\eta \left(\frac{\partial T}{\partial z}\right)$$

$$T_\upsilon = x_\upsilon \left(\frac{\partial T}{\partial x}\right) + y_\upsilon \left(\frac{\partial T}{\partial y}\right) + z_\upsilon \left(\frac{\partial T}{\partial z}\right)$$

where subscripts denote differentiation, develop an expression for the Jacobian of the transformation.

11.9. Two-dimensional energy equation in the x,y Cartesian coordinates of the physical domain is given in the form

$$\frac{\partial T}{\partial t} + u \frac{\partial T}{\partial x} + v \frac{\partial T}{\partial y} = \alpha_t \left(\frac{\partial^2 T}{\partial x^2} + \frac{\partial^2 T}{\partial y^2}\right)$$

where α_t is the thermal diffusivity. Transform this equation from the x,y coordinates into the ξ,η coordinates of the computational plane.

11.10. Write the finite difference form of the two coupled partial differential equations (11.31) and (11.32) for the determination of x,y over a rectangular computational ξ,η domain with square mesh $\Delta \xi = \Delta \eta = 1$.

11.11. Consider a two-layer, irregular, doubly connected composite region with insulated inner surface as shown in the accompanying figure. By applying one branch cut along the interface and one branch cut across the layers, map this composite region into two simply connected rectangular regions.

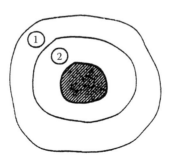

11.12. By applying the orthogonality condition given by equation (11.50) to the boundary η = J in the computational domain shown in Figure 11.3, develop the finite difference expression at the node i,J for the determination of $x_{i,J}$.

11.13. Derive equation (11.16), which gives the Laplacian transformed from a Cartesian two-dimensional (x,y) domain

$$\nabla^2 T = \frac{\partial^2 T}{\partial x^2} + \frac{\partial^2 T}{\partial y^2}$$

into the computational domain (ξ,η).

11.14. Repeat Problem 11.13 for cylindrical axisymmetric coordinates (see Note 1 at the end of this Chapter).

11.15. Repeat Problem 11.13 for polar coordinates (see Note 2 at the end of this chapter).

11.16. Use the elliptic scheme in Cartesian coordinates presented in this chapter to numerically generate a grid in the region shown in the accompanying figure. Use first-kind boundary conditions for all boundaries and no grid control (i.e., P = Q = 0).

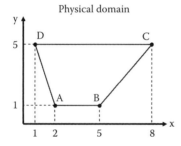
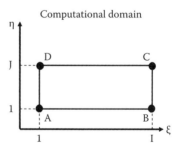

11.17. Repeat Problem 11.16 by considering orthogonal grid lines within the domain (i.e., β = 0 inside the domain) and at the boundaries. Do not use grid control, that is, make P = Q = 0.

11.18. Repeat Problem 11.16 and examine the effects of the control functions on the grids generated by using the boundary lines (ξ = 1, ξ = I, η = 1, and η = J) as attracting lines.

11.19. Repeat Problem 11.17 and examine the effects of the control functions on the grids generated by using the boundary lines (ξ = 1, ξ = I, η = 1, and η = J) as attracting lines.

11.20. Solve the transient heat conduction problem in the region of Problem 11.16, given in dimensionless form by

$$\frac{\partial T}{\partial t} = \frac{\partial^2 T}{\partial x^2} + \frac{\partial^2 T}{\partial y^2} \quad \text{in the region for } t > 0$$

$T = 0$ at all boundaries, for $t > 0$

$T = 10$ at $t = 0$ in the region

Use the grids generated in Problems 11.16 through 11.19 for the solution of the problem.

11.21. Numerically generate a grid with the elliptic scheme presented in this chapter, for the transformation presented in the accompanying figure in Cartesian coordinates. Use orthogonal grid lines at all boundaries. Use the control functions to generate grids with points evenly distributed.

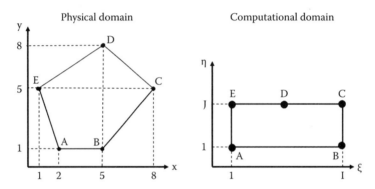

11.22. Numerically generate a grid with the elliptic scheme presented in this chapter for the transformation presented in the accompanying figure in Cartesian coordinates. Use orthogonal grid lines at all boundaries. Use the control functions to generate grids with points evenly distributed.

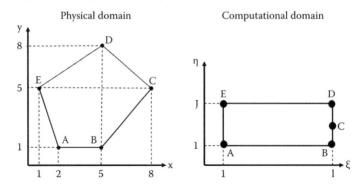

11.23. Solve the following dimensionless heat conduction problem with the transformation used in Problem 11.21:

$$\frac{\partial T}{\partial t} = \frac{\partial^2 T}{\partial x^2} + \frac{\partial^2 T}{\partial y^2} \quad \text{in the region, for } t > 0$$

$$\frac{\partial T}{\partial \mathbf{n}} = 0 \text{ at the boundary AB, for } t > 0$$

T = 0 at the other boundaries, for t > 0

T = 10 at t = 0 in the region

11.24. Repeat Problem 11.23 with the transformation used in Problem 11.22. Compare the solution obtained here with that obtained in Problem 11.23.

NOTES

1. Laplacian and Grid-Generation Equations Transformed from Axisymmetric Physical Domains in Cylindrical Coordinates

The Laplacian in two-dimensional cylindrical coordinates (R,z), in a physical domain with axial symmetry is given by

$$\nabla^2 T = \frac{\partial^2 T}{\partial z^2} + \frac{\partial^2 T}{\partial R^2} + \frac{1}{R}\frac{\partial T}{\partial R} \qquad (a)$$

Let's say that the Laplacian given by equation (a) applies to an irregular geometry in the physical domain (R,z) shown by Figure 11.27a. Such geometry in the physical domain is transformed into a regular geometry in the computational domain illustrated by Figure 11.27b. The Laplacian in the axisymmetric coordinates (ξ,η) in the computational domain can be obtained by directly applying the chain rule to the derivatives in equation (a). The following nonconservative form results:

$$\nabla^2 T = \frac{1}{J^2}[\alpha T_{\xi\xi} - 2\beta T_{\xi\eta} + \gamma T_{\eta\eta}] + [(\nabla^2 \xi)T_\xi + (\nabla^2 \eta)T_\eta] \qquad (b)$$

where

$$\alpha = \left(\frac{\partial R}{\partial \eta}\right)^2 + \left(\frac{\partial z}{\partial \eta}\right)^2 \qquad (c)$$

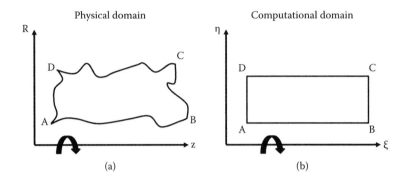

FIGURE 11.27
Transformation in cylindrical axisymmetric coordinates: (a) Physical domain; (b) Computational domain.

$$\beta = \frac{\partial R}{\partial \xi}\frac{\partial R}{\partial \eta} + \frac{\partial z}{\partial \xi}\frac{\partial z}{\partial \eta} \tag{d}$$

$$\gamma = \left(\frac{\partial R}{\partial \xi}\right)^2 + \left(\frac{\partial z}{\partial \xi}\right)^2 \tag{e}$$

$$J = \frac{\partial z}{\partial \xi}\frac{\partial R}{\partial \eta} - \frac{\partial z}{\partial \eta}\frac{\partial R}{\partial \xi} \tag{f}$$

$$\nabla^2 \xi = \frac{\partial^2 \xi}{\partial z^2} + \frac{\partial^2 \xi}{\partial R^2} + \frac{1}{R}\frac{\partial \xi}{\partial R} \tag{g}$$

$$\nabla^2 \eta = \frac{\partial^2 \eta}{\partial z^2} + \frac{\partial^2 \eta}{\partial R^2} + \frac{1}{R}\frac{\partial \eta}{\partial R} \tag{h}$$

The Laplacian in nonconservative form given by equation (b) can be written in conservative form by appropriate manipulation of the partial derivatives.

The equations for numerical elliptic grid generation in cylindrical axisymmetric coordinates are given by

$$\alpha \frac{\partial^2 z}{\partial \xi^2} - 2\beta \frac{\partial^2 z}{\partial \xi \partial \eta} + \gamma \frac{\partial^2 z}{\partial \eta^2} + J^2 \left[P(\xi, \eta)\frac{\partial z}{\partial \xi} + Q(\xi, \eta)\frac{\partial z}{\partial \eta} \right] = 0 \tag{i}$$

$$\alpha \frac{\partial^2 R}{\partial \xi^2} - 2\beta \frac{\partial^2 R}{\partial \xi \partial \eta} + \gamma \frac{\partial^2 R}{\partial \eta^2} + J^2 \left[P(\xi, \eta)\frac{\partial R}{\partial \xi} + Q(\xi, \eta)\frac{\partial R}{\partial \eta} - \frac{1}{R} \right] = 0 \tag{j}$$

where α, β, γ, and J are defined above by Equations c–f, respectively. $P(\xi, \eta) = \nabla^2\xi$ and $Q(\xi, \eta) = \nabla^2\eta$ are the grid control functions given, for example, by equations (11.39) and (11.40).

2. Laplacian and Grid-Generation Equations Transformed from Physical Domains in Polar Coordinates

The Laplacian in two-dimensional polar coordinates (R,θ) is given by:

$$\nabla^2 T = \frac{\partial^2 T}{\partial R^2} + \frac{2}{R}\frac{\partial T}{\partial R} + \frac{1}{R^2}\frac{\partial^2 T}{\partial \theta^2} \tag{k}$$

Consider that the Laplacian given by equation (k) applies to an irregular geometry in the physical domain shown by Figure 11.28a. Such geometry in the physical domain is transformed into a regular geometry in the computational domain illustrated by Figure 11.28b. The Laplacian in the coordinates (ξ,η) in the computational domain can be obtained by directly applying the chain rule to the derivatives in equation (k). The following non-conservative form results:

$$\nabla^2 T = \frac{1}{J^2}[\alpha T_{\xi\xi} - 2\beta T_{\xi\eta} + \gamma T_{\eta\eta}] + [(\nabla^2\xi)T_\xi + (\nabla^2\eta)T_\eta] \tag{l}$$

where

$$\alpha = \frac{1}{R^2}\left(\frac{\partial R}{\partial \eta}\right)^2 + \left(\frac{\partial \theta}{\partial \eta}\right)^2 \tag{m}$$

$$\beta = \frac{1}{R^2}\frac{\partial R}{\partial \xi}\frac{\partial R}{\partial \eta} + \frac{\partial \theta}{\partial \xi}\frac{\partial \theta}{\partial \eta} \tag{n}$$

$$\gamma = \frac{1}{R^2}\left(\frac{\partial R}{\partial \xi}\right)^2 + \left(\frac{\partial \theta}{\partial \xi}\right)^2 \tag{o}$$

$$J = \frac{\partial R}{\partial \xi}\frac{\partial \theta}{\partial \eta} - \frac{\partial R}{\partial \eta}\frac{\partial \theta}{\partial \xi} \tag{p}$$

$$\nabla^2\xi = \frac{\partial^2 \xi}{\partial R^2} + \frac{2}{R}\frac{\partial \xi}{\partial R} + \frac{1}{R^2}\frac{\partial^2 \xi}{\partial \theta^2} \tag{q}$$

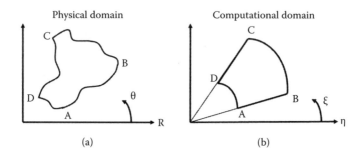

FIGURE 11.28
Transformation in polar coordinates: (a) Physical domain; (b) Computational domain.

$$\nabla^2 \eta = \frac{\partial^2 \eta}{\partial R^2} + \frac{2}{R}\frac{\partial \eta}{\partial R} + \frac{1}{R^2}\frac{\partial^2 \eta}{\partial \theta^2} \qquad (r)$$

The Laplacian in nonconservative form given by equation (l) can be written in conservative form by appropriate manipulation of the partial derivatives.

The equations for numerical elliptic grid generation in polar coordinates are given by

$$\alpha \frac{\partial^2 \theta}{\partial \xi^2} - 2\beta \frac{\partial^2 \theta}{\partial \xi \partial \eta} + \gamma \frac{\partial^2 \theta}{\partial \eta^2} + J^2 \left[P(\xi, \eta) \frac{\partial \theta}{\partial \xi} + Q(\xi, \eta) \frac{\partial \theta}{\partial \eta} \right] = 0 \qquad (s)$$

$$\alpha \frac{\partial^2 R}{\partial \xi^2} - 2\beta \frac{\partial^2 R}{\partial \xi \partial \eta} + \gamma \frac{\partial^2 R}{\partial \eta^2} + J^2 \left[P(\xi, \eta) \frac{\partial R}{\partial \xi} + Q(\xi, \eta) \frac{\partial R}{\partial \eta} - \frac{1}{R} \right] = 0 \qquad (t)$$

where α, β, γ, and J are defined above by equations (m–p), respectively. $P(\xi, \eta) = \nabla^2 \xi$ and $Q(\xi, \eta) = \nabla^2 \eta$ are the grid control functions given, for example, by equations (11.39) and (11.40).

12

Hybrid Numerical–Analytical Solutions

After the computer boom in the 1960s, and the consequent and necessary advancements on numerical analysis and algorithms, purely discrete methods for the solution of partial differential equations (PDEs) became dominant over analytical approaches, such as the finite differences approach studied here. The classical analytical methods for solving PDEs were mostly developed during the 19th century and the first half of the 20th century and were, in general, limited in applicability to certain classes of linear problems. Naturally, after the numerical methods reached a wide enough public, such analytical approaches became less popular and were mostly employed in verification of numerical codes and/or in the solution of simplified models.

Nevertheless, as the progress on discrete methods led to a reasonable maturity level, the practical limitations of this class of methods also became more evident, especially in light of the remarkable increase in computational cost for nonlinear, multidimensional, transient situations, and on the associated difficulties in automatically controlling relative errors in such computations. Motivated not only by the continuous search for more complete benchmark solutions but also by the perception of the opportunity for new developments on more robust, accurate, and cost-effective approaches, a number of hybrid numerical–analytical techniques have been proposed in the literature, particularly since the 1980s. With the use of hybrid methods of solution, the powers of numerical and analytical approaches have been combined, and new possibilities have been opened for solving PDEs in the physical sciences and engineering, including the heat and fluid flow problems in the present context.

One such hybrid approach is well-known as the generalized integral transform technique (GITT) (Cotta 1990, 1993, 1994a, 1994b, 1998; Cotta and Serfaty 1992; Cotta and Mikhailov 1997, 2006; Cotta et al. 2016a, 2016c), which combines the analytical ideas behind the Classical Integral Transform Method (Özişik 1968, 1993; Mikhailov and Özişik 1984b), and the use of numerical techniques for solving systems of ordinary differential equations (ODEs) or PDEs, such as the finite difference method. Integral transforms have been successfully used in different branches of the physical, mathematical, and engineering sciences for about 200 years. Its introduction might be attributed to Fourier, after the publication of his treatise on the analytical theory of heat (Fourier 1822). Fourier at that time advanced the idea of separation of variables, so as to handle and interpret the solutions of the newly derived heat conduction equation, after proposing the constitutive equation known nowadays as Fourier's law. According to Luikov (1968, 1980), it was not until the work of

Koshlyakov (1936) that the path was provided in handling diffusion problems with nonhomogeneous equations and boundary conditions by the integral transforms method, overcoming part of the limitations of the original work of Fourier. The theory of such integral transforms was also further developed by Grinberg (1948). Further limitations on the classical analytical approach were later on independently discussed by Özişik and Murray (1974) and Mikhailov (1975) in attempts of approximately handling linear diffusion problems with time-variable coefficients, respectively, at the boundary conditions and at the diffusion equation. These early works on approximate analytical solutions offered the clue to the advancement of the hybrid numerical–analytical method, first described as a complete solution of the coupled transformed system in Cotta (1986), soon extended to the complex domain (Cotta and Özişik 1986a), and afterward to nonlinear diffusion problems (Cotta 1990; Serfaty and Cotta 1990). The transformed ordinary differential systems typical of such integral transformations are likely to present significant stiffness, especially for larger truncation orders, but by the 1980s, reliable solvers for stiff initial value problems were already widely available, allowing for error-controlled solutions of the transformed potentials. Only then, would the coined GITT become a full-hybrid numerical–analytical solution of non-transformable diffusion or convection–diffusion problems.

A few years later (Cotta and Gerk 1994), the opportunity of further combining analytical and numerical techniques was explored, by introducing a partial transformation scheme, as opposed to the total transformation above discussed, which appeared to be particularly convenient for convection–diffusion problems with a preferential convective direction. It was then perceived that the integral transformation process could be already quite advantageous to the PDE solution, in eliminating one or more spatial variables, while yielding a system of transformed PDEs to be numerically solved, after removing those coordinates more closely associated with the diffusion effects. This hybrid partial transformation procedure not only modifies stability criteria and robustness with respect to a purely numerical scheme applied to the original problem but also enhances accuracy and reduces computational cost, offering an interesting alternative to purely discrete approaches and to the total transformation hybrid solution as well. In Cotta and Gerk (1994), a second-order accurate modified upwind scheme (Cotta et al. 1986) was employed in the numerical solution of the transformed one-dimensional partial differential system, but variable order and variable step size finite difference schemes with automatic error control, available in different subroutines libraries, were also considered in various applications that followed (Guedes and Özişik 1994a, 1994b; Cotta 1996; Santos et al. 1998; Kakaç et al. 2001; Cotta et al. 2003, 2005; Castellões and Cotta 2006; Castellões et al. 2007; Knupp et al. 2013a, 2013b, 2014, 2015a, 2015b).

In recent years, besides the continuous search for more challenging problems and different application areas, emphasis has also been placed on unifying and simplifying the use of the GITT. Hybrid methods become even more powerful and applicable when symbolic manipulation systems, which were also widely disseminated along the last three decades, are employed. The effort to integrate the knowledge on the GITT application into a symbolic-numerical algorithm resulted in the so-called UNIT code (UNified Integral Transforms) (Cotta et al. 2010, 2013, 2014; Sphaier et al. 2011) built on the *Mathematica* symbolic-numerical platform (Wolfram 2015). The open-source UNIT algorithm is intended to bridge the gap between simple problems that allow for a straightforward analytical solution, and more complex and involved situations that almost unavoidably require purely discrete approaches and specialized software systems. It is thus an implementation and development platform for researchers and engineers interested in integral transform solutions of diffusion and convection–diffusion problems. Both the total and partial transformation schemes have been implemented in the automatic UNIT algorithm (Cotta et al. 2014), in combination with the built-in routine for finite difference solution of either ODEs or PDEs, NDSolve, readily available in the *Mathematica* system (Wolfram 2015).

This chapter first provides an introduction to the combined use of finite differences and integral transforms in the hybrid numerical–analytical solution of convection–diffusion problems by considering an example taken from transient forced convection in channels (Guedes and Özişik 1994a, 1994b). In fact, the solution of a class of transient forced convection problems with a dominant convective direction, as reviewed in Kakaç et al. (2001) and Cotta et al. (2003, 2005), was the motivation for the proposition of this hybrid approach (Cotta and Gerk 1994). In the sections that follow, a more advanced and unified view of this modern hybrid approach (GITT) is presented, first describing the formal solution through both the total and partial transformation schemes. Afterward, an overview of the computational algorithm is provided based on the UNIT algorithm (Cotta et al. 2013, 2014). In the sequence, test cases with Burgers-type equations are illustrated for inspection of convergence characteristics. Special emphasis is given to an example that involves the partial transformation scheme, in light of the crucial role played by the finite difference method in such cases. Finally, an alternative integral transform solution for this same class of transient convection–diffusion problems is briefly considered, through which the convective effects can be fully or partially incorporated into the eigenfunction expansion basis, by obtaining a generalized diffusive formulation via an algebraic transformation in the original problem formulation coefficients.

12.1 Combining Finite Differences and Integral Transforms

In this section, we introduce the hybrid approach combining the GITT and finite differences. An application is illustrated, chosen from the area of transient forced convection in laminar flow inside channels. The numerical part of this approach utilizes the material already covered in the previous chapters, and hence requires no additional background; however, the analytical part of the method requires some background in the theory and application of the integral transform technique. The reader may consult the books by Özişik (1968, 1993) and Mikhailov and Özişik (1984) for the theory and application of the classical integral transform technique, and the books by Cotta (1993, 1998), Cotta and Mikhailov (1997), Santos et al. (2001), and Cotta et al. (2016a) for the theory and applications on the GITT.

12.1.1 The Hybrid Approach

The reason for choosing GITT for the analytic part of the problem lies in the fact that the integral transform technique provides a unified and systematic approach for removing the spatial operators from the PDEs. In addition, the inversion formula is available at the onset of the problem. Thus, considering a convection–diffusion problem formulation for the transient multidimensional potential, $T(x,y,z,t)$, the basic steps and salient features of the hybrid method of solution, utilizing GITT for the analytic part of the solution and finite differences for the numerical part, are summarized below:

1. **The Auxiliary Problem:** Choose a related auxiliary problem of the Sturm–Liouville type in the space variables to be eliminated by integral transformation (x,y for instance). This may be the natural eigenvalue problem for the system if the separation of variables is possible or the most appropriate eigenvalue problem if the natural eigenvalue problem cannot be obtained by the classical separation of variables.

2. **The Integral Transform Pair:** By utilizing the auxiliary problem chosen in Step 1, construct the integral transform pair (i.e., the integral transform and the inversion formula) in the eliminated space variables (x,y).

3. **Taking the Integral Transform:** Apply the integral transformation and remove the space partial derivatives from the PDE with respect to the x and y variables. During this procedure, the potential $T(x,y,z,t)$ is converted into the transformed potentials $\overline{T}(\mu_i,z,t)$ after replacing it with the inversion formula. The problem is thus transformed to an infinite system of first-order PDEs for the transformed potentials $\overline{T}(\mu_i,z,t)$. Eventually, a reliable two-dimensional numerical solver might be available, and the user might be interested in promoting one single integral transformation, thus eliminating only one spatial variable and yielding a two-dimensional transformed system of PDEs, also within the scope of the present hybrid approach.

4. **Truncation:** Truncate the infinite system obtained in Step 3 to a sufficiently large number of N coupled PDEs for the transformed potentials, $\overline{T}(\mu_i,z,t)$.

5. **Finite Difference Solution:** Solve the system of N PDEs for the transformed potentials, $\overline{T}(\mu_i,z,t)$, via finite differences.

6. **Inversion:** Invert the solution for the transformed potentials, $\overline{T}(\mu_i,z,t)$, with respect to the x and y variables using the inversion formula developed in Step 2, and recover the potential distribution T(x,y,z,t).

In some situations, under a suitable choice of the auxiliary problem, the coupling terms which are the nondiagonal elements in the coefficients matrix for the $\overline{T}(\mu_i,z,t)$ equations might be considered negligible in comparison with the diagonal elements of the coefficients matrix. In such a case, the nondiagonal terms of the coefficients matrix may be neglected, and the equations for the transformed potentials become uncoupled, offering a fairly simplified treatment of the coupled system (Cotta and Özişik 1986b, Cotta and Özişik 1987). Having established the formalism of the hybrid method, we now proceed to illustrating the details of this procedure with a specific example, taken from an application in transient forced convection in channels.

12.1.2 Hybrid Approach Application: Transient Forced Convection in Channels

Consider transient forced convection in thermally developing, hydrodynamically developed laminar flow between two parallel plates, subjected to a sinusoidally varying inlet temperature. We assume constant physical properties, neglect viscous dissipation, free convection, and axial conduction in the fluid. The mathematical formulation of this problem, in dimensionless form, is given by (Guedes and Özişik 1994a):

$$\frac{\partial \Theta(R,Z,\tau)}{\partial \tau} + W(R)\frac{\partial \Theta(R,Z,\tau)}{\partial Z} = \frac{\partial^2 \Theta(R,Z,\tau)}{\partial R^2}, \quad \text{in } 0 < R < 1,\ Z > 0,\ \tau > 0 \tag{12.1a}$$

$$\Theta(R,Z,0) = 0, \quad 0 \le R \le 1,\ Z \ge 0 \tag{12.1b}$$

$$\Theta(R,0,\tau) = \sin(\Omega\tau), \quad 0 \le R \le 1,\ \tau > 0 \tag{12.1c}$$

$$\frac{\partial \Theta(0,Z,\tau)}{\partial R} = 0, \quad Z > 0,\ \tau > 0 \tag{12.1d}$$

$$\theta(1,Z,\tau) = 0, \quad Z > 0,\ \tau > 0 \tag{12.1e}$$

where various quantities are defined as:

D_h	hydraulic diameter ($= 4r_w$), m
r_w	half the spacing between parallel plates, m
r	normal coordinate, m
R	dimensionless normal coordinate ($= r/r_w$)
t	time, s
T_i	initial temperature, °C
T(r,z,t)	fluid temperature, °C
ΔT	reference temperature difference, °C
u_m	mean flow velocity, m/s
$u(r) = \frac{3}{2} u_m [1 - (r/r_w)^2]$,	flow velocity distribution, m/s
W(R)	dimensionless flow velocity ($= u(r)/16\, u_m$)
λ	discretization parameter ($= \Delta\tau/\Delta Z$)
z	axial coordinate, m
Z	dimensionless axial coordinate ($= \alpha z/u_m D_h^2$)
α	thermal diffusivity of the fluid, m²/s
$\theta(R,Z,\tau)$	dimensionless temperature ($= [T(r,z,t) - T_i]/\Delta T$)
γ	Courant number
τ	dimensionless time ($= \alpha t/r_w^2$)
Ω	dimensionless frequency of oscillations ($= \omega r_w^2/\alpha$)
ω	frequency of oscillations, Hz

To solve this problem with the hybrid approach, we follow the formalism described above and detailed below:

1. **The Auxiliary Problem:** The complete separation of the governing PDE being impossible, we choose the following closely related Sturm–Liouville type eigenvalue problem as the auxiliary problem:

$$\frac{d^2\psi(\mu_i, R)}{dR^2} + \mu_i^2 \psi(\mu_i, R) = 0, \quad 0 < R < 1 \tag{12.2a}$$

$$\frac{d\psi(\mu_i, R)}{dR} = 0, \quad R = 0 \tag{12.2b}$$

$$\psi(\mu_i, R) = 0, \quad R = 1 \tag{12.2c}$$

which is readily solved to yield

$$\psi(\mu_i, R) = \cos \mu_i R, \quad i = 1, 2, \ldots \tag{12.3a}$$

$$\mu_i = \frac{(2i-1)}{2}\pi, \quad i = 1, 2, \ldots \tag{12.3b}$$

and the normalization integral N_i is determined from its definition

$$N_i = \int_{R=0}^{1} \psi^2(\mu_i, R)\,dR \tag{12.3c}$$

The integral in equation (12.3c) can readily be performed, but for generality we prefer to carry out the analysis by using the symbol N_i in order to illustrate the role of the normalization integral in the analysis.

2. **The Integral Transform Pair:** By utilizing the eigenfunctions, $\psi(\mu_i, R)$, the integral transform pair in the R variable for the temperature function $\theta(R, Z, \tau)$ is determined by the procedure described Cotta (1993) as

$$\Theta(R, Z, \tau) = \sum_{i=1}^{\infty} \frac{\psi(\mu_i, R)}{N_i^{1/2}} \overline{\Theta}_i(Z, \tau), \quad \text{inversion} \tag{12.4a}$$

$$\overline{\Theta}_i(Z, \tau) = \int_{R=0}^{1} \frac{\psi(\mu_i, R)}{N_i^{1/2}} \Theta(R, Z, \tau)\,dR, \quad \text{transform} \tag{12.4b}$$

The symmetric form of the kernel is preferred here by splitting the normalization integral as $N^{1/2}$, whereas in Özişik (1993), in the context of the classical integral transforms method, the norm appears only in the inversion formula.

3. **Taking the Integral Transform:** We now take the integral transform of equation (12.1a) by the application of the integral transform defined in equation (12.4b). That is, we operate on equation (12.1a) with the operator

$$\int_{R=0}^{1} \frac{\psi(\mu_i, R)}{N_i^{1/2}}\,dR$$

and obtain

$$\frac{\partial \overline{\Theta}_i(Z, \tau)}{\partial \tau} + \int_{R=0}^{1} W(R)\frac{\psi(\mu_i, R)}{N_i^{1/2}}\frac{\partial \Theta(R, Z, \tau)}{\partial Z}\,dR + \mu_i^2 \overline{\Theta}_i(Z, \tau) = 0 \tag{12.5}$$

where the bar denotes the transform with respect to the R variable. To obtain equation (12.5), we performed integration by parts twice in the term containing $\partial^2\Theta/\partial R^2$ and utilized the boundary conditions [equations (12.1d) and (12.1e)]. We note that every $\Theta(R,Z,\tau)$ in equation (12.1a) could not be transformed; as a result, equation (12.5) still contains $\Theta(R,Z,\tau)$. However, $\Theta(R,Z,\tau)$ can be expressed in terms of $\overline{\Theta}_i(Z,\tau)$ by replacing $\Theta(R,Z,\tau)$ with its equivalent inversion formula [equation (12.4a)]. Since the inversion is in the form of an infinite series, equation (12.5) transforms into an infinite system of first-order PDEs for the transformed temperature $\overline{\Theta}_i(Z,\tau)$. We obtain:

$$\frac{\partial \overline{\Theta}_i(Z,\tau)}{\partial \tau} + \sum_{k=1}^{\infty} A_{ik} \frac{\partial \overline{\Theta}_k(Z,\tau)}{\partial Z} + \mu_i^2\, \overline{\Theta}_i(Z,\tau) = 0,$$
$$i = 1, 2, \ldots,\quad Z > 0,\quad \tau > 0 \tag{12.6a}$$

where

$$A_{ik} = A_{ki} = \frac{1}{N_i^{1/2} N_k^{1/2}} \int_{R=0}^{1} W(R)\psi(\mu_i, R)\psi(\mu_k, R)\, dR \tag{12.6b}$$

The boundary condition at $Z = 0$ and the initial condition for $\tau = 0$, needed for the solution of the system of equations (12.6a–d), are obtained by taking the integral transform of equations (12.1c) and (12.1b), respectively, to yield

$$\overline{\theta}_i(Z, 0) = 0 \quad \tau = 0,\ Z > 0 \tag{12.6c}$$

$$\overline{\theta}_i(0, \tau) = \frac{1}{N_i^{1/2}} \int_{R=0}^{1} \sin(\Omega \tau)\psi(\mu_i, R)\, dR \equiv \overline{f}_i, \tag{12.6d}$$
$$i = 1, 2, \ldots \quad \text{for } Z = 0,\ \tau > 0$$

The chosen eigenvalue problem and the associated integral transform pair allow for the exact integral transformation of both the diffusion and transient terms in equation (12.1a), while the convective term introduces the coupling between the transformed potentials, due to the R-variable velocity coefficient, $W(R)$, as in equation (12.6a). Alternatively, an eigenvalue problem could have been chosen that accounts for the velocity coefficient in the weighting function, which would allow for the exact integral transformation of the convective term. However, in this case, the transient term would lead to the coupling among the transformed potentials. These two possibilities have been critically compared by Castellões et al. (2006, 2007).

4. **Truncation:** The infinite system of hyperbolic equations (12.6a–d) is truncated to a sufficiently large number of terms, N.
5. **Solution:** The system of N first-order hyperbolic PDEs (12.6a–d) is solved with finite differences by using a second-order accurate explicit finite difference scheme (Cotta et al. 1986; Cotta and Gerk 1994), based on an extension of the Warming and Beam (1975, 1976) upwind scheme, given by

Predictor:

$$\overline{\Theta}_{i,j}^{\overline{n+1}} = \overline{\Theta}_{i,j}^n - \lambda \sum_{k=1}^{N} A_{ik}(\overline{\Theta}_{k,j}^n - \overline{\Theta}_{k,j-1}^n) - \mu_i^2 \Delta\tau \, \overline{\Theta}_{i,j}^n \qquad (12.7a)$$

Corrector:

$$\overline{\Theta}_{i,j}^{n+1} = \frac{1}{2}\left[\overline{\Theta}_{i,j}^n + \overline{\Theta}_{i,j}^{\overline{n+1}} - \lambda \sum_{k=1}^{N} A_{ik}\left(\overline{\Theta}_{k,j}^{\overline{n+1}} - \overline{\Theta}_{k,j-1}^{\overline{n+1}}\right) \right.$$
$$\left. - \lambda \sum_{k=1}^{N} A_{ik}\left(\overline{\Theta}_{k,j}^n - 2\overline{\Theta}_{k,j-1}^n + \overline{\Theta}_{k,j-2}^n\right) - \mu_i^2 \Delta\tau \overline{\Theta}_{i,j}^{\overline{n+1}}\right] \qquad (12.7b)$$

where

$$\lambda = \frac{\Delta\tau}{\Delta Z} \quad \text{and} \quad \overline{\Theta}_{i,j}^n \equiv \overline{\Theta}(n\,\Delta\tau,\ j\,\Delta Z) \qquad (12.7c,d)$$

and the superscript $\overline{n+1}$ denotes evaluation at an intermediate time. The stability analysis of this system, by using Fourier stability analysis, leads to the following stability criterion

$$0 \le \gamma_i \le 2 \quad i = 1,2,\ldots,N \qquad (12.8)$$

where $\gamma_i = c_i \Delta\tau/\Delta Z$ is the Courant number, and the c_i are the eigenvalues of the matrix defined in the appendix of the article by Guedes and Özişik (1994a).

To solve this system, one needs to march along Z and τ up to a final axial position and time. Then, the unknown temperature $\Theta(R, Z, \tau)$ is recovered at any desired normal position R, for the axial positions Z and time τ specified in the numerical solution, by the application of the inversion formula [equation (12.4a)].

6. **Inversion:** The transforms $\overline{\Theta}_i(Z, \tau)$ are inverted by the inversion formula [equation (12.4a)] to recover the solution for the temperature field, $\Theta(R, Z, \tau)$.

Knowing the dimensionless temperature profile $\Theta(R, Z, \tau)$, the fluid bulk temperature $\Theta_b(Z, \tau)$ can be evaluated from its definition

$$\Theta_b(Z,\tau) = 16 \int_0^1 W(R)\Theta(R,Z,\tau)\, dR \qquad (12.9a)$$

An alternative form of this expression, which is more convenient for computational purposes, is obtained by replacing $\Theta(R,Z,\tau)$ with its equivalent inversion formula [equation (12.4a)] as

$$\Theta_b(Z,\tau) = 16 \sum_{i=1}^{N} \frac{\bar{g}_i}{N_i^{1/2}} \overline{\Theta}_i(Z,\tau) \qquad (12.9b)$$

where

$$\bar{g}_i = \int_{R=0}^{1} W(R)\psi(\mu_i, R)\, dR \qquad (12.9c)$$

Since the transforms $\overline{\Theta}_i(Z,\tau)$ are available from the solution of the finite difference equations (12.7a–d), the bulk temperature $\Theta_b(Z,\tau)$ is readily evaluated from equation (12.9b).

Also of interest is the evaluation of the dimensionless heat flux at the outer boundary, $\partial\Theta(1,Z,\tau)/\partial R$, for which an explicit analytical expression is readily obtained according to the inversion formula [equation (12.4a)] as

$$\frac{\partial\Theta(1,Z,\tau)}{\partial R} = \sum_{i=1}^{N} \frac{1}{N_i^{1/2}} \frac{d\psi(\mu_i,1)}{dR} \overline{\Theta}_i(Z,\tau) \qquad (12.10)$$

We now present numerical results for the dimensionless fluid bulk temperature and wall heat flux, and compare the present hybrid method of solution with purely numerical methods reported in the literature. Table 12.1 shows a comparison of fluid bulk temperature distributions given by the present hybrid approach and the purely numerical solution by Hatay et al. (1991). The agreement between the purely numerical and hybrid approaches is very good, the maximum deviation being 2.6% for large times near the inlet. However, the limitation to the accuracy of the purely numerical solution technique for the calculation of heat flux, resulting from the stability restriction imposed on the size of normal steps, should be recognized. Since the evaluation of heat flux distributions depends on numerical computation of derivatives with respect to the normal variable, the normal mesh size, ΔR, has to be small enough to ensure accuracy, especially near the channel inlet where gradients are steeper.

TABLE 12.1

Comparison of the Hybrid Solution θ_{hyb} with the Purely Numerical Finite Difference Solution θ_{num} (Hatay et al. 1991) for $\Omega = 0.05$ and 0.50 ($\gamma = 0.625$, $\eta = 0.18$)

τ	Z	5×10^{-3}	1×10^{-2}	2×10^{-2}	3×10^{-2}	5×10^{-2}	1×10^{-1}
				$\Omega = 0.05$			
1	Θ_{hyb}	0.0397	0.0338	0.0249	0.0184	0.0101	0
	Θ_{num}	0.0399	0.0340	0.0251	0.0185	0.0101	0
5	Θ_{hyb}	0.1986	0.1689	0.1247	0.0922	0.0504	0.0112
	Θ_{num}	0.2043	0.1731	0.1273	0.0943	0.0516	0.0114
10	Θ_{hyb}	0.3850	0.3275	0.2416	0.1787	0.0977	0.0216
	Θ_{num}	0.3950	0.3345	0.2466	0.1824	0.0998	0.0221
				$\Omega = 0.5$			
τ	Z	5×10^{-3}	1×10^{-2}	2×10^{-2}	3×10^{-2}	5×10^{-2}	1×10^{-1}
1	Θ_{hyb}	0.3810	0.3244	0.2393	0.1770	0.0968	0
	Θ_{num}	0.3827	0.3257	0.2403	0.1777	0.0972	0
5	Θ_{hyb}	0.4805	0.4087	0.3014	0.2229	0.1219	0.0270
	Θ_{num}	0.4931	0.4175	0.3079	0.2277	0.1246	0.0276
10	Θ_{hyb}	−0.7701	−0.6550	−04831	−0.3573	−0.1954	−0.0432
	Θ_{num}	−0.7901	−0.6690	−0.4933	−0.3649	−0.1996	−0.0442

On the contrary, with the hybrid approach, heat flux distributions at any specified location within the duct cross section can be evaluated *a posteriori* using the explicit analytic expression given by the derivative of the inversion formula [equation (12.4a)].

For the situation shown in Table 12.1, the execution times for both schemes (hybrid and numerical) are practically the same; however, the execution time can be further reduced and dispersive errors virtually eliminated with the hybrid scheme by increasing the Courant number toward unity. Such flexibility is limited when dealing with the purely numerical approach because an increase in the Courant number requires a decrease in the value of the discretization parameter in the radial direction, $\eta = \Delta\tau/(\Delta R)^2$, in order to satisfy the more restrictive stability criterion. Since ΔR has to remain small, a decrease in the parameter η requires a decrease in $\Delta\tau$ as well. The stability of the present hybrid approach depends only on the value of the Courant number, whereas in the purely numerical scheme, two different stability criteria need to be satisfied. The present results with the hybrid method were fully converged to the digits presented with a maximum of N = 15 terms in the series, and the Courant number value was maintained as 0.93.

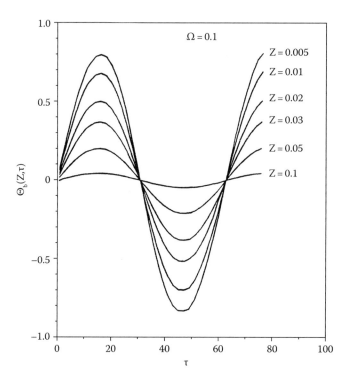

FIGURE 12.1
Variation of fluid bulk temperature with time inside parallel-plate duct. (From Guedes, R.O.C. & Özişik, M.N., *Int. J. Heat Fluid Flow*, 15(2), 116–121, 1994a.)

Figure 12.1 shows the variation of the bulk temperature as a function of time at different axial positions for $\Omega = 0.1$, for sinusoidal oscillation of inlet temperature. As expected, the amplitude becomes smaller with increasing distance from the inlet.

Figure 12.2 shows the variation of the heat flux at the outer boundary of the channel as a function of time at different axial positions for the frequency $\Omega = 0.1$. A comparison of the results in Figures 12.1 and 12.2 reveals that when the fluid bulk temperature is greater than the wall temperature, the heat flux is negative, which implies that the heat flows from the fluid to the wall and vice-versa. Hatay et al.'s (1991) solution presented no heat flux results to allow for comparisons.

The hybrid numerical–analytical method has advantages over conventional purely numerical approaches, in that the method is less restricted by stability considerations, thus allowing more flexibility to improve the accuracy of computations. The stability analysis of the combined integral transforms-finite difference approach, in the

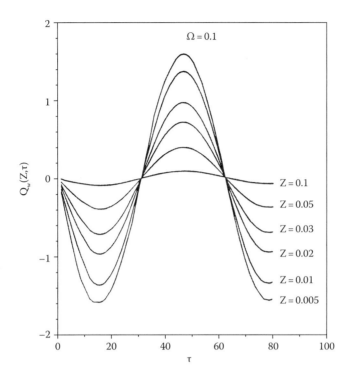

FIGURE 12.2
Variation of wall heat flux with time inside parallel-plate duct. (From Guedes, R.O.C. & Özişik, M.N., *Int. J. Heat Fluid Flow*, 15(2), 116–121, 1994a.)

realm of transient convection problems, has been discussed by (Cotta and Gerk 1994), in comparison with the fully discrete modified upwind method for the same problem (Cotta et al. 1986). Also, the heat flux anywhere in the medium can be computed *a posteriori* by using analytical expressions.

12.2 Unified Integral Transforms

After introducing the hybrid approach concept and illustrating its use, step by step, through an application, it is now possible to provide a unified view of the methodology, as applied to a fairly general class of diffusion and convection–diffusion problems. The total integral transformation scheme is first described when all the independent variables of the original formulation are eliminated, except one, normally the time variable in transient problems, yielding a transformed system of ODEs to be numerically solved. Then, the partial integral transformation scheme, more relevant in the present context, is generalized by retaining an additional variable without integral transformation, usually the space variable along which the convective effects predominate.

Thus, a general transient convection–diffusion problem of n coupled potentials is considered, defined in the region V, with boundary surface S, and position vector **r** (Cotta et al. 2014):

$$w_k(\mathbf{r})L_{k,t}T_k(\mathbf{r},t) = G_k(\mathbf{r},t,\mathbf{T}), \quad \mathbf{r} \in V, \ t > 0, \ k = 1,2,\ldots,n \quad (12.11a)$$

where the t operator, $L_{k,t}$, for a parabolic or parabolic–hyperbolic formulation may be given by

$$L_{k,t} \equiv \frac{\partial}{\partial t} \quad (12.11b)$$

while for an elliptic formulation, it is written as

$$L_{k,t} \equiv -a_k(t)\frac{\partial}{\partial t}\left[b_k(t)\frac{\partial}{\partial t}\right] \quad (12.11c)$$

and

$$G_k(\mathbf{r},t,\mathbf{T}) = \nabla \cdot [K_k(\mathbf{r})\nabla T_k(\mathbf{r},t)] - d_k(\mathbf{r})T_k(\mathbf{r},t) - \mathbf{u}(\mathbf{r},t,\mathbf{T}) \cdot \nabla T_k(\mathbf{r},t) + g_k(\mathbf{r},t,\mathbf{T}) \quad (12.11d)$$

with initial or boundary conditions in the t variable given, respectively, by

$$T_k(\mathbf{r},0) = f_k(\mathbf{r}), \quad \mathbf{r} \in V, \text{ for the initial value problem} \quad (12.11e)$$

or

$$\left[\lambda_{k,l}(t) + (-1)^{l+1}\gamma_{k,l}(t)\frac{\partial}{\partial t}\right]T_k(\mathbf{r},t) = f_{k,l}(\mathbf{r}), \quad \text{at } t = t_l, \ l = 0,1, \ \mathbf{r} \in V,$$

for the boundary value problem $\quad (12.11f)$

and the boundary conditions in the remaining coordinates

$$\left[\alpha_k(\mathbf{r}) + \beta_k(\mathbf{r})K_k(\mathbf{r})\frac{\partial}{\partial \mathbf{n}}\right]T_k(\mathbf{r},t) = \phi_k(\mathbf{r},t,\mathbf{T}), \quad \mathbf{r} \in S, \ t > 0 \quad (12.11g)$$

where **n** denotes the outward-drawn normal to the surface S, and the potential vector is given by

$$\mathbf{T} = \{T_1, T_2, \ldots, T_k, \ldots, T_n\} \quad (12.11h)$$

Equations (12.11a–h) are quite general since nonlinear and convection terms may be grouped into the equations and boundary condition source terms. It may be highlighted that in the case of decoupled linear source terms, that is, $g \equiv g(\mathbf{r},t)$ and $\phi \equiv \phi(\mathbf{r},t)$, this example is reduced to a class I linear diffusion problem for each potential, according to the classification by Mikhailov and Özişik (1984b), and exact analytical solutions are readily available via the classical integral transform technique. Otherwise, this problem shall not be *a priori* transformable, except for a few linear coupled situations also illustrated in Mikhailov and Özişik (1984). However, the formal solution procedure provided by the GITT (Cotta 1990, 1993) may be invoked in order to provide hybrid numerical–analytical solutions for nontransformable problems.

As mentioned in the introduction to the chapter, this approach has been recently consolidated into an algorithm via symbolic-numerical computation, known as the UNIT code (Cotta et al. 2010, 2013, 2014; Sphaier et al. 2010), and built on the *Mathematica* symbolic-numerical platform (Wolfram 2015). The formal solution regarding the standard procedure of the UNIT algorithm is known as the total transformation scheme, described in Cotta et al. (2010, 2013), in which all spatial variables are integral transformed. Here, we specially highlight the partial integral transformation scheme option of the UNIT algorithm as an alternative solution path to problems with a predominant convective direction, which is not eliminated through integral transformation but kept within the transformed system.

12.2.1 Total Transformation

Following the formal solution procedure for nonlinear convection–diffusion problems through integral transforms, the proposition of eigenfunction expansions for the associated potentials is first required. The preferred eigenvalue problem choice appears from the direct application of the separation of variables methodology to the linear, homogeneous, purely diffusive version of the proposed problem. Thus, the recommended set of decoupled auxiliary problems is given here by:

$$\nabla \cdot [K_k(\mathbf{r})\nabla \psi_{ki}(\mathbf{r})] + [\mu_{ki}^2 w_k(\mathbf{r}) - d_k(\mathbf{r})]\psi_{ki}(\mathbf{r}) = 0, \quad \mathbf{r} \in V \quad (12.12a)$$

$$\left[\alpha_k(\mathbf{r}) + \beta_k(\mathbf{r}) K_k(\mathbf{r}) \frac{\partial}{\partial \mathbf{n}}\right] \psi_{ki}(\mathbf{r}) = 0, \quad \mathbf{r} \in S \quad (12.12b)$$

where the eigenvalues, μ_{ki}, and associated eigenfunctions, $\psi_{ki}(\mathbf{r})$, are assumed to be known from exact analytical expressions, for instance, obtained through symbolic computation (Wolfram 2015) or application of the GITT itself (Cotta 1993; Sphaier and Cotta 2000; Naveira-Cotta et al. 2009). One should notice that equations (12.11a–h) are presented in such a form which already reflects this choice of eigenvalue problems, given by equations (12.12a,b), with the adoption of linear coefficients in both the equations and boundary conditions, and incorporating the remaining terms (coupling, nonlinear, and convective terms) into the source terms, without loss of generality.

Making use of the orthogonality property of the eigenfunctions, one defines the following integral transform pairs:

$$\overline{T}_{ki}(t) = \int_V w_k(\mathbf{r})\tilde{\psi}_{ki}(\mathbf{r})T_k(\mathbf{r},t)dv, \quad \text{transforms} \qquad (12.13a)$$

$$T_k(\mathbf{r},t) = \sum_{i=1}^{\infty}\tilde{\psi}_{ki}(\mathbf{r})\overline{T}_{k,i}(t), \quad \text{inverses} \qquad (12.13b)$$

where the symmetric kernels $\tilde{\psi}_{ki}(\mathbf{r})$ are given by

$$\tilde{\psi}_{ki}(\mathbf{r}) = \frac{\psi_{ki}(\mathbf{r})}{\sqrt{N_{ki}}}; \quad N_{ki} = \int_V w_k(\mathbf{r})\psi_{ki}^2(\mathbf{r})dv \qquad (12.13c,d)$$

with N_{ki} being the normalization integral.

The integral transformation of equation (12.11a) is accomplished by applying the operator $\int_V \tilde{\psi}_{ki}(\mathbf{r})(\cdot)dv$ and making use of the boundary conditions given by equations (12.11g) and (12.12b), yielding:

$$L_{k,t}\overline{T}_{ki}(t) + \mu_{ki}^2\overline{T}_{ki}(t) = \overline{g}_{ki}(t,\overline{T}) + \overline{b}_{ki}(t,\overline{T}), \quad i=1,2,..., \quad t>0, \quad k=1,2,...,n \qquad (12.14a)$$

where the first transformed source term $\overline{g}_{ki}(t,\overline{T})$ is due to the integral transformation of the equation source terms, including the convective term, and the second one, $\overline{b}_{ki}(t,\overline{T})$, is due to the contribution of the boundary source terms:

$$\overline{g}_{ki}(t,\overline{T}) = \int_V \tilde{\psi}_{ki}(\mathbf{r})[-\mathbf{u}(\mathbf{r},t,\overline{T})\cdot\nabla T_k(\mathbf{r},t) + g_k(\mathbf{r},t,\overline{T})]dv \qquad (12.14b)$$

$$\overline{b}_{ki}(t,\overline{T}) = \int_S K_k(\mathbf{r})\left[\tilde{\psi}_{ki}(\mathbf{r})\frac{\partial T_k(\mathbf{r},t)}{\partial n} - T_k(\mathbf{r},t)\frac{\partial \tilde{\psi}_{ki}(\mathbf{r})}{\partial n}\right]ds \qquad (12.14c)$$

The boundary conditions contribution may also be expressed in terms of the boundary source terms, after manipulating equations (12.11g) and (12.12b), to yield:

$$\overline{b}_{ki}(t,\overline{T}) = \int_S \phi_k(\mathbf{r},t,\overline{T})\left[\frac{\tilde{\psi}_{ki}(\mathbf{r}) - K_k(\mathbf{r})\frac{\partial \tilde{\psi}_{ki}(\mathbf{r})}{\partial n}}{\alpha_k(\mathbf{r}) + \beta_k(\mathbf{r})}\right]ds \qquad (12.14d)$$

The initial or boundary conditions in the t variable given by equations (12.11d,e) are transformed through the operator $\int_V w_k(\mathbf{r})\tilde{\psi}_{ki}(\mathbf{r})(\cdot)dv$, to provide:

$$\overline{T}_{ki}(0) = \overline{f}_{ki} \equiv \int_V w_k(\mathbf{r})\tilde{\psi}_{ki}(\mathbf{r})f_k(\mathbf{r})dv \quad \text{for the initial value problem}$$
(12.14e)

$$\left[\lambda_{k,l}(t) + (-1)^{l+1}\gamma_{k,l}(t)\frac{d}{dt}\right]\overline{T}_{ki}(t) = \overline{f}_{k,li} \equiv \int_V w_k(\mathbf{r})\tilde{\psi}_{ki}(\mathbf{r})f_{k,l}(\mathbf{r})dv$$
(12.14f)

at $t = t_l$, $l = 0, 1$, for the boundary value problem

For the solution of the infinite coupled system of nonlinear ODEs given by equations (12.14a–f), one must make use of numerical algorithms after the truncation of the system to a sufficiently large finite order. For instance, the built-in routine of the *Mathematica* system (Wolfram 2015), NDSolve, may be employed, which is able to provide reliable solutions under automatic absolute and relative errors control. After the transformed potentials have been numerically computed, the *Mathematica* routine automatically provides an interpolating function object that approximates the t variable behavior of the solution in a continuous form. Then, the inversion formula can be recalled to yield the potential field representation at any desired position **r** and t.

12.2.2 Partial Transformation

An alternative hybrid solution strategy to the above-described total transformation is described, which is of particular interest in the treatment of transient convection–diffusion problems with a predominant convective direction. In such cases, the partial integral transformation in all but one space coordinate may offer an interesting combination of relative advantages between the eigenfunction expansion approach and the selected numerical method for handling the coupled system of one-dimensional PDEs that results from the transformation procedure, as summarized in Figure 12.3.

To illustrate this partial integral transformation procedure, again a transient convection–diffusion problem of n coupled potentials is considered, but this time separating the preferential direction that is not to be integral transformed. The position vector **r** now includes the space coordinates that will be eliminated through integral transformation, here denoted by **r***, as well as the space variable to be retained in the transformed partial differential system. Thus, consider a general three-dimensional problem with $\mathbf{r} = \{x_1, x_2, x_3\}$, where only the coordinates $\mathbf{r}^* = \{x_1, x_2\}$ are intended to be eliminated by the integral transformation process, while the remaining space variable x_3 shall be retained in the transformed system to be numerically solved. The problem to be solved, taking only the initial value problem for the sake of simplicity, is written in the following form (Cotta et al. 2014):

$$w_k(\mathbf{r}^*)\frac{\partial T_k(\mathbf{r}, t)}{\partial t} = G_k(\mathbf{r}, t, \mathbf{T}), \quad \mathbf{r} \in V, \ t > 0, \ k = 1, 2, \ldots, n$$
(12.15a)

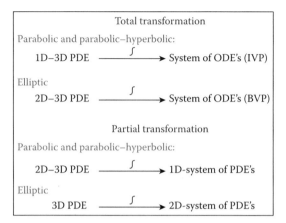

FIGURE 12.3
Schematic summary of total and partial transformation schemes.

with

$$G_k(\mathbf{r}, t, \mathbf{T}) = \nabla^* \cdot [K_k(\mathbf{r}^*)\nabla^* T_k(\mathbf{r}, t)] - d_k(\mathbf{r}^*)T_k(\mathbf{r}, t) \\ - \mathbf{u}(\mathbf{r}, t, \mathbf{T}) \cdot \nabla T_k(\mathbf{r}, t) + g_k(\mathbf{r}, t, \mathbf{T}) \quad (12.15b)$$

where the operator ∇^* refers only to the coordinates to be integral transformed, $\mathbf{r}^*$, and with initial and boundary conditions given, respectively, by

$$T_k(\mathbf{r}, 0) = f_k(\mathbf{r}), \quad \mathbf{r} \in V \quad (12.15c)$$

$$\left[\lambda_k(x_3) + (-1)^{l+1}\gamma_k(x_3)\frac{\partial}{\partial x_3}\right] T_k(\mathbf{r}, t) = \varphi_k(\mathbf{r}, t, \mathbf{T}), \quad x_3 \in S_3 = \{x_{3,l}\}, \ l = 0, 1, t > 0 \quad (12.15d)$$

$$\left[\alpha_k(\mathbf{r}^*) + \beta_k(\mathbf{r}^*)K_k(\mathbf{r}^*)\frac{\partial}{\partial \mathbf{n}^*}\right] T_k(\mathbf{r}, t) = \phi_k(\mathbf{r}, t, \mathbf{T}), \quad \mathbf{r}^* \in S^*, \ t > 0 \quad (12.15e)$$

where $\mathbf{n}^*$ denotes the outward-drawn normal to the surface S^* formed by the coordinates $\mathbf{r}^*$, and S_3 refers to the boundary values of the coordinate x_3.

The coefficients $w_k(\mathbf{r}^*)$, $d_k(\mathbf{r}^*)$, $K_k(\mathbf{r}^*)$, $\alpha_k(\mathbf{r}^*)$, and $\beta_k(\mathbf{r}^*)$ in equations (12.15a–e) inherently carry the information on the auxiliary problem that will be considered in the eigenfunction expansion, and all the remaining terms from this rearrangement are collected into the source terms, $g_k(\mathbf{r},t,\mathbf{T})$ and $\phi_k(\mathbf{r}, t, \mathbf{T})$, including the existing nonlinear terms and diffusion terms with respect to the independent variable x_3. Also, the coefficients $\lambda_3(x_3)$ and $\gamma_3(x_3)$ provide any combination of first- to third-kind boundary conditions in the untransformed coordinate, while the x_3 boundary source terms, $\varphi_k(\mathbf{r},t,\mathbf{T})$, collect the rearranged information that is not contained in the left-hand side of equation (12.15d).

Hybrid Numerical–Analytical Solutions

Following the solution path previously established, the formal integral transform solution of the posed problem requires the proposition of eigenfunction expansions for the associated potentials. The recommended set of uncoupled auxiliary problems is given by

$$\nabla \cdot [K_k(\mathbf{r}^*)\nabla \psi_{ki}(\mathbf{r}^*)] + [\mu_{ki}^2 w_k(\mathbf{r}^*) - d_k(\mathbf{r}^*)]\psi_{ki}(\mathbf{r}^*) = 0, \quad \mathbf{r}^* \in V^* \quad (12.16a)$$

$$\left[\alpha_k(\mathbf{r}^*) + \beta_k(\mathbf{r}^*)K_k(\mathbf{r}^*)\frac{\partial}{\partial \mathbf{n}^*}\right]\psi_{ki}(\mathbf{r}^*) = 0, \quad \mathbf{r}^* \in S^* \quad (12.16b)$$

The problem indicated by equations (12.16a,b) allows, through the associated orthogonality property of the eigenfunctions, for the definition of the following integral transform pairs:

$$\bar{T}_{ki}(x_3, t) = \int_{V^*} w_k(\mathbf{r}^*)\tilde{\psi}_{ki}(\mathbf{r}^*)T_k(\mathbf{r}, t)dv^*, \quad \text{transforms} \quad (12.17a)$$

$$T_k(\mathbf{r}, t) = \sum_{i=1}^{\infty} \tilde{\psi}_{ki}(\mathbf{r}^*)\bar{T}_{ki}(x_3, t), \quad \text{inverses} \quad (12.17b)$$

where the symmetric kernels $\tilde{\psi}_{ki}(\mathbf{r}^*)$ are given by

$$\tilde{\psi}_{ki}(\mathbf{r}^*) = \frac{\psi_{ki}(\mathbf{r}^*)}{\sqrt{N_{ki}}}; \quad N_{ki} = \int_{V^*} w_k(\mathbf{r}^*)\psi_{ki}^2(\mathbf{r}^*)dv^* \quad (12.17c,d)$$

with N_{ki} being the normalization integral.

The integral transformation of equation (12.15a) is accomplished by applying the operator $\int_{V^*} \tilde{\psi}_{ki}(\mathbf{r}^*)(\cdot)dv^*$ and making use of the boundary conditions given by equations (12.15e) and (12.16b), yielding:

$$\frac{\partial \bar{T}_{ki}(x_3, t)}{\partial t} + \mu_{ki}^2 \bar{T}_{ki}(x_3, t) = \bar{g}_{ki}(x_3, t, \bar{T}) + \bar{b}_{ki}(x_3, t, \bar{T}), \quad (12.18a)$$

$$i = 1, 2, \ldots, \quad k = 1, 2, \ldots, n, \quad x_3 \in V_3, \quad t > 0,$$

where the transformed source term $\bar{g}_{ki}(x_3, t, \bar{T})$ is due to the integral transformation of the equation source terms, and the other, $\bar{b}_{ki}(x_3, t, \bar{T})$, is due to the contribution of the boundary source terms corresponding to the directions being integral transformed:

$$\bar{g}_{ki}(x_3, t, \bar{T}) = \int_{V^*} \tilde{\psi}_{ki}(\mathbf{r}^*)[-\mathbf{u}(\mathbf{r}, t, \mathbf{T}) \cdot \nabla T_k(\mathbf{r}, t) + g_k(\mathbf{r}, t, \mathbf{T})]dv^* \quad (12.18b)$$

$$\bar{b}_{ki}(x_3, t, \bar{T}) = \int_{S^*} K_k(\mathbf{r}^*)\left[\tilde{\psi}_{ki}(\mathbf{r}^*)\frac{\partial T_k(\mathbf{r}, t)}{\partial \mathbf{n}^*} - T_k(\mathbf{r}, t)\frac{\partial \tilde{\psi}_{ki}(\mathbf{r}^*)}{\partial \mathbf{n}^*}\right]ds^* \quad (12.18c)$$

The contribution of the boundary conditions at the directions being transformed may also be expressed in terms of the boundary source terms:

$$\overline{b}_{ki}(x_3, t, \overline{T}) = \int_{S^*} \phi_k(r, t, T) \left[\frac{\tilde{\psi}_{ki}(r^*) - K_k(r^*) \frac{\partial \tilde{\psi}_{ki}(r^*)}{\partial n^*}}{\alpha_k(r^*) + \beta_k(r^*)} \right] ds^* \quad (12.18d)$$

The initial conditions given by equation (12.5c) are transformed through the operator $\int_{V^*} w_k(r^*) \tilde{\psi}_{ki}(r^*)(\cdot) dv^*$, to provide:

$$\overline{T}_{ki}(x_3, 0) = \overline{f}_{1,ki}(x_3) \equiv \int_{V^*} w_k(r^*) \tilde{\psi}_{ki}(r^*) f_k(r) dv^* \quad (12.18e)$$

Finally, the boundary conditions with respect to the direction x_3 are also transformed through the same operator, yielding:

$$\left[\lambda_k(x_3) + (-1)^{l+1} \gamma_k(x_3) \frac{\partial}{\partial x_3} \right] \overline{T}_{ki}(x_3, t) = \overline{\phi}_{ki}(x_3, t, \overline{T}), \quad x_3 \in S_3, \ l = 0, 1, \ t > 0$$

$$(12.18f)$$

with

$$\overline{\phi}_{ki}(x_3, t, \overline{T}) \equiv \int_{V^*} w_k(r^*) \tilde{\psi}_{ki}(r^*) \phi_k(r, t, T) dv^*, \quad x_3 \in S_3, \ t > 0 \quad (12.18g)$$

Equations (12.18a–g) form an infinite coupled system of nonlinear one-dimensional PDEs for the transformed potentials, $\overline{T}_{ki}(x_3, t)$, which is unlikely to be analytically solvable. Nonetheless, reliable algorithms are readily available to numerically handle this PDE system after truncation to a sufficiently large finite order. For instance, the *Mathematica* system provides the built-in routine NDSolve (Wolfram 2015), which employs the Method of Lines based on finite difference formulae of variable order and step size, and can handle this system under automatic absolute and relative errors control. The Method of Lines is a numerical technique for PDEs that involves the finite difference approximation of all the space coordinates differential operators, and thus transforming the original PDE into an ODE system as an initial value problem in the time variable, either in an actual transient problem or in a pseudo-transient formulation. One interesting aspect of this approach is that reliable and automatic solvers for initial value problems can then be readily employed. In our partial transformation hybrid approach, once the transformed potentials have been numerically computed, the *Mathematica* routine automatically provides an interpolating function object that approximates the x_3 and t variables behavior of the solution in a continuous form. Then, the inversion formula [equation (12.17b)] can be recalled to yield the potential field representation at any desired position r and time t.

12.2.3 Computational Algorithm

The formal solutions derived above provide the basic working expressions for the integral transform method in either the total or partial transformation schemes. However, for an improved computational performance, it is always recommended to reduce the importance of the equation and boundary source terms so as to enhance the eigenfunction expansions convergence behavior (Cotta and Mikhailov 1997). One possible approach for achieving this goal is the proposition of analytical filtering solutions, which essentially remove information from the source terms into a preferably simple analytical expression. Several different alternative filters may be proposed for the same problem. For instance, analytical filters for the above formulations are written in general form as:

$$T_k(\mathbf{r},t) = \theta_k(\mathbf{r},t) + T_{f,k}(\mathbf{r};t) \qquad (12.19a)$$

$$T_k(\mathbf{r}^*,x_3,t) = \theta_k(\mathbf{r}^*,x_3,t) + T_{f,k}(\mathbf{r}^*;x_3,t) \qquad (12.19b)$$

where the second term on the right-hand side represents a filter solution which is generally sought in analytic form. The first term on the right-hand side represents the filtered potentials that are obtained through integral transformation. Once the filtering problem formulation is chosen, equations (12.19a,b) are substituted back into equations (12.11a–h) or (12.15a–e), respectively, to obtain the resulting formulation for the filtered potential. Thus, the previously established formal solutions are directly applicable to the filtered problems, once the initial conditions, the filtered equation, and boundary source terms have been adequately substituted. It is desirable that the filtering solution contains as much information on the operators of the original problem as possible. This information may include, for instance, linearized versions of the source terms, so as to reduce their influence on convergence of the final eigenfunction expansions.

In multidimensional applications, the final integral transform solution for the related potential could in principle be expressed as double or triple infinite summations for two- or three-dimensional transient problems, respectively. However, if one just truncates each individual summation to a certain prescribed finite order, computations become quite ineffective. Some still important information to the final result can be disregarded due to the fixed summations limits, while other terms are accounted for that have essentially no contribution to convergence of the potential in the relative accuracy required. Therefore, for an efficient computation of these expansions, the infinite multiple summations should first be converted to a single sum representation with the appropriate reordering of terms according to their individual contribution to the final numerical result (Mikhailov and Cotta 1996; Correa et al. 1997). Then, the minimal number of eigenvalues and related derived quantities required to reach the user-prescribed accuracy target shall be evaluated. The most common choice of reordering strategy is based on arranging in increasing order the sum of the squared eigenvalues in each spatial coordinate. However, individual applications may require more elaborate reordering that accounts for the influence of transformed initial conditions and transformed nonlinear source terms in the ODE system.

To more clearly understand alternative reordering schemes, let us start from the formal solution of the transformed potential equations (12.14a–f), for the initial value problem case, which is written as:

$$\overline{T}_{ki}(t) = \overline{f}_{ki}\exp(-\mu_i^2 t) + \int_0^t \hat{\overline{g}}_{ki}(t', T)\exp[-\mu_i^2(t-t')]dt' \qquad (12.20)$$

where the nonlinear transformed source term $\hat{\overline{g}}_{ki}(t, T)$ includes the contributions of both the equation and boundary conditions source terms.

$$\hat{\overline{g}}_{ki}(t, T) = \overline{g}_{ki}(t, T) + \overline{b}_{ki}(t, T) \qquad (12.21)$$

Integration by parts of equation (12.20) provides an alternative expression that allows the understanding of the influence of the transformed initial conditions and source terms in the choice of reordering scheme, in rewriting the multiple series as a single one:

$$\overline{T}_{ki}(t) = \overline{f}_{ki}\exp(-\mu_i^2 t) + \frac{1}{\mu_i^2}[\hat{\overline{g}}_{ki}(t, T) - \hat{\overline{g}}_{ki}(0, T)\exp(-\mu_i^2 t)]$$

$$- \frac{1}{\mu_i^2}\int_0^t \frac{d\hat{\overline{g}}_{ki}}{dt'}\exp[-\mu_i^2(t-t')]dt' \qquad (12.22)$$

It is evident that the squared eigenvalues, which involve the combination of the eigenvalues in each spatial coordinate, play the most important role in the decay of the absolute values of the infinite summation components, both through the exponential term $\exp(-\mu_i^2 t)$ and, at a lower convergence rate, through the inverse of the squared eigenvalues, $1/\mu_i^2$. Therefore, this traditionally employed reordering scheme should usually be able to account for the most important terms in the adequate reordering of the expansion. Nevertheless, supposing that the last integral term in equation (12.22) plays a less important role in the reordering choice, and in fact it vanishes when the source term is not time dependent, one concludes that the decay of the transformed initial condition and the transformed source term evolution from its initial value play a complementary role in the selection of terms in the eigenfunction expansion for a fixed truncation order. Thus, a more robust selection can be proposed, based on adding to the initially reordered terms, according to the squared eigenvalues criterion, a few extra terms that are of possible contribution to the final result under the analysis of the initial condition decay and/or the transformed source term behavior. In the first case, for the lowest time value of interest, $t = t_{min}$, the criterion that reorders the terms based on the decay of the initial conditions is based on sorting in decreasing order from the expression $\bar{f}_{ki}\exp(-\mu_i^2 t_{min})$. In the second case, for the general situation of a nonlinear transformed source term, the estimation of the term's importance is more difficult, since the source terms are not known *a priori*. One possible approach is to consider the limiting situation of a uniform unitary source term, representing for instance its normalized maximum value, and analyzing the reordering of terms in descending absolute value based on the expression $\frac{1}{\mu_i^2}\int_V \tilde{\varphi}_{ki}(\mathbf{r})dv$. Therefore, combining the three criteria, and eliminating the duplicates with respect to the traditional reordering scheme based on the squared eigenvalues, a few extra terms may be added to the initially reordered terms that have still some relevant effect in the final truncated summation.

In order to computationally solve the problem defined by equations (12.11a–h) and (12.15a–e), a straightforward general algorithm can be described as follows:

- The filtering strategy is chosen and implemented. The simplest possible filtering solution is written as a linear function in the space variable that simultaneously satisfies both boundary conditions, which can be implemented progressively for the multidimensional situation (Cotta et al. 2013). The option of not providing a filtering solution is also feasible, either because it might not be actually necessary or as a solution strategy to be complemented, such as by an integral balance acceleration *a posteriori* (Cotta and Mikhailov 1997).

- The auxiliary eigenvalue problem (12.12) or (12.16) is solved for the eigenvalues and related normalized eigenfunctions, either in analytic explicit form, when applicable, as obtained for instance by the symbolic routine DSolve (Wolfram 2015), after separation of variables in multidimensional applications, or through the GITT itself (Cotta 1993; Mikhailov and Cotta 1994; Sphaier and Cotta 2000; Naveira-Cotta et al. 2009; Cotta et al. 2016d).

- The transformed initial conditions are computed, either analytically or through symbolic computation, for instance, function *Integrate* (Wolfram 2015), or with a general-purpose procedure through adaptive numerical integration, such as function *NIntegrate* (Wolfram 2015). Two additional options are also readily available, namely a semi-analytical evaluation where the analytical integration of the eigenfunction oscillatory behavior is preserved (Cotta and Mikhailov 2005), and a simplified and cost-effective numerical integration with Gaussian quadrature, automatically exploiting the frequency of oscillation of the eigenfunctions in the choice of subintervals for integration. Similarly, the transformed source term in the transformed ODE system in equations (12.14a) and (12.18a), once not dependent on the transformed potentials, can be evaluated in advance.

For the more general situation of nonlinear coefficients, there are some computational savings in grouping them into one single integrand, as represented in equations (12.14b) and (12.18b). The alternative semi-analytical integration procedure is particularly convenient in nonlinear formulations that might require costly numerical integration. For instance, the integral transformation of the equation source term 12.14b would then be evaluated as:

$$\overline{g}_{ki}^*(t, T) = \int_V \tilde{\psi}_{ki}(\mathbf{r}) g_k(\mathbf{r}, t, T) dv = \sum_{m=1}^{M} \int_{V_m} \tilde{\psi}_{ki}(\mathbf{r}) \hat{g}_{k,m}(\mathbf{r}, t, T) dv_m \quad (12.23)$$

where $\hat{g}_{k,m}(\mathbf{r}, t, T)$ are simpler representations of the source term, defined in subregions V_m, for which analytical integration of the eigenfunctions is still obtainable. The simplest choice would be the adoption of uniform values of the source terms within the subdomains (zeroth-order approximation), but linear and quadratic representations of the source terms' behavior can also be implemented for improved accuracy and cost-effectiveness (Cotta et al. 2013, 2014).

- The truncated ODE or PDE system [Equation (12.14) or (12.18)] is then numerically solved, preferably through an adaptive algorithm with automatic error control, such as through the NDSolve routine of the *Mathematica* system (Wolfram 2015). In general, for the initial value problem solvers, it is recommended to work under the automatic selection of a stiff system situation, since the resulting system is likely to become stiff, especially for increasing truncation orders in the expansion.
- Once all the intermediate numerical tasks are accomplished within user-prescribed accuracy, one is left with the need of reaching convergence in the eigenfunction expansions and controlling the truncation order N for the requested accuracy in the final solution. The analytical nature of the inversion formula allows for a direct error testing procedure at each specified position within the medium, and the truncation order N can also be controlled to fit the user global error requirements over the entire solution domain (Cotta 1993). The tolerance testing formulae employed in the total and partial transformations are, respectively:

$$\varepsilon = \max_{\mathbf{r} \in V} \left| \frac{\sum_{i=N^*}^{N} \tilde{\psi}_{ki}(\mathbf{r}) \overline{T}_{k,i}(t)}{T_{f,k}(\mathbf{r};t) + \sum_{i=1}^{N} \tilde{\psi}_{ki}(\mathbf{r}) \overline{T}_{k,i}(t)} \right| \quad (12.24a)$$

$$\varepsilon = \max_{\mathbf{r}^* \in V^*} \left| \frac{\sum_{i=N^*}^{N} \tilde{\psi}_{ki}(\mathbf{r}^*) \overline{T}_{k,i}(x_3, t)}{T_{f,k}(\mathbf{r}^*; x_3, t) + \sum_{i=1}^{N} \tilde{\psi}_{ki}(\mathbf{r}^*) \overline{T}_{k,i}(x_3, t)} \right| \quad (12.24b)$$

The numerator in equations (12.24a,b) represents those terms (from orders N* to N) that in principle might be abandoned in the evaluation of the inverse formula, without disturbing the final result to within the user-requested accuracy target, once convergence has been achieved. Therefore, this testing can be implemented by choosing the value of N* so as to have a small odd number of terms in the numerator sum, then offering error estimations at any of the selected test positions within the domain.

12.2.4 Test Case

Burgers-type equations provide valuable test cases for transient convection–diffusion problems, including nonlinear formulations with different relative importance of the advective and diffusive terms. The mathematical dimensionless formulation of the three-dimensional nonlinear Burgers' equation, here considered as test case, is given by (Cotta et al. 2014):

$$\frac{\partial T(x,y,z,t)}{\partial t} + u(T)\frac{\partial T(x,y,z,t)}{\partial x} = \nu\left[\frac{\partial^2 T(x,y,z,t)}{\partial x^2} + \frac{\partial^2 T(x,y,z,t)}{\partial y^2} + \frac{\partial^2 T(x,y,z,t)}{\partial z^2}\right],$$
$$0 < x < 1, \quad 0 < y < 1, \quad 0 < z < 1, \quad t > 0 \qquad (12.25a)$$

with initial and boundary conditions given by

$$T(x,y,z,0) = 1, \quad 0 \leq x \leq 1, \ 0 \leq y \leq 1, \ 0 \leq z \leq 1 \qquad (12.25b)$$

$$T(0,y,z,t) = \phi; \quad T(1,y,z,t) = 0, \quad t > 0 \qquad (12.25c,d)$$

$$\frac{\partial T(x,0,z,t)}{\partial y} = 0; \quad T(x,1,z,t) = 0, \quad t > 0 \qquad (12.25e,f)$$

$$\frac{\partial T(x,y,0,t)}{\partial z} = 0; \quad T(x,y,1,t) = 0, \quad t > 0 \qquad (12.25g,h)$$

and, for the present application, the nonlinear function u(T) is taken as:

$$u(T) = u_0 + cT \qquad (12.25i)$$

The inherent choice of the eigenvalue problem is made when the user establishes the equivalence between the general problem, given by equations (12.11a–h) or 12.15a–e, and the considered application. For the total transformation scheme, equations (12.25a–i) are recovered through the following parameters:

$$\mathbf{r} \to \{x,y,z\}; \quad n \to 1; \quad T_k(\mathbf{r},t) \to T(x,y,z,t); \quad f_k(\mathbf{r}) \to 1; \quad w_k(\mathbf{r}) \to 1;$$
$$K_k(\mathbf{r}) \to \nu; \quad d_k(\mathbf{r}) \to 0; \quad \mathbf{u}(\mathbf{r},t,T) \to \{u(T),0,0\}; \quad g_k(\mathbf{r},t,T) \to 0;$$
(12.26a–i)

$$\alpha_k(\mathbf{r}) \to 1; \quad \beta_k(\mathbf{r}) \to 0 \text{ for } x=0;$$
$$\alpha_k(\mathbf{r}) \to 0; \quad \beta_k(\mathbf{r}) \to 1 \text{ for } y,z=0; \qquad (12.26\text{j–o})$$
$$\alpha_k(\mathbf{r}) \to 1; \quad \beta_k(\mathbf{r}) \to 0 \text{ for } x,y,z=1$$

whereas the boundary source terms are given by

$$\phi_k(\mathbf{r},t,T) \to \phi, \text{ for } x=0, \text{ with } \phi=0 \text{ (homogeneous)}, \phi=1 \text{ (nonhomogeneous)}$$
$$\phi_k(\mathbf{r},t,T) \to 0, \text{ for } y,z=0, \; x,y,z=1 \qquad (12.27\text{a,b})$$

For the partial transformation scheme, equations (12.25a–i) are recovered through the following parameters:

$$\mathbf{r} \to \{x,y,z\}; \quad \mathbf{r}^* \to \{y,z\}; \quad x_3 \to x; \quad n \to 1; \quad T_k(\mathbf{r},t) \to T(x,y,z,t);$$
$$f_k(\mathbf{r}) \to 1; \quad w_k(\mathbf{r}^*) \to 1; \quad K_k(\mathbf{r}^*) \to \nu; \quad d_k(\mathbf{r}^*) \to 0;$$
$$\mathbf{u}(\mathbf{r},t,T) \to \{u(T),0,0\}; \quad g_k(\mathbf{r},t,T) \to \nu \frac{\partial^2 T(x,y,z,t)}{\partial x^2}; \quad \phi_k(\mathbf{r},t,T) \to 0;$$
(12.28a–l)

$$\lambda_k(x_3) \to 1; \quad \gamma_k(x_3) \to 0 \text{ for } x=0;$$
$$\lambda_k(x_3) \to 1; \quad \gamma_k(x_3) \to 0 \text{ for } x=1;$$
$$\alpha_k(\mathbf{r}^*) \to 0; \quad \beta_k(\mathbf{r}^*) \to 1 \text{ for } y,z=0; \qquad (12.28\text{m–t})$$
$$\alpha_k(\mathbf{r}^*) \to 1; \quad \beta_k(\mathbf{r}^*) \to 0 \text{ for } y,z=1$$

with the x_3 boundary source terms as:

$$\varphi_k(\mathbf{r},t,T) \to \phi, \text{ for } x=0, \text{ with } \phi=0 \text{ (homogeneous)}, \phi=1 \text{ (nonhomogeneous)}$$
$$\varphi_k(\mathbf{r},t,T) \to 0, \text{ for } x=1 \qquad (12.29\text{a,b})$$

Here, results are presented for both the case with homogeneous ($\phi = 0$) and nonhomogeneous boundary conditions ($\phi = 1$). In the analysis of the three-dimensional Burgers' equations (12.25a–i), the following governing parameters values have been adopted (Cotta et al. 2013): $u_0 = 1$, $c = 5$, and $\nu = 1$. The Burgers' equation formulation with homogenous boundary conditions is investigated first. In order to offer a benchmark solution, the GITT with total transformation was first employed with analytical integration of the transformed initial condition and nonlinear source term (Cotta et al. 2013). Then, high truncation orders could be achieved with reduced computational cost so as to offer a reliable hybrid solution with four fully converged significant digits, as presented in

Table 12.2, where convergence is reached at truncation orders around $N = 300$, with the traditional squared eigenvalues reordering criterion.

Table 12.3 illustrates the convergence behavior of the GITT solution with the partial transformation scheme, in which the direction x has been chosen not to be integral transformed, and Gaussian quadrature numerical integration was preferred so as to illustrate the overall accuracy of the solution under this most automatic and straightforward option for the transformed system coefficients determination. The results are presented for different number of terms in the eigenfunction expansion, $N = 30, 35, 40$, and 45, and fixed number of points in the Gaussian quadrature integration in each coordinate, $M = \{19,19\}$. In Table 12.3, the total transformation solution with $N = 300$ terms, obtained with analytical integration, is repeated as reference results. A fully discrete solution is also presented, obtained by the Method of Lines from the NDSolve routine of the *Mathematica* system (Wolfram 2015), under the same precision control with mesh refinement, as in the options employed in the partial transformation solution. One should notice that at the selected positions, the partial transformation results are converged to the third significant digit for $t = 0.02$ and to all four digits shown for $t = 0.1$, with the expected improved convergence for larger values of t. It is also observed that the partial transformation results agree well with the NDSolve numerical results in three significant digits for $t = 0.02$ and $t = 0.1$. These results, in comparison with those obtained by the total transformation scheme in Table 12.2, indicate a faster convergence rate of the partial transformation solution, as expected, since three eigenfunction expansions are represented in the total transformation solution, while only two are represented in the partial transformation solution. Nevertheless, it should be remembered that the partial transformation scheme demands additional numerical effort per equation, since a system of one-dimensional PDEs is being solved, while the total transformation scheme leads to a transformed ODE system. In the partial transformation, even considerably reducing the numerical effort when transforming from a three-dimensional partial differential system to a one-dimensional one, the numerical solution of the coupled transformed PDEs system still yields some effect on the final computed results.

Figure 12.4 depicts the graphical comparison of the partial and total transformation GITT solutions, at different times, against those obtained with the NDSolve routine, where an excellent agreement between the three solutions is observed throughout, to the graph scale.

Proceeding to the Burgers' equation problem with nonhomogeneous boundary conditions, Table 12.4 illustrates the convergence behavior of the GITT solution with partial transformation, by varying the truncation order, $N = 55$, 60, 65, and 70, with a fixed number of points in the Gaussian quadrature integration, $M = \{19,19\}$. We point out that no filtering procedure was performed in this solution. Since the direction with nonhomogeneous boundary conditions, x, is the one chosen not to be integral transformed, an excellent

TABLE 12.2
Convergence of GITT Solution for Three-Dimensional Burgers' Equation ($N = 40$ to 300 Terms and Analytical Integration)

x	$N = 40$	$N = 70$	$N = 100$	$N = 140$	$N = 180$	$N = 220$	$N = 260$	$N = 280$	$N = 300$
				$t = 0.02$, $y = 0.5$ and $z = 0.5$					
0.1	0.2601	0.2868	0.2868	0.2782	0.2814	0.2813	0.2780	0.2797	0.2798
0.3	0.7416	0.7349	0.7361	0.7402	0.7414	0.7415	0.7398	0.7395	0.7396
0.5	0.9208	0.9355	0.9361	0.9351	0.9331	0.9329	0.9330	0.9331	0.9331
0.7	0.9208	0.9157	0.9151	0.9128	0.9140	0.9141	0.9155	0.9157	0.9156
0.9	0.4815	0.4857	0.4858	0.4893	0.4898	0.4898	0.4905	0.4905	0.4905
				$t = 0.1$, $y = 0.5$ and $z = 0.5$					
x	$N = 40$	$N = 70$	$N = 100$	$N = 140$	$N = 180$	$N = 220$	$N = 260$	$N = 280$	$N = 300$
0.1	0.04898	0.04981	0.04983	0.04920	0.04938	0.04938	0.04923	0.04922	0.04922
0.3	0.1491	0.1491	0.1491	0.1493	0.1495	0.1495	0.1492	0.1492	0.1492
0.5	0.2221	0.2233	0.2233	0.2231	0.2231	0.2231	0.2231	0.2231	0.2231
0.7	0.2237	0.2236	0.2235	0.2231	0.2233	0.2233	0.2235	0.2235	0.2235
0.9	0.1017	0.1024	0.1024	0.1028	0.1030	0.1030	0.1031	0.1031	0.1031

TABLE 12.3

Convergence of GITT Solution with the Partial Transformation Scheme for Three-Dimensional Burgers' Equation with Homogeneous Boundary Conditions (GITT with N = 30, 35, 40, or 45 Terms, and M = {19,19} in the Gaussian Quadrature Integration)

	$t = 0.02$, $y = 0.5$, $z = 0.5$					
x	N = 30	N = 35	N = 40	N = 45	GITT TT[a]	NDSolve[b]
0.1	0.2780	0.2803	0.2807	0.2808	0.2798	0.2807
0.3	0.7354	0.7401	0.7407	0.7409	0.7396	0.7406
0.5	0.9291	0.9333	0.9336	0.9337	0.9331	0.9336
0.7	0.9127	0.9155	0.9153	0.9152	0.9156	0.9155
0.9	0.4898	0.4898	0.4892	0.4891	0.4905	0.4896
	$t = 0.1$, $y = 0.5$, $z = 0.5$					
x	N = 30	N = 35	N = 40	N = 45	GITT TT[a]	NDSolve[b]
0.1	0.04943	0.04945	0.04946	0.04946	0.04922	0.04947
0.3	0.1494	0.1494	0.1494	0.1494	0.1492	0.1494
0.5	0.2232	0.2232	0.2232	0.2232	0.2231	0.2233
0.7	0.2236	0.2236	0.2236	0.2236	0.2235	0.2236
0.9	0.1031	0.1031	0.1031	0.1031	0.1031	0.1031

[a] N = 300 terms with analytical integration and total transformation (TT), from Table 12.1.
[b] Mathematica v.7.

convergence behavior is observed at the selected positions, the results being converged to the fourth significant digit throughout.

Figure 12.5 brings the comparison between the solutions obtained with the partial transformation without filter and the total transformation with user provided polynomial filtering (Cotta et al. 2014), against the fully discrete NDSolve routine solution, along the x coordinate. In these results, one may again confirm the excellent adherence between the GITT solutions as well as with the NDSolve solution curves throughout the x-variable domain.

12.3 Convective Eigenvalue Problem

In both the total and partial transformation schemes presented above for the integral transform method, the choices of eigenvalue problems that form the basis of the eigenfunction expansions did not account for the convective effects, which were in fact fully incorporated into the equation source terms, and thus only appear in the transformed system and their respective solutions, the transformed potentials. In order to offer another GITT alternative approach for convection–diffusion problems, the present section considers incorporating

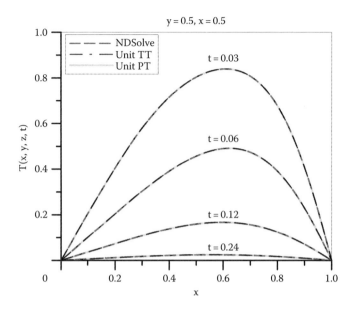

FIGURE 12.4
Comparison between the partial transformation (PT) and total transformation (TT) solutions and the NDSolve routine solution for the three-dimensional Burgers' equation problem with homogeneous boundary conditions. (From Cotta, R.M., et al., *Comput. Therm. Sci.*, 6(6), 507–524, 2014. With permission.)

the convective effects, either fully or partially, into the chosen eigenvalue problem that forms the basis of the proposed eigenfunction expansion. The aim is to markedly improve convergence behavior of the eigenfunction expansions, especially in the case of highly convective formulations, in comparison against the traditional approach via a purely diffusive eigenvalue problem, by directly accounting for the relative importance of convective and diffusive effects within the eigenfunctions themselves (Cotta et al. 2016b, 2016c, 2017). Through a straightforward algebraic transformation of the original convection–diffusion problem, basically by redefining the coefficients associated with the transient and diffusive terms, the convective term is merged, in linearized form, into a generalized diffusion term with a transformed space variable diffusion coefficient. The generalized diffusion problem then naturally leads to the eigenvalue problem to be adopted for deriving the eigenfunction expansion in the linear situation, as well as for the appropriate linearized version in the case of a nonlinear application. The resulting eigenvalue problem with space variable coefficients is then solved through the GITT itself, yielding the corresponding algebraic eigenvalue problem upon selection of a simple auxiliary eigenvalue problem of known analytical solution (Cotta 1993; Naveira-Cotta et al. 2009). The GITT is also applied in the solution of the generalized diffusion or convection–diffusion problem. In the case of

TABLE 12.4

Convergence of GITT Solution with the Partial Transformation Scheme for Three-Dimensional Burgers' Equation with Nonhomogenous Boundary Conditions (GITT with N = 55, 60, 65, or 70 Terms, and M = {19,19} in the Gaussian Quadrature Integration)

(x,y,z)	N = 55[a]	N = 60	N = 65	N = 70	NDSolve[b]
(0.2,0.5,0.5)	0.9407	0.9409	0.9408	0.9408	0.9408
(0.5,0.5,0.5)	0.8523	0.8523	0.8523	0.8523	0.8523
(0.8,0.5,0.5)	0.5945	0.5945	0.5945	0.5945	0.5948
(0.5,0.2,0.5)	0.9119	0.9119	0.9119	0.9119	0.9119
(0.5,0.8,0.5)	0.4897	0.4896	0.4896	0.4897	0.4896
(0.5,0.5,0.2)	0.9119	0.9119	0.9119	0.9119	0.9119
(0.5,0.5,0.8)	0.4897	0.4897	0.4897	0.4897	0.4896
(x,y,z)	N = 55	N = 60	N = 65	N = 70	NDSolve[b]
(0.2,0.5,0.5)	0.8798	0.8801	0.8799	0.8799	0.8798
(0.5,0.5,0.5)	0.6400	0.6400	0.6400	0.6400	0.6398
(0.8,0.5,0.5)	0.3365	0.3365	0.3365	0.3365	0.3364
(0.5,0.2,0.5)	0.7457	0.7457	0.7457	0.7457	0.7456
(0.5,0.8,0.5)	0.3282	0.3282	0.3282	0.3282	0.3281
(0.5,0.5,0.2)	0.7457	0.7457	0.7457	0.7457	0.7456
(0.5,0.5,0.8)	0.3282	0.3282	0.3282	0.3282	0.3281

[a] GITT solution with partial transformation and no filter (M = {19,19}).
[b] *Mathematica 7*.

multidimensional formulations, again the choice of a total or partial transformation scheme can be employed.

Here, the developed methodology is briefly illustrated for both linear and nonlinear applications, both in one-dimensional and multidimensional formulations, as represented by examples of Burgers' equation.

The adoption of convective eigenvalue problems in the integral transforms solution of transient convection–diffusion is analyzed first by considering a nonlinear one-dimensional formulation, a special case of Problem 12.11, given by

$$w(x)\frac{\partial T(x,t)}{\partial t} + u(x)\frac{\partial T(x,t)}{\partial x} = \frac{\partial}{\partial x}\left[k(x)\frac{\partial T(x,t)}{\partial x}\right] - d(x)T(x,t) + P(x,t,T),$$
$$x_0 < x < x_1, t > 0 \qquad (12.30)$$

where u(x) is a characteristic linear representation of the convective term coefficient, while the remaining of the nonlinear convective operator (or of any other operator) is incorporated into the nonlinear source term, P(x,t,T). Problem 12.30 can be readily rewritten as a generalized diffusion problem, through a simple transformation of the diffusive, dissipative, and transient terms as

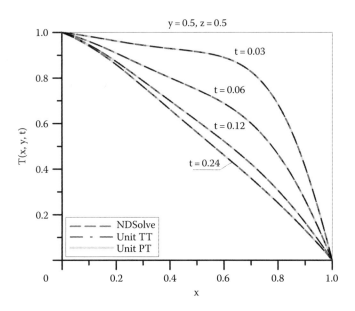

FIGURE 12.5
Comparison between the partial transformation (PT) and total transformation (TT) solutions against the NDSolve routine solution for the Burgers' equation problem with nonhomogeneous boundary conditions, along x variable. (From Cotta, R.M., et al., *Comput. Therm. Sci.*, 6, 507–524, 2014. With permission.)

$$\hat{w}(x)\frac{\partial T(x,t)}{\partial t} = \frac{\partial}{\partial x}\left[\hat{k}(x)\frac{\partial T(x,t)}{\partial x}\right] - \hat{d}(x)T(x,t) + \hat{P}(x,t,T), \quad x_0 < x < x_1, t > 0$$

(12.31)

where

$$\hat{w}(x) = \frac{w(x)\,\hat{k}(x)}{k(x)}; \quad \hat{d}(x) = \frac{d(x)\,\hat{k}(x)}{k(x)}; \quad \hat{P}(x,t,T) = \frac{P(x,t,T)\,\hat{k}(x)}{k(x)};$$

$$u^*(x) = \frac{1}{k(x)}\left[u(x) - \frac{dk(x)}{dx}\right]; \quad \text{and} \quad \hat{k}(x) = e^{-\int u^*(x)dx}$$

(12.32a–e)

Equation (12.31) is a special case of the nonlinear diffusion problems that have been extensively handled through the GITT, as described in Section 12.2, but is treated here through an eigenfunction expansion basis that includes convective effects through the characteristic convective term coefficient, u(x). The self-adjoint eigenvalue problem with space variable coefficients to be considered would then be given by the following equation:

$$\frac{d}{dx}\left[\hat{k}(x)\frac{d\psi(x)}{dx}\right] + [\mu^2\hat{w}(x) - \hat{d}(x)]\psi(x), \quad x_0 < x < x_1$$

(12.33)

which can be readily solved by the GITT itself (Cotta 1993; Naveira-Cotta et al. 2009; Cotta et al. 2016d), yielding the corresponding algebraic eigenvalue problem.

The same straightforward transformation can be employed to rewrite a more general multidimensional convection–diffusion formulation, given in the general region V with the position vector $\mathbf{r}$. Then, the following parabolic nonlinear multidimensional problem, already with the linear characteristic coefficients identified, is considered:

$$w(\mathbf{r})\frac{\partial T(\mathbf{r},t)}{\partial t} + \mathbf{u}(\mathbf{r})\cdot\nabla T(\mathbf{r},t) = \nabla\cdot[k(\mathbf{r})\nabla T(\mathbf{r},t)] - d(\mathbf{r})T(\mathbf{r},t) + P(\mathbf{r},t,T),$$
$$\mathbf{r}\in V, t > 0 \quad (12.34)$$

where the linear coefficients in each operator, dependent only on the spatial variables, already imply the choice of characteristic functional behaviors to be accounted for in the eigenfunction expansion basis, while the remaining nonlinearities are gathered into the redefined nonlinear source term $P(\mathbf{r},t,T)$. Considering that the convective term coefficient vector $\mathbf{u}$ can be represented in the three-dimensional situation by the three components $\{u_x, u_y, u_z\}$, illustrating here the transformation in the Cartesian coordinates system, $\mathbf{r} = \{x, y, z\}$, then equation (12.34) is rewritten in the generalized diffusive form as

$$\hat{w}(\mathbf{r})\frac{\partial T(\mathbf{r},t)}{\partial t} = \hat{k}_y(\mathbf{r})\hat{k}_z(\mathbf{r})\frac{\partial}{\partial x}\left[\hat{k}_x(\mathbf{r})\frac{\partial T(\mathbf{r},t)}{\partial x}\right] + \hat{k}_x(\mathbf{r})\hat{k}_z(\mathbf{r})\frac{\partial}{\partial y}\left[\hat{k}_y(\mathbf{r})\frac{\partial T(\mathbf{r},t)}{\partial y}\right]$$
$$+ \hat{k}_x(\mathbf{r})\hat{k}_y(\mathbf{r})\frac{\partial}{\partial z}\left[\hat{k}_z(\mathbf{r})\frac{\partial T(\mathbf{r},t)}{\partial z}\right] - \hat{d}(\mathbf{r})T(\mathbf{r},t) + \hat{P}(\mathbf{r},t,T),$$
$$\mathbf{r}\in V,\ t > 0 \quad (12.35a)$$

where

$$\hat{k}(\mathbf{r}) = \hat{k}_x(\mathbf{r})\hat{k}_y(\mathbf{r})\hat{k}_z(\mathbf{r});\quad \hat{w}(\mathbf{r}) = \frac{w(\mathbf{r})\,\hat{k}(\mathbf{r})}{k(\mathbf{r})};\quad \hat{d}(\mathbf{r}) = \frac{d(\mathbf{r})\,\hat{k}(\mathbf{r})}{k(\mathbf{r})};$$
$$\hat{P}(\mathbf{r},t,T) = \frac{P(\mathbf{r},t,T)\,\hat{k}(\mathbf{r})}{k(\mathbf{r})};\quad \mathbf{u}^*(\mathbf{r}) = \frac{1}{k(\mathbf{r})}[\mathbf{u}(\mathbf{r}) - \nabla k(\mathbf{r})]; \quad (12.35\text{b-i})$$
$$\text{and}\quad \hat{k}_x(\mathbf{r}) = e^{-\int u_x^*(\mathbf{r})dx};\quad \hat{k}_y(\mathbf{r}) = e^{-\int u_y^*(\mathbf{r})dy};\quad \hat{k}_z(\mathbf{r}) = e^{-\int u_z^*(\mathbf{r})dz}$$

For the general situation of equation (12.35a), the separation of variables as applied to the homogeneous version of the problem leads to a nonself-adjoint eigenvalue problem; hence, the eigenfunctions are not orthogonal, and the classical integral transformation approach is not directly applicable in the present form. However, the GITT can still be directly employed with an appropriate choice of a self-adjoint eigenvalue problem. On the contrary, when the transformed diffusion coefficients are functions of only the corresponding space coordinate, or $\hat{k}_x(\mathbf{r}) = \hat{k}_x(x)$, $\hat{k}_y(\mathbf{r}) = \hat{k}_y(y)$; $\hat{k}_z(\mathbf{r}) = \hat{k}_z(z)$, with the consequent restrictions on the choices of the characteristic linear coefficients k (r) and **u**(r), a generalized diffusion formulation is constructed, which leads to a self-adjoint eigenvalue problem and can be written in such special case as

$$\hat{w}(\mathbf{r})\frac{\partial T(\mathbf{r},t)}{\partial t} = \nabla.[\hat{k}(\mathbf{r})\nabla T(\mathbf{r},t)] - \hat{d}(\mathbf{r})T(\mathbf{r},t) + \hat{P}(\mathbf{r},t,T), \quad \mathbf{r} \in V, \ t > 0 \quad (12.36a)$$

where

$$\hat{k}(\mathbf{r}) = \hat{k}_x(x)\hat{k}_y(y)\hat{k}_z(z) \quad (12.36b)$$

for which the appropriate self-adjoint eigenvalue problem would be

$$\nabla.[\hat{k}(\mathbf{r})\nabla\psi(\mathbf{r})] + [\mu^2\hat{w}(\mathbf{r}) - \hat{d}(\mathbf{r})]\psi(\mathbf{r}) = 0, \quad \mathbf{r} \in V \quad (12.37)$$

which can be directly solved by the GITT itself (Cotta 1993; Naveira-Cotta et al. 2009; Cotta et al. 2016d) to yield the corresponding algebraic eigenvalue problem for the eigenvalues and eigenvectors that reconstruct the desired eigenfunction, ψ(**r**). Problem 12.37 again incorporates relevant information on the convective effects, as specified in the chosen linear convective term coefficients that undergo the exponential transformation, which can provide a desirable convergence enhancement effect in the integral transform solution of this multidimensional nonlinear problem. The remaining steps in the GITT solution of the nonlinear generalized diffusion problem 12.36 are well documented in Section 12.2 and are not repeated here.

The illustration of the proposed procedure is started with the analysis of the one-dimensional Burgers' equation, both in linear and nonlinear formulations, which allows for the analysis of the convective eigenvalue problem choice on the eigenfunction expansion convergence behavior. The convection–diffusion problem first analyzed is given in dimensionless form by

$$\frac{\partial T(x,t)}{\partial t} + U(T)\frac{\partial T(x,t)}{\partial x} = \frac{\partial^2 T(x,t)}{\partial x^2}, \quad 0 < x < 1, t > 0 \quad (12.38a)$$

$$T(x,0) = 1; \quad T(0,t) = 0; \quad T(1,t) = 0 \quad (12.38b\text{-d})$$

where the nonlinear velocity coefficient is taken as

$$U(T) = u_0 + b.T \quad (12.38e)$$

which readily allows for the separate consideration of the linear and nonlinear situations. From direct comparison with the general one-dimensional formulation in equation (12.30), choosing the characteristic convective coefficient as the linear portion of equation (12.38e), and moving the remaining nonlinear convective term to the source term, the following correspondence is reached:

$$w(x) = 1; \quad u(x) = u_0; \quad k(x) = 1; \quad d(x) = 0; \quad P(x, t, T) = -b\, T(x,t)\frac{\partial T}{\partial x} \quad (12.39\text{a--e})$$

where for the special case of a linear situation, the nonlinear source term vanishes. For this particular linear situation, it can be anticipated that the choice of the linear portion of the convective coefficient shall lead to an exact integral transform solution. On the contrary, for the more general nonlinear situation, other choices of the characteristic coefficient can offer improved convergence with respect to the present choice, by for instance including linearized information on the second nonlinear portion of the original coefficient.

The resulting problem with transformed coefficients, after incorporation of the characteristic convective term into the generalized diffusion term, is obtained with the following correspondence:

$$u(x) = u_0; \quad \hat{k}(x) = e^{-u_0 x}$$
$$\hat{w}(x) = \hat{k}(x); \quad \hat{d}(x) = 0; \quad \hat{P}(x, t, T) = P(x, t, T)\, \hat{k}(x) \quad (12.40\text{a--e})$$

and the corresponding eigenvalue problem is given by

$$\frac{d}{dx}\left[\hat{k}(x)\frac{d\psi(x)}{dx}\right] + \mu^2 \hat{k}(x)\psi(x) = 0, \quad 0 < x < 1 \quad (12.41\text{a})$$

with boundary conditions

$$\psi(0) = 0; \quad \psi(1) = 0 \quad (12.41\text{b,c})$$

First, the linear one-dimensional convection–diffusion problem is considered, with $b = 0$ in equation (12.38e). A maximum truncation order of $N = 100$ terms has been considered in the solution of the generalized diffusive eigenvalue problem, and $N < 100$ in computing the $T(x,t)$ expansion. Table 12.5 illustrates the convergence behavior of the eigenfunction expansions for $T(x,t)$, with $u_0 = 10$ and $b = 0$ (linear problem), while Table 12.6 provides results for the case with $u_0 = 10$ and $b = 5$ (nonlinear problem). In Table 12.5 for the linear situation, from those columns associated with the use of the convective eigenvalue problem (labeled "Conv."), one may clearly observe the marked gain in convergence rates in comparison with the diffusive alternative basis (labeled "Diff."), when the convective eigenvalue problem fully accounts for the influence of the linear convection term. From the first two columns, for $x = 0.5$ and $t = 0.05$, it can be observed that the results through the convective eigenvalue problem are already fully converged to four significant digits at truncation orders as low as $N = 4$, while the solution with the diffusive eigenvalue problem needs around $N = 40$ terms to achieve the same level of precision. For $x = 0.9$ and $t = 0.01$, slightly larger truncation orders are required, as expected for smaller values of t. For the convective basis, full convergence to four significant digits has already been achieved for $N = 10$, while truncation orders of around $N = 80$ are required through

TABLE 12.5

Convergence Analysis of Eigenfunction Expansions with Convective and Diffusive Eigenvalue Problems in the Solution of the One-Dimensional Linear Burgers' Equation

	$u_0 = 10$, $b = 0$ (Linear Problem)			
N	T(0.5,0.05) Conv.	T(0.5,0.05) Diff.	T(0.9,0.01) Conv.	T(0.9,0.01) Diff.
2	0.3865	0.4624	−1.814	0.4302
4	0.3797	0.3562	0.1994	0.6538
6	0.3797	0.3888	0.7033	0.7263
8		0.3754	0.7369	0.7484
10		0.3820	0.7374	0.7508
12		0.3783	0.7374	0.7464
14		0.3806		0.7411
16		0.3791		0.7372
18		0.3801		0.7353
20		0.3794		0.7350
30		0.3798		0.7382
40		0.3797		0.7371
50		0.3797		0.7376
60				0.7373
70				0.7375
80				0.7374
90				0.7374

the purely diffusive basis. The two solutions perfectly match each other, but in this linear case, the computational cost of the convective basis solution, which is fully explicit and analytical, is indeed negligible in comparison to the coupled transformed system solution with the usual diffusive eigenvalue problem choice.

For the nonlinear, one-dimensional problem, a maximum truncation order of N = 30 terms has been considered in the solution of the generalized diffusive eigenvalue problem and N < 30 in computing the potential expansion. From Table 12.6 for the nonlinear situation, one concludes that similar observations can be drawn with respect to the comparative behavior of the convective and diffusive basis, with a somehow less marked difference in this case, since for the nonlinear case, the convective eigenvalue problem does not fully account for the convective term influence but only for a characteristic behavior of the velocity coefficient. For instance, at x = 0.5 and t = 0.05, convergence to four significant digits is achieved for N as low as 10 for the convective basis, while the diffusive basis requires N = 30, while for x = 0.9 and t = 0.01, the convective basis yields four significant digits at N = 18, and the diffusive basis requires at least N = 30.

TABLE 12.6

Convergence Analysis of Eigenfunction Expansions with Convective and Diffusive Eigenvalue Problems in the Solution of the One-Dimensional Nonlinear Burgers' Equation

	$u_0 = 10$, $b = 5$ (Nonlinear Problem)			
N	T(0.5,0.05) Conv.	T(0.5,0.05) Diff.	T(0.9,0.01) Conv.	T(0.9,0.01) Diff.
2	0.2862	0.3606	−1.437	0.4504
4	0.2762	0.2496	0.4146	0.6881
6	0.2770	0.2877	0.7687	0.7808
8	0.2768	0.2718	0.7980	0.8125
10	0.2769	0.2795	0.8009	0.8162
12	0.2769	0.2753	0.8006	0.8104
14		0.2778	0.8001	0.8035
16		0.2762	0.7998	0.7984
18		0.2773	0.7996	0.7960
20		0.2765	0.7996	0.7956
22		0.2771		0.7965
24		0.2766		0.7977
26		0.2770		0.7988
28		0.2767		0.7995
30		0.2769		0.7996

In order to illustrate the proposed approach in a multidimensional situation, we borrow an example of the two-dimensional Burgers' equation, again allowing for direct comparison against previously published results through the GITT with a purely diffusive eigenfunction basis (Cotta et al. 2013). The problem here considered is written in dimensionless form as

$$\frac{\partial T(x,y,t)}{\partial t} + U_x(T)\frac{\partial T(x,y,t)}{\partial x} + U_y(T)\frac{\partial T(x,y,t)}{\partial y} = \frac{\partial^2 T(x,y,z,t)}{\partial x^2} + \frac{\partial^2 T(x,y,z,t)}{\partial y^2},$$
$$0 < x < 1, \quad 0 < y < 1, \quad t > 0 \tag{12.42a}$$

with initial and boundary conditions given by

$$T(x,y,0) = 1, \quad 0 \le x \le 1, \quad 0 \le y \le 1 \tag{12.42b}$$

$$T(0,y,t) = 0; \quad T(1,y,t) = 0, \quad t > 0 \tag{12.42c,d}$$

$$T(x,0,t) = 0; \quad T(x,1,t) = 0, \quad t > 0 \tag{12.42e,f}$$

and for the present application, the nonlinear functions $U_x(T)$ and $U_y(T)$ are taken as

$$U_x(T) = u_0 + b_x T; \quad U_y(T) = v_0 + b_y T \quad (12.42\text{g,h})$$

Again, from direct comparison against the multidimensional formulations in equations (12.34) and (12.35)a–i, taking the linear portion of the velocity coefficients to represent the characteristic convective terms and transporting the remaining terms to the nonlinear source term, the following coefficients correspondence can be reached:

$$u_x(\mathbf{r}) = u_0; \quad v_x(\mathbf{r}) = v_0; \quad w(\mathbf{r}) = 1; \quad k(\mathbf{r}) = 1; \quad d(\mathbf{r}) = 0;$$

$$P(\mathbf{r},t,T) = -\left(b_x T \frac{\partial T}{\partial x} + b_y T \frac{\partial T}{\partial y}\right); \quad \hat{k}_x(\mathbf{r}) = e^{-u_0 x}; \quad \hat{k}_y(\mathbf{r}) = e^{-v_0 y};$$

$$\hat{k}(\mathbf{r}) = \hat{k}_x(x)\hat{k}_y(y) = e^{-(u_0 x + v_0 y)}; \quad \hat{w}(\mathbf{r}) = \hat{k}(\mathbf{r}); \quad \hat{d}(\mathbf{r}) = 0;$$

$$\hat{P}(\mathbf{r},t,T) = P(\mathbf{r},t,T)\,\hat{k}(\mathbf{r})$$

$$(12.43\text{a–l})$$

The resulting transformed equation with the generalized diffusive representation becomes

$$\hat{k}(x,y)\frac{\partial T(x,y,t)}{\partial t} = \frac{\partial}{\partial x}\left[\hat{k}(x,y)\frac{\partial T}{\partial x}\right] + \frac{\partial}{\partial y}\left[\hat{k}(x,y)\frac{\partial T}{\partial y}\right] + \hat{P}(\mathbf{r},t,T),$$

$$0 < x < 1, \quad 0 < y < 1, \quad t > 0 \quad (12.44)$$

which leads to the following eigenvalue problem

$$\frac{\partial}{\partial x}\left[\hat{k}(x,y)\frac{\partial \psi}{\partial x}\right] + \frac{\partial}{\partial y}\left[\hat{k}(x,y)\frac{\partial \psi}{\partial y}\right] + \mu^2 \hat{k}(x,y)\psi(x,y) = 0, \quad 0 < x < 1, \quad 0 < y < 1$$

$$(12.45\text{a})$$

with boundary conditions

$$\psi(0,y) = 0; \quad \psi(1,y) = 0, \quad t > 0 \quad (12.45\text{b,c})$$

$$\psi(x,0) = 0; \quad \psi(x,1) = 0, \quad t > 0 \quad (12.45\text{d,e})$$

This two-dimensional eigenvalue problem with space variable coefficients is also readily solved by applying the GITT itself, based on a simpler auxiliary eigenvalue problem, as described in different sources (Cotta 1993; Naveira-Cotta et al. 2009; Cotta et al. 2016d).

Table 12.7 provides a convergence analysis for $T(x,y,t)$ in the two-dimensional Burgers' equation, with $u_0 = 10$, $v_0 = 1$, $b_x = 0$, $b_y = 0$ (linear problem). For this linear two-dimensional problem, a maximum truncation order of $N = 140$ terms has been considered in the solution of the generalized diffusive eigenvalue problem, and $N < 140$ in computing the potential expansion. The solution of the eigenvalue problems, either for the convective or diffusive basis, employs a reordering scheme based on the sum of the squared eigenvalues of the auxiliary problem. The results in Table 12.7 again

TABLE 12.7

Convergence Analysis of Eigenfunction Expansions with Convective and Diffusive Eigenvalue Problems in the Solution of the Two-Dimensional Linear Burgers' Equation

			$u_0 = 10, v_0 = 1, b_x = 0, b_y = 0$ (Linear Problem)			
N	T(0.5,0.1,0.01) Conv.	T(0.5,0.1,0.01) Diff.	T(0.1,0.1,0.01) Conv.	T(0.1,0.1,0.01) Diff.	T(0.9,0.1,0.05) Conv.	T(0.9,0.1,0.05) Diff.
10	0.5185	0.4857	0.1113	0.1334	0.1082	0.09789
20	0.4936	0.5014	0.1356	0.1551	0.1083	0.1023
30	0.4950	0.4978	0.1410	0.1381	0.1083	0.1062
40	0.4946	0.4924	0.1419	0.1452		0.1085
50	0.4947	0.4914	0.1421	0.1395		0.1094
60	0.4947	0.4941	0.1421	0.1408		0.1098
70		0.4951		0.1409		0.1099
80		0.4953		0.1410		0.1098
90		0.4936		0.1405		0.1095
100		0.4929		0.1414		0.1091
110		0.4931		0.1419		0.1092
120		0.4942		0.1411		0.1087
130		0.4947		0.1406		0.1088
140		0.4947		0.1419		0.1082

reconfirm the excellent convergence behavior of the expansions that follow the convective basis proposal, with four converged significant digits at truncation orders as low as $N = 50$ ($x = 0.5$, $y = 0.1$, $t = 0.01$), $N = 50$ ($x = 0.1$, $y = 0.1$, $t = 0.01$), and $N = 20$ ($x = 0.9$, $y = 0.1$, $t = 0.05$), as expected, with decreasing required truncation orders for larger values of the variable t. The results achieved by the diffusive basis are only converged for the largest truncation order here adopted, $N = 140$, but it is clearly noticeable that this solution presents an oscillatory behavior to reach convergence at the fourth significant digit. Again, it should be recalled that a fully analytical solution is achieved through the use of the convective basis in this linear example.

Thus, it has been demonstrated that the solutions based on the convective eigenvalue problem choice lead to markedly improved convergence rates, in comparison to the traditional solutions with the purely diffusive basis and with the convective terms fully incorporated into the source terms. Clearly, the adoption of a more informative eigenfunction expansion basis, including at least part of the convective term contribution, while at the same time reducing the importance of the source term that incorporates the convective terms, significantly contributes to the convergence acceleration. This alternative can be directly combined with the partial transformation scheme emphasized here, especially for nonlinear situations, when the convective information on the directions to be eliminated by integral transformation can be fully or partially casted into the generalized diffusion formulation. The approach also opens new perspectives for future work, as research advances toward the integral transform solution of nonlinear problems through the adoption of nonlinear eigenvalue problems (Cotta et al. 2016e).

PROBLEMS

12.1. The equation that governs the transient transpiration cooling of a slab can be written in dimensionless form as

$$\frac{\partial T(x,t)}{\partial t} + u\frac{\partial T(x,t)}{\partial x} = \frac{\partial^2 T(x,t)}{\partial x^2}, \quad 0 < x < 1, \; t > 0$$

Or according to the transformation in Section 2.5, for $u = $ constant as

$$e^{-ux}\frac{\partial T(x,t)}{\partial t} = \frac{\partial}{\partial x}\left[e^{-ux}\frac{\partial T(x,t)}{\partial x}\right], \quad 0 < x < 1, \; t > 0$$

with initial and boundary conditions given by

$$T(x,0) = 0, \quad 0 \le x \le 1$$
$$u\,T(0,t) - \frac{\partial T(0,t)}{\partial x} = 0; \quad \delta\,T(1,t) + \gamma\frac{\partial T(1,t)}{\partial x} = \varphi, \; t > 0$$

Obtain through GITT the hybrid solution of the transpiration cooling problem, in its original form with the convection term, by proposing a purely diffusive eigenvalue problem. Then, find the exact solution of the same problem, after the transformation that eliminates the convection term, through classical integral transforms.

12.2. Propose a solution via GITT of the nonself-adjoint eigenvalue problem below, adopting an auxiliary eigenvalue problem with constant coefficients. Employ the transformation of Section 12.5 to rewrite the problem as a self-adjoint formulation.

$$\frac{d}{dy}\left[k(y)\frac{d\varphi}{dy}\right] - v(y)\frac{d\varphi}{dy} + \mu^2 w(y)\varphi(y) = 0, \quad 0 < y < H$$

$$\left.\frac{d\varphi}{dy}\right|_{y=0} = 0; \quad \varphi(H) = 0$$

12.3. The bio-heat transfer equation for heterogeneous tissues, including the blood perfusion term and the metabolic heat generation, $g(x,t)$, can be written as

$$\rho(x)C_p(x)\frac{\partial T(x,t)}{\partial t} = \frac{\partial}{\partial x}\left[k(x)\frac{\partial T}{\partial x}\right] - \omega(x)\rho_b C_b(T(x,t) - T_a) + g_m(x,t)$$

$$t > 0, \ 0 < x < L$$

$$T(x,0) = T_p(x), \quad 0 < x < L$$

$$-k(0)\frac{\partial T(x,t)}{\partial x}\bigg|_{x=0} = q_0(t) + h[T_\infty(t) - T(0,t)], \quad t > 0$$

$$\frac{\partial T(x,t)}{\partial x}\bigg|_{x=L} = 0, \quad t > 0$$

where $T(x,t)$ is the local tissue temperature and $q_0(t)$ is the irradiated heat absorbed by the tissue surface. The remaining parameters are: ρ = specific mass of the tissue, C_p = specific heat of tissue, k = thermal conductivity of tissue, ω = blood perfusion rate, ρ_b = specific mass of blood, C_b = specific heat of blood, T_a = arterial blood temperature, L = thickness of tissue slab, $T_p(x)$ = initial temperature in tissue, h = heat transfer coefficient at tissue surface, and T_∞ = temperature of external environment. Establish the correspondence between this problem and the general formulation in equations (12.1a–e) and (12.2a–c), and obtain the integral transforms solution of this problem without repeating the solution steps, but just recovering it from the general formal solution. Employ a linear boundary conditions filter only.

12.4. The energy equation and boundary conditions for the determination of the temperature distribution for transient conjugated heat transfer over a flat plate can be written as

$$w(y)\frac{\partial T}{\partial t} + w_f u(x,y)\frac{\partial T}{\partial x} + w_f v(x,y)\frac{\partial T}{\partial y} = k(y)\frac{\partial^2 T}{\partial x^2} + \frac{\partial}{\partial y}\left[k(y)\frac{\partial T}{\partial y}\right] + g(y,t)$$

$$t > 0, \quad 0 < y < H, \quad 0 < x < L$$

$$T(x,y,0) = T_\infty$$

$$T(0,y,t) = T_\infty; \quad \left.\frac{\partial T}{\partial x}\right|_{x=L} = 0$$

$$\left.\frac{\partial T}{\partial y}\right|_{y=0} = 0; \quad T(x,H,t) = T_\infty$$

where $T(x,y,t)$ is the temperature field over the fluid–solid regions and $g(y,t)$ is the heat generation within the plate, which may vary over the thickness y and along time t. The remaining parameters are: $w(y)$ = thermal capacitance in the solid–fluid regions, $k(y)$ = thermal conductivity in the solid–fluid regions, w_f = thermal capacitance in the fluid, H = total thickness of the solid–fluid regions, L = plate length, T_∞ = free stream temperature, and $u(x,y)$ and $v(x,y)$ = velocity field components. Establish the correspondence between the present problem and the general formulation in equations (12.5) and (12.6a-d), so as to build the hybrid integral transforms solution under the partial transformation scheme, by keeping the x variable in the transformed problem. The space variable coefficients $w(y)$ and $k(y)$ should be accounted for in the eigenvalue problem. Adopt just a linear filter for the boundary conditions.

12.5. The mathematical formulation of a two-dimensional transient Burgers' equation is written as

$$\frac{\partial T(x,y,t)}{\partial t} + u(T)\frac{\partial T(x,y,t)}{\partial x} + v(T)\frac{\partial T(x,y,t)}{\partial y}$$
$$= \frac{\partial^2 T(x,y,t)}{\partial x^2} + \frac{\partial^2 T(x,y,t)}{\partial y^2}, \quad 0 < x < 1, \ 0 < y < 1, t > 0$$

with initial and boundary conditions

$$T(x,y,0) = 0, \quad 0 \leq x \leq 1, \ 0 \leq y \leq 1$$

$$T(0,y,t) = 1; \quad T(1,y,t) = 0, \ t > 0$$
$$\frac{\partial T(x,0,t)}{\partial y} = 0; \quad T(x,1,t) = 0, \ t > 0$$

with given nonlinear convection term coefficients, u(T) and v(T). Establish the correspondence between the general problem in equations (12.5) and (12.6a-d) with the present application. Without repeating the solution steps, write the formal solution through GITT under the partial transformation scheme, by eliminating the y variable through integral transformation, in which the boundary conditions are homogeneous.

12.6. The mathematical formulation of a three-dimensional Burgers' equation may be written as

$$\frac{\partial T(x,y,z,t)}{\partial t} + u_0 \frac{\partial T(x,y,z,t)}{\partial x} = \nu \left[\frac{\partial^2 T(x,y,z,t)}{\partial x^2} + \frac{\partial^2 T(x,y,z,t)}{\partial y^2} + \frac{\partial^2 T(x,y,z,t)}{\partial z^2} \right],$$

$$0 < x < 1, \quad 0 < y < 1, 0 < z < 1, \quad t > 0$$

with initial and boundary conditions

$$T(x,y,z,0) = 1, \quad 0 \le x \le 1, \quad 0 \le y \le 1, \quad 0 \le z \le 1$$

$$T(0,y,z,t) = 1; \quad T(1,y,z,t) = 0, \quad t > 0$$

$$\frac{\partial T(x,0,z,t)}{\partial y} = 0; \quad T(x,1,z,t) = 0, \quad t > 0$$

$$\frac{\partial T(x,y,0,t)}{\partial z} = 0; \quad T(x,y,1,t) = 0, \quad t > 0$$

with a linear convection term coefficient, u_0.

Establish the correspondence between the general problem in equations (12.5) and (12.6a-d) with the present application. Without repeating the solution steps, write the formal solution through GITT under the partial transformation scheme, eliminating the y and z variables through integral transformation.

12.7. Consider steady-state heat transfer in thermally developing, hydrodynamically developed forced laminar flow inside a rectangular micro-channel of cross section dimensions $2x_1$ and $2y_1$, under the following additional assumptions:
- The flow is incompressible with constant physical properties.
- Free convection of heat and energy viscous dissipation is negligible.
- The entrance temperature distribution is uniform.
- The temperature of the channel wall is prescribed and uniform.
- Axial diffusion of heat is negligible.

The temperature $T(x,y,z)$ of a fluid with a nonseparable velocity profile $u(x,y)$, thermal diffusivity α, flowing along the channel in the region $0 \leq z \leq \infty$, $0 \leq y \leq y_1$, $0 \leq x \leq x_1$ is then described by the following problem:

$$u(x,y)\frac{\partial T(x,y,z)}{\partial z} = \alpha\left[\frac{\partial^2 T(x,y,z)}{\partial y^2} + \frac{\partial^2 T(x,y,z)}{\partial x^2}\right],$$

$$0 < y < y_1, \quad 0 < x < x_1, z > 0$$

The boundary conditions at $y = 0$ and at the surface $y = y_1$ are

$$\frac{\partial T(x,y,z)}{\partial y}\bigg|_{y=0} = 0, \quad \beta_t \lambda \frac{\partial T(x,y,z)}{\partial y}\bigg|_{y=y1} = T_s - T(x,y_1,z), \quad z > 0$$

The boundary conditions at $x = 0$ and at the surface $x = x_1$ are

$$\frac{\partial T(x,y,z)}{\partial x}\bigg|_{x=0} = 0, \quad \beta_t \lambda \frac{\partial T(x,y,z)}{\partial x}\bigg|_{x=x1} = T_s - T(x_1,y,z), \quad z > 0$$

where $\beta_t = [(2-\alpha_t)/\alpha_t)(2\gamma/(\gamma+1)]/Pr$, α_t is the thermal accommodation coefficient, λ is the molecular mean free path, and $\gamma = C_p/C_v$, while C_p is the specific heat at constant pressure, C_v is the specific heat at constant volume, T_s is the temperature at the channel wall, and the Knudsen number is defined as $Kn = \lambda/2y_1$. The temperature distribution at the channel entrance is considered uniform

$$T(x,y,0) = T_{in}, \quad 0 \leq y \leq y_1$$

(a) Establish the correspondence between this problem and the general formulation in equations (12.1a-e) and (12.2a-c) and find the GITT solution via the total transformation scheme;

(b) Establish the correspondence between this problem and the general formulation in equations (12.5) and (12.6a-d) and find the GITT solution via the partial transformation scheme. Considering that $y_1 \ll x_1$, the diffusion effects are predominant in the y direction, which will be the independent variable to be eliminated by integral transformation. Adopt the eigenvalue problem with a parabolic velocity profile in the y direction (parallel plates formulation).

12.8. Introduce the axial diffusion term, $\dfrac{\partial^2 T(x, y, z)}{\partial z^2}$, in the formulation of Problem 12.7, for low values of Péclét number, and apply GITT in the total transformation scheme to eliminate both transversal coordinates x and y. Adopt a boundary condition of zero flux for a sufficiently large axial distance z = L. Propose an analytical solution for the transformed boundary value problem. Employ a pseudo-transient formulation so as to rederive the solution of the transformed boundary value problem as a one-dimensional partial differential system.

12.9. Consider the one-dimensional transient heat conduction problem that governs the temperature field T(x,t) for a heterogeneous plate of length L_x. The plate material heterogeneity is characterized by an x-variable graded thermal conductivity, k(x), and thermal capacity w(x). In addition, the plate is thermally thin in the thickness, z direction, so as to incorporate the boundary conditions in the diffusion equation through lumping, in the form of a variable prescribed heat flux on one face, $q_w(x,t)$, and both a nonlinear radiation boundary condition, with emissivity ε, and a space variable convective heat transfer coefficient at the other face, h(x). The dimensional problem formulation is written as

$$w(x)\frac{\partial T(x,t)}{\partial t} = \frac{\partial}{\partial x}\left[k(x)\frac{\partial T(x,t)}{\partial x}\right] - \frac{h(x)[T(x,t) - T_\infty]}{L_z}$$
$$- \frac{\varepsilon\sigma[T^4(x,t) - T^4_\infty]}{L_z} + \frac{q_w(x,t)}{L_z},$$
$$0 < x < L_x, t > 0$$

with initial and boundary conditions

$$T(x, t=0) = T_\infty$$

$$\left.\frac{\partial T}{\partial x}\right|_{x=0} = 0, \quad \left.\frac{\partial T}{\partial x}\right|_{x=L_x} = 0$$

where L_z is the plate thickness and T_∞ is the external ambient temperature.

Following the approach described in this chapter, obtain the hybrid GITT solution of the proposed problem, fully presenting the transformed system to be solved. Consider an eigenvalue problem that incorporates the variable thermal conductivity and capacity in its formulation. Then, consider a second eigenvalue problem that accounts for the convective and radiative heat exchanges through an effective heat transfer coefficient, $h_{ef}(x) = h(x) + h_r(x)$ as a dissipation term, where $h_r(x)$ is a linearized radiative heat transfer coefficient. Assume the eigenvalue problems' solutions are known.

12.10. Consider the two-dimensional transient heat conduction problem that governs the temperature field T(x,y,t), for a heterogeneous plate of dimensions L_x and L_y. The plate material heterogeneity is characterized by an x-variable graded thermal conductivity, k(x), and thermal capacity, w(x). In addition, the plate is thermally thin in the thickness, z direction, so as to incorporate the boundary conditions in the diffusion equation through lumping, in the form of a variable prescribed heat flux on one face, $q_w(x,y,t)$, and a space variable effective heat transfer coefficient at the other face, $h_{ef}(x,y)$. The dimensional problem formulation is written as

$$w(x)\frac{\partial T(x,y,t)}{\partial t} = \frac{\partial}{\partial x}\left[k(x)\frac{\partial T}{\partial x}\right] + k(x)\frac{\partial^2 T}{\partial y^2} + \frac{q_w(x,y,t)}{L_z}$$

$$- \frac{h_{ef}(x,y)}{L_z}(T - T_\infty),$$

$$0 < x < L_x, 0 < y < L_y, t > 0$$

with initial and boundary conditions

$$T(x,y,t=0) = T_\infty$$

$$\left.\frac{\partial T}{\partial x}\right|_{x=0} = 0, \quad \left.\frac{\partial T}{\partial x}\right|_{x=L_x} = 0$$

$$\left.\frac{\partial T}{\partial y}\right|_{y=0} = 0, \quad \left.\frac{\partial T}{\partial y}\right|_{y=L_y} = 0$$

where L_z is the plate thickness, and T_∞ is the external ambient temperature.

Following the approach described in this chapter, obtain the total and partial transformation schemes solutions of the proposed problem, fully presenting the transformed systems. In the partial transformation scheme, eliminate only the x variable, retaining the y variable in the transformed problem. Consider eigenvalue problems, either two-dimensional for the total transformation or one-dimensional for the partial transformation that incorporates the variable thermal conductivity and capacity in its formulation. Assume the eigenvalue problem solution is known.

Appendix A. Subroutine Gauss

```fortran
!SUBROUTINE GAUSS uses Gauss direct
!elimination method to solve
![a][t]=[d]
!
!m=dimension of the matrices
!a=coefficient matrix
!t=solution vector

subroutine gauss(m,a,d,t)
  implicit none
  integer i,j,m,m1,kk,jj,ll,mm,k,l
  real(8) a(m,m),d(m),t(m)
  real(8) ad(m,m+1)
  real(8) eps,atemp,div,amult,sum

  eps=epsilon(0.d0)
  m1=m+1

!Combine [a] and [d] into [ad] matrix
  do i=1,m
    do j=1,m1
      if(j.ne.m1)then
        ad(i,j)=a(i,j)
      else
        ad(i,j)=d(i)
      end if
    end do
  end do

!Check and exchange rows for zero-diagonals
  kk=0
  jj=0
  do i=1,m
    jj=kk+1
    ll=jj
    kk=kk+1
20  if(dabs(ad(jj,kk))-eps)21,21,22
21  jj=jj+1
    goto 20
22  if(ll-jj)23,24,23
```

```fortran
23  do mm=1,m1
      atemp=ad(ll,mm)
      ad(ll,mm)=ad(jj,mm)
      ad(jj,mm)=atemp
    end do

!Simplify the matrix into an upper triangular system
24  div=ad(i,i)
    do j=1,m1
      ad(i,j)=ad(i,j)/div
    end do
    k=i+1
    if(k-m1)12,13,13
12  do l=k,m
      amult=ad(l,i)
      do j=1,m1
        ad(l,j)=ad(l,j)-ad(i,j)*amult
      end do
    end do
  end do

!Backward substitution
13  t(m)=ad(m,m1)
    l=m
    do j=2,m
      sum=0.d0
      i=m1+1-j
      do k=i,m
        sum=sum+ad(i-1,k)*t(k)
      end do
      l=l-1
      t(l)=ad(i-1,m1)-sum
    end do
end subroutine gauss
```

Appendix B. Subroutine Trisol

```
!SUBROUTINE TRISOL uses the Thomas algorithm
!to solve a tri-diagonal matrix equation
!
!m=diagonal do the matrix (number of elements in the diagonal)
!a=off-diagonal term (lower)
!b=diagonal term
!c=off-diagonal term (upper)
!d=on input - right-hand-side
!  on output - solution

subroutine trisol(m,a,b,c,d)
  implicit none
  integer i,j,m
  real(8) a(m),b(m),c(m),d(m)
  real(8) r

!Establish upper triangular matrix
  do i=2,m
    r=a(i)/b(i-1)
    b(i)=b(i)-r*c(i-1)
    d(i)=d(i)-r*d(i-1)
  end do

!Back substitution
  d(m)=d(m)/b(m)
  do i=2,m
    j=m-i+1
    d(j)=(d(j)-c(j)*d(j+1))/b(j)
  end do

!Solution stored in d

end subroutine trisol
```

Appendix C. Subroutine SOR

```fortran
!SUBROUTINE SOR uses sucessive over relaxation to solve
![a][t]=[d]
!
!a=coefficient matrix
!d=right-hand-side vector
!t=on input - initial guess
!  on output - solution
!w=relaxation factor
!  set w=1 for Gauss-Seidel iteration
!m=dimension of matrices
!eps=convergence criteria
!iter=number of iterations for convergence
!maxiter=maximum number of iterations

subroutine sor(m,a,d,t,w,eps,maxiter,iter)
   implicit none
   integer m,maxiter,iter,i,j,flag1,flag2
   real(8) a(m,m),d(m),t(m)
   real(8) eps,w,er,err

!Check a sufficient criteria for convergence
   flag1=0
   flag2=1
   do i=1,m
     er=0.d0
     do j=1,m
       if(i.ne.j)then
         er=er+dabs(a(i,j))
       end if
       if(dabs(a(i,i)).lt.er)then
         flag1=1
       end if
       if(dabs(a(i,i)).gt.er)then
         flag2=0
       end if
     end do
   end do

   if((flag1.eq.1).or.(flag2.eq.1))then
     write(3,*)"A *sufficient* criteria for convergence"
```

```
      write(3,*)"   was not satisfied"
      write(3,*)"Solution might not be correct"
      write(*,*)"A *sufficient* criteria for convergence"
      write(*,*)"   was not satisfied"
      write(*,*)"Solution might not be correct"
    end if

    iter=0
1   do i=1,m
      er=0.d0
      err=0.d0
      do j=1,m
        er=er+a(i,j)*t(j)
      end do
      er=w*(d(i)-er)/a(i,i)
      t(i)=er+t(i)
      err=dmax1(err,dabs(er))
    end do

!err gives the maximum relative errors for current iteration
!
!Increase nuber of iterations by one
!Check iter and err
    iter=iter+1
    if(iter.ge.maxiter) goto 50
    if(err.ge.eps) goto 1
    return

!Number of iterations exceeds the maximum number allowed
50  write(3,90)maxiter
90  format("*** Convergence not reached after",i4," it")
    end subroutine sor
```

Appendix D. Subroutine BICGM2

```fortran
!***********************************************************
!BICONJUGATE GRADIENT METHOD
!Based on the code presented in Press et al (1992)
!n=size of the matrix (input)
!a=matrix containing n x n elements (input)
!x2=solution vector (output)
!b2=right hand side of the system (input)
!maxit=maximum number of iterations (input)
!tol=tolerance for convergence (input)
!info=0 (code ran without problems); 1 (an error was found)
!***********************************************************
subroutine bicgm2(n,a,x2,b2,maxit,tol,info)
 implicit none
 integer k,maxit,it,n,n26,i,j,ndm,info,it2
 real(8),dimension(:,:), allocatable :: aux
 real(8),dimension(:),allocatable:: r2,z2,rr2,zz2,p2,&
                          pp2,vd,vd2,vd3,precond,x2tmp
 real(8) a(n,n),x2(n),b2(n)
 real(8) tol,error,errorold,bk,bknum,bkden,ak,akden,bnorm
 REAL time1,time2,time3,time4
 INTEGER itime1,itime2,itime3,itime4

 info=0

 if(any(b2.eq.1.d200))then
  info=1
  return
 end if

 allocate(aux(n,n),r2(n),z2(n),rr2(n), &
         zz2(n),p2(n),pp2(n),vd(n),vd2(n), &
         vd3(n),precond(n),x2tmp(n))

 do k=1,n
  precond(k)=1.d0
  if(a(k,k).ne.0.d0)precond(k)=a(k,k)
 end do

 do k=1,n
  x2(k)=0.0d0
 end do
```

```
  it = 0
1 it2 = 0

  do k = 1,n
    r2(k) = 0.d0
  end do

  do i = 1,n
    do j = 1,n
      aux(j,i) = a(i,j)*x2(j)
    end do
  end do

  do i = 1,n
    do j = 1,n
      r2(i) = r2(i) + aux(j,i)
    end do
  end do
  do k = 1,n
    r2(k) = b2(k) - r2(k)
  end do

  do k = 1,n
    rr2(k) = 0.d0
  end do

  do i = 1,n
    do j = 1,n
      aux(j,i) = a(i,j)*r2(j)
    end do
  end do

  do i = 1,n
    do j = 1,n
      rr2(i) = rr2(i) + aux(j,i)
    end do
  end do

  do k = 1,n
    z2(k) = (b2(k)/precond(k))
  end do
  do k = 1,n
    vd(k) = z2(k)*z2(k)
  end do
```

Subroutine BICGM2

```
bnorm = 0.d0
do k = 1, n
  bnorm = bnorm + vd(k)
end do
bnorm = dsqrt(bnorm)
if(bnorm.eq.0.d0)then
  info = 1
  deallocate(aux,r2,z2,rr2,zz2,p2,pp2, &
             vd,vd2,vd3,precond,x2tmp)
  return
end if
if(2.d0*bnorm.ne.2.d0*bnorm)then
   info = 1
  deallocate(aux,r2,z2,rr2,zz2,p2,pp2, &
             vd,vd2,vd3,precond,x2tmp)
  return
end if

it = it + 1
it2 = it2 + 1

do k = 1, n
  z2(k) = (r2(k)/precond(k))
  zz2(k) = (rr2(k)/precond(k))
end do

do k = 1, n
  vd(k) = z2(k)*rr2(k)
end do

bknum = 0.d0
do k = 1, n
  bknum = bknum + vd(k)
end do

do k = 1, n
  p2(k) = z2(k)
  pp2(k) = zz2(k)
end do

bkden = bknum

 do k = 1, n
  z2(k) = 0.d0
 end do
```

```
do i = 1, n
  do j = 1, n
    aux(j,i) = a(i,j) * p2(j)
  end do
end do

do i = 1, n
  do j = 1, n
    z2(i) = z2(i) + aux(j,i)
  end do
end do

do k = 1, n
  vd(k) = z2(k) * pp2(k)
end do

akden = 0.d0
do k = 1, n
  akden = akden + vd(k)
end do
if (akden.eq.0.d0) then
  info = 1
  deallocate(aux, r2, z2, rr2, zz2, p2, pp2, &
             vd, vd2, vd3, precond, x2tmp)
  return
end if

ak = bknum/akden

do k = 1, n
  zz2(k) = 0.d0
end do

do i = 1, n
  do j = 1, n
    aux(j,i) = a(j,i) * pp2(j)
  end do
end do

do i = 1, n
  do j = 1, n
    zz2(i) = zz2(i) + aux(j,i)
  end do
end do
```

Subroutine BICGM2

```
do k=1,n
  vd(k)=x2(k)+ak*p2(k)
  vd2(k)=r2(k)-ak*z2(k)
  vd3(k)=rr2(k)-ak*zz2(k)
end do

do k=1,n
  x2(k)=vd(k)
  r2(k)=vd2(k)
  rr2(k)=vd3(k)
end do

do k=1,n
  z2(k)=(r2(k)/precond(k))
end do

do k=1,n
  vd(k)=z2(k)*z2(k)
end do

error=0.d0
do k=1,n
  error=error+vd(k)
end do
if(error.lt.0.d0)then
  info=1
  deallocate(aux,r2,z2,rr2,zz2,p2,pp2, &
             vd,vd2,vd3,precond,x2tmp)
  return
end if
if(2.d0*error.ne.2.d0*error)then
  info=1
  deallocate(aux,r2,z2,rr2,zz2,p2,pp2, &
             vd,vd2,vd3,precond,x2tmp)
  return
end if
error=dsqrt(error)/bnorm
errorold=error

do while (((it.lt.maxit).and.(error.gt.tol) &
       .and.(error.le.1.d2*errorold))
  it2=it2+1
  !if(it2.eq.n)goto 1
  if(error.le.errorold)then
```

```
    errorold = error
    do k = 1, n
      x2tmp(k) = x2(k)
    end do
  end if
  it = it + 1
  do k = 1, n
    zz2(k) = (rr2(k)/precond(k))
    vd(k) = z2(k) * rr2(k)
  end do

  bknum = 0.d0
  do k = 1, n
    bknum = bknum + vd(k)
  end do

  if (bkden.eq.0.d0) then
    info = 1
    deallocate(aux, r2, z2, rr2, zz2, p2, pp2, &
               vd, vd2, vd3, precond, x2tmp)
    return
  end if
  bk = bknum/bkden

  do k = 1, n
    vd(k) = bk*p2(k) + z2(k)
    vd2(k) = bk*pp2(k) + zz2(k)
  end do

  do k = 1, n
    p2(k) = vd(k)
    pp2(k) = vd2(k)
  end do

  bkden = bknum

  do k = 1, n
    z2(k) = 0.d0
  end do

  do i = 1, n
    do j = 1, n
      aux(j,i) = a(i,j) * p2(j)
    end do
  end do
```

Subroutine BICGM2

```
do i = 1,n
 do j = 1,n
  z2(i) = z2(i) + aux(j,i)
 end do
end do

do k = 1,n
 vd(k) = z2(k)*pp2(k)
end do

akden = 0.d0
do k = 1,n
 akden = akden + vd(k)
end do

if(akden.eq.0.d0)then
 info = 1
 deallocate(aux,r2,z2,rr2,zz2,p2,pp2, &
            vd,vd2,vd3,precond,x2tmp)
 return
end if
ak = bknum/akden

do k = 1,n
 zz2(k) = 0.d0
end do

do i = 1,n
 do j = 1,n
  aux(j,i) = a(j,i)*pp2(j)
 end do
end do

do i = 1,n
 do j = 1,n
  zz2(i) = zz2(i) + aux(j,i)
 end do
end do

do k = 1,n
 vd(k) = x2(k) + ak*p2(k)
 vd2(k) = r2(k) - ak*z2(k)
 vd3(k) = rr2(k) - ak*zz2(k)
end do
```

```
  do k = 1, n
    x2(k) = vd(k)
    r2(k) = vd2(k)
    rr2(k) = vd3(k)
  end do

  do k = 1, n
    z2(k) = (r2(k)/precond(k))
  end do

  do k = 1, n
    vd(k) = z2(k) * z2(k)
  end do

  error = 0.d0
  do k = 1, n
    error = error + vd(k)
  end do
  if(error.lt.0.d0)then
    info = 1
    deallocate(aux, r2, z2, rr2, zz2, p2, pp2, &
               vd, vd2, vd3, precond, x2tmp)
    return
  end if
  if(2.d0 * error.ne.2.d0 * error)then
    info = 1
    deallocate(aux, r2, z2, rr2, zz2, p2, pp2, &
               vd, vd2, vd3, precond, x2tmp)
    return
  end if
  error = dsqrt(error)/bnorm
end do
tol = errorold
maxit = it
do k = 1, n
  x2(k) = x2tmp(k)
end do
deallocate(aux, r2, z2, rr2, zz2, p2, pp2, &
           vd, vd2, vd3, precond, x2tmp)
end subroutine bicgm2
```

Appendix E. Program to Solve Example 10.1

```
!This program reads the input data, calls trisol to solve
!tridiagonal matrix equations for 1-dimensional single-phase
!solidification problem defined by example 10-1 in chapter 10
!
!a=off-diagonal terms (lower)
!b=diagonal terms
!c=off-diagonal terms (upper)
!d=on input - right-hand-side vector
!  on output - solution
!alpha=thermal diffusivity of the solid
!bb=thickness of the region
!dt=time step
!dtnew=new time step
!dx=space interval
!h=heat transfer coefficient
!maxit=maximum number of iterations
!n=number of space intervals over the thickness of the region
!nit=number of iterations
!s=interface position
!t=temperature
!tinf-surrounding temperature
!tmelt=melting temperature
!tol=relative error in dt

program example_10_1
  implicit none
  integer n,maxit,i,nit,j
  real(8) a(100),b(100),c(100),d(100),t(100),s(100)
  real(8) tol,h,tinf,alpha,tmelt,bb,dx,dt,dtnew
  real(8) a1,b1,c1,a2,b2,c2,r,t02,t12,ratio
!Read input data
  write(*,*)"Enter input data"
  read(*,*)n
  read(*,*)tol
  read(*,*)maxit
  read(*,*)h
  read(*,*)tinf
  read(*,*)alpha
```

```
      read(*,*)tmelt
      read(*,*)bb

!Space interval
      dx=bb/dble(n)

!Interface position
      do i=1,n
        s(i)=dx*dble(i)
      end do

      write(3,1)

!Calculate numerical values (time step, surface temperature,
!                            number of iterations)
      do i=1,n
        nit=0
!for the time step dt0 (Eq. 10-26)
        if(i.eq.1)then
          dt=dx*(1.d0+h*dx)/(h*(tmelt-tinf))
          t(1)=(1.d0-tinf*dx)/(1.d0+h*dx)
          goto 15
        elseif(i.eq.2)then
          a1=-(1.d0+h*dx)
          b1=1.d0
          c1=tinf*dx
   13     r=dt/(dx*dx)
          a2=-alpha*r
          b2=2.d0*alpha*r+1.d0
          c2=1.d0+alpha*r
          t02=(c1*b2-c2*b1)/(a1*b2-a2*b1)
          t12=(a1*c2-a2*c1)/(a1*b2-a2*b1)
          dtnew=(dx*dx)/(1.d0-t12)
          ratio=dabs((dt-dtnew)/dt)
          if(ratio.le.tol)then
            t(1)=t02
            t(2)=t12
            goto 15
          elseif(nit.le.maxit)then
            nit=nit+1
            dt=dtnew
            goto 13
          else
            write(3,2)i
            stop
```

Program to Solve Example 10.1

```
      end if
!for the time steps dtn, n=2,3,...
    elseif(i.ge.3)then
14  r=dt/(dx*dx)
!set up the coefficients of tri-diagonal matrix
    b(1) =-(1.d0+h*dx)
    c(1) =1.d0
    d(1) =tinf*dx
    do j=2,i-1
      a(j) =-alpha*r
      b(j) =(2.d0*alpha*r+1.d0)
      c(j) =-alpha*r
      d(j) =t(j)
    end do
    a(i) =-alpha*r
    b(i) =(2.d0*alpha*r+1.d0)
    d(i) =1.d0+alpha*r
!call subroutine trisol
    call trisol2(i,a,b,c,d)
    dtnew=(dx*dx)/(1.d0-d(i))
    ratio=dabs((dt-dtnew)/dt)
!check convergence
    if(ratio.le.tol)then
      do j=1,i
        t(j) =d(j)
      end do
      goto 15
    elseif(nit.le.maxit)then
      nit=nit+1
      dt=dtnew
      goto 14
    else
      write(3,2)i
      stop
    end if
  end if
15  continue
    write(3,3)s(i),dt,t(1),nit
  end do
1 format("Interface position     Time step    ", &
        "T(0,t)   Number of iterations")
2 format(5x,"When i= ",i3," # of iterations > maxit ")
3 format(8x,f6.4,13x,f6.4,6x,f6.4,12x,i2)
end program example_10_1
```

```fortran
!SUBROUTINE TRISOL uses the Thomas algorithm
!to solve a tri-diagonal matrix equation
!
!m=diagonal do the matrix
!a=off-diagonal term (lower)
!b=diagonal term
!c=off-diagonal term (upper)
!d=on input - right-hand-side
!  on output - solution

subroutine trisol2(m,a,b,c,d)
  implicit none
  integer i,j,m
  real(8) a(m),b(m),c(m),d(m)
  real(8) r

!Establish upper triangular matrix
  do i=2,m
   r=a(i)/b(i-1)
   b(i)=b(i)-r*c(i-1)
   d(i)=d(i)-r*d(i-1)
  end do

!Back substitution
  d(m)=d(m)/b(m)
  do i=2,m
   j=m-i+1
   d(j)=(d(j)-c(j)*d(j+1))/b(j)
  end do

!Solution stored in d

end subroutine trisol2
```

Bibliography

Allada, S. R. and D. Quon. (1966). A Stable, Explicit Numerical Solution of the Conduction Equation for Multidimensional Nonhomogeneous Media, *Heat Transf. Los Angeles Chem. Eng. Symp. Ser.*, 62, 151–156.

Ames, W. F. (1977). *Numerical Methods for Partial Differential Equations*, 2nd ed., Academic Press, New York.

Amsden, A. A. and C. W. Hirt. (1973). A Simple Scheme for Generating Curvilinear Grids, *J. Comput. Phys.*, 11, 348–359.

Anderson, J. (1990). *Modern Compressible Flow with Historical Perspective*, McGraw-Hill, New York.

Anderson, J. (1995). *Computational Fluid Dynamics: The Basics with Applications*, McGraw-Hill, Singapore.

André, S. and A. Degiovanni. (1995). A Theoretical Study of the Transient Coupled Conduction and Radiation Heat Transfer in Glass: Phonic Diffusivity Measurements by the Flash Technique, *Int. J. Heat Mass Transf.*, 38, 3401–3412.

Aparecido, J. B. and R. M. Cotta. (1990). Laminar Flow Inside Hexagonal Ducts, *Comp. Mech.*, 6, 93–100.

Aparecido, J. B., R. M. Cotta and M. N. Özişik. (1989). Analytical Solutions to Two-Dimensional Diffusion Type Problems in Irregular Geometries, *J. Franklin Inst.*, 326, 421–434.

Arpaci, V. S. (1966). *Conduction Heat Transfer*, Addison-Wesley, Reading, MA.

ASME V&V 20-2009. (2009). *Standard for Verification and Validation in Computational Fluid Dynamics and Heat Transfer*, ASME, New York.

ASTM Standard E1461-01. (2001). *Standard Test Method for Thermal Diffusivity by the Flash Method*, ASTM, West Conshohocken, PA.

Atkinson, K. E. (1978). *An Introduction to Numerical Analysis*, Wiley, New York.

Barakat, H. Z. and J. A. Clark. (1966). On the Solution of Diffusion Equation by Numerical Methods, *J. Heat Transf.*, 88, 421–427.

Barichello, L. B. (2011). Explicit Formulations for Radiative Transfer Problems, in *Thermal Measurements and Inverse Techniques*, H. Orlande, O. Fudym, D. Maillet, and R. Cotta (Eds.), CRC Press, Boca Raton, FL, pp. 541–562.

Barichello, L. B. and C. E. Siewert. (1999). A Discrete-Ordinates Solution for a Non-Grey Model with Complete Frequency Redistribution, *J. Quant. Spectrosc. Radiat. Transf.*, 62, 665–675.

Barichello, L. B. and C. E. Siewert. (2002). A New Version of the Discrete-Ordinates Method, *Proceedings: Computational Heat and Mass Transfer-2001*, Rio de Janeiro, pp. 340–347.

Bayazitoglu, Y. and M. N. Özişik. (1980). On the Solution of Graetz Type Problems with Axial Conduction, *Int. J. Heat Mass Transf.*, 23, 1399–1402.

Bejan, A. (1984). *Convection Heat Transfer*, Wiley-Interscience, New York.

Benjapiyaporn, C., V. Timchenko, E. Leonardi, G. de Vahl Davis and H. C. de Groh, III. (2000). Effects of Space Environment on Flow and Concentration during Directional Solidification, *Int. J. Fluid Dyn.*, 4, 1–27.

Bennacer, R. and D. Gobin. (1996). Cooperating Thermosolutal Convection in Enclosures—I. Scale Analysis and Mass Transfer, *Int. J. Heat Mass Transf.*, 39, 2671–2681.

Benson, A. R. and D. S. McRae. (1990). A Three-Dimensional Dynamic Solution—Adaptive Mesh Algorithm, AIAA Paper No. 90-1566, *AIAA 21st Fluid Dynamics, Plasma Dynamics and Lasers Conference*, June 18–20, Seattle, WA.

Berezin, I. S. and N. P. Zhidkov. (1965). *Computing Methods*, Vol. 1, Addison-Wesley, Reading, MA, pp. 210–215.

Bertrand, O., B. Binet, H. Combeau, S. Couturier, Y. Delanny, D. Gobin, M. Lacroix, P. L. Quéré, M. Médale, J. Mencinger, H. Sadat and G. Vieira. (1999). Melting Driven by Natural Convection—A Comparison Exercise: First Results, *Int. J. Therm. Sci.*, 38, 5–26.

Beyeler, E. P., B. A. Yost and S. I. Guceri. (1987). *Two-Dimensional Solidification in Irregularly Shaped Domains*, Report # CCM-87-01, Center for Composite Materials, University of Delaware, Newark, DE.

Bogado Leite, S. Q., M. N. Özişik and K. Verghese. (1980). On the Solution of Linear Diffusion Problems in Media with Moving Boundaries, *Nucl. Sci. Eng.*, 76, 345–350.

Bonacina, C., G. Comini, A. Fasano and M. Primicerio. (1973). Numerical Solutions of Phase Change Problems, *Int. J. Heat Mass Transf.*, 16, 1825–1832.

Brackbill, J. U. (1982). Coordinate System Control: Adaptive Meshes, in *Numerical Grid Generation*, Joe F. Thompson (Ed.), Elsevier, New York, pp. 277–293.

Brackbill, J. U. and J. S. Saltzman. (1982). Adaptive Zoning for Singular Problems in Two Dimensions, *J. Comput. Phys.*, 46, 342–368.

Brazhnikov, A. M., V. A. Karpychev and A. V. Luikov. (1975). One Engineering Method of Calculating Heat Conduction Processes, *Inzhenerno Fizicheskij Zhurnal*, 28, 677–680.

Brian, P. L. T. (1961). A Finite Difference Method of High-Order Accuracy for the Solution of Three-Dimensional Transient Heat Conduction Problems, *Am. Inst. Chem. Eng. J.*, 7, 367–370.

Briley, W. R. and H. McDonald. (1973). Solution of Three-Dimensional Compressible Navier-Stokes Equations by an Implicit Technique, *Proceedings of the Fourth International Conference on Numerical Methods in Fluid Dynamics*, Boulder, CO, Lecture Notes in Physics, Vol. 35, Springer-Verlag, New York, pp. 105–110.

Buzbee, B. L., G. H. Golub and C. W. Nielson. (1970). On Direct Methods for Solving Poisson's Equation, *SIAM J. Numer. Anal.*, 7, 627–656.

Camaréro, R., B. Ozzel, H. Yang, H. Zhang and C. Dupvis. (1981). Computed Aided Grid Design, in *Numerical Grid Generation in Computational Fluid Dynamics*, J. Hauser and C. Taylor (Eds.), Pineridge Press, Swansea, UK, pp. 15–34.

Caretto L. S., A. D. Gosman, S. V. Patankar and D. B. Spalding. (1973). Two calculation procedures for steady, three-dimensional flows with recirculation, in *Proceedings of the Third International Conference on Numerical Methods in Fluid Mechanics*. Lecture Notes in Physics, H. Cabannes and R. Temam (Eds.), Vol. 19, Springer, Berlin, Heidelberg.

Carey, G. (1997). *Computational Grids: Generation, Adaption and Solution Strategies*, Taylor & Francis, New York.

Carey, G. F. and M. Tsai. (1982). Hyperbolic Heat Transfer with Reflection, *Numer. Heat Transf.*, 5, 309–327.

Carslaw, H. S. and J. C. Jaeger. (1959). *Conduction of Heat in Solids*, 2nd ed., Oxford University Press, New York.

Carter, J. E. (1971). *Numerical Solutions of the Supersonic, Laminar Flow Over a Two-Dimensional Compression Corner*, Ph.D. Thesis, Virginia Polytechnic Institute and State University, Blacksburg, VA.
Carvalho, T. M. B., R. M. Cotta and M. D. Mikhailov. (1993). Flow Development in the Entrance Region of Ducts, *Comm. Numer. Meth. Eng.*, 9, 503–509.
Castellões, F. V., C. R. Cardoso, P. Couto and R. M. Cotta. (2007). Transient Analysis of Slip Flow and Heat Transfer in Microchannels, *Heat Transf. Eng.*, 28, no. 6, 549–558.
Castellões, F. V. and R. M. Cotta. (2006). Analysis of Transient and Periodic Convection in Microchannels via Integral Transforms, *Progress in Computational Fluid Dynamics*, 6, no. 6, 321–326.
Cattaneo, C. (1958). A Form of Heat Conduction Equation Which Eliminates the Paradox of Instantaneous Propagation, *Comptes Rendus Hebdomadaires des Seances de 'l'Academie des Sciences*, 247, 431–433.
Cebeci, T. and P. Bradshaw. (1977). *Momentum Transfer in Boundary Layers*, Hemisphere/McGraw-Hill, New York.
Cebeci, T. and P. Bradshaw. (1984). *Physical and Computational Aspects of Convective Heat Transfer*, Springer-Verlag, New York.
Cebeci, T. and A. M. O. Smith. (1974). *Analysis of Turbulent Boundary Layers*, Academic Press, New York.
Chandrasekhar, S. (1960). Radiative Transfer, Dover Publications, New York.
Chester, M. (1963). Second Sound in Solids, *Phys. Rev.*, 131, 2013–2015.
Cho, S. and S. Krishnan. (2013). *Cancer Nanotechnology*, CRC Press, Boca Raton, FL.
Chow, C. Y. (1979). *An Introduction to Computational Fluid Dynamics*, Wiley, New York.
Chu, W. H. (1971). Development of a General Finite Difference Approximation for the General Domain, *J. Comput. Phys.*, 8, 392–408.
Chui, E. H. and Raithby, G. D. (1993). Computation of Radiant Heat Transfer on a Nonorthogonal Mesh Using the Finite-Volume Method, *Numer. Heat Transf. Part B*, 23, 269–288.
Churchill, R. V. (1948). *Introduction to Complex Variables*, McGraw-Hill, New York.
Colaço, M. J. and G. Dulikravich. (2006). A Multilevel Hybrid Optimization of Magnetohydrodynamic Problems in Double-Diffusive Fluid Flow, *J. Phys. Chem. Solids*, 67, 1965–1972.
Colaço, M. J. and G. S. Dulikravich. (2007). Solidifcation of Double-Diffusive Flows using Thermo-Magneto-Hydrodynamics and Optimization, *Mater. Manuf. Process.*, 22, 594–606.
Colaço, M. J., G. S. Dulikravich., T. J. Martin, and S. Lee (2003). An Inverse Method Allowing User-Specified Layout of Magnetized Micro-Fibers in Solidifying Composites, *Journal of Composite Materials*, UK, 37, pp. 1351–1365.
Colaço, M. J., G. S. Dulikravich and T. J. Martin. (2004). Optimization of Wall Electrodes for Electro-Hydrodynamic Control of Natural Convection during Solidification, *Mater. Manuf. Process.*, 19, 719–736.
Colaço, M. J., G. S. Dulikravich and T. J. Martin. (2005). Control of Unsteady Solidification via Optimized Magnetic Fields, *Mater. Manuf. Process.*, 20, 435–458.
Colaço, M. J. and H. R. B. Orlande. (2001a). Inverse Forced Convection Problem of Simultaneous Estimation of Two Boundary Heat Fluxes in Irregularly Shaped Channels, *Numer. Heat Transf. Part A, App.*, 39, 737–760.
Colaço, M. J. and H. R. B. Orlande. (2001b). Inverse Problem of Simultaneous Estimation of Two Boundary Heat Fluxes in Parallel Plate Channels, *J. Braz. Soc. Mech. Sci. Eng.*, XXIII, 201–215.

Colaço, M. J. and H. R. B. Orlande. (2002). Inverse Convection Problems in Irregular Geometries, *21st Southeastern Conference on Theoretical and Applied Mechanics*, Orlando, FL, pp. 423–432.
Colaço, M. J. and H. R. B. Orlande. (2004). Inverse Natural Convection Problem of Simultaneous Estimation of Two Boundary Heat Fluxes in Irregular Cavities, *Int. Heat Mass Transf.*, 47, 1201–1215.
Colaço, M. J., H. R. B. Orlande and G. S. Dulikravich. (2006). Inverse and Optimization Problems in Heat Transfer, *J. Braz. Soc. Mech. Sci. Eng.*, 28, 1–24.
Colaço, M. J., C. V. Teixeira and L. M. Dutra. (2010). Thermal Analysis of a Diesel Engine Operating with Diesel-Biodiesel Blends, *Fuel*, 89, 3742–3752.
Collatz, L. (1960). *The Numerical Treatment of Differential Equations*, Springer-Verlag, Berlin.
Conte, S. D. and C. de Boor. (1972). *Elementary Numerical Analysis, An Algorithm Approach*, 2nd ed., McGraw-Hill, New York.
Correa, E. J., R. M. Cotta and H. R. B. Orlande. (1997). On the Reduction of Computational Costs in Eigenfunction Expansions of Multidimensional Diffusion Problems, *Int. J. Num. Meth. Heat Fluid Flow*, 7, no. 7, 675–695.
Cotta, R. M. (1986). Diffusion in Media with Prescribed Moving Boundaries: Application to Metals Oxidation at High Temperatures, *Proceedings of the 2nd Latin American Congress of Heat & Mass Transfer*, Vol. 1, Sao Paulo, Brazil, pp. 502–513.
Cotta, R. M. (1990). Hybrid Numerical-Analytical Approach to Nonlinear Diffusion Problems, *Numer. Heat Transf. B.*, 127, 217–226.
Cotta, R. M. (1993). *Integral Transforms in Computational Heat and Fluid Flow*, CRC Press, Boca Raton, FL.
Cotta, R. M. (1994a). The Integral Transform Method in Computational Heat and Fluid Flow, Special Keynote Lecture, *Proceedings of the 10th International Heat Transfer Conference*, Vol. 1, Brighton, UK, SK-3, pp. 43–60, August.
Cotta, R. M. (1994b). Benchmark Results in Computational Heat and Fluid Flow: The Integral Transform Method, *Int. J. Heat Mass Transf.* (Invited Paper), 37, 381–394.
Cotta, R.M., (1996). *Integral Transforms in Transient Convection: Benchmarks and Engineering Simulations*, Invited Keynote Lecture, ICHMT International Symposium on Transient Convective Heat Transfer, Turkey, pp. 433–453.
Cotta, R. M. (Ed.). (1998). *The Integral Transform Method in Thermal and Fluids Sciences and Engineering*, Begell House, New York.
Cotta, R. M. and J. E. V. Gerk. (1994). Mixed Finite Difference/Integral Transform Approach for Parabolic-Hyperbolic Problems in Transient Forced Convection, *Numer. Heat Transf. B Fund.*, 25, 433–448.
Cotta, R. M., S. Kakaç, M. D. Mikhailov, F. V. Castellões and C.R. Cardoso. (2005). Transient Flow and Thermal Analysis in Microfluidics, in *Microscale Heat Transfer—Fundamentals and Applications*, S. Kakaç, L. Vasiliev, Y. Bayazitoglu and Y. Yener (Eds.), NATO ASI Series, Kluwer Academic Publishers, Dordrecht, The Netherlands, pp. 175–196.
Cotta, R. M., D. C. Knupp and C. P. Naveira-Cotta. (2016a). *Analytical Heat and Fluid Flow in Microchannels and Microsystems*, Mechanical Engineering Series, Springer, New York.
Cotta, R. M., D. C. Knupp and C. P. Naveira-Cotta. (2016b). Integral Transforms in Linear or Nonlinear Convection-Diffusion Through Convective Eigenvalue Problems, *24th International Congress on Theoretical and Applied Mechanics, XXIV ICTAM*, Montreal, Canada, pp. 21–26, August.

Cotta, R. M., D. C. Knupp, C. P. Naveira-Cotta, J. L. Z. Zotin and P. C. Pontes. (2016c). Eigenfunction Expansions for Coupled Nonlinear Convection-Diffusion Problems in Complex Physical Domains, Invited Plenary Lecture, *7th European Thermal Sciences Conference, EUROTHERM 2016*, Krakow, Poland, June 2016; also, *J. Phys. Conf. Ser.*, 745, 022001, 1–20.

Cotta, R. M., D. C. Knupp and C. P. Naveira-Cotta. (2016d). Enhanced Eigenfunction Expansions in Convection-Diffusion Problems with Multiscale Space Variable Coefficients, *Num. Heat Transf. A. Appl.*, 70, no. 5, 492–512.

Cotta, R. M., D. C. Knupp and C. P. Naveira-Cotta. (2016e). Nonlinear Eigenvalue Problem in the Integral Transforms Solution of Convection-Diffusion with Nonlinear Boundary Conditions, *Int. J. Num. Meth. Heat Fluid Flow* (Invited Paper, 25th Anniversary Special Issue), 26, nos. 3&4, 767–789.

Cotta, R. M., D. C. Knupp, C. P. Naveira-Cotta, L. A. Sphaier, and J. N. N. Quaresma. (2013). Unified Integral Transform Algorithm for Solving Multidimensional Nonlinear Convection-Diffusion Problems, *Num. Heat Transf. A—Appl.*, 63, no. 11, 840–866.

Cotta, R. M., D. C. Knupp, C. P. Naveira-Cotta, L. A. Sphaier and J. N. N. Quaresma. (2014). The Unified Integral Transforms (UNIT) Algorithm with Total and Partial Transformation, *Comput. Therm. Sci.*, 6, no. 6, 507–524.

Cotta, R. M. and M. D. Mikhailov. (1997). *Heat Conduction: Lumped Analysis, Integral Transforms, Symbolic Computation*, Wiley-Interscience, Chichester, UK.

Cotta, R. M. and M. D. Mikhailov. (2005). Semi-Analytical Evaluation of Integrals for the Generalized Integral Transform Technique, *4th Workshop on Integral Transforms and Benchmark Problems*, Rio de Janeiro, August.

Cotta, R. M. and M. D. Mikhailov. (2006). Hybrid Methods and Symbolic Computations, in *Handbook of Numerical Heat Transfer*, 2nd ed., W. J. Minkowycz, E. M. Sparrow and J. Y. Murthy (Eds.), Wiley, New York. pp. 493–522.

Cotta, R. M., M. D. Mikhailov and M. N. Özişik. (1987). Transient Conjugated Forced Convection in Ducts with Periodically Varying Inlet Temperature, *Int. J. Heat Mass Transf.*, 30, no. 10, 2073–2082.

Cotta, R.M., C.P. Naveira-Cotta and D.C. Knupp (2017). Convective Eigenvalue Problems for Convergence Enhancement of Eigenfunction Expansions in Convection-Diffusion Problems, *ASME J. Thermal Science and Eng. Appl.*, (in press).

Cotta, R. M., H. R. B. Orlande, M. D. Mikhailov and S. Kakaç. (2003). Experimental and Theoretical Analysis of Transient Convective Heat and Mass Transfer: Hybrid Approaches, Invited Keynote Lecture, *ICHMT International Symposium on Transient Convective Heat And Mass Transfer in Single and Two-Phase Flows*, Cesme, Turkey, 17–22 August.

Cotta, R. M. and M. N. Özişik. (1986a). Laminar Forced Convection in Ducts with Periodic Variation of Inlet Temperature, *Int. J. Heat Mass Transf.*, 29, no. 10, 1495–1501.

Cotta, R. M. and M. N. Özişik. (1986b). Transient Forced Convection in Laminar Channel Flow with Stepwise Variations of Wall Temperature, *Can. J. Chem. Eng.*, 64, 734–742.

Cotta, R. M. and M. N. Özişik. (1987). Diffusion Problems with General Time-Dependent Coefficients, *J. Braz. Assoc. Mech. Sci.*, 9, no. 4, 269–292.

Cotta, R. M., M. N. Özişik and D. S. McRae. (1986). Transient Heat Transfer in Channel Flow with Step Change in Inlet Temperature, *Numer. Heat Transf.*, 9, 619–630.

Cotta, R. M., J. N. N. Quaresma, L. A. Sphaier and C. P. Naveira-Cotta. (2010). Unified Integral Transform Approach in the Hybrid Solution of Multidimensional

Nonlinear Convection-Diffusion Problems, *14th International Heat Transfer Conference*, Washington, DC, August.

Cotta, R. M. and C. A. C. Santos. (1992). Transient Diffusion Problems with Time-Dependent Boundary Condition Coefficients, *Inzh. Fizich. Z.*, 61, no. 5, 829–837 (in Russ); also *J. Eng. Phys.*, 61, no. 5, 1411–1418.

Cotta, R. M. and R. Serfaty. (1992). Hybrid Analysis of Transient Nonlinear Convection-Diffusion Problems, *Int. J. Num. Meth. Heat Fluid Flow*, 2, 55–62.

Coulter, J. P., S. D. Gilmore and S. I. Guceri. (1986). "TGFLOW" a Software Package for the Analysis of Laminar Fluid Flow, in *Numerical Grid Generation in Computational Fluid Dynamics*, J. Häuser and C. Taylor (Eds.), Pineridge Press, Swansea, UK, pp. 515–526.

Coulter, J. P. and S. I. Guceri. (1987). Laminar and Turbulent Natural Convection within Irregularly Shaped Enclosures, *Numer. Heat Transf.*, 12, 211–227.

Coulter, J. P. and S. I. Guceri. (1988). Resin Impregnation during the Manufacturing Composite Materials Subject to Prescribed Injection Rate, *J. Reinforc. Plast. Compos.*, 7, 200–219.

Courant, R. (1956). *Differential and Integral Calculus*, Blackie & Son, Ltd., London, p. 133.

Courant, R., E. Isaacson and M. Rees. (1952). On the Solution of Non-Linear Hyperbolic Differential Equations by Finite Differences, *Comm. Pure Appl. Math.*, 5, 243.

Crandal, S. H. (1956). *Engineering Analysis*, McGraw-Hill, New York.

Crank, J. (1957). Two Methods for the Numerical Solution of Moving Boundary Problems in Diffusion and Heat How, *J. Mech. Appl. Math.*, 10, 220–231.

Crank, J. (1984). *Free and Moving Boundary Problems*, Oxford University Press, New York.

Crank, J. and R. S. Gupta. (1972). A Method of Solving Moving Boundary Problems in Heat Flow Using Cubic Splines or Polynomials, *J. Inst. Math. Appl.*, 10, 296–304.

Crank, J. and P. Nicolson. (1947). A Practical Method for Numerical Evaluation of Solution of Partial Differential Equations of the Heat Conduction Type, *Proc. Camb. Phil. Soc.*, 43, 50–67.

Crowley, A. B. (1978). Numerical Solution of Stefan Problems, *Int. J. Heat Mass Transf.*, 21, 215–219.

Davis, R. T. (1979). 4th Computational Fluid Dynamics Conference, *Numerical Methods for Coordinate Generation Based on Schwarz-Christoffel Transformations*, AIAA Paper 79-1463, Williamsburg, VA.

Doormaal, J. P. V. and G. D. Raithby. (1984). Enhancements of the Simple Method for Predicting Incompressible Fluid Flow, *Numer. Heat Transf.*, 7, 147–163.

Dorr, F. W. (1970). The Direct Solution of the Discrete Poisson Equation on a Rectangle, *SIAM Rev.*, 12, 248–263.

Douglas, J. (1955). On the Numerical Integration of $\partial^2 u/\partial x^2 + \partial^2 u/\partial y^2 = \partial u/\partial t$ by Implicit Methods, *J. Soc. Ind. Appl. Math.*, 3, 42–65.

Douglas, J. (1962). Alternating Direction Methods for Three Space Variables, *Numer. Math.*, 4, 41–63.

Douglas, J. and T. M. Gallie. (1955). On the Numerical Integration of a Parabolic Differential Equation Subject to a Moving Boundary Condition, *Duke Math. J.*, 22, 557–570.

Douglas, J. and J. E. Gun. (1964). A General Formulation of Alternating Direction Methods-Part I, Parabolic and Hyperbolic Problems, *Numer. Math.*, 6, 428–453.

Douglas, J. and H. H. Rachford. (1956). On the Numerical Solution of Heat Conduction Problems in Two and Three Space Variables, *Trans. Am. Math. Soc.*, 82, 421–439.
DuFort, E. C. and S. P. Frankel. (1953). Stability Conditions in the Numerical Treatment of Parabolic Differential Equations, *Math. Tables Other Aids Comput.*, 7, 135–152.
Dulikravich, G. S. and M. J. Colaço. (2006). Convective Heat Transfer Control Using Magnetic and Electric Fields, *J. Enhanc. Heat Transf.*, 13, 139–155.
Dulikravich, G. S., M. J. Colaço, B. H. Dennis, T. J. Martin, I. N. Egorov and S. Lee. (2004). Optimization of Intensities and Orientations of Magnets Controlling Melt Flow During Solidification, *Mater. Manuf. Process.*, 19, 695–718.
Dulikravich, G. S., M. J. Colaço, T. J. Martin and S. Lee. (2003). An Inverse Method Allowing User-Specified Layout of Magnetized Micro-Fibers in Solidifying Composites, *J. Compo. Mater.*, 37, 1351–1365.
Dupont, T., G. Fairweather and J. P. Johnson. (1974). Three-Level Galerkin Methods for Parabolic Equations, *SIAM J. Numer. Anal.*, 11, 392–410.
Egerton, P., J. A. Howarth and G. Poots. (1979). A Theoretical Investigation of Heat Transfer in a Ladle of Molten Steel during Pouring, *Int. J. Heat Mass Transf.*, 22, 1525–1532.
Eibner, S., R. Jaime, B. Lamien, R. Basto, H. Orlande and O. Fudym. (2014). Near Infrared Light Heating of Soft Tissue Phantoms Containing Nanoparticles, *Therm. Eng.*, 13, 13–18.
Eiseman, P. R. (1982). Automatic Algebraic Coordinate Generation, in *Numerical Grid Generation*, J. F. Thompson (Ed.), North-Holland, Amsterdam, pp. 447–463.
Eiseman, P. R. (1985). Grid Generation for Fluid Mechanics Computations, *Rev. Fluid Mech.*, 17, 487.
Ekrlick, L. W. (1958). A Numerical Method of Solving a Heat Flow Problem with Moving Boundary, *J. Assn. Comp. Math.*, 5, 161–176.
Elshamy, M. M., M. N. Özişik and J. P. Coulter. (1990). Correlation for Natural Convection between Confocal Horizontal Elliptical Cylinders, *Numer. Heat Transf.*, 18, 95–112.
Evans, D. J. and G. Avdelas. (1978). Fast Methods for the Iterative Solution of Linear Elliptic and Parabolic Partial Differential Equations Involving 2 Space Dimensions, *Int. J. Comput. Math. Sec. B*, 6, 335–358.
Farnia, K. and J. V. Beck. (1977). Numerical Solution of Transient Heat Conduction Equation for Heat-Treatable Alloys Whose Thermal Properties Change with Times and Temperature, *J. Heat Transf.*, 99, 471–478.
Farraye, A. and S. I. Guceri. (1985). *George W. Laird Computer-Aided Engineering Laboratory*, Report No. CAE Report-4/85, Mechanical Engineering Department, University of Delaware, Newark, DE.
Forsythe, G. E. and W. R. Wasow. (1967). *Finite-Difference Methods for Partial Differential Equations*, Wiley, New York.
Fourier, J. B. (1822). *Theorie Analytique de la Chaleur*, Paris, (English translation by Freeman, A. (1955), Dover Publications, Inc., New York.
Fox, L. (1962). *Numerical Solution of Ordinary and Partial Differential Equations*, Addison-Wesley, Reading, MA.
Frankel, J., B. Vick and M. N. Özişik. (1986). Hyperbolic Heat Conduction in Composite Regions, *International Heat Transfer Conference*, Vol. 2, San Francisco, CA, pp. 615–620, 17–22 August.

Frankel, S. P. (1950). Convergence Rates of Iterative Treatments of Partial Differential Equations, *Math. Tables Other Aids Comput.*, 4, 65–75.
Frank-Kamenetskii, D. A. (1969). *Diffusion and Heat Transfer in Chemical Kinetics*, 2nd ed., Plenum Press, New York. (Translated by S. P. Appleton).
Friedman, M. (1970). Flow in a Circular Pipe with Recessed Walls, *J. Appl. Mech.*, 37, 5–8.
Fromm, J. E. and F. H. Harlow. (1963). Numerical Solution of the Problem of Vortex Street Development, *Phys. Fluid*, 6, 975–982.
Furzeland, R. M. (1980). A Comparative Study of Numerical Methods for Moving Boundary Problems, *J. Inst. Math. Appl.*, 26, 411–429.
Gear, C. W. (1971). *Numerical Initial Value Problems in Ordinary Differential Equations*, Prentice-Hall, Englewood Cliffs, NJ.
Gerald, C. F. and P. D. Wheatley. (1984). *Applied Numerical Analyses*, 3rd ed., Addison-Wesley, Reading, MA.
Gerges, H. and J. A. McCorquodale. (1997). Modelling of Flow in Rectangular Sedimentation Tanks by an Explicit Third-Order Upwinding Technique, *Int. J. Numer. Meth. Fluids*, 24, 537–561.
Ghosh, A. (1990). *Principles of Secondary Processing and Casting of Liquid Steel*, Oxford and IBH, New Delhi.
Ghosh, A. (2001). *Segregation in cast products*, Sadhana, 26, 5–24.
Glass, D. E., M. N. Özişik, D. S. McRae and B. Vick. (1985a). On the Numerical Solution of Hyperbolic Heat Conduction, *Numer. Heat Transf.*, 8, 497–504.
Glass, D. E., M. N. Özişik and B. Vick. (1985b). Hyperbolic Heat Conduction with Surface Radiation, *Int. J. Heat Mass Transf.*, 28, 1823–1830.
Glass, D. E., M. N. Özişik and D. S. McRae. (1986). Hyperbolic Heat Conduction with Temperature Dependent Thermal Conductivity, *J. Appl. Phys.*, 59, 1861–1865.
Glass, D. E., M. N. Özişik and D. S. McRae. (1987). Hyperbolic Heat Conduction with Radiation in an Absorbing and Emitting Medium, *Numer. Heat Transf.*, 12, 321–333.
Glass, D. E., M. N. Özişik and B. Vick. (1987). Non-Fourier Effects on Transient Temperature Resulting from Periodic On-Off Heat Flux, *Int. J. Heat Mass Transf.*, 30, 1623–1631.
Gobin, D. and Bennacer, R. (1996). Cooperating Thermosolutal Convection in Enclosures—II. Scale Analysis and Mass Transfer, *Int. J. Heat Mass Transf.*, 39, 2683–2697.
Goldman, A. and Y. C. Kao. (1981). Numerical Solution to a Two-Dimensional Conduction Problem Using Rectangular and Cylindrical Body-Fitted Coordinate Systems, *J. Heat Transf.*, 103, 753–758.
Golub, G. H. and C. F. Van Loan. (1996). *Matrix Computations*, 3rd ed., The Johns Hopkins University Press, Baltimore, Maryland.
Goodling, J. S. and M. S. Khader. (1974a). One-Dimensional Inward Solidification with a Convective Boundary Condition, *AFS Cast Metals Res. J.*, 10, 26–29.
Goodling, J. S. and M. S. Khader. (1974b). Inward Solidification with Radiation-Convection Boundary Condition, *J. Heat Transf.*, 96, 114–115.
Goodrich, L. E. (1978). Efficient Numerical Technique for One-Dimensional Thermal Problems with Phase Change, *Int. J. Heat Mass Transf.*, 21, 615–621.
Gosman, A. D., W. M. Pun, A. K. Runchal, D. B. Spalding and M. W. Wolfstein. (1969). *Heat and Mass Transfer in Recirculating Flows*, Academic Press, London.
Graetz, L. (1883). Über di Wärmeleitung von Flü ssigkeiten. Part I, *Ann. Phys. Chem.*, 18, 70–94; Part II. *Ann. Phys. Chem.*, 25, 337–357, 1885.

Greenspan, D. (1969). Numerical Studies of Prototype Cavity Flow Problem, *Comput. J.*, 12, 88–93.
Grinberg, G. A. (1948). *Selected Problems of Mathematical Theory of Electrical and Magnetic Effects*, Akademii Nauk SSSR, Leningrad.
Guceri, S. I. (1988). Finite Difference Methods in Polymer Processing, in *Fundamentals of Computer Modelling for Polymer Processing*, C. L. Tucker (Ed.), Hanser, Munchen.
Guedes, R. O. C. and R. M. Cotta. (1991). Periodic Laminar Forced Convection within Ducts Including Wall Heat Conduction Effects, *Int. J. Eng. Sci.*, 29, no. 5, 535–547.
Guedes, R. O. C., R. M. Cotta and N. C. L. Brum. (1991). Heat Transfer in Laminar Tube Flow with Wall Axial Conduction Effects, *J. Thermophys. Heat Transf.*, 5, no. 4, 508–513.
Guedes, R. O. C. and M. N. Özişik. (1994a). Hybrid Approach for Solving Unsteady Laminar Forced Convection Inside Ducts with Periodically Varying Inlet Temperature, *Int. J. Heat & Fluid Flow*, 15, no. 2, 116–121.
Guedes, R. O. C. and M. N. Özişik. (1994b). Transient Heat Transfer in Simultaneously Developing Channel Flow with Step Change in Inlet Temperature, *Int. J. Heat & Mass Transfer*, 37, no. 17, 2699–2706.
Guedes, R. O. C., M. N. Özişik and R. M. Cotta. (1992). Conjugated Periodic Turbulent Forced Convection in a Parallel Plate Channel, *1992 National Heat Transfer Conference*, Vol. 201, HTD, San Diego, CA, pp. 63–70.
Guedes, R. O. C., M. N. Özişik and R. M. Cotta. (1994). Conjugated Periodic Turbulent Forced Convection in a Parallel Plate Channel, *J. Heat Transf.*, 116, 40–46.
Gupta, R. S. (1974). Moving Grid Method without Interpolations, *Comput. Meth. Appl. Mech. Eng.*, 4, 143–152.
Gupta, R. S. and D. Kumar. (1980). A Modified Variable Time Step Method for One-Dimensional Stefan Problem, *Comp. Meth. Appl. Mech. Eng.*, 23, 101–109.
Gupta, R. S. and D. Kumar. (1981). Variable Time Step Methods for One-Dimensional Stefan Problem with Mixed Boundary Condition, *Int. J. Heat Mass Transf.*, 24, 251–259.
Hageman, L. A. and D. M. Young. (1981). *Applied Iterative Methods*, Academic Press, New York.
Hamming, R. W. (1962). *Numerical Methods for Scientists and Engineers*, McGraw-Hill, New York.
Hatay, F. F., W. Li, S. Kakaç and F. Mayinger. (1991). Numerical and Experimental Analysis of Unsteady Laminar Forced Convection in Channels, *Int. Comm. Heat Mass Transf.*, 18, no. 4, 407–417.
Hauser, J. and C. Taylor. (1986). *Numerical Grid Generation in Computational Fluid Dynamics*, Pineridge Press, Swansea, UK.
Hellwig, G. (1977). *Partial Differential Equations*, B. G. Teubner, Stuttgard.
Hirsch, C. (1988). *Numerical Computation of Internal and External Flows*, Vol. 1, Wiley, New York.
Hirsch, C. (1990). *Numerical Computation of Internal and External Flows—Volume 2*, Wiley, Chichester.
Hogge, M. A. (1981). A Comparison of Two-and-Three-Level Integration Schemes for Non-linear Heat Conduction, in *Numerical Methods in Heat Transfer*, R. W. Lewis, K. Morgan and O. C. Zienkiewicz (Eds.), Wiley, New York, pp. 75–90.
Isenberg, J. and G. de Vahl Davis. (1975). Finite Difference Methods in Heat and Mass Transfer, in *Topics in Transport Phenomena*, C. Gutfinger (Ed.), Hemisphere, New York, pp. 457–553.

Jaime, R. A. O., R. L. Q. Basto, B. Lamien, H. R. B. Orlande, S. Eibner and O. Fudym. (2013). Fabrication Methods of Phantoms Simulating Optical and Thermal Properties, *Procedia Eng.*, 59, 30–36.

Jaluria, Y. and K. E. Torrance. (1986). *Computational Heat Transfer*, Hemisphere, New York.

Johnson, B. H. (1982). Numerical Modelling of Estuarine Hydrodynamics on a Boundary-Fitted Coordinate System, in *Numerical Grid Generation*, J. F. Thompson (Ed.), Elsevier Science, Amsterdam, pp. 409–436.

Kakaç, S. and R. M. Cotta. (1993). Experimental and Theoretical Investigation on Transient Cooling of Electronic Systems, *Proceedings of the NATO Advanced Study Institute on Cooling of Electronic Systems*, Invited Lecture, NATO ASI Series E: Applied Sciences, Vol. 258, Turkey, pp. 239–275, June/July.

Kakaç, S., W. Li and R. M. Cotta. (1990). Unsteady Laminar Forced Convection in Ducts with Periodic Variation of Inlet Temperature, *J. Heat Transf.*, 112, 913–920.

Kakaç, S., C. A. C. Santos, M. R. Avelino and R. M. Cotta. (2001). Computational Solutions and Experimental Analysis of Transient Forced Convection in Ducts (Invited Paper), *Int. J. Transp. Phenom.*, 3, 1–17.

Kaminski, W. (1990). Hyperbolic Heat Conduction Equation for Materials with a Nonhomogeneous Inner Structure, *J. Heat Transf.*, 112, 555–560.

Katnasis, T. (1967). A Computer Program for Calculating Velocities and Streamlines for Two-Dimensional Incompressible Flow in Axial Blade Rows, *NASA-TN D-3762*, January.

Kays, W. M. and M. E. Crawford. (1980). *Convective and Mass Transfer*, McGraw-Hill, New York.

Keller, H. B. (1970). A New Difference Scheme for Parabolic Problems, in *Numerical Solution of Partial Differential Equations*, Vol. 2, J. Bramble (Ed.), Academic Press, New York, pp. 327–350.

Kim, W. S., R. M. Cotta and M. N. Özişik. (1990). Laminar Internal Forced Convection with Periodically Varying, Arbitrarily Shaped Inlet Temperature, *Proceedings of the 9th International Heat Transfer Conference*, Israel, Paper # 16-TR-17, pp. 383–388.

Kim, S. H. and K. Y. Huh. (2000). A New Angular Discretization Scheme of the Finite Volume Method for 3-D Radiative Heat Transfer in Absorbing, Emitting and Anisotropically Scattering Media, *Int. J. Heat Mass Transf.*, 43, 1233–1242.

Kim, W. S. and M. N. Özişik. (1987). Transient Laminar Forced Convection in Ducts with Suddenly Applied Uniform Wall Heat Flux, *Int. J. Heat Mass Transf.*, 30, 1753–1756.

Kim, W. S. and M. N. Özişik. (1989). Turbulent Forced Convection Inside a Parallel-Plate Channel with Periodic Variation of Inlet Temperature, *J. Heat Transf.*, 111, 882–888.

Knupp, D. C., R. M. Cotta and C. P. Naveira-Cotta. (2013b). Heat Transfer in Microchannels with Upstream–Downstream Regions Coupling and Wall Conjugation Effects, *Num. Heat Transf. B Fund.*, 64, 365–387.

Knupp, D. C., R. M. Cotta, C. P. Naveira-Cotta and S. Kakaç. (2015a). Transient Conjugated Heat Transfer in Microchannels: Integral Transforms with Single Domain Formulation, *Int. J. Therm. Sci.*, 88, 248–257.

Knupp, D. C., C. P. Naveira-Cotta and R. M. Cotta. (2012). Theoretical Analysis of Conjugated Heat Transfer with a Single Domain Formulation and Integral Transforms, *Int. Comm. Heat Mass Transf.*, 39, no. 3, 355–362.

Knupp, D. C., C. P. Naveira-Cotta and R. M. Cotta. (2013a). Conjugated Convection-Conduction Analysis in Microchannels with Axial Diffusion Effects and a Single Domain Formulation, *ASME J. Heat Transf.*, 135, 091401.

Knupp, D. C., C. P. Naveira-Cotta and R. M. Cotta. (2014). Theoretical-Experimental Analysis of Conjugated Heat Transfer in Nanocomposite Heat Spreaders with Multiple Microchannels, *Int. J. Heat Mass Transf.*, 74, 306–318.

Knupp, D. C., C. P. Naveira-Cotta, A. Renfer, M. K. Tiwari, R. M. Cotta and D. Poulikakos. (2015b). Analysis of Conjugated Heat Transfer in Micro-Heat Exchangers via Integral Transforms and Non-Intrusive Optical Techniques, *Int. J. Num. Meth. Heat Fluid Flow*, 25, no. 6, 1444–1462.

Knupp, P. and S. Steinberg. (1993). *Fundamentals of Grid Generation*, CRC Press, Boca Raton, FL.

Kober, H. (1957). *Dictionary of Conformal Representation*, 2nd ed., Dover Publications, New York.

Koshlyakov, N. S. (1936). *Fundamental Differential Equations of Mathematical Physics*, 4th edition, ONTI, Moscow.

Kubiček, M. and V. Hlavacek. (1983). *Numerical Solution of Nonlinear Boundary Value Problems with Applications*, Prentice Hall, Englewood Cliffs, NJ.

Lage, P. L. C. (1995). Interpolation Functions for Convection-Diffusion Problems: Approximations of Exponential-Based Functions and Solution Accuracy, *Proceedings of the 15th Brazilian Congress of Mechanical Engineering*, Aguas de Lindóia, São Paulo.

Lamien, B., H. Orlande and G. Eliçabe. (2014). Computer Simulation of Laser Heating of Soft Tissue Phantoms Loaded with Plasmonic Nanoparticles, *15th Brazilian Congress of Thermal Sciences and Engineering*, Belém, PA, Brazil, 10–13 November.

Lamien, B., H. Orlande and G. Eliçabe. (2016a). Inverse Problem in the Hyperthermia Therapy of Cancer with Laser Heating and Plasmonic Nanoparticles, *Inv. Prob. Sci. Eng*, 25, 1–24.

Lamien, B., H. Orlande and G. Eliçabe. (2016b). Particle Filter and Approximation Error Model for State Estimation in Hyperthermia, *ASME J. Heat Transf*, 139, 012001–012012.

Lamien, B., H. Orlande, G. Eliçabe and L. Varon. (2016c). State Estimation in Bioheat Transfer: A Comparison of Particle Filter Algorithms, *Int. J. Num. Meth. Heat Fluid Flow*, 27, 615–638.

Lamien, B., H. Orlande and G. Eliçabe. (2017). Inverse problem in the hyperthermia therapy of cancer with laser heating and plasmonic nanoparticles. Inverse Problems in Science and Engineering, 25, 608–631.

Lancaster, P. (1970). Explicit Solutions of Linear Matrix Equations, *SIAM Rev.*, 12, 544–566.

Laney, C. (1998). *Computational Gasdynamics*, Cambridge University Press, Cambridge.

Lapidus, L. and G. F. Pinder. (1982). *Numerical Solution of Partial Differential Equations in Science and Engineering*, Wiley-Interscience, New York.

Larkin, B. K. (1964). Some Stable Explicit Difference Approximations to the Diffusion Equation, *Math. Comput.*, 18, 196–202.

Lax, P. D. (1954). Weak Solutions of Nonlinear Hyperbolic Equations and their Numerical Computations, *Comm. Pure Appl. Math.*, 7, 159–193.

Lee, J. S. and U. C. Fung. (1970). Flow in a Locally Constricted Tubes at Low Reynolds Numbers, *J. Appl. Mech.*, 37, 9–17.

Lentini, M. and V. Pereyra. (1968). *SIAM J. Num. Anal.*, 15, 59.

Leonard, B. P. (1979). A Stable and Accurate Convective Modelling Procedure Based on Quadratic Upstream Interpolation, *Comput. Meth. Appl. Mech. Eng.*, 19, 59–98.

Leonard, B. P. (1991). The Ultimate Conservative Difference Scheme Applied to Unsteady One-Dimensional Advection, *Comput. Meth. Appl. Mech. Eng.*, 88, 17–74.

Leonard, B. P. (1997). Bounded High-Order Upwind Multidimensional Finite-Volume Convection-Diffusion Algorithms, in *Advances in Numerical Heat Transfer*, W. J. Minkowycz and E. M. Sparrow (Eds.), Vol. 1, Taylor & Francis, New York, pp. 1–57.

Leonard, B. P., M. K. MacVean and A. P. Lock. (1995). The Flux Integral Method for Multidimensional Convection and Diffusion, *Appl. Math. Model.*, 19, 333–342.

Leonard, B. P. and S. Mokhtari. (1990). Beyond First-Order Upwinding: The Ultra-Sharp Alternative for Non-Oscillatory Steady-State Simulation of Convection, *Int. J. Num. Meth. Eng.*, 30, 729–766.

Leonard, B. P. and H. S. Niknafs. (1991). Sharp Monotonic Resolution of Discontinuities without Clipping of Narrow Extrema, *Comput. Fluids*, 19, 141–154.

Liseikin, V. (2010). *Grid Generation Methods*, 2nd ed., Springer, Dordrecht.

Liu, S.-L. (1967). Numerical Solution of Two-Point Boundary Value Problems in Simultaneous Second Order Nonlinear Ordinary Differential Equations, *Chem. Eng. Sci.*, 22, 871.

Luikov, A. V. (1968). *Analytical Heat Diffusion Theory*, Academic Press, New York.

Luikov, A. V. (1980). *Heat and Mass Transfer*, Mir Publishers, Moscow.

MacCormack, R. and B. Baldwin. (1975). *A Numerical Method for Solving the Navier-Stokes Equations with Application to Shock-Boundary Layer Interactions*, AIAA Paper 75-1, AIAA 13th Aerospace Sciences Meeting, Pasadena, CA.

MacCormack, R. W. (1969). *The Effect of Viscosity on Hypervelocity Impact Conference*, AIAA Paper 69–354, 4th Aerodynamic Testing Conference, Cincinnati, OH.

MacCormack, R. W. (1971). Numerical Solution of the Interaction of a Shock Wave with a Laminar Boundary Layer, *Proceedings Second International Conference Numerical Methods Fluid Dynamics*, Lecture Notes in Physics, Vol. 8, Springer-Verlag, New York, pp. 151–163.

Maliska, C. R. (1981). *A Solution Method for Three-Dimensional Parabolic Fluid Flow Problems in Nonorthogonal Coordinates*, Ph.D. Thesis, University of Waterloo, Waterloo, Canada.

Mallison, G. D. and D. G. De Vahl. (1973). The Method of False Transient for the Solution of Coupled Elliptic Equations, *J. Comput. Phys.*, 12, 435.

Masiulaniec, K. C., T. G. Keith, Jr. and K. J. Dewitt. (1984). *Finite Difference Solution of Heat Conduction Problems in Multi-layered Bodies with Complex Geometries*, ASME Paper 84-HT-58, National Heat Transfer Conference, Niagara Falls, NY.

Mastanaiah, K. (1976). On the Numerical Solution of Phase Change Problems in Transient Non-Linear Heat Conduction, *Int. J. Numer. Meth. Eng.*, 10, 833–844.

Mastin, C. W. (1982). Error Induced by Coordinate Systems, in *Numerical Grid Generation*, J. F. Thompson (Ed.), New York and Amsterdam, North-Holland, pp. 31–40

Mastin, C. W., C. Wayne and J. F. Thompson. (1978). Elliptic Systems and Numerical Transformations, *J. Math. Anal. Appl.*, 62, 52.

Maxwell, J. C. (1867). On the Dynamical Theory of Gases, *Phil. Trans. Roy. Soc.*, 157, 49–88.

McWhorter, J. C. and M. H. Sadd. (1980). Numerical Anisotropic Heat Conduction Solutions Using Boundary-Fitted Coordinate System, *J. Heat Transf.*, 102, 308–311.

Meyer, G. H. (1973). Multidimensional Stefan Problems, *SIAM J. Numer. Anal.*, 10, 522–528.
Mikhailov, M. D. (1975). On the Solution of the Heat Equation with Time Dependent Coefficient, *Int. J. Heat Mass Transf.*, 18, 344–345.
Mikhailov, M. D. and R. M. Cotta. (1994). Integral Transform Method for Eigenvalue Problems, *Commum. Num. Meth. Eng.*, 10, 827–835.
Mikhailov, M. D. and R. M. Cotta. (1996). Ordering Rules for Double and Triple Eigenseries in the Solution of Multidimensional Heat and Fluid Flow Problems, *Int. Commun. Heat Mass Transf.*, 23, 299–303.
Mikhailov, M. D. and M. N. Özişik. (1984a). An Alternative General Solution of the Steady-State Heat Diffusion Equation, *Int. J. Heat Mass Transf.*, 23, 609–612.
Mikhailov, M. D. and M. N. Özişik. (1984b). *Unified Analysis and Solutions of Heat and Mass Diffusion*, Wiley, New York.
Milne-Thompson, L. M. (1950). *Theoretical Hydrodynamics*, 2nd ed., MacMillan, New York.
Minkowycz, W. J., E. M. Sparrow, G. E. Schneider and R. H. Plecher. (1988). *Handbook of Numerical Heat Transfer*, Wiley-Interscience, New York.
Mitchell, A. R. (1969). *Computational Methods in Partial Differential Equations*, Wiley, New York.
Modest, M. (2013). *Radiative Heat Transfer—Third Edition*, Elsevier, Amsterdam.
Moreira Filho, E., M. Mejias, H. Orlande and A. Leiroz. (2002). Computational Aspects of Metrics Evaluation for the Finite Volume Method, *9th Brazilian Congress of Thermal Engineering and Sciences*, Paper CIT02–0623, Caxambu.
Murray, W. D. and F. Landis. (1959). Numerical and Machine Solutions of Transient Heat Conduction Problems Involving Melting or Freezing, *J. Heat Transf.*, 81, 106–112.
Naveira, C. P., M. Lachi, R. M. Cotta and J. Padet. (2009). Hybrid Formulation and Solution for Transient Conjugated Conduction-External Convection, *Int. J. Heat Mass Transf.*, 52, no. 1–2, 112–123.
Naveira-Cotta, C. P., R. M. Cotta, H. R. B. Orlande and O. Fudym. (2009). Eigenfunction Expansions for Transient Diffusion in Heterogeneous Media, *Int. J. Heat Mass Transf.*, 52, 5029–5039.
Nóbrega, P., H. R. B. Orlande and J.-L. Battaglia. (2011). Bayesian Estimation of Thermophysical Parameters of Thin Metal Films Heated by Fast Laser Pulses, *Int. Commun. Heat Mass Transf.*, 38, 1172–1177.
Nunes, J. S., R. M. Cotta, M. R. Avelino and S. Kakaç. (2010). Conjugated Heat Transfer in Microchannels, in *Microfluidics Based Microsystems: Fundamentals and Applications*, S. Kakaç, B. Kosoy and A. Pramuanjaroenkij (Eds.), Vol. 1, NATO Science for Peace and Security Series A: Chemistry and Biology, Springer, Dordrecht, The Netherlands, pp. 61–82.
Oleinik, O. A. (1960). A Method of Solution of the General Stefan Problem, *Sov. Math. Dokl.*, 1, 1350–1354.
Orivouri, S. (1979). Efficient Method for Solution of Nonlinear Heat Conduction Problems, *Int. J. Numer. Meth. Eng.*, 14, 1461–1476.
Orlande, H. R. B., M. N. Özişik and D. Y. Tzou. (1995). Inverse Analysis for Estimating the Electron–Phonon Coupling Factor, *J. Appl. Phys.*, 78, 1843–1899.
Ortega, J. M. (1989). *Introduction to Parallel and Vector Solution of Linear Systems*, Plenum Press, New York, pp. 160–163.
Ortega, J. M. and W. C. Reinholdt. (1970). *Iterative Solution of Nonlinear Equations in Several Variables*, Academic Press, New York.

Özişik, M. N. (1968). *Boundary Value Problems of Heat Conduction*, International Textbook, Scranton, PA.
Özişik, M. N. (1973). *Radiative transfer and interactions with conduction and convection*, Wiley, New York.
Özişik, M. N. (1977). *Basic Heat Transfer*, McGraw-Hill, New York.
Özişik, M. N. (1985). *Heat Transfer a Basic Approach*, McGraw-Hill, New York.
Özişik, M. N. (1993). *Heat Conduction*, Wiley, New York.
Özişik, M. N. and R. L. Murray. (1974). On the Solution of Linear Diffusion Problems with Variable Boundary Condition Parameters, *J. Heat Transf.*, 96c, 48–51.
Özişik, M. N. and D. Y. Tzou. (1994). On the Wave Theory in Heat Conduction, *ASME J. Heat Transf.*, 116, 526–535.
Özişik, M. N. and B. Vick. (1984). Propagation and Reflection of Thermal Waves in a Finite Medium, *Int. J. Heat Mass Transf.*, 27, 1845–1854.
Pao, Y. H. and R. J. Daugherty. (1969). *Time Dependent Viscous Incompressible Flow Past a Finite Flat Plate*, Boeing Scientific Research Laboratories, Technical Report DI-82-0822, Boeing Sci. Res. Lab. (Flight Sciences Lab), January.
Paris, J. and S. Whitaker. (1965). Confined Wakes—A Numerical Solution of Navier-Stokes Equations, *Am. Inst. Chem. Eng.*, 11, 1033–1041.
Park, T. S. and J. H. Kwon. (1996). An Improved Multistage Time Stepping for Second-Order Upwind TVD Schemes, *Comput. Fluids*, 25, 629–645.
Patankar, S. V. (1975). Numerical Prediction of Three-Dimensional Flows, in *Studies in Convection: Theory, Measurement and Applications*, Vol. 1, ed., B. E. Launder, Academic Press, New York.
Patankar, S. V. (1979). A Calculation Procedure for Two-Dimensional Elliptic Situations, *Numer. Heat Transf*, 4, 409–425.
Patankar, S. V. (1980). *Numerical Heat Transfer and Fluid Flow*, Hemisphere Publishing Corp., New York.
Patankar, S. V. (1988). Parabolic Systems: Finite-Difference Method I, in *Handbook of Numerical Heat Transfer*, W. J. Minkowycz, E. M. Sparrow, G. E. Schneider and R. H. Pletcher (Eds.), Wiley, New York.
Patankar, S. V. and D. B. Spalding. (1967a). A Finite-Difference Procedure for Solving the Equations of the Two-Dimensional Boundary Layers, *Int. J. Heat Mass Transf.*, 10, 1369–1412.
Patankar, S. V. and D. B. Spalding. (1967b). *Heat and Mass Transfer in Boundary Layers*, Morgan-Grampian Books, London.
Patankar, S. V. and D. B. Spalding. (1970). Heat and Mass Transfer in Boundary Layers, 2nd edn, Intertext, London.
Patankar, S. V. and D. B. Spalding. (1972). A Calculation Procedure for Heat and Momentum Transfer in Three-Dimensional Parabolic Flows, *Int. J. Heat Mass Transf.*, 15, 1787–1806.
Peaceman, D. W. and H. H. Rachford. (1955). The Numerical Solution of Parabolic and Elliptic Differential Equation, *J. Soc. Ind. Appl. Math.*, 3, 28–41.
Pereira, L. M., R. M. Cotta and J. S. Pérez-Guerrero. (2000). Forced and Natural Convection in Annular Concentric Channels and Cavities by Integral Transforms, *8th Brazilian Congress of Thermal Engineering and Sciences*, Porto Alegre, Brazil, October 2000.
Peyret, R. and T. D. Taylor. (1983). *Computational Methods for Fluid Flow*, Springer-Verlag, New York.

Pham, Q. (1985). A Fast Unconditionally Stable Finite-Difference Scheme for Heat Conduction with Phase Change, *Int. J. Heat Mass Transf.*, 28, 2079–2084.
Piperno, S. and S. Depeyre, S. (1996). Criteria for the Design of Limiters Yielding Efficient High Resolution TVD Schemes, *Comput. Fluids*, 27, 183–197.
Pletcher, R. H., J. C. Tannehill and D. A. Anderson. (2012). *Computational Fluid Mechanics and Heat Transfer—Third Edition*, CRC Press, Boca Raton, FL.
Poirier, D. and M. Salcudean. (1988). On Numerical Methods Used in Mathematical Modelling of Phase Change in Liquid Metals, *J. Heat Transf.*, 110, 562–570.
Press, W. H., S. A. Teukolsky, W. T. Vetterling and B. P. Flannery. (1992). *Numerical Recipes in Fortran 77: The Art of Scientific Computing*, 2nd ed., Cambridge University Press, Cambridge, UK.
Price, R. H. and M. R. Slack. (1954). The effect of Latent Heat on Numerical Solutions of the Heat Flow Equation, *Br. J. Appl. Phys.*, 5, 285–287.
Projahn, V., H. Rieger and H. Beer. (1981). Heat Conduction in Anisotropic Composites of Arbitrary Shape (A Numerical Analysis), *Wärme Stoffübertr.*, 15, 223–232.
Qiu, T. Q., T. Juhasz, C. Suarez, W. E. Bron and C. L. Tien. (1994). Femtosecond Laser Heating of Multi-Layer Metals—II. Experiments, *Int. J. Heat Mass Transf.*, 37, 2799–2808.
Qiu, T. Q. and C. L. Tien. (1992). Short-Pulse Laser Heating on Metals, *Int. J. Heat Mass Transf.*, 35, 719–726.
Qiu, T. Q. and C. L. Tien. (1993). Size Effects on Nonequilibrium Laser Heating of Metal Films, *ASME J. Heat Transf.*, 115, 842–847.
Qiu, T. Q. and C. L. Tien. (1994). Femtosecond Laser Heating of Multi-Layer Metals—I. Analysis, *Int. J. Heat Mass Transf.*, 37, 2789–2797.
Quaresma, J., E. Macedo, H. Fonseca and H. Orlande. (2010). An Analysis of Heat Conduction Models for Nanofluids, *Heat Trans. Eng.*, 31, 1126–1136.
Raithby, G. D. (1976). Skew Upstream Differencing Schemes for Problems Involving Fluid Flow, *Comput. Meth. Appl. Mech. Eng.*, 9, 153–164.
Raithby, G. D. (1999). Discussion of the Finite-Volume Method for Radiation, and Its Application Using 3D Unstructured Meshes, *Numer. Heat Transf. B*, 35, 389–405.
Raithby, G. D. and E. H. Chui. (1990). A Finite-Volume Method for Predicting a Radiant Heat Transfer in Enclosures with Participating Media, *ASME J. Heat Transf.*, 112, 415–423.
Raithby, G. D. and G. E. Schneider. (1988). Elliptic Systems: Finite-Difference Method II, in *Handbook of Numerical Heat Transfer*, eds., W. J. Minkowycz, E. M. Sparrow, G. E. Schneider, and R. H. Pletcher, Wiley, New York, pp. 241–291.
Raithby, G. D. and K. E. Torrance. (1974). Upstream-Weighted Differencing Schemes and Their Application to Elliptic Problems Involving Fluid Flow, *Comput. Fluids*, 2, 191–206.
Ralston, A. and P. Rabinowitz. (1978). *A First Course in Numerical Analysis*, 2nd ed., McGraw-Hill, New York.
Ranchal, A. K. and M. Wolfstein. (1969). *J. Mech. Eng. Sci.*, 11, 445–453.
Rappaz, M. (1989) Modelling of Microstructure Formation in Solidification Processes, *Int. Mater. Rev.*, 34, 93–123.
Richardson, L. F. (1910). The Approximate Arithmetical Solution by Finite Differences of Physical Problems Involving Differential Equations, with Applications to the Stresses in a Masonry Dam, *Philos. Trans. R. Soc. London, Ser. A*, 210, 307–357.

Richtmeyer, R. D. and K. W. Morton. (1967). *Difference-Methods in Initial-Value Problems*, 2nd ed., Interscience, New York.
Rieger, H., U. Projahn and H. Beer. (1982). Analysis of the Heat Transfer Mechanisms During Melting Around a Horizontal Cylinder, *Int. J. Heat Mass Transf.*, 25, 137–147.
Roache, P. (1975). The LAD, NOS and SPLIT NOS Methods for the Steady-State Navier-Stokes Equations, *Comput. Fluids*, 3, 179–195.
Roache, P. J. (1976). *Computational Fluid Dynamics*, Hermosa Publishers, Albuquerque, NM.
Roberts, G. O. (1971). Computational Meshes for Boundary Layer Problems, *Proceedings of the Second International Conference Numerical Methods Fluid Dynamics*, Vol. 8, Lecture Notes in Physics, Springer-Verlag, New York, pp. 171–177.
Rose, M. E. (1960). A Method for Calculating Solutions of Parabolic Equations with a Free Boundary, *Math. Comput.*, 14, 249–256.
Runchal, A. K., D. B. Spalding and Wolfstein. (1969). Numerical Solution of the Elliptic Equations for Transport for Vorticity, Heat and Matter in Two-Dimensional Flow, *Phys. Fluids*, 12, Suppl. II, 1121–1128.
Saltzman, J. and J. Brackbill. (1982). Application and Generalization of Variational Methods for Generating Adaptive Meshes, in *Numerical Grid Generation*, J. F. Thompson (Ed.), Elsevier, New York, pp. 865–884.
Santos, C. A. C., R. M. Cotta and M. N. Özişik. (1988). Laminar Forced Convection Inside Externally Finned Tubes, *Proceedings of the 2nd National Thermal Sciences Meeting*, Sao Paulo, Brazil, ENCIT-88, pp. 87–90.
Santos, C. A. C., R. M. Cotta and M. N. Özişik. (1991). Heat Transfer Enhancement in Laminar Flow within Externally Finned Tubes, *Int. J. Heat Technol.*, 9, 46–68.
Santos, C. A. C., M. J. Medeiros, R. M. Cotta and S. Kakaç. (1998). Theoretical Analysis of Transient Laminar Forced Convection in Simultaneous Developing Flow in Parallel-Plate Channel, *7th AIAA/ASME Joint Themophysics and Heat Transfer Conference*, AIAA Paper #97-2678, Albuquerque, NM, June.
Santos, C. A. C., J. N. N. Quaresma and J. A. Lima. (2001). *Convective Heat Transfer in Ducts: The Integral Transform Approach*, ABCM Mechanical Sciences Series, Editora E-Papers, Rio de Janeiro, Brazil.
Saul'yev, V. K. (1957). On a Method of Numerical Integration of a Diffusion Equation, *Dokl. Akad. Nauk SSSR*, 115, 1077–1079. (In Russian).
Schlichting, H. (1979). *Boundary Layer Theory*, 7th ed., McGraw-Hill, New York.
Seidel, P. L. (1874). Uber ein Verfahren, die Gleinchungen, auf welche die Methode der kleinsten quadrate furht, sowie lineare Gleinchungen uberhaupt, durch successive Annaherung aufzulosen, *Abh. Bayer. Akad*, 11, 81–108.
Serfaty, R. and R. M. Cotta. (1990). Integral Transform Solutions of Diffusion Problems with Nonlinear Equation Coefficients, *Int. Comm. Heat Mass Transf.*, 17, no. 6, 851–864.
Shamsundar, N. and E. M. Sparrow. (1975). Analysis of Multi-dimensional Conduction Phase Change via the Enthalpy Method, *J. Heat Transf.*, 97, 333–340.
Shih, T. M. (1984). *Numerical Heat Transfer*, Hemisphere/McGraw-Hill, New York.
Shih, T. M. and H. J. Huang. (1981). A Method of Solving Nonlinear Differential Equations for Boundary-Layer Flows, *Numer. Heat Transf.*, 4, 159–178.
Siegel, R. and J. Howell. (2002). *Thermal Radiation Heat Transfer—4th Edition*, Taylor & Francis, New York.

Sieniutycz, S. (1977). The Variational Principle of Classical Type for Non-Coupled Non-Stationary Irreversible Transport Processes with Convective Motion and Relaxation, *Int. J. Heat Mass Transf.*, 20, 1221–1231.

Sieniutycz, S. (1981). Thermodynamics of Coupled Heat, Mass and Momentum Transport with Finite Wave Speed I—Basic Ideas of Theory, *Int. J. Heat Mass Transf.*, 24, 1723–1732.

Singh, P. and Liburdy, J. A. (1986), Effect of Plate Inclination on Natural Convection From a Plate to Its Cylindrical Enclosure, *J. Heat Transfer.*, 108, 770–775.

Smith, G. D. (1978). *Numerical Solution of Partial Differential Equations: Finite Difference Methods*, 2nd ed., Oxford University Press, London.

Smith, R. E. (1982). Algebraic Grid Generation, in *Numerical Grid Generation*, J. F. Thompson (Ed.), North-Holland, New York, pp. 137.

Sommeljer, B. P., P. J. van der Houwen and J. G. Verwer. (1981). On the Treatment of Time-Dependent Boundary Conditions in Splitting Methods for Parabolic Differential Equations, *Int. J. Numer. Methods Eng.*, 17, 335–346.

Sorenson, R. L. (1980). *A Computer Program to Generate Two-Dimensional Grids about Airfoils and Other Shapes by the Use of Poisson's Equations*, NASA-Technical-Memorandum 81198, NASA AMES Research Center, Moffett Field, CA.

Sottos, N. R. and S. I. Güceri. (1986). Residual and Transient Thermal Stresses in Laminar Orthotopic Composites, in Numerical Grid Generation in Computational Fluid Dynamics, Hau ser and Taylor (Eds.), Pineridge Press, UK, pp. 741–754.

Spalding, D. B. (1972). A Novel Finite-Difference Formulation for Differential Expressions Involving Both First and Second Derivatives, *Int. J. Numer. Method Eng.*, 4, 551.

Spalding, D. B. (1977). *Genmix: A General Computer Program for Two-Dimensional Parabolic Phenomena*, Pergamon Press, Oxford.

Sparrow, E. M. and Cess, R. D. (1978). Radiation Heat Transfer. Brooks/Cole, Belmont, Calif., 1966 and 1970., New Augmented Edition.

Sphaier, L. A. and R. M. Cotta. (2000). Integral Transform Analysis of Multidimensional Eigenvalue Problems within Irregular Domains, *Numer. Heat Transf. Part B Fund.*, 38, 157–175.

Sphaier, L. A., R. M. Cotta, C. P. Naveira-Cotta and J. N. N. Quaresma. (2011). The UNIT Algorithm for Solving One-Dimensional Convection-Diffusion Problems via Integral Transforms, *Int. Commun. Heat Mass Transf.*, 38, 565–571.

Spirou, G. M., A. A. Oraevsky, I. A. Vitkin and W. M. Whelan. (2005). Optical and Acoustic Properties at 1064 nm of Polyvinyl Chloride-Plastisol for Use as a Tissue Phantom in Biomedical Optoacoustics, *Phys. Med. Biol.*, 50, N141–N143.

Stefan, J. (1891). Uber die Theorie der Eisbildung, insbesondere fiber die Eisbildung im Polarmeere, *Ann. Phys. Chemie (Wiedemannsche Annalen)*, 42, 269–286.

Subbiah, S., D. L. Trafford and S. I. Guceri. (1989). Non-Isothermal Flow of Polymers into Two-Dimensional Thin Cavity Molds: A Numerical Grid Generation Approach, *Int. J. Heat Mass Transf.*, 32, 415–434.

Swaminathan, C. R. and Voller, V. R. (1997). Towards a General Numerical Scheme for Solidification Systems, *Int. J. Heat Mass Transf.*, 40, 2859–2868.

Szekely, J. and R. G. Lee. (1968). The Effects of Slag Thickness on Heat Loss from Ladles Holding Molten Steel, *Trans. Am. Inst. Min. Engrs.*, 242, 961.

Tacke, K. H. (1985). Discretization of the Explicit Enthalpy Method for Planar Phase Change, *Int. J. Num. Methods Eng.*, 21, 543–554.

Tafti, D. (1996). Comparison of Some Upwind-Biased High-Order Formulations with a Second-Order Central-Difference Scheme for Time Integration of the Incompressible Navier-Stokes Equations, *Comput. Fluids*, 25, no. 7, 647–665.

Tanaka, T. (1981). Gels. *Sci. Am.*, 244, 124–136, 138.

Tannehill, J. C., T. L. Hoist and J. V. Rakish. (1975). *Numerical Computation of Two-Dimensional Viscous Blunt Body Hows with an Impinging Shock*, AIAA Paper 75-154, AIAA 13th Aerospace Sciences Meeting, Pasadena, CA.

Temperton, C. (1979). Direct Methods for the Solution of the Discrete Poisson's Equation: Some Comparisons, *J. Comput. Phys.*, 31, 1–20.

Thoman, D. C. and A. A. Szewczyk. (1966). *Numerical Solutions of Time Dependent Two-Dimensional Flow of a Viscous, Incompressible Fluid Over Stationary and Rotating Cylinders*, Tech. Rep., 66–14, Heat Transfer and Fluid Mechanics Laboratory, Department of Mechanical Engineering, University of Notre Dame, Notre Dame, IN.

Thomas, L. H. (1949). Elliptic Problems in Linear Difference Equations Over a Network, *Sci. Comp. Lab. Rept.*, Columbia University, New York.

Thomson, J., B. Soni and N. Weatherill. (1999). *Handbook of Numerical Grid Generation*, CRC-Press, Boca Raton, FL.

Thompson, J. F. (1982). *Elliptic Grid generation*, Elsevier Publishing Co. Inc., New York, pp. 79–105.

Thompson, J. F. (1983). A Survey of Grid Generation Techniques in Computational Fluid Dynamics, AIAA Paper No. 83-0447, *AIAA 21st Aerospace Sciences Meeting*, January, Reno, Nevada.

Thompson, J. F. (1984). Grid Generation Techniques in Computational Fluid Dynamics, *AIAA J.*, 22, 1505–1523.

Thompson, J. F. (1985). A Survey of Dynamically Adaptive Grids in the Numerical Solution of Partial Differential Equations, *Appl. Numer. Math.*, 1, 3–29.

Thompson, J. F., F. C. Thames and C. W. Mastin. (1974). Automatic Numerical Generation of Body-Fitted Curvilinear Coordinate System for Field Containing any Number of Arbitrary Two-Dimensional Bodies, *J. Comput. Phys.*, 15, 299–319.

Thompson, J. F., F. C. Thames and C. W. Mastin. (1977). TOMCAT—A Code for Numerical Grid Generation of Boundary-Fitted Coordinate System of Fields Containing any Number of Arbitrary Two Dimensional Bodies, *J. Comput. Phys.*, 24, 274–302.

Thompson, J. F., Z. U. A. Varsi and C. W. Mastin. (1982). Boundary-Fitted Coordinate Systems for Numerical Solution of Partial Differential Equations—A Review, *J. Comput. Phys.*, 48, 1–108.

Thompson, J. F., Z. U. A. Varsi and C. Mastin. (1985). *Numerical Grid Generation Foundations and Applications*, North-Holland, Elsevier Science Publishers, Amsterdam, The Netherlands.

Toro, E. F. (1999). *Riemann Solvers and Numerical Methods for Fluid Dynamics*, 2nd ed., Berlin, Springer.

Traub, J. (1964). *Iterative Methods for the Solution of Equations*, Prentice-Hall, Englewood Cliffs, NJ.

Trefethen, L. N. (1980). Numerical computation of the Schwarz–Christoffel transformation, SIAM Journal on Scientific and Statistical Computing, 1, 82–102.

Tucker, C. L. and E. C. Bernhardt (Ed.). (1989). *Fundamentals of Computer Modelling for Polymer Processing*, Hanser Publishers, Munich, Germany.

Tzou, D. Y. (1996). *Macro- To Micro-Scale Heat Transfer: The Lagging Behavior*, CRC Press, Boca Raton, FL.

Tzou, D. Y., R. J. Chiffelle and M. N. Özişik. (1994). On the Wave Behavior of Heat Conduction in the Two-Step Model for Microscale Heat Transfer, *ASME J. Heat Transf.*, 116, 1034–1041.

Ushikawa, S. and R. Takeda. (1985). Use of Boundary-fitted Coord. Transformation for Unsteady Heat Cond. Prob. in Multiconnected Regions with Arbitrarily Shaped Boundary, *J. Heat Transf.*, 107, 494–498.

Van Wylen, G., R. Sonntag and C. Borgnakke. (1994). *Fundamentals of Classical Thermodynamics*, Wiley, New York.

Varon, L., H. Orlande, G. Eliçabe and L. Varon. (2015). Estimation of State Variables in the Hyperthermia Therapy of Cancer with Heating Imposed by Radiofrequency Electromagnetic Waves, *Int. J. Therm. Sci.*, 98, 228–236.

Varon, L., H. Orlande, G. Eliçabe and L. Varon. (2016). Combined Parameter and State Estimation in the Radiofrequency Hyperthermia Treatment of Cancer, *Numer. Heat Transf. Part A*: Applications, 70, 581–594.

Vernotte, P. (1958). Les Panadoxes de la Theorie Continue de L'equation de la Chaleauv, *C. r. Acad. Sci. Paris*, 246, 3154–3155.

Vick, B. and M. N. Özişik. (1983). Growth and Decay of a Thermal Pulse Predicted by the Hyperbolic Heat Conduction Equation, *J. Heat Transf.*, 105, 902–907.

Vick, B. and R. G. Wells. (1986). Laminar Flow with an Axially Varying Heat Transfer Coefficient, *Int. J. Heat Mass Transf.*, 29, no. 12, 1881–1889.

Voller, V. and M. Cross. (1981). Accurate Solutions of Moving Boundary Problems Using she Enthalpy Method, *Int. J. Heat Mass Transf.*, 24, 545–556.

Voller, V. and M. Cross. (1983). An Explicit Numerical Method to Track a Moving Phase-Change Front, *Int. J. Heat Mass Transf.*, 26, 147–150.

Voller, V. R. (2004). An Explicit Scheme for Coupling Temperature and Concentration Fields in Solidification Models, *Appl. Math. Model.*, 28, 79–94.

Voller, V. R., A. D. Brent and C. Prakash. (1989). The Modeling of Heat, Mass and Solute Transport in Solidification Systems, *Int. J. Heat Mass Transf.*, 32, 1719–1731.

Vuik, C. (1993). Some Historical Notes about the Stefan Problem, *Nieuw Arch. voor Wiskunde*, 11, 157–167.

Wang, L., X. Zhou and X. Wei. (2010). *Heat Conduction: Mathematical Models and Analytical Solutions*, Springer, New York.

Warming, R. F. and R. M. Beam. (1975). Upwind Second-Order Difference Schemes and Applications in Unsteady Aerodynamic Flow, *Proceedings of the AIAA 2nd Computational Fluid Dynamics Conference*, Hartford, CT, pp. 17–28.

Warming, R. F. and R. M. Beam. (1976). Upwind Second-Order Difference Schemes and Applications in Aerodynamic Flows, *AIAA J.*, 14, 1241–1249.

Warming, R. F. and B. S. Hyett. (1974). The Modified Equation Approach to the Stability and Accuracy Analysis of Finite-Difference Methods, *J. Comput. Phys.*, 14, 159–179.

Webb, B. W. and R. Viskanta. (1986). Analysis of Heat Transfer During Melting the Pure Metal from an Isothermal Vertical Wall, *Num. Heat Transf.*, 9, 539–558.

Weinstock, R. (1952). *Calculus of Variations with Applications to Physics and Engineering*, McGraw-Hill, New York.

Wellele, O., H. Orlande, N. Ruperti, Jr., M. Colaço and A. Delmas. (2006). Coupled Conduction–Radiation in Semi-Transparent Materials at High Temperatures, *J. Phys. Chem. Solid.*, 67, 2230–2240.

Westlake, J. R. (1968). *A Handbook of Numerical Matrix Inversion and Solution of Linear Equations*, Wiley, New York.
Weymann, H. D. (1967). Finite Speed of Propagation in Heat conduction, Diffusion and Viscous Shear Motion, *Am. J. Phys.*, 35, 488–496.
White, F. M. (1974). *Viscous Fluid Flows*, McGraw-Hill, New York.
Wilhemson, R. B. and J. H. Ericksen. (1977). Direct Solutions for Poisson's Equation in Three Dimensions, *J. Comput. Phys.*, 25, 319–331.
Williams, S. D. and D. M. Curry. (1977). An Implicit-Iterative Solution of the Heat Conduction Equation with a Radiation Boundary Condition, *Int. J. Numer. Method Eng.*, 11, 1605–1619.
Wolfram, S. (2015). *The Mathematica Book*, Wolfram, version 10, Champaign, IL.
Wu, J. C. (1961). On the Finite-Difference Solution of Laminar Boundary Layer Problem, *Proceedings of the Heat Transfer and Fluid Mechanics Institute*, Stanford University Press, Stanford, CA.
Xiu, D. (2010). *Numerical Methods for Stochastic Computations: A Spectral Method Approach*, Princeton University Press, Princeton, NJ.
Xu, T., C. Zhang, X. Wang, L. Zhang and J. Tian. (2003). Measurement and Analysis of Light Distribution in Intralipid-10% at 650nm, *Appl. Opt.*, 42, 5777–5784.
Yanenko, N. N. (1971). *The Method of Fractional Steps: The Solution of Problem of Mathematical Physics in Several Variables*, M. Holt (Ed.), Springer-Verlag, New York, pp. 160.
Yee, H. C. (1981). *Numerical Approximation of Boundary Conditions with Applications to Inviscid Equations of Gas Dynamics*, NASA Technical Memorandum 81265, NASA Ames Research Center, Moffett Field, CA.
Young, D. (1954). Iterative Methods for Solving Partial Differential Equations of Elliptic Type, *Trans. Am. Math. Soc.*, 76, 92–111.
Young D. M. (1971). *Iterative Solution of Large Linear Systems*, Academic Press, New York.
Yuen, W. W. and L. W. Wong. (1980). An Efficient Algorithm for the Numerical Solution of the Transient Diffusion Equation with an Implicit Formulation, *Numer. Heat Transf.*, 3, 373–380.
Zabaras, N. and D. Samanta. (2004). A Stabilized Volume-Averaging Finite Element Method for Flow in Porous Media and Binary Alloy Solidification Processes, *Int. J. Numer. Method Eng.*, 60, 1103–1138.
Zachmanoglou, E. C. and D. W. Thoe. (1976). *Introduction to Partial Differential Equations with Applications*, Williams & Wilkinson, Baltimore, MD.

Index

A

Absolute convergence, 77
Adaptive grid generation methods, 362
Advective–diffusive systems
 advection–diffusion equation, 179–184
 conservation equation, 418–419
 explicit scheme, 179–182
 hyperbolic heat conduction equation, 185–189
 implicit finite volume method, 182–184
 implicit scheme, 182–184
 MacCormack's method, 172–178
 metrics evaluation, computational aspects, 461–465
 purely advective (wave) equation, 169–178
 stability, 179–181
 upwind method, 170–172, 174–178, 181–182
 Warming and Beam's method, 173–178
Algebraic equations, solving systems of
 biconjugate gradient method, 83–84
 direct methods, 70–74
 Gauss elimination method, 71–72
 Gauss–Seidel iteration, 75–79
 iterative methods, 75–84
 LU decomposition with iterative improvement, 83
 nonlinear systems, 84–88
 red-black ordering scheme, 81–83
 reduction to algebraic equations, 65–70
 successive overrelaxation, 79–81
 systems of equations, 9–11
 Thomas algorithm, 72–74
Alternating direction explicit (ADE) methods, 207, 224–228
Alternating direction implicit (ADI) methods, 207, 220–224
Alternative reordering schemes, 498–501
Amplification factor, 140–141, 172, 179–180
Auxiliary problem, 480, 482–483
Axisymmetric physical domains, cylindrical coordinates, 473–474

B

Back substitution, 73–75
Backward differencing schemes
 boundary conditions, control volumes, 43
 diffusive–advective systems, 118
 hyperbolic heat conduction, 188
 MacCormack's method, 173
 mesh size, changing, 31–32
 reduction, algebraic equations, 67
 Taylor series, 24–25, 27–28
 two-dimensional steady laminar boundary layer flow, 331
 upwind differencing, 170
 upwind scheme, 181
 Warming and Beam's method, 174
Backward first order approximation, 147
Backward sweep, 223, 224
Barakat and Clark scheme, 224–228
Beer–Lambert's law, 230
Bessel function, zero-order, 163
Biconjugate gradient method, 75, 83–84
Binary alloy application, 405–409
Black-body intensity, 275, 277
Boundary and initial conditions
 basic relations, 12–15
 control volume approach, 42–46
 convection, explicit method, 132–133, 145–146
 convective eigenvalue problem, 513–514
 enthalpy method, 383, 386, 391
 explicit method, 136–137, 145–146, 158–159
 first kind, 41
 hollow cylinder and sphere, 106–108

mathematical formulation, physical problems, 38
nonlinear diffusion, 253
numerical grid generation, 437–439
prescribed flux, explicit method, 133–134
prescribed potential, explicit method, 131
pressure, vorticity-stream function formulation, 307
primitives variables formulation, 321–329
reduction, algebraic equations, 66–70
second kind, 41–42
sharp corners, 305–306
single-phase solidification, variable time step approach, 370
solid cylinder and sphere, 100–101
steady-state heat conduction, irregular geometry, 441–443
steady-state laminar free convection, irregular enclosures, 449
symmetry, 306–307
Taylor series, 40–42
third kind, 41–42
three-dimensional equation, radiative transfer, 276
two-phase solidification, variable time step approach, 376
unified integral transforms, 492, 494–496
velocity, 302
vorticity-stream function formulation, 302–309
Boundary-fitted coordinate method, 411–412
Boundary grid point control, 434–435
Boundary nodes, 259–261
Boundary value problem, 429–439
Boussinesq approximation, 446, 457
Branch cut, 431–432
Burgers-type equations, 479, 501–503, 507, 510, 513–514

C

Cancer, hyperthermia treatment, 228–242
Cartesian coordinates
  boundary conditions, 12

boundary value problem, 429
code verification, 53
comparisons, 23
connected region, 422
control volume approach, 37
convective eigenvalue problem, 509
multiple connected region, 428
one-dimensional heat conduction, 39
primitive variables, 457
solution verification, 55
Taylor series, 28
transformations, 419
vorticity-stream function, 292
Cattaneo–Vernotte's constitutive equation, 185–186
Center of control volumes, 318–320
Central differencing
  boundary conditions, control volumes, 43
  diffusive–advective systems, 117–118
  explicit method, 132, 158–159
  finite difference operators, 33–34
  implicit method, 165–166
  mixed partial derivatives, 29–30
  pressure, solution method, 302
  solid cylinder and sphere, 100
  Taylor series, 25, 27
  two-dimensional steady laminar boundary layer flow, 331
  vorticity-stream function, 296
Chain rule of differentiation, 413
Channels, transient forced convection, 482–489
Classical integral transform technique (CITT), 50
Classification, second-order partial differential equations, 4–5
Code verification, 50–54
Columnar dendritic zone, 394
Compressible flow
  MacCormack's method, 342–348
  quasi-one-dimensional flow, 339–354
  subsonic/supersonic inlet/outlet, 341, 343
  two-dimensional flow, 354–358
  WAF-TVD method, 348–354
Computational derivatives, 414
Concave corners, 305
Connected regions, 422–429

Index   567

Conservation equations and properties
  control volume approach, 35–36
  convection, explicit method, 145–146
  primitive variables, 311–312
  SIMPLEC method, 326–328
Conservative properties, parabolic
    systems, 5
Consistency, 48–49
Continuity equations
  coordinate transformation relations,
    414–415
  primitive variables formulation,
    309–310
  transformations, 420–421
  two-dimensional steady laminar
    boundary layer flow, 329–330
  vorticity-stream function, 292, 294
Control volume and approach
  advective–diffusive systems, 169
  derivatives, 34–38
  discrete approximation, derivatives, 23
  discretization, 43–46
  enthalpy, 402
  evaluation of metrics, 463
  interpolation functions, 121, 123
  primitive variables, 311–312
  SIMPLEC method, 317–320
  three-dimensional equation, radiative
    transfer, 279
  two-dimensional diffusion, 209
  two-dimensional transient
    convection-diffusion, 215–219
  upwind scheme, 182–183
Convection, 449, 450–452
Convection boundary conditions
  Crank–Nicolson method, 150–151,
    167–168
  cylindrical and spherical
    properties, 161
  explicit method, 132–133,
    145–146, 159
  hollow cylinder and sphere, 163
  implicit method, 165
  solid steel bar, 101–103
Convective–diffusive problems, 392–409
Convective eigenvalue problem, 505–516
Convective flux, 45, 326–329
Convergence, 48, 77–78, 298–299
Convex corners, 305–306

Coordinate-average approach, 463–469
Coordinate transformation relations,
    413–419
Correction equations, 315–317
Coupled conduction and radiation, 253,
    268–285
Courant–Friedrichs–Lewy condition, 171
Courant number
  hyperbolic heat conduction, 188–189
  MacCormack's method, 173
  transient forced convection,
    channels, 487
  upwind differencing, 170
  WAF-TVD method, 350
Cramer's rule, 71, 414
Crank–Nicolson method
  ADI method, 220, 222
  application, 166–169
  combined methods, 152–155, 220
  diffusive systems, 148–152, 166–169
  lagging properties, 254
  overview, 148–151
  prescribed heat flux boundary
    condition, 151–152
  transient multidimensional
    systems, 207
  two-time-level scheme, 257
Cylindrical properties
  hollow, explicit method, 161–164
  hyperthermia treatment, cancer,
    229–242
  implicit method, 164, 258–259
  numerical grid generation, 473–474
  solid, explicit method, 156–160
  stability, 160–161
  symmetry, diffusive systems,
    155–156
  two-dimensional diffusion, 211

D

Darcy's law, 394
Derivatives, coordinate transformation
    relations, 416–419
Derivatives, discrete approximation of
  boundary and initial conditions,
    38–46
  code verification, 50–54
  consistency, 48–49

control volume approach, 34–38, 42–46
differencing via polynomial fitting, 28–29
discretization, 40–47
errors, 46–49
first derivative, 25–27
mesh size, changing, 31–33
mixed partial derivative, 29–31
overview, 23–24
round-off errors, 46
second derivative, 27–28
solution verification, 54–58
stability, 48
Taylor series, 24–34, 40–42
total errors, 47
truncation errors, 46–47
verification and validation, 49–58
Diagonal dominance, 70, 75
Differencing via polynomial fitting, 28–29
Differential equation, 65–70, 370
Diffusion
false transient, 267–268
irregular enclosures, 449, 452–453
one-dimensional problem, 268–272
parabolic systems, 5
Diffusive–advective systems, 116–124
Diffusive flux, 325–329
Diffusive systems
boundary conditions, effects of, 136–137
combined method, 152–155
convection boundary conditions, 132–133
Crank–Nicolson method, 148–152, 166–169
cylinders, 156–166
explicit method, 130–146, 156–164
Fourier method, 138–146
heat conduction through fins, 110–116
hollow cylinder and sphere, 105–110
implicit method, 146–148, 164–166
prescribed flux boundary conditions, 133–134
prescribed potential at boundaries, 131
slab, 97–98
solid cylinder and sphere, 98–105
solution, stability of, 160–161
spheres, 156–160
spherical symmetry, 155–156
stability, 134–148, 160–161
symmetry, cylindrical, 155–156
truncation error, 137–138
Dimensionless variables, 366–367
Direct methods, 70–74
Dirichlet boundary conditions, 12
Discretization
with control volumes, 42–46
grid-generation differential equation, 436–437
index, method, stability analysis, 138
with Taylor series, 23, 40–42
Dissipation function, 308
Divergence, 384, 415
Double-grid approach, 463–467, 469
Doubly connected region, 426
Downstream boundary condition, 303, 305
Dufour effects, 393
Dupont-II scheme, 256–261

E

Eckert number, 308–309
Elliptic systems and properties
boundary value problem, 429
characterization, 10–11
classification, second-order partial differential equations, 4–5
fluid flow, 8
overview, 2–5
physical significance, 4–5
primitive variables, 459–460
steady-state advection-diffusion, 7–8
steady-state diffusion, 7
unified integral transforms, 490
vorticity-stream function, 293
Energy conservation equation, 36, 276–279
Energy equations
convective–diffusive problems, 398–409
diffusive systems, 97
hyperbolic systems, 9
primitive variables, 309–310, 325–329
vorticity-stream function, 308–309
Enthalpy method, 362, 383–392

Index 569

Equilibrium solidification, 396
Errors; *see also* Truncation
 ADE method, 226–227
 code verification, 53
 Crank–Nicolson, combined method, 154
 Dupont-II scheme, 258
 explicit method, 137–138
 Fourier method, stability analysis, 138–141
 freezing problem, 389
 implicit method, 147–148
 LU decomposition, iteration, 83
 in numerical solutions, 46–49
Euler's equations, 170, 354–355
Explicit method and scheme
 advection–diffusion equation, 179–182
 advective–diffusive systems, 179–182
 application, 156–164
 boundary conditions, effects of, 136–137
 considerations, 134–136
 convection boundary conditions, 132–133
 enthalpy method, 385–389, 392
 false transient, 266
 Fourier method, 138–146
 hollow cylinder and sphere, 161–164
 lagging properties, 254
 phase change problems, 385–389
 prescribed flux boundary conditions, 133–134
 prescribed potential at boundaries, 131
 solid cylinder and sphere, 156–160
 solution, stability of, 160–161
 stability, 134–146, 160–161
 transient multidimensional systems, 207–219
 truncation error, 137–138
 two-dimensional diffusion, 208–212
 two-dimensional transient convection-diffusion, 213–219
Extended surfaces, 110–116
Extinction coefficient, 269, 275

F

False transient technique, 264–268
Fictitious elements
 boundary conditions, Taylor series, 41–42

computational domain, 445
Crank–Nicolson method, 150, 167–168
derivative boundary conditions, 209
explicit method, 132, 158–159
false transient, 265
heat conduction through fins, 113
hollow cylinder and sphere, 106–107, 163
reduction, algebraic equations, 67
second-order accurate finite-difference scheme, 69
single-phase solidification, 370
solid cylinder and sphere, 99–100
two-phase solidification, 382
Field equation transformation, 441, 448–449
Finite difference approximation, 369–371, 376–377
Finite difference method (FDM), 1–2
Finite difference operators, 33–34
Finite difference representation, 295–298, 436–439, 443–445, 449–450
Finite differences/integral transforms combination, 481
Finite difference solution, 481
Finite-element method (FEM), 1
Finite volume method, 119–124
Finite volume method (FVM), 1
Fins, 110–116
First derivative, 25–27, 31–32
First-order accurate central difference, 130
First-order accurate differencing scheme, 101–103
First-order accurate finite difference, 113, 141–142
First-order central differencing scheme, 32
First-order linear (wave) equation, 9
First-order partial differential equations, 10, 186
First-order wave equation, 169
Fixed grid method, 361
Flash method, 273, 283
"Floating point" form, 46
Fluid flow, elliptic systems, 8
FORTRAN subroutines, 72, 73, 81, 84, 374

Forward differencing schemes
  boundary conditions, control
    volumes, 44–45
  hyperbolic heat conduction, 188
  MacCormack's method, 173
  mesh size, changing, 31–32
  reduction, algebraic equations, 67
  solution verification, 57–58
  Taylor series, 24–25, 27–28
  truncation errors, 47
  two-dimensional steady laminar
    boundary layer flow, 331
  upwind differencing, 170
  upwind scheme, 181
Forward sweep, 223–224
Forward time and central space (FTCS),
  208, 211, 213–214
Fourier methods
  ADE method, 226, 228
  advective–diffusive systems, 179
  background, 477–478
  boundary conditions, 14
  control volume approach, 35
  coupled conduction and
    radiation, 269
  cylindrical and spherical properties, 161
  explicit method, 138–146
  hyperbolic heat conduction, 185
  implicit method, 147
  stability analysis, 138
  transient forced convection,
    channels, 485
  two-dimensional diffusion, 209
  upwind differencing, 170–172
Four-point fomulae, 26
Freezing/frost problem, 389–392
Fresnel coefficient, 229–230
Front fixing method, 362
Full width half maximum (FWHM), 231,
  233, 235–236

G

Gauss divergence theorem, 279–280
Gauss elimination method, 71–72, 103
Gaussian distribution, 281
Gaussian laser beam, 229–236
Gaussian quadrature numerical
  integration, 503

Gauss–Seidel iteration, 75–81, 438
Generalization to multidimensions,
  365–366
Generalized integral transform technique
  (GITT), 50, 477–479; *see also*
  Hybrid numerical–analytical
  solutions
GITT, *see* Generalized integral transform
  technique
Gradient, coordinate transformation
  relations, 415
Gray media, 273, 278
Grid control functions, 429, 432–434
Grid convergence analysis, 52–53
Grid-generation differential equation,
  discretization, 436–437
Grid-generation equations
  transformation, 473–476
Grid-generation problem, 440–441
Grid lines orthogonality, 431–432, 438–439

H

Harmonic means, 38, 218
Henyey–Greenstein phase function, 281
Heuristic stability analysis, 171, 263–264
HLL/HLLC scheme, 351
Hollow cylinder and sphere, 105–110,
  161–164, 465–469
Homogeneous boundary conditions, 12
Hybrid numerical–analytical solutions
  application, 481–489
  auxiliary problem, 480, 482–483
  background, 477–479
  convective eigenvalue problem,
    505–516
  finite differences/integral transforms
    combination, 479–489
  finite difference solution, 481
  integral transform pair, 480, 483
  integral transforms, 480
  inversion, 481, 485–489
  solution, 485
  taking the integral transform, 480,
    483–484
  transient forced convection, channels,
    481–489
  truncation, 481, 485
  unified integral transforms, 489–505

Hyperbolic heat conduction, 4–5, 185–189
Hyperbolic systems
  basic relations, 8–9
  characterization, 10, 11
  classification, second-order partial
    differential equations, 4–5
  overview, 2–5
  wave equation, 11
Hyperthermia treatment, cancer,
    228–242
Hypoeutectic alloy, 393–394

I

Identity matrix, 11
Impermeable wall condition, 303, 304,
    306–307
Implicit method and scheme
  advection–diffusion equation,
    182–184
  advective–diffusive systems, 182–184
  application, 164–166
  enthalpy, 389–392
  false transient, 267
  finite volume method, 182–184
  hollow cylinder and sphere, 165–166
  lagging properties, 254
  overview, 146–148
  solid cylinder and sphere, 164–165
  stability analysis, 147–148
Incompatibility, see Consistency
Incompressible flow, see
    Multidimensional
    incompressible laminar flow
Inflow boundary condition, 303–305
Initial conditions, see Boundary and
    initial conditions
In-scattering contribution, 281
Integral transforms, 480, 483–484
Integrate function, 499
Interface, control volumes, 319–320
Interface conditions
  mathmathical formulation, phase
    change problems, 364–365
  single-phase solidification, variable
    time step approach, 371
  two-phase solidification, variable
    time step approach, 377
Internal nodes, 257–258

Interpolation functions, 121–124
Inversion, 481, 485–489
Iron rod application, 114–116
Iterative methods, 70, 75–84, 299–301

J

Jacobian properties
  boundary value problem, 430, 436–437
  coordinate transformation relations,
    413–414
  nonlinear systems, 86
  systems of equations, 10
Jacobi preconditioner, 84

K

Keller's box method, 207
Kozeny–Carman equation, 394

L

Lagging properties
  enthalpy method, 386
  hyperbolic heat conduction, 185
  nonlinear diffusion, 254
  nonlinear systems, 85
  one time step, nonlinear diffusion,
    254–256
Laminar flow, see Multidimensional
    incompressible laminar flow
Laplacian operator
  coordinate transformation relations,
    416–418
  irregular geometry and enclosures,
    441, 448
  numerical grid generation, 473–476
  truncation errors, 47
  two-dimensional diffusion, 212
Larkin scheme, 224
Lax's equivalence theorem, 48, 348
Lax–Wendroff methods, 173
Lever rule, 397, 403
L'Hospital's rule, 98, 259
Liebman iteration, see Gauss–Seidel
    iteration
LINBCG subroutine, 83
Linear boundary conditions, 12, 38
Linearization, nonlinear diffusion, 261–264

Liquid phase equation, 377
LUBKSB subroutine, 83
LUDCMP subroutine, 83
LU decomposition with iterative improvement, 83

M

MacCormack's method
    application, 174–178
    dispersion, 171
    hyperbolic heat conduction, 186–189
    overview, 169
    quasi-one-dimensional flow, 342–348
    upwind method, 172–173
Manufactured solutions, code verification, 50–52
Mapping, 422–429, 440, 447
Mass advection–diffusion, see Advection–diffusion systems
Mathemathical formulation, phase change problems, 363–367
Mathematica system, 479, 491, 496, 500
Matrix method, stability analysis, 137
Mesh
    distorted, 466–469
    finite difference operators, 33–34
    size, explicit method, 135–136
    Taylor series formulation, 31–33
Method of Lines, 496, 503
Metrics evaluation, computational aspects, 461–469
Mixed partial derivative, finite difference approximation, 29–31
Mixed partial derivatives, 29–31
"Model" defined, 49
Modified equations, 152, 171–172
Modified heat transfer coefficient, 45
Modified variable time step (MVTS) method, 368–374
Momentum equations
    parabolic systems, 6
    pressure, solution method, 301
    primitive variables, 309–310, 325–329
    SIMPLEC method, 314–315
    two-dimensional steady laminar boundary layer flow, 329–331
    vorticity-stream function, 292–294

Monotonic properites, 56, 348, 422, 424–425, 427
Moving wall conditions, 304–307
MPROVE subroutine, 83
Multidimensional incompressible laminar flow
    boundary conditions, 302–309, 321–329
    energy equation, 308–309, 325–329
    finite difference representation, 295–298
    inflow boundary condition, 305
    initial condition, 307
    momentum equation, 325–329
    outflow boundary condition, 305
    overview, 291–292
    Poisson's equation, 297–298
    pressure, 298, 301–302, 307, 321–325
    primitives variables formulation, 309–329
    sharp corners boundary condition, 305–306
    SIMPLEC method, 314–320
    steady-state problem solution, 299–303
    stream function, 292–295, 297
    symmetry boundary condition, 306–307
    transient problem solution, 298–299
    two-dimensional steady laminar boundary layer flow, 329–331
    velocity, boundary conditions treatment, 302
    velocity field determination, 314–320
    vorticity function, 292–295
    vorticity-stream function formulation, 291–309
    vorticity transport equation, 296–297
Multidimensions, generalization to, 365–366
Mushy zone, 394, 396–397, 399–400

N

Navier–Stokes equation, 119, 182, 293
NDSolve routine, 496, 500, 503, 505
Net radiative heat flux, 283–284
Neumann boundary conditions, 12
Newtonian properties, 8, 393

Index

Newton–Raphson iterative method, 85–88
NIntegrate function, 499
Nonlinear diffusion
   boundary nodes, 259–261
   coupled conduction/radiation, participating media, 268–285
   cylinder and sphere, limiting case, 258–259
   diffusion equations, set of, 267–268
   explicit scheme, 266
   false transient, 264–268
   implicit scheme, 267
   internal nodes, 257–258
   lagging properties, one time step, 254–256
   linearization, 261–264
   one-dimensional problem, diffusion approximation, 268–272
   overview, 253–254
   stability criterion, 263–264
   three-dimensional equation, radiative transfer, 272–285
   three-time-level implicit scheme, 256–261
Nonlinear systems, 84–88
Nonorthogonality, 435
No-slip at the wall condition, 303, 304
Numerical grid generation
   advection–diffusion equation, 461–465
   axisymmetric physical domains, 473–474
   boundary conditions, 437–439
   boundary grid point control, 434–435
   boundary value problem, 429–439
   branch cut, 432
   coordinate transformation relations, 413–419
   cylindrical coordinates, 473–474
   finite difference representation, 436–439
   grid control functions, 432–434
   grid-generation differential equation, discretization, 436–437
   grid-generation equations transformation, 473–476
   grid lines orthogonality, 431–432
   hollow sphere, 465–469
   irregular enclosures and geometry, 439–460
   Laplacian equation transformations, 473–476
   mapping, 422–429
   metrics evaluation, computational aspects, 461–469
   multiply connected region, 426–429
   nonorthogonality, effects of, 435
   one-dimensional advection–diffusion equation, 461–465
   overview, 411–412
   physical domains, 475–476
   polar coordinates, 475–476
   simple transformations, 419–421
   simply connected region, 422–426
   steady-state heat conduction, 439–445
   steady-state laminar free convection, 445–456
   transient laminar free convection, 457–460
   two-dimensional heat conduction, 465–469
Nusselt number, 454–456, 460

O

Oberbeck–Boussomesq equation, 393, 398
One-dimensional advection–diffusion equation, 461–465
One-dimensional diffusion equation, 146
One-dimensional heat conduction, 37, 39
One-dimensional problem, diffusion approximation, 268–272
One-dimensional steady-state systems
   diffusive–advective systems, 116–124
   diffusive systems, 97–116
   finite volume method, 119–124
   heat conduction through fins, 110–116
   hollow cylinder and sphere, 105–110
   interpolation functions, 121–124
   iron rod application, 114–116
   overview, 97
   slab, 97–98
   solid cylinder and sphere, 98–105
   stability, 118–119

One-dimensional transient systems
  advection–diffusion, 169–184
  combined method, 152–155
  Crank–Nicolson method, 148–152, 166–169
  cylindrical symmetry, 155–156
  diffusive systems, 129–169
  explicit method, 130–146, 156–164
  finite difference representation, 186–189
  hyperbolic heat conduction equation, 185–189
  implicit method, 146–148, 164–166
  overview, 129
  purely advective (wave) equation, 169–178
  spherical symmetry, 155–156
One-sided differencing, 170
One-sided three-point formula, 302
One time step, lagging properties, 254–256
Operators, finite difference, 33–34
"Order of notation," 25
Orthogonality, grid lines, 431–432
Outflow boundary condition, 303, 305
Overrelaxation, 80–81

P

Parabolic systems and properties, 2–6, 207–219, 322, 490
Parallelepiped properties, 273, 412
Partial lumping formulation, 111
Partial transformation, 493–496
Passive scalar transport equation model, 395–398
PDE, *see* Partial differential equations
Peclet number
  advective–diffusive systems, 181
  diffusive–advective systems, 118
  energy equation, 308–309
  interpolation functions, 121–123
  two-dimensional transient convection-diffusion, 218
  upwind scheme, 182
  vorticity-stream function, 297
Phase change problems
  adaptive grid generation methods, 362
  boundary conditions, 370, 376
  convective–diffusive problems, 392–409

differential equation, 370
dimensionless variables, 366–367
energy equation model, 398–409
enthalpy method, 362, 383–392
explicit enthalpy method, 385–389, 392
finite difference approximation, 369–371, 376–377
fixed grid method, 361
front fixing method, 362
generalization to multidimensions, 365–366
implicit enthalpy method, 389–392
interface conditions, 364–365, 371, 377
liquid phase equation, 377
mathemathical formulation, 363–367
overview, 361–362
passive scalar transport equation model, 395–398
single melting temperature, 385–392
single-phase solidification, 368–374
solid phase equation, 376
temperature range, 392
time steps determination, 371–374, 377–383
two-phase solidification, 374–383
variable grid method, 361–362
variable time step approaches, 368–383
Physical domains, polar coordinates, 475–476
"Pivot" equation, 71–73
Point-iterative procedure, 75–77
Poiseuille flow, 305
Poisson's equation
  boundary value problem, 429
  elliptic systems, 7
  energy equation, 309
  multidimensional incompressible laminar flow, 291
  pressure, solution method, 301
  SIMPLEC method, 321
  stream function, boundary conditions, 303
  vorticity-stream function, 293–295, 298–300
  vorticity-stream function formulation, 297–298, 307
Polar coordinates, 475–476

Index

Polynomial fitting, differencing via, 28–29
Predictor-corrector procedure, 173, 186–189, 342–348
Prescribed flux boundary conditions
  Crank–Nicolson method, 168–169
  explicit method, 133–134
  hollow cylinder and sphere, 163
  implicit method, 165
Prescribed heat flux boundary condition, 151–152
Prescribed potential at boundaries, 131
Pressure
  primitives variables formulation, 321–325
  SIMPLEC method, 317, 321–325
  solution, vorticity-stream function formulation, 301–302
  vorticity-stream function, 293–294, 298, 301–302, 307
Primitive variables formulation, 309–329
Purely advective (wave) equation, 169–178

Q

Quadratic upstream interpolation for convective kinematics (QUICK), 122
Quadratic upstream interpolation for convective kinematics with estimated streaming terms (QUICKEST), 122
Quadrilateral properties, 412
Quasilinear systems of equations, 10
Quasi-one-dimensional flow, 170, 339–354

R

Radial heat conduction, 108–109
Radiative transfer, three-dimensional equation, 272–285
Rayleigh numbers, 455–456, 460
"Real world" defined, 49–50
Red-black ordering scheme, 75, 81–83
Reduction to algebraic equations, 65–70
Refractive index, 275
Relative convergence criterion, 77

Relaxation formula, 300
Relaxation parameter, 80
Relaxation time, 9
Reynolds number, 121, 295, 297, 309
Riemann solver, 351
Robin boundary conditions, 12
Round-off errors, 46, 83

S

Scheil's model, 397, 403–404
Schwarz–Christoffel formula, 419
Second derivative, 27–28, 32–34
Second-order accurate central difference
  Crank–Nicolson method, 167
  diffusive–advective systems, 117
  explicit method, 130, 132
  heat conduction through fins, 113
  hollow cylinder and sphere, 106–108
  solid cylinder and sphere, 99
Second-order accurate differencing, 101–103, 150
Second-order accurate finite difference
  Crank–Nicolson method, 168
  explicit method, 133–134
  implicit method, 166
  reduction, algebraic equations, 69–70
  transient heat conduction, 141–142, 144–145
Second-order central differencing scheme
  Crank–Nicolson method, 150
  mesh size, changing, 32
  solid cylinder and sphere, 100
  Taylor series, 25
Second-order linear (wave) equation, 9
Second-order partial differential equations, 186
Self-adjoint eigenvalue problem, 508–510
Semi-elliptical physical domain, 439–445
Sharp corners, 305–306
SIMPLEC method, 314–320, 405, 459
Simply connected region, 422–426
Single melting temperature, 385–392
Single-phase solidification, variable time step approach, 368–374
Skew third-order upwinding scheme (STOUS), 123

Skew weighted upstream differencing scheme (SWUDS), 122
Slab, diffusive systems, 97–98
Smith analysis, 161
Solid cylinder and sphere, 98–105, 156–160
Solid phase equation, 376
Solution verification, 54–58
SOR, see Successive overrelaxation
Soret effects, 393
Space variable index, 148
Spherical properties
    hollow, explicit method, 161–164
    implicit method, 164
    solid, explicit method, 156–160
    stability, 160–161
    symmetry, 155–156
    three-time-level implicit scheme, 258–259
Stability
    ADE method, 226
    ADI method, 220
    advection–diffusion equation, 179–181
    advective–diffusive systems, 179
    analysis, matrix method, 137
    combined method, 154–155
    Crank–Nicolson method, 152–155
    criterion, nonlinear diffusion, 263–264
    diffusive–advective systems, 118–119
    enthalpy method, 386
    errors in numerical solutions, 48
    explicit method, 134–146, 160–161, 266
    implicit method, 147–148
    two-dimensional diffusion, 209–210
    two-dimensional transient convection-diffusion, 214, 218
    upwind method, 170
    vorticity-stream function, 297
Steady-state advection-diffusion, 7–8
Steady-state diffusion, 7, 251–252
Steady-state heat conduction, irregular geometry, 440–445
Steady-state laminar free convection, irregular enclosures, 445–454
Steady-state one-dimensional heat conduction equation, 37
Steady-state problem solution, 299–303
Steady-state subsonic forced compressible flow, 8
Stefan–Boltzmann constant, 14, 269, 253, 273
Stefan number, 367
Stream function, 291–295, 297
Stretching parameter, 420
Sturm–Liouville problem, 480, 482
Subsonic inlet/outlet, 341, 343
Successive overrelaxation (SOR), 75, 79–81, 438
Successive-substitution formula, 444, 447–448
Supersonic inlet/outlet, 341, 343
Systems of equations, 9–11; see also Algebraic equations, solving systems of

T

Taking the integral transform, 480, 483–484
Tangential derivatives, 417–419
Taylor series
    advective–diffusive systems, 169
    changing mesh size, 31–33
    Crank–Nicolson, combined method, 153
    Crank–Nicolson method, 149
    differencing via polynomial fitting, 28–29
    discrete approximation, derivatives, 23
    discretization with, 40–42
    Dupont-II scheme, 260
    finite difference operators, 33–34
    first derivative, 25–27, 31–32
    linearization, 262
    mesh size, changing, 31–33
    mixed partial derivative, finite difference approximation, 29–31
    nonlinear systems, 85
    overview, 23, 24–25
    second derivative, 27–28, 32–33
    upwind differencing, 170
    vorticity-stream function formulation, 307

Temperature, 385–392
Test case, 501–505
Tetrahedral properties, 412
Thomas algorithm, 72–74, 154, 222–224, 267
Thompson's approach, 411
Three-dimensional equation, radiative transfer, 272–285
Three-point fomulae, 26
Three-time-level implicit scheme, 256–261
Time steps, 371–374, 377–383
Total errors, 47; see also Errors
Total transformation, 491–493
Transient convection–diffusion, 493
Transient forced convection, channels, 482–489
Transient heat conduction, 6, 141–145
Transient laminar free convection, irregular enclosures, 457–460
Transient mass advection–diffusion, 6; see also Advection–diffusion
Transient multidimensional systems
  ADE method, 224–228
  ADI method, 220–224
  combined method, 219–220
  explicit method, 207–219
  hyperthermia treatment, cancer, 228–242
  overview, 207
  steady-state diffusion, 251–252
  two-dimensional diffusion, 208–212
  two-dimensional transient convection-diffusion, 213–219
Transient problem solution, 298–299
Transient radial heat conduction, 163–164
Transport equation, 296–297
Two-dimensional diffusion, 208–212
Two-dimensional flows, 329–331, 354–358
Two-dimensional heat conduction, hollow sphere, 465–469
Two-dimensional steady laminar boundary layer flow, 329–331
Two-dimensional transient convection-diffusion, 6, 213–219
Two-dimensional transient heat conduction, 6

Two-dimensional transient mass advection–diffusion, 6
Two-phase solidification, variable time step approach, 374–383
Two-point fomulae, 26

U

Uncertainty, 50
Underrelaxation, 80
Unified integral transforms
  computational algorithm, 497–501
  overview, 489–491
  partial transformation, 493–496
  test case, 501–505
  total transformation, 491–493
Unified Integral Transforms (UNIT code), 479
Uniformly third-order polynomial interpolation algorithm (UTOPIA), 122–123
Uniqueness of the solution, 15–17
Universal limiter, 122
Unknown potential, 5
Upper diagonalizing equation, 73–74
Upstream boundary condition, 303, 304, 305
Upstream differencing, 117, 122, 217, 312, 451–452
Upwind differencing
  application, 174–178
  diffusive–advective systems, 117–118, 119
  overview, 169
  two-dimensional transient convection-diffusion, 214
  vorticity transport equation, 296
  Warming and Beam's method, 174
Upwind method
  advective–diffusive systems, 170–172, 181–182
  application, 174–178
  interpolation functions, 121–124
  MacCormack's method, 172–173
  modified equation, 171–172
  purely advective (wave) equation, 170–172
  stability criteria, 170
  Warming and Beam's method, 173–178

V

Variable grid method, 361–362
Variables, dimensionless, 366–367
Variable time step approach, 368–383
Velocity, 302
Velocity field determination, 314–320
Verification and validation (V&V), 49–58
Volumetric energy generation, 35
Volumetric source, 5
von Neumann method, 138
Vorticity function, 292–295
Vorticity-stream function
    boundary conditions, 302–309
    energy equation, 308–309
    finite difference representation, 295–298
    inflow boundary condition, 305
    initial condition, 307
    irregular enclosures, 445–456
    outflow condition, 305
    overview, 291–292
    Poisson's equation, 297–298
    pressure, 298, 301–302, 307
    sharp corners, 305–306
    steady-state laminar free convection, 445–456
    steady-state problem solution, 299–303
    stream function, 292–295, 297
    symmetry, 306–307
    transient problem solution, 298–299
    velocity, 302
    vorticity function, 292–295
Vorticity transport equation, 293–295, 299, 307

W

Warming and Beam's method, 169, 171, 173–178
Wave equation, 11
Weighted average flux - total variation diminishing (WAF-TVD) method, 348–358
Weighted upstream differencing scheme (WUDS)
    evaluation of metrics, 462
    interpolation functions, 122
    primitive variables, 312, 459
    SIMPLEC method, 326
    two-dimensional transient convection-diffusion, 217

Z

Zero-order Bessel function, 163